Achieving sustainable cultivation of soybeans

Volume 1: Breeding and cultivation techniques

It is widely recognised that agriculture is a significant contributor to global warming and climate change. Agriculture needs to reduce its environmental impact and adapt to current climate change whilst still feeding a growing population, i.e. become more 'climate-smart'. Burleigh Dodds Science Publishing is playing its part in achieving this by bringing together key research on making the production of the world's most important crops and livestock products more sustainable.

Based on extensive research, our publications specifically target the challenge of climate-smart agriculture. In this way we are using 'smart publishing' to help achieve climate-smart agriculture.

Burleigh Dodds Science Publishing is an independent and innovative publisher delivering high quality customer-focused agricultural science content in both print and online formats for the academic and research communities. Our aim is to build a foundation of knowledge on which researchers can build to meet the challenge of climate-smart agriculture.

For more information about Burleigh Dodds Science Publishing simply call us on +44 (0) 1223 839365, email info@bdspublishing.com or alternatively please visit our website at www.bdspublishing.com.

Related titles:

Achieving sustainable cultivation of soybeans Volume 2: Diseases, pests, food and other uses
Print (ISBN 978-1-78676-116-3); Online (ISBN 978-1-78676-119-4, 978-1-78676-118-7)

Achieving sustainable cultivation of grain legumes Volume 1:
Advances in breeding and cultivation techniques
Print (ISBN 978-1-78676-136-1); Online (ISBN 978-1-78676-139-2, 978-1-78676-138-5)

Achieving sustainable cultivation of grain legumes Volume 2:
Improving cultivation of particular grain legumes
Print (ISBN 978-1-78676-140-8); Online (ISBN 978-1-78676-142-2, 978-1-78676-143-9)

Chapters are available individually from our online bookshop: https://shop.bdspublishing.com

climate•SMART•publishing
IN AGRICULTURAL SCIENCE

BURLEIGH DODDS SERIES IN AGRICULTURAL SCIENCE

NUMBER 29

Achieving sustainable cultivation of soybeans

Volume 1: Breeding and cultivation techniques

Edited by Professor Henry T. Nguyen
University of Missouri, USA

burleigh dodds
SCIENCE PUBLISHING

Published by Burleigh Dodds Science Publishing Limited
82 High Street, Sawston, Cambridge CB22 3HJ, UK
www.bdspublishing.com

Burleigh Dodds Science Publishing, 1518 Walnut Street, Suite 900, Philadelphia, PA 19102-3406, USA

First published 2018 by Burleigh Dodds Science Publishing Limited

Library of Congress Control Number: 2017956486

British Library Cataloguing in Publication Data
A catalogue record for this book is available from the British Library

ISBN 978-1-78676-112-5 (print)
ISBN 978-1-78676-114-9 (online)
ISBN 978-1-78676-115-6 (online)
ISSN 2059-6936 (print)
ISSN 2059-6944 (online)

Typeset by Deanta Global Publishing Services, Chennai, India
Printed by Lightning Source

Contents

Series list

Title	Series number
Achieving sustainable cultivation of maize - Vol 1 From improved varieties to local applications Edited by: Dr Dave Watson, CGIAR Maize Research Program Manager, CIMMYT, Mexico	001
Achieving sustainable cultivation of maize - Vol 2 Cultivation techniques, pest and disease control Edited by: Dr Dave Watson, CGIAR Maize Research Program Manager, CIMMYT, Mexico	002
Achieving sustainable cultivation of rice - Vol 1 Breeding for higher yield and quality Edited by: Prof. Takuji Sasaki, Tokyo University of Agriculture, Japan	003
Achieving sustainable cultivation of rice - Vol 2 Cultivation, pest and disease management Edited by: Prof. Takuji Sasaki, Tokyo University of Agriculture, Japan	004
Achieving sustainable cultivation of wheat - Vol 1 Breeding, quality traits, pests and diseases Edited by: Prof. Peter Langridge, The University of Adelaide, Australia	005
Achieving sustainable cultivation of wheat - Vol 2 Cultivation techniques Edited by: Prof. Peter Langridge, The University of Adelaide, Australia	006
Achieving sustainable cultivation of tomatoes Edited by: Dr Autar Mattoo, USDA-ARS, USA & Prof. Avtar Handa, Purdue University, USA	007
Achieving sustainable production of milk - Vol 1 Milk composition, genetics and breeding Edited by: Dr Nico van Belzen, International Dairy Federation (IDF), Belgium	008
Achieving sustainable production of milk - Vol 2 Safety, quality and sustainability Edited by: Dr Nico van Belzen, International Dairy Federation (IDF), Belgium	009
Achieving sustainable production of milk - Vol 3 Dairy herd management and welfare Edited by: Prof. John Webster, University of Bristol, UK	010
Ensuring safety and quality in the production of beef - Vol 1 Safety Edited by: Prof. Gary Acuff, Texas A&M University, USA & Prof.James Dickson, Iowa State University, USA	011
Ensuring safety and quality in the production of beef - Vol 2 Quality Edited by: Prof. Michael Dikeman, Kansas State University, USA	012
Achieving sustainable production of poultry meat - Vol 1 Safety, quality and sustainability Edited by: Prof. Steven C. Ricke, University of Arkansas, USA	013
Achieving sustainable production of poultry meat - Vol 2 Breeding and nutrition Edited by: Prof. Todd Applegate, University of Georgia, USA	014
Achieving sustainable production of poultry meat - Vol 3 Health and welfare Edited by: Prof. Todd Applegate, University of Georgia, USA	015

Acknowledgements

We wish to acknowledge the following for their help in reviewing particular chapters:

- Chapter 2: Dr. Badara Gueye, IITA, Nigeria
- Chapter 3: Dr Amelio Dall'Agnol, EMBRAPA, Brazil
- Chapter 4: Dr Soon-Chun Jeong, Korea Research Institute of Bioscience and Biotechnology
- Chapter 6: Dr Tom Jacobs, Cornell University, USA
- Chapter 8: Professor Matthew Gilliham, University of Adelaide, Australia
- Chapter 12: Dr Amanda Ashworth, USDA-ARS Soil Scientist, USA
- Chapter 13: Dr Esther Ronner, Wageningen University, The Netherlands

Introduction

Soybeans are one of the most widely grown crops in the world. As the world's main source of vegetable protein, they have a wide range of food and non-food uses. Current yields need to increase significantly to meet growing demand but in a way that reduces input use, does not damage the environment and is resilient to climate change. These challenges are addressed in the two volumes of *Achieving sustainable cultivation of soybeans*:

- Volume 1: Breeding and cultivation techniques
- Volume 2: Diseases, pests, food and other uses

Volume 1 reviews the wealth of research addressing this challenge, focussing particularly on breeding and cultivation techniques. The volume reviews the latest research on crop physiology and genetic diversity, providing detailed coverage of key advances in breeding and their application to produce more resilient drought, salt and flood-tolerant varieties of soybean. The volume also summarises research on good agricultural practices to optimise soybean cultivation and make it more sustainable.

Part 1 Plant physiology and breeding

The chapters in the first part of the volume review our understanding of soybean physiology and genetic diversity. They then discuss advances in conventional and marker-assisted breeding, as well as transgenic techniques, and their use to produce more stress-resistant varieties. The subject of Chapter 1 is advances in understanding soybean physiology and growth. The chapter covers biological nitrogen fixation in soybean and classical techniques for understanding soybean physiology. It then moves on to look at soybean mutagenesis methods and soybean mutant lines, as well as modern experimental techniques applicable to soybean physiology research. Finally, the chapter examines the soybean genome and associated bioinformatics resources.

Chapter 2 moves on to look at the challenge of maintaining and utilizing the genetic diversity of soybeans. The aim of plant breeding is to create genetically diverse gene pools and then devise strategies to consolidate favorable alleles whilst eliminating those that impede improvement. The chapter describes the major soybean germplasm collections of the world as sources of available genetic diversity, then reviews the formation of the current commercially-used gene pool for soybean breeding in North America. It then considers how adding new genetic diversity can improve the rate of gain for seed yield and how various sources of exotic germplasm can be utilized within different segments of a national soybean program.

Complementing the preceding chapter's focus on soybean genetic diversity, Chapter 3 examines advances in conventional soybean breeding techniques. The chapter provides an overview of the evolution of modern soybean varieties. The chapter examines the genetic architecture of soybean and its germplasm diversity, as well as describing the process of seed certification for plant breeding. The chapter looks at the challenge of minimizing phenotypic variation due to non-genetic factors, examining field and laboratory phenotyping, breeding to reduce losses from biotic and abiotic stress, and the management and analysis of phenotyping data. Finally, the chapter looks at soybean cultivation in developing countries and includes a detailed case study of the work of the USAID Soybean Innovation Lab.

Continuing the preceding chapters' focus on genetics, the subject of Chapter 4 is mapping the soybean genome. Soybean is one of the most important grain legumes, but genetic mapping has been less advanced for soybean than for other crops, which limits modern soybean molecular breeding. The chapter discusses the significant progress in mapping the soybean genome that has now been made. The chapter examines the use of restriction fragment length polymorphism molecular markers, high-throughput genotyping markers, SLAF-Seq (specific-locus amplified fragment sequencing) mapping, whole-genome re-sequencing mapping and comparative genome mapping. This has allowed the identification of quantitative trait loci (QTLs) or genes to improve yield, nutritional quality, as well as biotic and abiotic stress tolerance. Finally, the chapter includes two detailed genomics case studies on comparing wild and cultivated soybean varieties and identifying salinity tolerance in soybean.

Chapter 5 continues the focus on soybean genetics, discussing advances in marker-assisted breeding of soybeans. It begins by discussing types of molecular markers such as simple sequence repeats (SSRs) and single nucleotide polymorphism (SNPs). It then discusses marker assays and genotyping platforms for marker-assisted selection, including SSR marker genotyping and developments in high-throughput genotyping. The chapter then reviews applications of marker-assisted selection in soybean breeding to deal with biotic threats such as nematodes, rusts and aphids, abiotic threats such as salinity, as well as aspects of quality such as fatty acid composition. The chapter concludes by looking at developments in genomic selection.

The following chapter, Chapter 6 considers advances in genetic modification of soybeans. Transgenic soybean holds a significant share of the market. However, the traits improved by genetic modification of soybean so far remain limited, with herbicide tolerance the primary trait exhibited by genetically modified (GM) soybean. Several other agronomic traits could be usefully improved, such as insect resistance, oil composition and amino acid composition. Valuable genes and efficient transformation systems are the key requirements for the improvement of GM soybean. The chapter describes recent advances in the genetic modification of important agronomic traits of soybean, and reviews the genome editing technology available.

Moving on to focus on the effects of genetics on specific characteristics of soybean, Chapter 7 examines advances in improving drought and heat resistance in soybean. Drought combined with heat is the major abiotic stress that threatens soybean crop production, particularly in the context of climate change. The success of soybean improvement under drought and heat stress depends on the discovery and utilization of genetic variations present in germplasm. The chapter discusses the identification of quantitative trait loci (QTL) associated with the traits contributing to drought and heat tolerance. In addition, the chapter discusses genomic resources that can facilitate a better understanding of phenotype-genotype association and formulate genomic-assisted breeding strategies.

As Chapter 8 indicates, the production of soybean is severely restricted by salt stress. The chapter describes different approaches to developing salt tolerance in soybean, including a study of germplasm resources for salt tolerance, 'omic' approaches and molecular breeding of salt-tolerant soybean varieties. The final chapter in Part 1, Chapter 9, deals with advances in flood-tolerant varieties of soybean. Flooding is a serious abiotic stress that can cause drastic soybean yield loss around the world. As a result, a large number of studies have been conducted to understand soybean responses to flooding on agronomic, morphological, physiological, and molecular levels. The chapter summarises current understanding of the impact of flooding and the natural adaptation of soybean to flooding. The chapter evaluates current and future flooding survival strategies, including the use of genetic engineering to improve flooding tolerance, and various agronomic techniques for the management of flooding stress. The chapter concludes that an integrated approach is likely to prove most effective in preventing and mitigating the effects of flooding on soybean growth and yield.

Part 2 Cultivation techniques

The second part of the volume reviews key advances in cultivation techniques to make the most of these new varieties. The subject of Chapter 10 is modelling the effects of temperature and photoperiod on soybean reproductive development. It has long been recognized that the development rate and yield of soybean is sensitive to temperature and photoperiod. The chapter considers a range of experimental evidence concerning the precise effects of these factors on soybean reproductive development, and reviews methods of modelling these effects. The chapter concludes that integrated modelling and experimentation at gene level may hold the key to improving the modelling of temperature and photoperiod effects on soybean development.

Chapter 11 moves on to the subject of defining and implementing best management practices in soybean production. In order to optimise the profitability of soybean production while minimizing adverse effects on the environment, it is important to define and implement the most effective management, agronomic and technological practices. The chapter considers the following components of soybean production: tillage systems, soil fertility, variety selection, agronomic practices and pest management. Such practices will, of course, vary between geographic regions; for instance, best tillage practice may be determined by soil type and environmental conditions. However, in each case, the chapter describes general practices that transcend geographic areas, giving consideration to how these practices might be adjusted to accommodate region-specific factors. The chapter concludes with recommendations for how these practices could be implemented within an overall management plan.

As Chapter 11 points out, good agricultural practices (GAPs) vary within and among production systems. Complementing the themes of the preceding two chapters on modelling and best management practices, Chapter 12 examines contrasting case studies from three soybean production areas to illuminate sustainable soybean production research and practice. The chapter focuses on examples from the USA (Nebraska), Brazil, and Uganda to compare and contrast current production practices and ways they can be optimized. As the chapter shows, in all three cases abundant opportunities exist to overcome production constraints while improving sustainability.

The final chapter of the volume, Chapter 13, looks at supporting smallholders in soybean cultivation, with a particular focus on the example of Zimbabwe. Low crop productivity due to soil fertility decline is a major challenge, constraining smallholder farming systems in sub-Saharan Africa. The decline in soil fertility is largely due to continuous cropping without soil replenishment and reduced nitrogen fixation. In Zimbabwe, soybean has emerged as a key crop that has the potential to help to improve soil fertility and food and income security among smallholder farmers. The chapter reviews collaborative projects that have promoted soybean cultivation by smallholder farmers, and makes recommendations regarding the training, financial and marketing systems required to support smallholder farmers further in improving soybean cultivation.

Part 1

Plant physiology and breeding

Chapter 1

Advances in understanding soybean physiology and growth

M. B. Zhang[1] and X. T. Chu[1], Centre for Integrative Legume Research, University of Queensland, Australia; H. N. Su, University of Queensland, Australia, and National Navel Orange Engineering Research Center, Gannan Normal University, China; and A. H. Hastwell, P. M. Gresshoff and B. J. Ferguson[2], Centre for Integrative Legume Research, University of Queensland, Australia

1 Introduction

Soybean (*Glycine max* (L.) Merr.) is the world's foremost legume crop and represents a multi-billion dollar industry worldwide (Singh and Singh, 1992; Specht et al., 2001; Norman, 2012). Global soybean production is roughly 320 million tonnes per year, with over 80% grown in three dominant countries: the United States (USA), Brazil and Argentina (USDA-FAS 2016). Due to its high seed protein and oil content, much of the soybean yield undergoes industrial processing to become crushed soybean meal (for human consumption and for livestock and poultry feed), soybean oil or biodiesel (Asbridge, 1995; Goldsmith, 2008; Norman, 2012; Gresshoff et al., 2015; Orf, 2016). Indeed, only a fraction of the soybean yield is used in other products for human consumption, such as soy sauce, tofu, edamame and soymilk (Goldsmith, 2008; Norman, 2012).

Soybean represents a source of high-quality protein, providing the full spectrum of essential amino acids, and has been incorporated into the daily diets of the Eastern

1. Authors contributed equally.
2. Corresponding author: b.ferguson1@uq.edu.au

http://dx.doi.org/10.19103/AS.2017.0034.01

world since ancient times (Wilson, 2008; Orf, 2016). Soybean oil is especially rich in unsaturated fatty acids compared with other vegetable oils, making it an affordable and healthy option to help meet human requirements (Asbridge, 1995; Goldsmith, 2008; Wilson, 2008). Coupling these health benefits with the need to replace fossil fuels with renewable energies, such as soybean biodiesel, is likely to further drive the demand for soybean production (Wilson, 2008). Apart from its health benefits and economic value, using soybean as a rotation crop can benefit agricultural sustainability and improve crop profitability by reducing farming costs (Tanner and Hume, 1978).

2 Biological nitrogen fixation in soybean

Some of the most important properties of soybean are achieved via a symbiotic association with nitrogen-fixing soil bacteria (broadly called 'rhizobium') that infect the roots of soybean plants and reside in specialised organs, called 'nodules' (Fig. 1), which form on the host plant roots. The atmospheric di-nitrogen (N_2) is initially fixed in the form of ammonium by the bacteria and subsequently converted into organic nitrogenous compounds (mainly ureides such as allantoin and allantoic acid, as well as glutamine) that benefit the host plant's nitrogen metabolism and improve soil conditions (Gresshoff et al., 2015). The rhizosphere benefits since the soil nitrogen content is replenished once the plant dies and decomposes. It has been estimated that soybean alone can produce up to 200 kg N ha^{-1} in a single season, of which up to just under 70% is derived from symbiotic nitrogen fixation (Jensen et al., 2012). As a result, legumes such as soybeans are widely used in crop rotations to improve soil health.

The global significance of the soybean market has resulted in a strong research effort to better understand the genetic and physiological basis of soybean development, with the goal of optimising its growth and yield in agriculture (Norman, 2012). Soybean research is largely facilitated by the continued development and refinement of experimental techniques and modern molecular and breeding approaches (Ferguson and Gresshoff 2009; Gresshoff et al., 2015). Advances from these efforts result in new varieties with improved traits, such as drought/pathogen resistance or enhanced nutritional qualities that meet the ever-evolving market needs. In particular, using soybean as a model legume species can assist in improving food security, soil health and agricultural sustainability via the enhanced understanding of legume nodulation and nitrogen fixation. Optimising

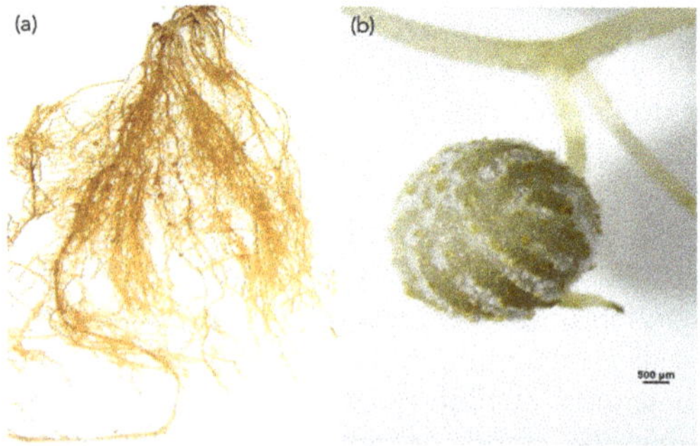

Figure 1 Soybean roots and nodules. (a) Soybean root system inoculated with *Bradyrhizobium diazoefficiens* exhibiting mature nodule structures. (b) A two-week-old soybean nodule.

these processes can help in reducing the use of expensive, often polluting, nitrogen-based fertilisers in agriculture.

The symbiotic relationship between legumes such as soybean and rhizobia is mediated by sophisticated signalling mechanisms (Ferguson, 2013). The process is initiated when the roots secrete flavonoids into the soil, attracting rhizobia which are stimulated to synthesise a signal molecule called Nod factor. LysM receptors on the plant root perceive Nod factor, which triggers a signalling cascade and drives nodule establishment (Ferguson, 2010; Indrasumunar et al., 2010; Indrasumunar et al., 2011). Following the attachment of bacteria to the emerging root hairs, specialised structures known as infection threads form, allowing bacteria to enter the roots. The infection causes cortical cells in the roots to divide, eventually forming the nodule. Dividing bacteria within nodules differentiate into specialised mature bacteroids which use a nitrogenase enzyme complex to fix the di-nitrogen into ammonia and other forms that the soybean plant can use. The ammonia is converted into compounds such as glutamine or ureides that can then be transported throughout the plant. To regulate nodule numbers the plant uses the Autoregulation of Nodulation (AON) pathway to achieve an appropriate balance between nodule formation and nitrogen acquisition. This broad introduction to nodule formation cannot, of course, capture the complexity of the processes involved which remain the subject of ongoing research using a range of techniques.

Research on nodule formation and other physiological processes in soybean has benefited from a wide range of analytical techniques. This chapter reviews the role of these different techniques in improving our understanding of soybean physiology, using nodule formation as one example of what different techniques can contribute.

3 Classical techniques for understanding soybean physiology

Fast germination, good size and stature, relatively short lifespan, large tissues and the ability to enter into symbioses are all traits that make soybean a great model species for both physiological and biochemical research. These features also enable soybean to be amenable to a number of classical physiological techniques (Ferguson and Gresshoff, 2009; Ferguson, 2013).

3.1 Grafting

As a conventional horticultural technique, grafting has its practical application in forward genetics to examine the contribution of different genetic backgrounds to a defined trait (Fig. 2 and 3). This is especially true when there is involvement of non-cell-autonomous signals travelling across the scion and rootstock junction.

In soybean, grafting has been a critical tool in understanding two prominent research areas: nodulation regulation and photoperiod/flowering studies (Mirzaei et al., 2017). The late flowering phenotype of scions from certain cultivars can be substantially reduced when grafted to early flowering rootstocks, indicating the existence of a systemic transmissible flowering stimulus (Heinze et al., 1942; Kiihl et al., 1977). These long-standing findings were subsequently shown in *Arabidopsis*, confirming the movement of a potent flowering-promoting factor, called FLOWERING LOCUS T protein (Sun et al., 2011).

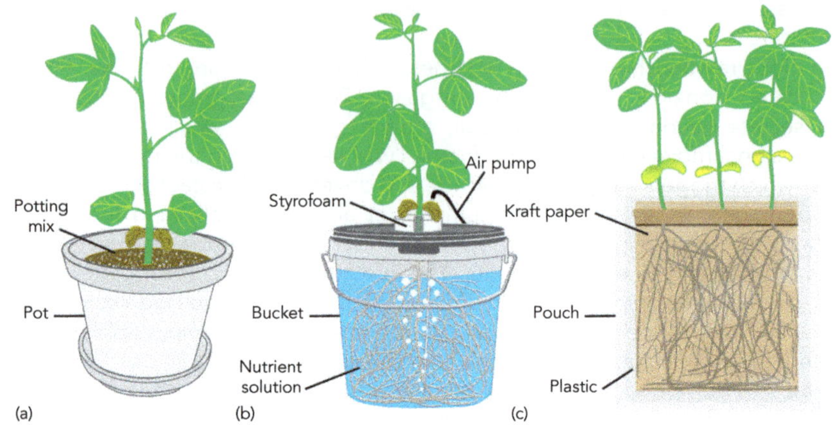

Figure 2 Different growing strategies for soybean research. (a) Potting media, such as soil, vermiculite, etc. (b) Hydroponic systems, where the growing solution requires regular replacement to avoid contamination. (c) Plastic growth pouches containing with unbleached kraft paper.

Similarly, reciprocal grafting between wild type and nitrate-tolerant symbiosis/supernodulation soybean mutants demonstrated the critical involvement of both root and shoot factors in nodulation regulation. Indeed, the supernodulation phenotype was determined by the shoot genotype, regardless of the genotype of the rootstock, indicating the existence of a systemic signal exchange pathway (Delves et al., 1986). This resulted in the identification of the AON process, an inbuilt mechanism restricting excessive nodulation in legumes (reviewed in Ferguson et al., 2010; Reid et al., 2011b).

3.2 Split root

Split root systems allow for the differential treatment of separated root systems on the same plant (*i.e.*, two root systems sharing a common shoot; Fig. 3a and 3b). Seedling taproots are severed and encouraged to grow into two segments that are separated into independent pots (Fig. 3a; Koch and Johnson, 1984; Olsson et al., 1989). Split root systems are useful for studying the effects of various factors on soybean growth and development, such as low soil pH (Lin et al., 2012), salinity (Singleton, 1983), drought (Marino et al., 2006) and the complex regulatory mechanisms induced by pathogens and symbionts such as rhizobia and mycorrhizae (Koch and Johnson, 1984; Olsson et al., 1989). Essentially, the split root system enables the identification and discrimination of systemic and local regulatory mechanisms, similarly to grafting (Fig. 3c).

3.3 Pouch growth

Growing plants in plastic growth pouches enables the visualisation of root system development (Fig. 2c). Soybean seeds or seedlings can be placed into commercially purchased pouches, consisting of unbleached filter paper surrounded by transparent

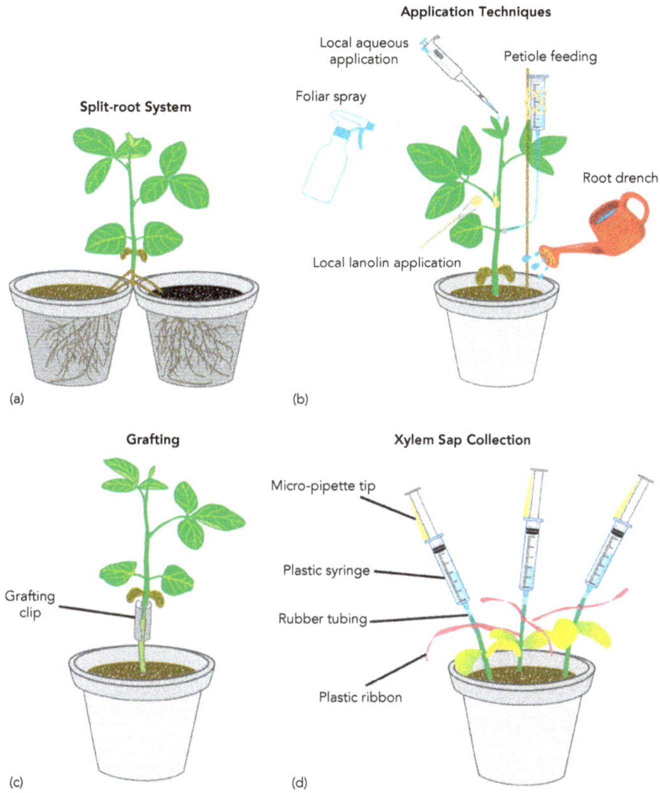

Figure 3 Established methods for soybean studies. (a) Split-root system enabling different treatments to be administered to the two sides of the root system. (b) Various application techniques of aqueous solutions. Aqueous solutions can be applied by root drenching, petiole feeding, micropipettes, foliar spray or diffusion from lanolin. (c) Grafting system. A heterologous scion and rootstock are secured at the wound site using grafting clips or rubber tubing. (d) Xylem sap extraction. Syringes connected to decapitated stems via short pieces of rubber tubing collect sap due to negative pressure created by placing a micropipette tip between the plunger and the barrel. The junction of the stem and rubber tubing is tightened using non-slip string or ribbon.

plastic (Fig. 2c). The technique is non-destructive, allows for the feeding of aqueous solutions and enables root systems to be continuously monitored throughout their growth (Hayashi et al., 2012; Corcilius et al., 2017). Limitations to pouch growth analyses include the limited space for root growth, and the risk that inhibitory gaseous compounds, such as ethylene, could build up within the pouch that inhibit processes such as legume nodulation and lateral root growth (Ferguson and Mathesius, 2014).

3.4 Hydroponics

Growing plants in soilless media, such as hydroponically, has become a widespread approach (Fig. 2b; Jones Jr, 1982; Parker and Norvell, 1999). It can minimise the possibility of pathogen infection and provides control over the constituents in the growing solution, including nutrients, signalling compounds and inhibitory ions (Kopittke et al., 2015). Drawbacks include lack of physical components for roots and root hairs to adhere to mechanically, thus representing an imperfect simulation of plant growth. Furthermore, the exchange of exudates at the root surface may be altered due to the root system being fully submerged.

3.5 Petiole feeding

To better understand the function of bioactive molecules in plant growth, a simple and effective petiole feeding bioassay has been developed to introduce aqueous solutions into the vascular stream and apoplast of various plants, including soybean (Lin et al., 2010). An advantage of petiole feeding is that relatively large volumes of solutions can be quickly fed directly into plants. The technique essentially consists of removing the plant petiole and replacing it with a syringe-based feeding apparatus that acts to directly feed solutions into the recipient plant (Fig. 2b; Lin et al., 2011). Petiole feeding represents a reliable method to study the role of signalling compounds in plant growth and development.

This approach is particular useful for compounds acting over long distances, such as those acting in source-sink allocation, hormone action, *etc.* (Ferguson et al., 2010; Ferguson and Mathesius 2014). It also adds an extra method to the toolbox used to distinguish the effects of signalling molecules on plant development, including foliar application (Fig. 2b; Carroll and Gresshoff, 1983), root irrigation (Fig. 2b; Ferguson et al., 2005; Reid et al., 2011a) and supplemented agar plates (Ridge et al., 1992).

3.6 Xylem sap collection

Xylem and phloem are two essential transport tissues in vascular plants; they act synergistically to transport and partition water, metabolites and signal molecules amongst plant organs (Alexou and Peuke, 2013). Xylem sap collection is a widely applied technique to help characterise the presence and function of root-to-shoot signalling molecules (Fig. 3d; Lin et al., 2010). It can also be used to determine the impact of various environmental parameters or exogenously applied compounds on the type and level of constituents being transported within the plant (Reid et al., 2012; Netting et al., 2012).

In addition to the systemic feeding, there are a few more classical and easily achievable techniques for applying chemicals of interest to soybean plants (Fig. 3b), including foliar spray (Garcia and Hanway, 1976), root drench (Heggestad, 1988), and aqueous (Wright and Nicell, 1999) and lanolin application (Huff and Dybing, 1980).

4 Mutagenesis methods and soybean mutant lines

A classical and powerful approach for determining a gene's function is the use of mutants to identify altered phenotypes and physiological responses. Indeed, the induction of mutations has been a driving force in plant science and soybean genetics since the 1980s (Carroll et al., 1985b; Day et al., 1987; Pracht et al., 1994; Meixner et al., 2005; Tripathi and Khare, 2016). Forward and reverse genetics are both widely employed to understand the function of genes in plant development, including soybean.

Forward genetics is the approach used to identify the gene responsible for a change in a mutant's phenotype (*i.e.*, to discover the mutated gene). One of the tremendous benefits of this approach is that genes are not specifically targeted but are identified based on function.

As an example, a series of soybean mutant lines were generated, including supernodulators *nts382* and *nts1007*, hypernodulator *nts1116* (Carroll et al., 1985a) and non-nodulators *nod49*, *nod772* and *nod139* (Carroll et al., 1986). Using forward genetics, all of the lines exhibiting excess nodulation were found to have mutations in a key membrane receptor protein acting in the nodulation control mechanism, AON, whilst *nod49* and *nod139* have mutations in the Nod factor receptor encoding genes, *NFR1* and *NFR5*, respectively (reviewed in Ferguson et al., 2010; Indrasumunar et al., 2010; Indrasumunar et al., 2011).

Reverse genetics begins with the selection of a gene or an array of genes, having a known nucleotide sequence and a proposed biochemical function, but the biological/cellular functions are undetermined. With the recent advance of deep genome sequencing and associated large-scale sequence data, the analysis of multi-gene families has particularly benefited from reverse genetics due to functional redundancy making forward genetics difficult to achieve (Alonso and Ecker, 2006). Indeed, many genes that are integral to a developmental process are unlikely to be identified by traditional forward genetic approaches due to functional redundancy, small gene size (such as peptide encoding genes) or because they are located in a region of the genome that is less amenable to mutagenesis. One drawback is that reverse genetics can be both costly and labour intensive (Alonso et al., 2003). However, it enables the targeted characterisation of a gene of interest, which forward genetic approaches cannot always achieve. For example, a number of CLE peptide encoding genes involved in AON and nitrate regulation of nodulation were identified using reverse genetic approaches in soybean (Reid et al., 2011a) and common bean (Ferguson et al., 2014).

A range of mutagenesis approaches are available, including chemical, irradiation and insertion mutagenesis. Each alters the nucleic acid sequence and has both advantages and disadvantages, which are detailed further here.

4.1 Ethyl methanesulfonate mutagenesis

Ethyl methanesulfonate (EMS) mutagenesis is the most commonly used chemical mutagen for plants, and can induce a number of mutations in the genome (Carroll et al., 1985a; Gresshoff and Delves, 1986; Greene et al., 2003). At an efficiency rate of 99%, EMS induces cytosine (C) to thymine (T) changes, resulting in C/G (guanine) to T/A (adenine) substitutions, with a mutation rate of 10^{-8} to 10^{-7} events per base pair per generation (Kovalchuk et al., 2000).

Baoge et al. (1995) demonstrated that higher doses (over 0.9%) of EMS reduce the frequency of successful mutations in soybean; therefore, an optimal dose is required to achieve high level of mutations.

Several attributes make EMS attractive for plant with large genomes, such as soybean and pea. Firstly, EMS can be applied to essentially any species of interest. Secondly, mutation frequency appears to be high regardless of genome size (Caldwell et al., 2004). Finally, EMS mutated lines are considered non-GMO, and thus can be grown in the field.

The single base substitutions caused by EMS fall into three categories: silent mutations (a base pair change that does not alter the encoded amino acid), missense mutations (a base pair change that causes an amino acid replacement), and nonsense mutations (a base pair change that introduces a stop codon) (McCallum et al., 2000a). Mutations also occur in non-coding sequences, sometimes affecting regulatory elements or splicing signals that control gene expression.

4.2 Fast neutron mutagenesis

Deletion mutants can also be obtained using fast neutron irradiation (Koornneeff et al., 1982; Eon et al., 2001). Results from fast neutron bombardment can delete DNA segments ranging from a few base pairs to several megabases (Li et al., 2001). As a result, one or multiple genes can be deleted per line, following which genes can be identified and subsequently cloned using forward genetics screening approaches (Meinke et al., 2003). Recent revelations in whole-genome technologies have increased the capacity for mutant screening. Its use in non-model species can be limited by not knowing the sequence of a targeted gene(s) (Men et al., 2002). A soybean fast neutron mutant population available for screening seed composition, maturity, morphology, pigmentation and nodulation can be found at http://www.soybase.org/mutants (Bolon et al., 2011), providing a resource for soybean research.

4.3 Transferred DNA (T-DNA)

Random insertional mutagenesis can be induced by transferred DNA (T-DNA) mediated by transposons. This technique has been quite effective for generating mutant populations in *Medicago truncatula*, *Lotus japonicus*, *Arabidopsis thaliana*, *Oryza sativa* (rice) and *Zea mays* (maize; *e.g.*, Mathieu et al., 2009). Biological agents, such as *Agrobacterium tumefaciens* and *Agrobacterium rhizogenes*, which can mediate the transfer of DNA or RNA into plants can also be employed as mutagens as outlined below.

4.4 Targeting Induced Local Lesions in Genomes

Targeting Induced Local Lesions in Genomes (TILLING) is a powerful reverse genetics technique that was developed to detect induced or natural DNA base pair changes (Alonso and Ecker, 2006). After the mutagen is applied, DNA samples from individual offspring are pooled and utilised as PCR templates to amplify the region of interest. To detect mutations, PCR templates are denatured and re-natured to allow for heteroduplex formation between wild type and mutated fragments. Base changes are subsequently detected around mismatches through digestion. After a substitution is detected, DNA from individual plants is amplified and PCR is used to confirm the sequence (McCallum et al., 2000a,b).

TILLING enables a wide range of substitutions to be screened in the genome. It offers advantages over other reverse genetic techniques, including its applicability to virtually any organism and its facility for high-throughput or less-expensive processes when in combination with high-performance liquid chromatography (Gilchrist and Haughn, 2005) or a mismatch-specific celery nuclease, CEL1, to identify SNPs (Colbert et al., 2001). The method does have drawbacks as it is relatively labour intensive and is based on random mutagenesis, so desired mutants might not be generated. TILLING has been successfully applied to a variety of plant species, including *Arabidopsis* (Greene et al., 2003), *L. japonicus* (Perry et al., 2003), maize (Till et al., 2004) and soybean (Cooper et al., 2008; Dierking and Bilyeu, 2009).

5 Modern experimental techniques applicable to soybean physiology research

With revolutionary advances achieved over the last few decades, molecular genetics has reached new heights and gradually become an integral part of modern biological research. Employment of appropriate techniques can largely facilitate context-dependent functional characterisation of various genes and dissection of complex signalling pathways. As a model legume, soybean is amenable to a wide array of experimental methodologies, ranging from the classical techniques mentioned above to the more sophisticated genome editing approaches discussed below.

Agrobacterium-mediated transformation is a well-established and cost-effective technique that is highly applicable to soybeans (Hinchee et al., 1988; Gelvin, 2003; Olhoft et al., 2003; Kereszt et al., 2007). It is based on the pathogenetic mechanism of *A. tumefaciens* or *A. rhizogenes*, which induce the formation of tumour- or root-like structures in a multitude of plant species by semi-randomly integrating a characteristic DNA segment (T-DNA) from its native Ti or Ri (tumour/root-inducing) plasmid into the host genome (Martinell et al., 2002). Genetic engineering of exogenous DNA sequences specified as T-DNA cassettes can enable the insertion of transgenes into the plant genome (Martinell et al., 2002). This transformation technique is commonly coupled with modern molecular research methods to study the functionality of root-specific genes, such as RNA interference (RNAi). *A. tumefaciens* infection of cotyledonary nodes is reported to be useable for the regeneration of stable transgenic soybean lines via tissue culturing (Ko et al., 2003; Olhoft et al., 2003; Somers et al., 2003). In contrast, infection of *A. rhizogenes* leads to the growth of transgenic hairy roots on an otherwise non-transgenic plant, resulting in the formation of a chimeric plant (Kereszt et al., 2007). The technique involves a relatively short generation time with reasonable accuracy, and is inexpensive in soybean (Kereszt et al., 2007). It provides significant advantages in soybean research relating to root architecture, soil-borne pathogen invasion and symbiosis establishment (Somers et al., 2003; Lozovaya et al., 2004; Kobae et al., 2010; Reid et al., 2011a, 2013; Liu et al., 2012; Wang et al., 2014).

Gene overexpression, RNAi and promoter-reporter gene fusions are all readily achievable in soybean, particularly in combination with hairy root transformation. For example, soybean hairy root plants harbouring an RNAi construct for an isoflavone synthase encoding gene were used to demonstrate the role of isoflavone in host plant signalling with their rhizobium partners (Subramanian et al., 2006). Furthermore, RNAi-based knockdown of a leucine-rich-repeat receptor-like kinase (*GmSARK*) helped to establish the role of this receptor in regulating leaf senescence (Li et al., 2006). Constitutive expression of *AtDREB1A* in soybean indicated that this factor can downregulate gibberellin levels (Suo et al., 2012). Moreover, overexpression of nodulation-suppressive CLE peptides, *GmRIC1* and *GmRIC2*, demonstrated that these signals act in the suppression of nodulation in the AON pathway (Reid et al., 2011a). Reporter gene expression driven by endogenous promoters has been critical to understand tissue-specific roles of key factors acting in various signalling pathways. Common reporter systems, such as β-glucuronidase (GUS), DsRed (a red fluorescent protein) and green fluorescent protein, are all commonly used for molecular visualisation studies using soybean (Kobae et al., 2010; Lin et al., 2011; Liu et al., 2012).

Recently, genome editing techniques have become increasingly prevalent. One common advantage conferred by these modern technologies, including ZFNs (zinc-finger nucleases), TALENs (Transcription Activator-Like Effectors Nucleases) and CRISPR/Cas9 (Clustered

Regularly Interspaced Short Palindromic Repeats/CRISPR-associated protein 9), is precise sequence modification, which helps maintain genomic integrity (Urnov et al., 2010; Joung and Sander, 2013; Sun et al., 2015). However, RNA/DNA sequence complementarity-based recognition of the cleavage sites allows the Type-II CRISPR-Cas9 system to have better targeting specificity compared with ZFNs and TALENS, and its utility and convenience have been experimentally demonstrated in soybean (Fauser et al., 2014; Sun et al., 2015; Du et al., 2016).

The CRISPR/Cas9 system is derived from a bacterial defensive mechanism against bacteriophages, as bacteria preserve immunological memories of ancestral viral invasions by incorporating short stretches of bacteriophage and infectious plasmid sequences ('spacer') into their CRISPR loci, separated by 23–47 base pair palindromic repeats (Weinberger et al., 2012). Integration of the scaffold regions (repeats) of CRISPR single-stranded guide RNAs (sgRNAs) into the Cas9 protein-based enzymatic machineries enables spacer sequence-specific cleavage of foreign plasmids or viral genomes (Weinberger et al., 2012).

The Cas9 protein-sgRNA complex generates double strand breakage at the target sites which can be subsequently repaired by either non-homologous end joining or homologous recombination (HR) (Cai et al., 2015; Jacobs et al., 2015; Sun et al., 2015). The former is particularly useful for inducing gene knockouts due to the resulting random mutations that often disrupt gene function. HR offers the possibility of precise sequence manipulation at favourable genomic sites given the provision of a donor source, such as insertion of exogenous DNA fragments and generation of specific point mutations (Cai et al., 2015; Li et al., 2015). Site-directed mutagenesis of the soybean acetolactate synthase1 gene was successfully conducted using CRISPR-Cas9, resulting in a single amino acid substitution of P178S (Li et al., 2015). The versatility of the CRISPR-Cas9 system offers tremendous advantages and considerably reduces issues associated with unwanted background mutations (typical of most other mutagenesis techniques) or a lack of desired mutants (McCallum et al., 2000b; Henikoff and Comai, 2003).

Simultaneous targeting of homologous gene pairs can be achieved with a single CRISPR-Cas9 construct, which is extremely valuable when investigating the highly duplicated soybean genome (Cai et al., 2015; Sun et al., 2015). Although mutagenesis efficiency varies enormously between species, it is relatively high in soybean using hairy root transformation, but the rate of simultaneous modification of gene homologs is still low (Cai et al., 2015; Jacobs et al., 2015; Sun et al., 2015). Compared with RNAi, one major limitation of this technique is restricted choice of DNA sequences that can be selected as sgRNAs, which have to be specific to the targets and followed by a proto-spacer adjacent motif, usually 5′-NGG-3′ depending on the species (Sun et al., 2015; Du et al., 2016).

Viral cDNA-based gene delivery system is another viable tool to study plant molecular genetics and there are a number of viruses that have been engineered to become disarmed viral vectors (Zhang and Ghabrial, 2006; Nagamatsu et al., 2007; Seo et al., 2009; Zhang et al., 2010). For instance, vectors developed from *bean pod mottle virus* were shown to be effective for virus-induced gene silencing and heterologous gene expression in soybean (Zhang and Ghabrial, 2006; Zhang et al., 2010; Liu et al., 2012). Recently, it has been suggested that a newly modified *soybean mosaic virus* vector has the capacity for cellular co-expression of two transgenes, providing a technical means to examine *in vivo* protein-protein interactions (Seo et al., 2016).

6 The soybean genome and associated bioinformatics resources

Due in part to its considerable economic significance, soybean was one of the first plant species to have its genome sequenced at a high level, and many associated bioinformatics resources are now available to the soybean research community (Schmutz et al., 2010). Soybean is an allotetraploid (2n = 4x = 40) resulting from two major genomic duplication events occurring approximately 59 and 13 MYA (Schmutz et al., 2010). As a result, homeologous (*i.e.*, duplicate) counterparts are found for nearly three-quarters of all predicted genes. Moreover, repetitive heterochromatic sequences constitute more than 50% of the soybean genome (Schmutz et al., 2010).

There are a number of user-friendly and reliable databases associated with the soybean genome, with each having slightly different foci. Almost all platforms support multiple querying options for information retrieval, including sequence identifier/call, gene/locus/QTL name, ontological description, family name, domain name, individual metabolite, pathway name, species and keyword (Severin et al., 2010; Cannon et al., 2012; Goodstein et al., 2012; Joshi et al., 2012; Dash et al., 2016). Another prevalent feature available with most soybean genome websites is the provision of a specialised set of online analytic tools designed to fulfil the established objective of each database, with BLAST and BLAT commonly encompassed (Dash et al., 2012, 2016; Goodstein et al., 2012; Joshi et al., 2012; Li et al., 2012).

SoyBase (http://soybase.org) is a host site for an array of soybean data and was designed to be seamlessly accessible through other online portals. In this database, reference genome sequences (*Glycine max* Wm82.a2.v1) can be viewed using Gbrowse, which enables the selective display of user-customised informative tracks that correspond to diverse genetic features within the selected genomic region, such as gene models (from sequence assembly versions, Glyma.Wm82.a1.v1.1 and Glyma.Wm82.a2.v1), genetic markers, and intraspecific and interspecific regions sharing similar microsynteny. Alternatively, searches for single genes will bring up a descriptive summary of the request. An RNA-Seq atlas generated in 2010 is also hosted in SoyBase, enabling users to query tissue-specific expression profiles of target genes, which is useful for exploring evolutionary outcomes and functional specialisations of gene family members (Severin et al., 2010). The same set of expression data can also be intuitively visualised in the eFP (electronic Fluorescent Pictograph) browser (http://bar.utoronto.ca/efpsoybean/cgi-bin/efpWeb.cgi?dataSource = soybean), which employs a colour gradient to scale the transcriptional level of an input gene onto graphical representations of different plant tissues. Transcriptome atlases of other tissue types and *Bradyrhizobium diazoefficiens*-inoculated samples are also available for viewing in the eFP browser (Winter et al., 2007; Libault et al., 2010a,b; Severin et al., 2010). For researchers and soybean breeders conducting genome-wide association studies, genetic linkage maps created using numerous genetic markers and QTLs can be browsed from CMap, with an option for users to jump to the Gbrowse view (Severin et al., 2010; Cannon et al., 2012). A dataset of soybean mutant lines generated via fast neutron irradiation is also incorporated in SoyBase (Cannon et al., 2012). Those harbouring mutations of interest can be filtered out by specifying expected genomic locations of indels or desired phenotypic traits.

Phytozome (https://phytozome.jgi.doe.gov/pz/portal.html) is a comparative genomics resource currently comprised of sixty-five fully annotated plant genomes (latest release v11). Data of single genes can be retrieved by identifier-, keyword- or sequence-homology-based searches (*e.g.*, Hastwell et al., 2015, 2017). The returned interactive interface offers selective access to categorised information with respect to ontology annotations, gene models, colour-coded sequences of different types, ranked protein homologs across all species housed in Phytozome, gene ancestry and available gene expression atlases, while users have the choice of shuttling to the genome browser viewing mode (Goodstein et al., 2012). The entire dataset of the query gene can be reorganised in the form of a textual summarising report automatically generated by PhytoMine. For the gene family view, there are four main panels, a full list of family members with schematic illustrations of the respective gene structure, PFAM domain positioning and genomic synteny, a list of all functional annotations found in the family, a multiple sequence alignment tab and a table displaying the evolutionary history of the gene family (Goodstein et al., 2012).

The Legume Information System (LIS) (http://legumeinfo.org) is a database established to aid interspecific comparative studies by integrating the genomic information of nine fully sequenced legume species using elaborately designed analytic tools (Cannon et al., 2012). LIS provides hyperlinks to the QTL maps of three species, including soybean, and direct access to the genetic linkage maps of twelve legumes (Dash et al., 2016). Unlike the web portals described above, the result page for single gene searches begins with a table concisely showing the names of the organism, ontologies, Phytozome code of the corresponding gene family, and mRNA and protein identifiers, followed by a link leading the users to the genomic browser for further information. An advantage of LIS is the interactive phylogenetic tree viewer for Phytozome-based gene families, which can be easily converted to the genomic context viewer at internal nodes (Dash et al., 2016). The programmable feature of the LIS genomic context viewer permits horizontal display of more than ten genes flanking upstream and downstream the query sequence on a track, with gene icons of the same colour representing genes of the same family (Goodstein et al., 2012; Dash et al., 2016). Currently, the phylogram is constructed of nine legume species, plus seven phylogenetically or scientifically significant outgroups, such as rice and *Arabidopsis*. The multiple sequence alignment of all constituent members is also provided on top of the phylogenetic tree (Dash et al., 2016). Importantly, with respect to individual genes, identifying those with similar genomic syntenies across the genomes available in LIS, or having fluent traversal to other bioinformatics resources (*e.g.*, SoyBase), can be easily achieved (Cannon et al., 2012; Dash et al., 2016).

Soybean Knowledge Base (SoyKB) (http://soykb.org) is a soybean-specific database developed for -omics data management and compilation, ensuring that massive high-throughput data generated by modern technologies can be delivered in a reasonable manner (Joshi et al., 2012, 2014). The gene card interface in SoyKB can provide an exhaustive description of each soybean gene, which is designed in an attempt to cover all the available genomics, transcriptomics, proteomics, metabolomics and epigenomics data, along with associated regulatory miRNA and 3D protein structure. In addition, information regarding SNPs, miRNAs/sRNAs and metabolic compounds is rendered searchable with multiple querying methods. The results are presented as similar entity cards containing data like the tissue-specific transcript abundance of miRNAs, and GCMS- and LCMS-based metabolite profiling in sample tissues of various treatments (Joshi et al., 2012). However, a portion of data in this repository is classified as 'access restricted', requiring registration (Joshi et al., 2014). There are a number of useful web tools uniquely offered by SoyKB, such as the pathway viewer, 3D protein structure modeller and the motif sampler (Joshi et al., 2012, 2014).

Most published soybean gene expression datasets and their respective analyses can be browsed and downloaded through PLEXdb (Plant Expression Database: http://www.plexdb.org/plex.php?database=Soybean; Wise et al., 2007; Dash et al., 2012). There is an overview provided for each experiment conducted to generate the data, primarily introducing the experimental design and key findings. Databases, such as Legume Base (https://www.legumebase.brc.miyazaki-u.ac.jp), are also available that distribute research resources, including seeds of different soybean varieties and mutant lines (Kai et al., 2010). Other online information resources with a soybean focus include:

- http://bioinformatics.psb.ugent.be/plaza/versions/plaza_v3_dicots/organism/view/Glycine+max – Plaza 3.0 (Proost et al., 2015).
- http://plantgrn.noble.org/LegumeIPv2/index.jsp – LegumeIP (Li et al., 2012).
- http://www.cilr.uq.edu.au – Centre for Integrative Legume Research homepage.

7 Conclusion

Soybean is a tremendously important crop species. It has unique agronomic and biological characteristics that are not fully represented by *Arabidopsis* or other model legume systems. Advances in gene identification and characterisation as well as signalling mechanism elucidation is greatly aided by the range of classical and modern manipulation techniques that can be utilised in conjunction with the available bioinformatics resources. Further genomic and physiological-based research will enhance the understanding of soybean growth, reproduction, nodulation and abiotic/biotic stress resistance, with new findings potentially aiding in the selection of superior varieties and the optimisation of current fertiliser regimes and farming practices. As an example, research identifying the genes and signals involved in nodule formation and AON provide potential targets for plant breeders seeking to improve the ability of soybean and other legumes to acquire nitrogen.

8 Where to look for further information

1. http://www.cilr.uq.edu.au – Centre for Integrative Legume Research.
2. http://www.phytozome.net – Phytozome.
3. http://sci-wikibook.bacs.uq.edu.au – Plants in Action.
4. http://soybase.org – Soybase.
5. http://www.bar.utoronto.ca/efpsoybean/cgi-bin/efpWeb.cgi – Soybean eFP browser.

9 References

Alexou, M. and Peuke, A. D. (2013). Methods for xylem sap collection. *Plant Mineral Nutrients: Methods and Protocols*, 195–207.

Alonso, J. M. and Ecker, J. R. (2006). Moving forward in reverse: genetic technologies to enable genome-wide phenomic screens in *Arabidopsis*. *Nature Reviews Genetics*, 7, 524–36.

Alonso, J. M., Stepanova, A. N., Leisse, T. J., Kim, C. J., Chen, H., Shinn, P., Stevenson, D. K., Zimmerman, J., Barajas, P., Cheuk, R., Gadrinab, C., Heller, C., Jeske, A., Koesema, E., Meyers, C. C., Parker, H., Prednis, L., Ansari, Y., Choy, N., Deen, H., Geralt, M., Hazari, N., Hom, E., Karnes, M., Mulholland, C., Ndubaku, R., Schmidt, I., Guzman, P., Aguilar-Henonin, L., Schmid, M., Weigel, D., Carter, D. E., Marchand, T., Risseeuw, E., Brogden, D., Zeko, A., Crosby, W. L., Berry, C. C. and Ecker, J. R. (2003). Genome-wide insertional mutagenesis of *Arabidopsis thaliana*. *Science*, 301, 653–7.

Asbridge, D. D. (1995). *Soybeans vs. Other Vegetable Oils as a Source of Edible Oil Products*. AcoS Press.

Bolon, Y. T., Haun, W. J., Xu, W. W., Grant, D., Stacey, M. G., Nelson, R. T., Gerhardt, D. J., Jeddeloh, J. A., Stacey, G., Muehlbauer, G. J., Orf, J. H., Naeve, S. L., Stupar, R. M. and Vance, C. P. (2011). Phenotypic and genomic analyses of a fast neutron mutant population resource in soybean. *Plant Physiology*, 156, 240–53.

Cai, Y., Chen, L., Liu, X., Sun, S., Wu, C., Jiang, B., Han, T. and Hou, W. (2015). CRISPR/Cas9-mediated genome editing in soybean hairy roots. *PLoS ONE*, 10, e0136064.

Caldwell, D. G., McCallum, N., Shaw, P., Muehlbauer, G. J., Marshall, D. F. and Waugh, R. (2004). A structured mutant population for forward and reverse genetics in barley (*Hordeum vulgare* l.). *The Plant Journal*, 40, 143–50.

Cannon, S. B., Crow, J. A. and Grant, D. (2012). Soybase and the legume information system: accessing information about the soybean and other legume genomes. *Designing Soybeans for 21st Century Markets*, 49–62.

Carroll, B. J. and Gresshoff, P. M. (1983). Nitrate inhibition of nodulation and nitrogen fixation in white clover. *Zeitschrift für Pflanzenphysiologie*, 110, 77–88.

Carroll, B. J., McNeil, D. L. and Gresshoff, P. M. (1985a). Isolation and properties of soybean [*Glycine max* (L.) Merr.] mutants that nodulate in the presence of high nitrate concentrations. *Proceedings of the National Academy of Sciences*, 82, 4162–6.

Carroll, B. J., McNeil, D. L. and Gresshoff, P. M. (1985b). A supernodulation and nitrate-tolerant symbiotic (nts) soybean mutant. *Plant Physiology*, 78, 34–40.

Carroll, B. J., McNeil, D. L. and Gresshoff, P. M. (1986). Mutagenesis of soybean [*Glycine max* (L.) Merr.] and the isolation of non-nodulating mutants. *Plant Science*, 47, 109–14.

Colbert, T., Till, B. J., Tompa, R., Reynolds, S., Steine, M. N., Yeung, A. T., McCallum, C. M., Comai, L. and Henikoff, S. (2001). High-throughput screening for induced point mutations. *Plant Physiology*, 126, 480–4.

Cooper, J. L., Till, B. J., Laport, R. G., Darlow, M. C., Kleffner, J. M., Jamai, A., El-Mellouki, T., Liu, S., Ritchie, R. and Nielsen, N. (2008). Tilling to detect induced mutations in soybean. *BMC Plant Biology*, 8, 9.

Corcilius, L., Hastwell, A. H., Zhang, M., Williams, J., Mackay, J., Gresshoff, P. M., Ferguson, B. J. and Payne, R. (2017). Arabinosylation modulates the growth-regulating activity of the peptide Hormone CLE40a from Soybean. *Cell Chemical Biology*, 24, http://dx.doi.org/10.1016/j.chembiol.2017.08.014.

Dash, S., Campbell, J. D., Cannon, E. K., Cleary, A. M., Huang, W., Kalberer, S. R., Karingula, V., Rice, A. G., Singh, J., Umale, P. E., Weeks, N. T., Wilkey, A. P., Farmer, A. D. and Cannon, S. B. (2016). Legume information system (Legumeinfo.Org): A key component of a set of federated data resources for the legume family. *Nucleic Acids Research*, 44(D1), D1181–8.

Dash, S., Van Hemert, J., Hong, L., Wise, R. P. and Dickerson, J. A. (2012). PlEXdb: gene expression resources for plants and plant pathogens. *Nucleic Acids Research*, 40(D1), D1194–201.

Day, A., Price, G., Schuller, K. and Gresshoff, P. M. (1987). Nodule physiology of a supernodulating soybean (*Glyine max*) mutant. *Functional Plant Biology*, 14, 527–38.

Delves, A. C., Mathews, A., Day, D. A., Carter, A. S., Carroll, B. J. and Gresshoff, P. M. (1986). Regulation of the soybean-rhizobium nodule symbiosis by shoot and root factors. *Plant Physiology*, 82, 588–90.

Dierking, E. C. and Bilyeu, K. D. (2009). New sources of soybean seed meal and oil composition traits identified through TILLING. *BMC Plant Biology*, 9, 89.

Du, H., Zeng, X., Zhao, M., Cui, X., Wang, Q., Yang, H., Cheng, H. and Yu, D. (2016). Efficient targeted mutagenesis in soybean by TALENs and CRISPR/Cas9. *Journal of Biotechnology*, 217, 90–7.

Eon, S., Culard, F., Sy, D., Charlier, M. and Spotheim-Maurizot, M. (2001). Radiation disrupts protein-DNA complexes through damage to the protein. The lac repressor-operator system. *Radiation Research*, 156, 110–17.

Fauser, F., Schiml, S. and Puchta, H. (2014). Both CRISPR/Cas-based nucleases and nickases can be used efficiently for genome engineering in *Arabidopsis thaliana*. *Plant Journal*, 79, 348–59.

Ferguson, B. J. (2013). *The Development and Regulation of Soybean Nodules*: INTECH Open Access Publisher.

Ferguson, B. J. and Gresshoff, P. M. (2009). Soybean as a model legume. *Grain Legumes*, 53(7).

Ferguson, B. J., Indrasumunar, A., Hayashi, S., Lin, M. H., Lin, Y. H., Reid, D. E. and Gresshoff, P. M. (2010). Molecular analysis of legume nodule development and autoregulation. *Journal of Integrative Plant Biology*, 52, 61–76.

Ferguson, B. J., Li, D., Hastwell, A. H., Reid, D. E., Li, Y., Jackson, S., Gresshoff, P. M. (2014). The soybean (*Glycine max*) nodulation-suppressive CLE peptide, GmRIC1, functions interspecifically in common white bean (*Phaseolus vulgaris*), but not in a supernodulating line mutated in PvNARK. *Plant Biotechnology Journal* 12: 1085–97.

Ferguson, B. J. and Mathesius, U. (2014). Phytohormone regulation of legume-rhizobia interactions. *Journal of Chemical Ecology*, 40, 770–90.

Ferguson, B. J., Ross, J. J. and Reid, J. B. (2005). Nodulation phenotypes of gibberellin and brassinosteroid mutants of pea. *Plant Physiology*, 138, 2396–405.

Garcia, L. and Hanway, J. J. (1976). Foliar fertilization of soybeans during the seed-filling period. *Agronomy Journal*, 68, 653–7.

Gelvin, S. B. (2003). *Agrobacterium*-mediated plant transformation: The biology behind the 'Gene-Jockeying' tool. *Microbiology and Molecular Biology Reviews*, 67, 16–37.

Gilchrist, E. J. and Haughn, G. W. (2005). TILLING without a plough: a new method with applications for reverse genetics. *Current Opinion in Plant Biology*, 8, 211–15.

Goldsmith, P. D. (2008). Economics of soybean production, marketing and utilization. *Soybeans Chemistry, Production, Processing, and Utilization*, 117–50.

Goodstein, D. M., Shu, S., Howson, R., Neupane, R., Hayes, R. D., Fazo, J., Mitros, T., Dirks, W., Hellsten, U., Putnam, N. and Rokhsar, D. S. (2012). Phytozome: a comparative platform for green plant genomics. *Nucleic Acids Research*, 40 (Database Issue), D1178–86.

Greene, E. A., Codomo, C. A., Taylor, N. E., Henikoff, J. G., Till, B. J., Reynolds, S. H., Enns, L. C., Burtner, C., Johnson, J. E., Odden, A. R., Comai, L. and Henikoff, S. (2003). Spectrum of chemically induced mutations from a large-scale reverse-genetic screen in *Arabidopsis*. *Genetics*, 164, 731–40.

Gresshoff, P. M. and Delves, A. C. (1986). Plant genetic approaches to symbiotic nodulation and nitrogen fixation in legumes. *A Genetic Approach to Plant Biochemistry*. Springer Vienna, 159–206.

Gresshoff, P. M., Hayashi, S., Biswas, B., Mirzaei, S., Indrasumunar, A., Reid, D., Samuel, S., Tollenaere, A., van Hameren, B. and Hastwell, A. and Ferguson, B. J. (2015). The value of biodiversity in legume symbiotic nitrogen fixation and nodulation for biofuel and food production. *Journal of Plant Physiology*, 172, 128–36.

Hastwell, A. H., Gresshoff, P. M. and Ferguson, B. J. (2015). Genome-wide annotation and characterisation of CLAVATA/ESR (CLE) peptide hormones of soybean (*Glycine max*) and common bean (*Phaseolus vulgaris*), and their orthologues of *Arabidopsis thaliana*. *Journal of Experimental Botany*, 66: 5271–87.

Hastwell, A. H., de Bang, T., Gresshoff, P. M. and Ferguson, B. J. (2017). CLE peptide-encoding gene families in *Medicago truncatula* and *Lotus japonicus*, compared with those of soybean, common bean and Arabidopsis. *Scientific Reports*, 7: 9384.

Hayashi, S., Reid, D. E., Lorenc, M. T., Stiller, J., Edwards, D., Gresshoff, P. M. and Ferguson, B. J. (2012). Transient nod factor-dependent gene expression in the nodulation-competent zone of soybean (*Glycine max* [l.] merr.) roots. *Plant Biotechnology Journal*, 10, 995–1010.

Heggestad, H. E. (1988). Reduction in soybean seed yields by ozone air pollution. *JAPCA*, 38, 1040–1.

Heinze, P., Parker, M. and Borthwick, H. (1942). Floral initiation in biloxi soybean as influenced by grafting. *Botanical Gazette*, 103, 518–30.

Henikoff, S. and Comai, L. (2003). Single-nucleotide mutations for plant functional genomics. *Annual Review of Plant Biology*, 54, 375–401.

Hinchee, M. A., Connor-Ward, D. V., Newell, C. A., McDonnell, R. E., Sato, S. J., Gasser, C. S., Fischhoff, D. A., Re, D. B., Fraley, R. T. and Horsch, R. B. (1988). Production of transgenic soybean plants using *Agrobacterium*-mediated DNA transfer. *Bio/Technology*, 6, 915–22.

Huff, A. and Dybing, C. D. (1980). Factors affecting shedding of flowers in soybean (*Glycine max* (L.) Merrill). *Journal of Experimental Botany*, 31, 751–62.

Indrasumunar, A., Kereszt, A., Miyagi, M., Nguyen, C. D. T., Li, D.-X., Searle, I., Men, A., Carroll, B. J. and Gresshoff, P. M. (2010). Inactivation of duplicated Nod-Factor Receptor 5 (NFR5) genes in recessive loss-of-function non-nodulation mutants of allotetraploid soybean (*Glycine max* L. Merr.). *Plant Cell Physiology*, 51: 201–14.

Indrasumunar, A., Searle, I., Lin, M. H., Kereszt, A., Men, A., Carroll, B. J. and Gresshoff, P. M. (2011). Nodulation factor receptor kinase 1α controls nodule organ number in soybean (*Glycine max* L. Merr). *The Plant Journal*, 65(1), 39–50.

Jacobs, T. B., LaFayette, P. R., Schmitz, R. J. and Parrott, W. A. (2015). Targeted genome modifications in soybean with CRISPR/Cas9. *BMC Biotechnology*, 15, 16.

Jensen, E. S., Peoples, M. B., Boddey, R. M., Gresshoff, P. M., Hauggaard-Nielsen, H., Alves, B. J. R. and Morrison, M. J. (2012). Legumes for mitigation of climate change and the provision of feedstock for biofuels and biorefineries. A review. *Agronomy for Sustainable Development*, 32, 329–64.

Jones, Jr J. B. (1982). Hydroponics: its history and use in plant nutrition studies. *Journal of Plant Nutrition*, 5, 1003–30.

Joshi, T., Fitzpatrick, M. R., Chen, S., Liu, Y., Zhang, H., Endacott, R. Z., Gaudiello, E. C., Stacey, G., Nguyen, H. T. and Xu, D. (2014). Soybean knowledge base (SoyKB): A web resource for integration of soybean translational genomics and molecular breeding. *Nucleic Acids Research*, 42(D1), D1245–52.

Joshi, T., Patil, K., Fitzpatrick, M. R., Franklin, L. D., Yao, Q., Cook, J. R., Wang, Z., Libault, M., Brechenmacher, L., Valliyodan, B., Wu, X., Cheng, J., Stacey, G., Nguyen, H. T. and Xu, D. (2012). Soybean knowledge base (SoyKB): a web resource for soybean translational genomics. *BMC Genomics*, 13, S15.

Joung, J. K. and Sander, J. D. (2013). TALENs: a widely applicable technology for targeted genome editing. *Nature Reviews Molecular Cell Biology*, 14, 49–55.

Kai, S., Tanaka, H., Hashiguchi, M., Iwata, H. and Akashi, R. (2010). Analysis of genetic diversity and morphological traits of Japanese *Lotus japonicus* for establishment of a core collection. *Breeding Science*, 60, 436–46.

Kereszt, A., Li, D., Indrasumunar, A., Nguyen, C. D., Nontachaiyapoom, S., Kinkema, M. and Gresshoff, P. M. (2007). *Agrobacterium rhizogenes*-mediated transformation of soybean to study root biology. *Nature Protocols*, 2, 948–52.

Kiihl, R., Hartwig, E. and Kilen, T. (1977). Grafting as a tool in soybean breeding. *Crop Science*, 17, 181–3.

Ko, T. S., Lee, S., Krasnyanski, S. and Korban, S. S. (2003). Two critical factors are required for efficient transformation of multiple soybean cultivars: *Agrobacterium* strain and orientation of immature cotyledonary explant. *Theoretical and Applied Genetics*, 107, 439–47.

Kobae, Y., Tamura, Y., Takai, S., Banba, M. and Hata, S. (2010). Localized expression of arbuscular mycorrhiza-inducible ammonium transporters in soybean. *Plant and Cell Physiology*, 51, 1411–15.

Koch, K. E. and Johnson, C. R. (1984). Photosynthate partitioning in split-root citrus seedlings with mycorrhizal and nonmycorrhizal root systems. *Plant Physiology*, 75, 26–30.

Koornneeff, M., Dellaert, L. and Van der Veen, J. (1982). EMS-and relation-induced mutation frequencies at individual loci in *Arabidopsis thaliana* (L.) Heynh. *Mutation Research/Fundamental and Molecular Mechanisms of Mutagenesis*, 93, 109–23.

Kopittke, P. M., Moore, K. L., Lombi, E., Gianoncelli, A., Ferguson, B. J., Blamey, F. P. C., Menzies, N. W., Nicholson, T. M., McKenna, B. A., Wang, P., Gresshoff, P. M., Kourousias, G., Webb, R. I., Green, K. and Tollenaere, A. (2015). Identification of the primary lesion of toxic aluminum (Al) in plant roots. *Plant Physiology* 167: 1402–11.

Kovalchuk, I., Kovalchuk, O. and Hohn, B. (2000). Genome-wide variation of the somatic mutation frequency in transgenic plants. *The EMBO Journal*, 19, 4431–8.

Li, J., Dai, X., Liu, T. and Zhao, P. X. (2012). LegumeIP: an integrative database for comparative genomics and transcriptomics of model legumes. *Nucleic Acids Research*, 40(D1), D1221–9.

Li, X., Song, Y., Century, K., Straight, S., Ronald, P., Dong, X., Lassner, M. and Zhang, Y. (2001). A fast neutron deletion mutagenesis-based reverse genetics system for plants. *The Plant Journal*, 27, 235–42.

Li, X. P., Gan, R., Li, P. L., Ma, Y. Y., Zhang, L. W., Zhang, R., Wang, Y. and Wang, N. N. (2006). Identification and functional characterization of a leucine-rich repeat receptor-like kinase gene that is involved in regulation of soybean leaf senescence. *Plant Molecular Biology*, 61, 829–44.

Li, Z., Liu, Z. B., Xing, A., Moon, B. P., Koellhoffer, J. P., Huang, L., Ward, R. T., Clifton, E., Falco, S. C. and Cigan, A. M. (2015). Cas9-guide RNA directed genome editing in soybean. *Plant Physiology*, 169, 960–70.

Libault, M., Farmer, A., Brechenmacher, L., Drnevich, J., Langley, R. J., Bilgin, D. D., Radwan, O., Neece, D. J., Clough, S. J., May, G. D. and Stacey, G. (2010a). Complete transcriptome of the soybean root hair cell, a single-cell model, and its alteration in response to *Bradyrhizobium japonicum* infection. *Plant Physiology*, 152, 541–52.

Libault, M., Farmer, A., Joshi, T., Takahashi, K., Langley, R. J., Franklin, L. D., He. J., Xu, D., May, G. and Stacey, G. (2010b). An integrated transcriptome atlas of the crop model *Glycine max*, and its use in comparative analyses in plants. *Plant Journal*, 63, 86–99.

Lin, M. H., Gresshoff, P. M. and Ferguson, B. J. (2012). Systemic regulation of soybean nodulation by acidic growth conditions. *Plant Physiology*, 160, 2028–39.

Lin, M. H., Gresshoff, P. M., Indrasumunar, A. and Ferguson, B. J. (2011). pHairyRed: a novel binary vector containing the DsRed2 reporter gene for visual selection of transgenic hairy roots. *Molecular Plant*, 4, 537–45.

Lin, Y. H., Ferguson, B. J., Kereszt, A. and Gresshoff, P. M. (2010). Suppression of hypernodulation in soybean by a leaf-extracted, NARK-and Nod factor-dependent, low molecular mass fraction. *New Phytologist*, 185, 1074–86.

Lin, Y. H., Lin, M. H., Gresshoff, P. M. and Ferguson, B. J. (2011). An efficient petiole-feeding bioassay for introducing aqueous solutions into dicotyledonous plants. *Nature Protocols*, 6, 36–45.

Liu, S., Kandoth, P. K., Warren, S. D., Yeckel, G., Heinz, R., Alden, J., Yang, C., Jamai, A., El-Mellouki, T., Juvale, P. S., Hill, J., Baum, T. J., Cianzio, S., Whitham, S. A., Korkin, D., Mitchum, M. G. and Meksem, K. (2012). A soybean cyst nematode resistance gene points to a new mechanism of plant resistance to pathogens. *Nature*, 492, 256–60.

Lozovaya, V. V., Lygin, A. V., Zernova, O. V., Li, S., Hartman, G. L. and Widholm, J. M. (2004). Isoflavonoid accumulation in soybean hairy roots upon treatment with *Fusarium solani*. *Plant Physiology and Biochemistry*, 42, 671–9.

Marino, D., González, E. M. and Arrese-Igor, C. (2006). Drought effects on carbon and nitrogen metabolism of pea nodules can be mimicked by paraquat: evidence for the occurrence of two regulation pathways under oxidative stresses. *Journal of Experimental Botany*, 57, 665–73.

Martinell, B. J., Julson, L. S., Emler, C. A., Huang, Y., McCabe, D. E. and Williams, E. J. (2002). Soybean *Agrobacterium* transformation method. U.S. Patents and Trademark Office, No.6384301.

Mathieu, M., Winters, E. K., Kong, F., Wan, Wang, S., Eckert, H., Luth, D., Paz, M., Donovan, C., Zhang, Z., Somer, D., Wang, K., Nguyen, H., Shoemaker, R. C., Stacey, G. and Clemente, T. (2009). Establishment of a soybean (*Glycine max* Merr. L) transposon-based mutagenesis repository. *Planta*, 229, 279–89.

McCallum, C. M., Comai, L., Greene, E. A. and Henikoff, S. (2000a). Targeted screening for induced mutations. *Nature Biotechnology*, 18, 455–7.

McCallum, C. M., Comai, L., Greene, E. A. and Henikoff, S. (2000b). Targeting induced locallesions in genomes (TILLING) for plant functional genomics. *Plant Physiology*, 123, 439–42.

Meinke, D. W., Meinke, L. K., Showalter, T. C., Schissel, A. M., Mueller, L. A. and Tzafrir, I. (2003). A sequence-based map of *Arabidopsis* genes with mutant phenotypes. *Plant Physiology*, 131, 409–18.

Meixner, C., Ludwig-Müller, J., Miersch, O., Gresshoff, P. M., Staehelin, C. and Vierheilig, H. (2005). Lack of mycorrhizal autoregulation and phytohormonal changes in the supernodulating soybean mutant *nts1007*. *Planta*, 222, 709–15.

Men, A. E., Laniya, T. S., Searle, I. R., Iturbe-Ormaetxe, I., Gresshoff, I., Jiang, Q., Carroll, B. J. and Gresshoff, P. M. (2002). Fast neutron mutagenesis of soybean (*Glycine soja* L.) produces a supernodulating mutant containing a large deletion. *Genome Letters*, 3, 147–55.

Mirzaei, S., Batley, J., El-Mellouki, T., Liu, S., Meksem, K., Ferguson, B. J. and Gresshoff, P. M. (2017). Neodiversification of homeologous CLAVATA1-like receptor kinase genes in soybean leads to distinct developmental outcomes. *Scientific Reports*, 7: 8878.

Nagamatsu, A., Masuta, C., Senda, M., Matsuura, H., Kasai, A., Hong, J. S., Kitamura, K., Abe, J. and Kanazawa, A. (2007). Functional analysis of soybean genes involved in flavonoid biosynthesis by virus-induced gene silencing. *Plant Biotechnology Journal*, 5, 778–90.

Norman, A. G. Ed. (2012). *Soybean Physiology, Agronomy, and Utilization*. New York. Elsevier., pp. 1–17.

Olhoft, P. M., Flagel, L. E., Donovan, C. M. and Somers, D. A. (2003). Efficient soybean transformation using hygromycin B selection in the cotyledonary-node method. *Planta*, 216, 723–35.

Olsson, J. E., Nakao, P., Bohlool, B. B. and Gresshoff, P. M. (1989). Lack of systemic suppression of nodulation in split root systems of supernodulating soybean (*Glycine max* [l.] merr.) mutants. *Plant Physiology*, 90, 1347–52.

Orf, J. (2016). Introduction. In Bilyeu, K., Ratnaparkhe, M. B. and Kole, C. (Eds), *Genetics, Genomics, and Breeding of Soybean*. Boca Raton: CRC Press, pp. 1–18.

Parker, D. R. and Norvell, W. A. (1999). Advances in solution culture methods for plant mineral nutrition research. *Advances in Agronomy*, 65, 151–213.

Perry, J. A., Wang, T. L., Welham, T. J., Gardner, S., Pike, J. M., Yoshida, S. and Parniske, M. (2003). A TILLING reverse genetics tool and a web-accessible collection of mutants of the legume *Lotus japonicus*. *Plant Physiology*, 131, 866–71.

Pracht, J., Nickell, C., Harper, J. and Bullock, D. (1994). Agronomic evaluation of non-nodulating and hypernodulating mutants of soybean. *Crop Science*, 34, 738–40.

Proost, S., Van Bel, M., Vaneechoutte, D., Van de Peer, Y., Inze, D., Mueller-Roeber, B. and Vandepoele, K. (2015). Plaza 3.0: An access point for plant comparative genomics. *Nucleic Acids Research*, 43(Database issue), D974–81.

Reid, D. E., Ferguson, B. J. and Gresshoff, P. M. (2011a). Inoculation-and nitrate-induced CLE peptides of soybean control NARK-dependent nodule formation. *Molecular Plant-Microbe Interactions*, 24, 606–18.

Reid, D. E., Ferguson, B. J., Hayashi, S., Lin, Y. H. and Gresshoff, P. M. (2011b). Molecular mechanisms controlling legume autoregulation of nodulation. *Annals of Botany*, 108, 789–95.

Reid, D. E., Hayashi, S., Lorenc, M., Stiller, J., Edwards, D., Gresshoff, P. M. and Ferguson, B. J. (2012). Identification of systemic responses in soybean nodulation by xylem sap feeding and complete transcriptome sequencing reveal a novel component of the autoregulation pathway. *Plant Biotechnology Journal*, 10, 680–9.

Reid, D. E., Li, D., Ferguson, B. J. and Gresshoff, P. M. (2013). Structure–function analysis of the GmRIC1 signal peptide and CLE domain required for nodulation control in soybean. *Journal of Experimental Botany*, 64, 1575–85.

Ridge, R., Bender, G. and Rolfe, B. (1992). Nodule-like structures induced on the roots of wheat seedlings by the addition of the synthetic auxin 2, 4-dichlorophenoxyacetic acid and the effects of microorganisms. *Functional Plant Biology*, 19, 481–92.

Schmutz, J., Cannon, S. B., Schlueter, J., Ma, J., Mitros, T., Nelson, W., Hyten, D. L., Song, Q., Thelen, J. J., Cheng, J., Xu, D., Hellsten, U., May, G. D., Yu, Y., Sakurai, T., Umezawa, T., Bhattacharyya,

M. K., Sandhu, D., Valliyodan, B., Lindquist, E., Peto, M., Grant, D., Shu, S., Goodstein, D., Barry, K., Futrell-Griggs, M., Du, J., Tian, Z., Zhu, L., Gill, N., Joshi, T., Libault, M., Sethuraman, A., Zhang, X. C., Shinozaki, K., Nguyen, H. T., Wing, R. A., Cregan, P., Specht, J., Grimwood, J., Rokhsar, D., Stacey, G., Shoemaker, R. C. and Jackson, S. A. (2010). Genome sequence of the paleopolyploid soybean. *Nature*, 463, 178–83.

Seo, J. K., Lee, H. G. and Kim, K. H. (2009). Systemic gene delivery into soybean by simple rub-inoculation with plasmid DNA of a *soybean mosaic virus*-based vector. *Archives of Virology*, 154, 87–99.

Seo, J. K., Choi, H. S. and Kim, K. H. (2016). Engineering of *soybean mosaic virus* as a versatile tool for studying protein-protein interactions in soybean. *Scientific Report*, 6, 22436.

Severin, A. J., Woody, J. L., Bolon, Y. T., Joseph, B., Diers, B. W., Farmer, A. D, Muehlbauer, G. J., Nelson, R. T., Grant, D., Specht, J. E., Graham, M. A., Cannon, S. B., May, G. D., Vance, C. P. and Shoemaker, R. C. (2010). RNA-Seq atlas of *Glycine max*: A guide to the soybean transcriptome. *BMC Plant Biology*, 10, 1–16.

Singh, U. and Singh, B. (1992). Tropical grain legumes as important human foods. *Economic Botany*, 46, 310–21.

Singleton, P. (1983). A split-root growth system for evaluating the effect of salinity on components of the soybean *Rhizobium japonicum* symbiosis. *Crop Science*, 23(2), 259–62.

Somers, D. A., Samac, D. A. and Olhoft, P. M. (2003). Recent advances in legume transformation. *Plant Physiology*, 131, 892–9.

Specht, J., Chase, K., Macrander, M., Graef, G., Chung, J., Markwell, J., Germann, M., Orf, J. and Lark, K. (2001). Soybean response to water. *Crop Science*, 41, 493–509.

Subramanian, S., Stacey, G. and Yu, O. (2006). Endogenous isoflavones are essential for the establishment of symbiosis between soybean and *Bradyrhizobium japonicum. The Plant Journal*, 48, 261–73.

Sun, H., Jia, Z., Cao, D., Jiang, B., Wu, C., Hou, W., Liu, Y., Fei, Z., Zhao, D., and, Han, T. (2011). GmFT2a, a soybean homolog of FLOWERING LOCUS T, is involved in flowering transition and maintenance. *PLoS ONE*, 6, e29238.

Sun, X., Hu, Z., Chen, R., Jiang, Q., Song, G., Zhang, H. and Xi, Y. (2015). Targeted mutagenesis in soybean using the CRISPR-Cas9 system. *Science Report*, 5, 10342.

Suo, H., Ma, Q., Ye, K., Yang, C., Tang, Y., Hao, J., Zhang, Z. J., Chen, M., Feng, Y. and Nian, H. (2012). Overexpression of *AtDREB1A* causes a severe dwarf phenotype by decreasing endogenous gibberellin levels in soybean [*Glycine max* (L.) Merr.]. *PLoS ONE*, 7, e45568.

Tanner, J. W. and Hume, D. J. (1978). Management and production. In A. G. Norman (Ed.), *Soybean Physiology, Agronomy and Utilization* (pp. 157–217). New York: Academic Press.

Till, B. J., Reynolds, S. H., Weil, C., Springer, N., Burtner, C., Young, K., Bowers, E., Codomo, C. A., Enns, L. C., Odden, A. R., Greene, E. A., Comai, L. and Henikoff, S. (2004). Discovery of induced point mutations in maize genes by TILLING. *BMC Plant Biology*, 4, 12.

Tripathi, N. and Khare, D. (2016). Molecular approaches for genetic improvement of seed quality and characterization of genetic diversity in soybean: a critical review. *Biotechnology Letters*, 38, 1–10.

Urnov, F. D., Rebar, E. J., Holmes, M. C., Zhang, H. S. and Gregory, P. D. (2010). Genome editing with engineered zinc finger nucleases. *Nature Reviews Genetics*, 11, 636–46.

Wang, Y., Wang, L., Zou, Y., Chen, L., Cai, Z., Zhang, S., Zhao, F., Tian, Y., Jiang, Q., Ferguson, B. J., Gresshoff, P. M. and Li, X. (2014). Soybean miR172c targets the repressive AP2 transcription factor NNC1 to activate *ENOD40* expression and regulate nodule initiation. *The Plant Cell*, 26, 4782–801.

Weinberger, A. D., Sun, C. L., Pluciński, M. M., Denef, V. J., Thomas, B. C., Horvath, P., Barrangou, R., Gilmore, M. S., Getz, W. M. and Banfield, J. F. (2012). Persisting viral sequences shape microbial CRISPR-based immunity. *PLoS Computational Biology*, 8, e1002475.

Wilson, R. F. (2008). Soybean: market driven research needs. In G. Stacey (Ed.), *Genetics and Genomics of Soybean*. New York, Springer, pp. 3–15.

Winter, D., Vinegar, B., Nahal, H., Ammar, R., Wilson, G. V. and Provart, N. J. (2007). An 'Electronic Fluorescent Pictograph' browser for exploring and analyzing large-scale biological data sets. *PLoS ONE*, 2, e718.

Wise, R. P., Caldo, R. A., Hong, L., Shen, L., Cannon, E. and Dickerson, J. A. (2007). Barleybase/PLEXdb. *Plant Bioinformatics: Methods and Protocols*, 406, 347–63.

Wright, H. and Nicell, J. A. (1999). Characterization of soybean peroxidase for the treatment of aqueous phenols. *Bioresource Technology*, 70, 69–79.

Zhang, C., Bradshaw, J. D., Whitham, S. A. and Hill, J. H. (2010). The development of an efficient multi-purpose *bean pod mottle virus* viral vector set for foreign gene expression and RNA silencing. *Plant Physiology*, 153, 52–65.

Zhang, C. and Ghabrial, S. A. (2006). Development of *bean pod mottle virus*-based vectors for stable protein expression and sequence-specific virus-induced gene silencing in soybean. *Virology*, 344, 401–11.

Zhu, B., Gu, A., Deng, X. D., Deng, Y. X. and Lu, Z. X. (1995). Effects of caffeine or EDTA post-treatment on EMS mutagenesis in soybean. *Mutation Research/Environmental Mutagenesis and Related Subjects*, 334, 157–9.

Chapter 2

Maintaining and utilizing the genetic diversity of soybeans

Randall Nelson, USDA-ARS and University of Illinois, USA

1 Introduction

The goal of the art and science of plant breeding is to create genetically diverse gene pools and then devise strategies to consolidate the favourable alleles and eliminate those that are adverse to the objectives. Genetic diversity is the fuel that powers plant breeding but it is also the largest impediment to progress. Historically, most cultivar development programmes seem to operate with the assumption that the latter is more important than the former. In this chapter, the major soybean germplasm collections of the world which are the sources of available genetic diversity are briefly described. The formation of the current commercially used gene pool for soybean breeding in North America, some reasons why it has served us very well and some concerns about its limitations will be reviewed. This chapter also presents the perspective that adding new genetic diversity can improve the rate of gain for seed yield and how the various sources of exotic germplasm can be utilized within different segments of a national soybean programme.

2 Germplasm collections

Plant breeders access new genetic diversity from *ex situ* collections. Internationally, there are more than 40 soybean germplasm collections with at least 1000 accessions that, as of 2004, totalled more than 150 000 accessions (Carter et al., 2004). The most significant collections in terms of size and distinctiveness are in the countries of ancient cultivation as well as the United States. Table 1 lists the inventory of these five collections. The Institute of Crop Sciences in the Chinese Academy of Agriculture Science in Beijing, China, has

http://dx.doi.org/10.19103/AS.2017.0034.02

Table 1 Number of accessions in the major *Glycine* germplasm collections from the countries of ancient cultivation and the United States

Institution	Country	*G. max* accessions	*G. soja* accessions	Perennial *Glycine* species	Perennial *Glycine* accessions
Institute of Crop Germplasm Resources, Chinese Academy of Agricultural Sciences	China	25584	8426	–	–
Genetic Resources Center, National Agriculture and Food Research Organization	Japan	7879	1675	2	2
N. I. Vavilov Research Institute of Plant Genetic Resources	Russia	7055	330	–	–
Rural Development Administration Genebank	S. Korea	18445	3029	5	29
USDA Soybean Germplasm Collection	U. S.A	19622	1179	19	1005

the largest collection and it continues to grow. Since 2004 the number of *Glycine max* accessions has increased by 8% and the number of *G. soja* accessions has increased over 35% (Qiu Lijuan, pers. comm., 2016). China also has provincial collections including a major collection at the Jilin Academy of Agricultural Sciences and a large collection at Nanjing Agricultural University which may contain unique accessions. Rural Development Administration (RDA) Genebank in South Korea has recently expanded its facilities and has greatly increased the number of annual *Glycine* accessions in the past decade (http://genebank.rda.go.kr). The Genetic Resources Center at the National Agriculture and Food Research Organization (http://www.gene.affrc.go.jp) is the repository for *Glycine* germplasm in Japan. N.I. Vavilov Research Institute of Plant Genetic Resources (http://www.vir.nw.ru) has the soybean collection in Russia. The U.S. Department of Agriculture (USDA) Soybean Germplasm Collection (http://www.ars-grin.gov/npgs/) has had modest growth in the past decade with a 5% increase in the number of stored accessions. It is the primary source of *Glycine* germplasm in the world, distributing an average of 34 500 seed lots per year over the past decade. Commonwealth Scientific and Industrial Research Organization in Australia maintains the largest collection of perennial *Glycine* with over 2100 accessions (Carter et al., 2004) representing 26 species (Singh and Chung, 2007). There has been no universal inventory of the soybean germplasm collections but we know that there is considerable redundancy among all the collections. Using common identifiers that could be verified, at least 5000 accessions are in common between the USDA Soybean Germplasm Collection and the soybean collection in RDA Genebank in South Korea. Probably less than a third of the accessions held worldwide are unique, but it is currently not possible to verify that estimate (Carter et al., 2004). Whatever numbers of unique accessions are currently being preserved, compared to what is currently being used in cultivar development programmes, it represents an enormous untapped resource.

The global inventory of *G. max* is not likely to increase greatly. There are few places in the world where soybean can still be collected because primitive cultivars have been replaced with scientifically developed cultivars in almost all farmers' fields. Wild soybean accessions held in *ex situ* collections are only a very small percentage of the total *Glycine* holdings, and there is a need and significant opportunity to collect additional wild soybean throughout East Asia where it is native. Wild soybean is a vigorous colonizer and can be found in many habitats including disturbed landscapes. Because of these characteristics, populations of wild soybean are less vulnerable than some wild populations, but nevertheless they could be lost over time. Perhaps more importantly, wild soybean accessions not in *ex situ* collections are not available as plant breeding resources.

3 The U.S. gene pool

The records of the USDA Uniform Soybean Tests provide the best, although incomplete, information about how the current commercially used gene pool was established. The USDA-coordinated regional testing programmes began in the northern U.S. in 1939 and in the southern U.S. in 1943. Since the USDA Soybean Germplasm Collection was not established until 1949 and only 20% of the 8300 documented soybean introductions into the United States were still extant when the Collection was formed, we do not know exactly what was available to soybean breeders during these early years. We have complete records of what was included in the Uniform Tests but we do not know what was evaluated within states and not advanced to regional testing, so these records provide us with a conservative estimate of the range of diversity from which our commercially used gene pool was selected. From the beginning of the Uniform Tests through 1955, there were 86 plant introductions used as parents of the tested lines and about 60 plant introductions were evaluated directly as entries. Only 37 of the introductions used as parents became part of the genetic base of North American soybean breeding and these lines included 17 of the 36 ancestral lines that contribute 95% of the genes to the current U.S. genetic base (Gizlice et al., 1994). Most other members of the genetic base were selected as sources of disease resistance (Gizlice et al., 1994).

It is likely that some of these early parental lines did not become part of the U.S. gene pool because the number of progeny tested was insufficient to evaluate their potential or they were not crossed with the right lines to expose their genetic potential. There was obvious selection against some parental lines that were very extensively tested. For example, the introduction named Mandarin was in the pedigree of 50 entries in the Uniform Tests but none made any contribution to our current cultivars. The history of the formation of the U.S. gene pool is described in more detail by Carter et al. (2004).

Early breeding successes quickly dominated breeding programmes. Lincoln, derived from unknown parents (but presumably Chinese landraces), was released in 1943 and became the single most important contributor to the U.S. genetic base (Gizlice et al., 1994). Lincoln was much higher yielding and earlier in maturity than most entries to which it was compared. Chief was released in 1940 from an intentional cross made at the University of Illinois. In regional testing between 1942 and 1945, Lincoln yielded 7% more and was 8 days earlier than Chief. This large yield advantage made Lincoln a very popular parental line and at the same time reduced interest in using other lower-yielding lines as parents. This was the beginning of the closure of the North American gene pool.

The first lines with Lincoln as a parent were entered in the Uniform Tests in 1948 and during the next 8 years, 65% of all the lines tested in Maturity Groups (MGs) 0 to IV had Lincoln as a parent. Soybean breeding supported by state universities and USDA were highly collaborative programmes during these years. Decisions on releasing cultivars were made by the group and segregating populations were widely shared. There was efficiency in these collaborations but it also helped to homogenize the breeding efforts and many of the lines tested from different states during this period had the same pedigree. Seven states had Uniform Test entries in MG 0 to IV derived from the Lincoln (2) × Richland cross that resulted in five released cultivars (Bernard et al., 1988).

The release of the southern cultivar Lee in 1953 had a major effect on breeding in the South. Lee and its sister lines from the cross of S-100 × CNS dominated the pedigrees of the Southern Uniform Tests for a decade after the release of Lee. The cultivars Lincoln and Lee now constitute major pillars in the foundation of the current genetic base of U.S. cultivars.

Although the North American ancestral base was selected from a relatively few alternatives based on limited testing, it has served the breeding community very well. After breeding within this gene pool for approximately 80 years there is no indication that progress is still not occurring (Rincker et al., 2014) and discussions about a yield plateau seem less common now than 30 years ago. There may be several explanations for this phenomenon. Despite how it was selected and the lack of size, there is substantial evidence indicating this set of ancestral lines is quite diverse. Kisha et al. (1998) using RFLPs found that the 20 major North American ancestors were significantly more diverse than 14 introductions selected among northern MGs for being good parents in our breeding programme and similar to 28 introductions in the southern MGs selected for agronomic performance. Thompson and Nelson (1998a) characterized 18 major North American ancestors that by pedigree accounted for 86% of the genes in North American cultivars (Gizlice, 1994) and 17 introductions that had produced high-yielding progeny in their breeding programme using RAPDs. There were 10 introductions in common between this research and that of Kisha et al. (1998). Based on multiple clustering procedures the ancestral lines occurred in 7 of 8 clusters with CNS consistently classified as an outlier. Brown-Guedira et al. (2000) used the same 18 ancestral lines as Thompson and Nelson (1998a) but expanded the exotic germplasm to include 70 exotic accessions in the pedigrees of high-yielding experimental lines and 17 additional exotic accessions included in their breeding programme. These lines were characterized with both RAPDs and SSRs. These 18 ancestral lines were distributed among 8 of 11 clusters. CNS was again classified as an outlier. We cannot know for sure the proportions of good fortune and good breeding that were involved in creating this initial gene pool, but it has provided an excellent resource for decades of breeding progress.

Either the wise or fortuitous (or perhaps both) use of this diversity may have also contributed to the success of this gene pool. Thompson et al. (1998a) examined the pedigrees of publicly developed cultivars and determined that 70% of the gene pool described by Gizlice et al. (1994) came from just seven crosses. In each of these crosses, the parents were from different genetic clusters as defined by Brown-Guedira et al. (2000) and Thompson and Nelson (1998a). The genetic differences between the parents may have been one reason that these were the successful crosses, but almost all the major ancestors made their contribution through only one cross which greatly limits their contribution to the resulting gene pool.

Although the analyses demonstrate the notable diversity among the major ancestors of U.S. soybean breeding, they also raised concerns about the narrowness of the genetic base. A single cluster contained the three major ancestral lines – A.K. (Harrow), Illini and Lincoln (Brown-Guedira et al., 2000). These three genotypes together account for 34% genetic base of cultivars adapted to the northern growing region of North America. Although S-100 seems to be genetically unique among the major ancestral lines, it was most closely related to Illini and A.K. (Harrow) and accounts for 21% of the genetic base for cultivars developed for the southern U.S. S-100 is reported to be an MG V selection from the MG III line Illini. Thompson et al. (1998a) suggested that S-100 may be progeny of Illini rather than a maturity mutant from the line, given the relative large genetic distance between these two lines. This conclusion is also supported by Lorenzen and Shoemaker (1996) who found that S-100 and Illini differed at 7 of 48 RFLP loci. As good as this gene pool has been, it may still be imposing limits on current breeding progress.

Selection for agronomic traits including seed yield has made significant changes in the phenotypes of the commercially used gene pool. Rincker et al. (2014) were able to show by testing MGs II to IV cultivars released over an 80-year period that seed yields significantly increased, and lodging scores and plant height significantly decreased over this period. It should be expected that there would be corresponding genotypic differences between the ancestral lines and derived cultivars. Using RFLP markers Kisha et al. (1998) found less polymorphism in cultivars than in the ancestral pool. However, Hyten et al. (2006) sequenced fragments from 102 random genes and found no significant differences in nucleotide diversity between 17 major ancestral lines and 25 diverse U.S. cultivars released in the 1980s. Neither of these studies tells us what is happening to the commercially used gene pool regarding agronomically important alleles. Each type of measurement does provide some information but using DNA markers and sequences of random genes are sampling only a relatively few locations in the genome with unknown relationships to genes important for controlling yield.

We cannot dismiss the idea that we are continually creating new variation and that the variation observed in a gene pool is not all derived from the original parents. Rasmusson and Phillips (1997) made a strong case for this in malting barley breeding. One example was the cultivar Excel released in 1990 that exceeded the yield of its best parent by 6% and had improvements in malt extract and alpha amylase as well as other important characteristics. The estimated coefficient of parentage of Excel's parents is 0.87. Mating two lines this closely related would not be expected to produce these gains. Another example is the long-term oil and protein selection study in maize conducted at the University of Illinois with substantial variability still present after 100 cycles of selection (Dudley and Lambert, 2004). Rasmussen and Phillips (1997) suggested that intergenic recombination, unequal crossing over, transposable elements, DNA methylation and gene amplification could all be factors that generate variation *de novo*. More recently our knowledge of presence/absence variation (Zhang et al., 2010; Santos et al., 2016; McHale et al., 2012), copy number variation (McHale et al., 2012; Cook et al., 2012) and the fluidity of gene copy numbers (Lee et al., 2015b) provide evidence for other means of creating new variation. Although we currently have no specific evidence that any of the mechanisms are creating useful variation that would increase seed yield, it is certainly plausible that it could be occurring.

For soybean, there are examples of highly successful crosses coming from either very diverse or relatively closely related parents. Williams, released in 1971 (Bernard and Lindahl, 1972), may be the most important single cultivar in U.S. soybean breeding. It reached its peak production in 1979 when it was planted on 26% of the soybean acreage in seven North Central states (Illinois (32%), Indiana (30%), Iowa (7%), Kansas (30%), Missouri (50%), Nebraska (22%) and Ohio (28%)) (Crop Reporting Board, ESS, USDA, November 1979). It was used as the direct parent of 70 cultivars (Mikel, 2010). One of them was A3127, which was a direct parent of 63 cultivars (Mikel, 2010). Williams was the product of Wayne x L57-0034. By pedigree, Wayne was 69% Lincoln and 19% Richland and L57-0034 was 37% Lincoln and 13% Richland. Based on the coefficient of parentage, this is a cross that should not have been made.

The pedigree of Hardin (Fehr et al., 1983) is Corsoy (3) x Cutler 71. One objective of this cross was to transfer *Phytophthora* resistance from Cutler 71 but in the process, Hardin was developed which exceeded the yield of Corsoy by over 7% and was 3 days earlier in maturity. The donor parent of the final cross to Corsoy would have theoretically had 75% of the genes of Corsoy. Producing a large yield increase from this cross would not be expected.

In contrast, one of the most successful crosses in U.S. soybean breeding was Williams x Essex, which combines two very unrelated parents. Williams in MG III and Essex in MG V represent the two quite distinct northern and southern U.S. gene pools. The ancestral lines that contribute approximately three quarters of the genes to southern gene pool contribute fewer than 15% of the genes to the northern gene pool and the reverse is true for the northern gene pool (Gizlice et al., 1994). In this example the only ancestor these two lines have in common is CNS, which contributes over 30% to the pedigree of Essex and only 3% to the pedigree of Williams. Ten cultivars share this parentage (Stefaniak et al., 2006) including the very successful variety A3127 developed by Asgrow Seed Company. In comparing 10 released cultivars to a set of random lines derived from this cross, Stefaniak et al. (2006) reported higher rates of recombination in the cultivars indicating that these recombinations between the diverse parents were important in producing the higher yield. They also found some very highly conserved regions in the cultivars mostly contributed by Williams, implying that maintaining favourable linkages as well as creating new genetic combinations are critical for making progress in increasing yields.

Genetic changes are absolutely required for increasing yield. Plant breeding literature and plant breeders' experience both show that there are different strategies that can be used to create those genetic changes and we still have much to learn about the genetics of seed yield. Although Rincker et al. (2014) demonstrated that soybean breeders are still making progress in increasing yield, their data also showed that the rate of gain has not changed in at least the past 40 years despite enormous changes in soybean breeding. During this period, we have greatly increased the number of soybean breeders in the United States and technology has allowed a large expansion of each breeding programme, especially within commercial companies. Extensive use of winter nurseries to quickly develop inbred lines and do testing, and seed increases in the off-season has also greatly shortened the time between the initial cross and the cultivar release. It took 11 years between the time the cross was made and the release of Williams. Today, the process can be completed in approximately half that time. Despite all these changes that should be increasing the annual rate of gain for yield, Rincker et al. (2014) found no measureable change in rate over the past 40 years. Although new genetic diversity may be created through any of several ways during the breeding process, presently we have no control over any of these mechanisms. Over the long term, capturing spontaneous changes can be to our advantage, but it is a risky strategy to depend on these changes. Intentional introduction of new genetic diversity has been successful in the past and research in more recent years has confirmed the value of this approach.

S1346, a soybean cultivar released by the Northrup-King Company in the early 1980s, was derived from a cross of an experimental line with PI 257435. This MG 0 accession was introduced from Germany in 1959. Genotypically, it is nearly identical to two other lines received from Germany through Japan nearly 20 year later (Qijian Song, pers. comm., 2016), which would indicate that these lines are likely to be full sibs and probably developed from a cross between agronomic good lines within a German breeding programme. S1346 was widely used as a parent and Sneller (1994, 2003), using pedigree data, determined that it is a major factor in explaining clustering of cultivars from the northern U.S. PI 91110-1 contributes 25% to the pedigree of IVR 1120 through a three-way cross including the cultivar Provar. PI 91110-1 was introduced into the United States from Heilongjiang, China, in 1935. It is very likely that it predates scientific plant breeding in China. IVR 1120 is the parent of several cultivars developed by commercial companies (http://www.soybase.org/, 2016). PI 71506 was supposedly introduced from Nanjing, China, in 1927 but we know that the accession that was added to the Collection in 1949 had seeds that were unlike the original seeds so its origin is unknown. PI 71506 is a grandparent of Ripley (MG IV) and Hutcheson (MG V). Hutcheson, released in 1987, was one of the most successful publicly developed cultivars of that era and is the direct parent of more than ten cultivars (http://www.soybase.org/, 2016). The actual genetic contribution that each of these exotic parents made to these successful cultivars was never determined, which may be one reason why they did not serve as positive examples and provide an incentive for additional use of exotic parents.

There are many reasons for not using exotic germplasm in a cultivar breeding programme. The seed yield of these exotic parents will be lower, there may be detrimental agronomic characteristics, disease resistance may be absent and introducing new germplasm may disrupt favourable linkages. To make use of exotic germplasm attractive, there must be some countervailing positive effects. One justification that is often promoted is a reduction of genetic vulnerability. This vulnerability is usually described as susceptibility to pathogens or pests. With genetic uniformity equated to genetic vulnerability, increasing genetic diversity will decrease genetic vulnerability. There is some logic to this assumption and very dramatic examples from history bolster the argument. The devastation of the potato crop in Ireland by *Phytophthora infestans* that started in 1845 resulted in one million deaths due to starvation and disease in the following decade. The two million Irish that emigrated changed world history (Carefoot and Sprott, 1967). The southern corn leaf blight epidemic of 1970 was facilitated by susceptible male-sterile cytoplasm that made approximately 80% of crops vulnerable and changed a minor problem into a major disease (Tatum, 1971). This modern example changed our ideas about the importance of genetic diversity, which was a good thing; but it also reinforced the concept that diversity was the solution to vulnerability, which is a flawed concept. What is sometimes overlooked is that this type of genetic vulnerability requires a virulent pathogen or pest that is highly mobile. Until the advent of soybean rust on the North American continent we did not have such a soybean pathogen, but even this disease has proven to be not well adapted to this region. We have certainly seen the spread of yield-reducing pathogens such as *Fusarium virguliforme* causing sudden death syndrome and soybean cyst nematode (*Heterodera glycines*) but the time frame is generally measured in decades and not weeks. Perhaps more important is the knowledge that disease resistance is generally quite a rare phenomenon. Over 12000 soybean accessions have been screened for SCN resistance and slightly more than 100 have been found with usable resistance (http://www.ars-grin.gov/npgs/). Miles et al.

(2006) initially screened more than 16 000 soybean accessions for resistance to soybean rust and based on these initial screenings, Walker et al. (2011), after replicated testing, were able to identify 64 accessions with potentially useable resistance. The probability that random increases in genetic diversity will provide the relative rare alleles to resist disease is extremely low and that number decreases substantially if variation in the pathogen requires a specific allele for resistance. Promoting increased variation as a cure for vulnerability to pests and pathogens is not only a misleading genetic strategy but it can also undermine the productive reasons for expanding the commercially used gene pool. The lack of the genetic variation is not so much a threat of vulnerability but a genetic opportunity. An opportunity that can be exploited with targeted breeding objectives to improve the specific traits of interest. The trait of interest of this chapter is seed yield.

4 Categories of exotic germplasm

A major reason that exotic germplasm is excluded from most soybean breeding programmes is the perception that exotic germplasm carries a very large genetic load and extracting good phenotypes from a cross with an exotic parent is imagined to be akin to winning the lottery. For some crosses, this may be true but a more expansive view of exotic germplasm yields more opportunities. From a plant breeding perspective, defining exotic as anything that is not domestic is a practical strategy. With this definition, there are four major categories of exotic germplasm for the soybean breeder to consider: foreign released cultivars, primitive cultivars, wild soybean and perennial *Glycine*. These are listed in ascending order of challenges. The remainder of this chapter presents data that show the potential of each category to improve the seed yield of soybean cultivars.

Scientific soybean breeding began in China about the same time as it did in the United States The first recorded release in China from an intentional hybridization was in 1941 (Cui, 1999). Although Ogemaw, released in 1902, was the first cultivar released in the United States from a hybrid with known parents (Bernard et al., 1988), the parents of Ogemaw are not extant and little is known about this very early breeding work in Michigan. Modern soybean breeding began in North America with the release of Pagoda in 1939 in Ontario, Canada (Bernard et al., 1988). In 1940, Chief was released in Illinois and Ogden in Tennessee (Bernard et al., 1988). Using RAPD markers, Li et al. (2001) genetically characterized the major U.S. and Chinese ancestors. They found that even though the ancestral lines of these two oldest soybean breeding programmes in the world were selected from very similar regions at approximately the same time, they are genetically distinct. For nearly 90 years, these two national breeding programmes have been selecting for increased seed yield using divergent gene pools. The major soybean producing regions of China and the United States are in similar latitudes which facilitates transfer of breeding stock so that each county can take advantage of the accumulation of favourable alleles by the other country. China has been more adept in incorporating outside cultivars into their breeding programmes with the U.S. cultivar Amsoy occurring in the pedigree of Liao Dou 3 Hao that was released in 1983 (Cui et al., 1999). Soybean breeding in South Korea is relatively recent – pure line selections were released from landraces as early as 1913, but the first cultivar, Kwangkyo, produced by intentional hybridization was not released until 1969 (Lee et al., 2015a). Only 178 cultivars have been officially registered in South Korea (Lee et al., 2015a). Japan has been releasing cultivars since 1950, but its soybean breeding effort has also been quite limited (Zhou et al., 2000). Regardless of their size, all these Asian countries of ancient soybean cultivation have the potential of supplying unique genetic diversity to North American soybean breeding and examples of these contributions will be documented in the next section.

The next step up both in terms of increasing variation and difficulty in use involves primitive soybean cultivars. Soybean is a domesticated species so by definition, the entire species is 'improved' since they are all products of human intervention. However, compared to released cultivars, primitive cultivars will have lower seed yield and more undesirable agronomic traits. Within the USDA Soybean Germplasm Collection and worldwide, this group represents the largest number of germplasm accessions available, so it should not be avoided even though there are challenges in using these lines.

The wild soybean is within the primary gene pool (Harlan and deWet, 1971) of soybean, so obtaining fertile progeny is not an issue with these interspecific crosses; however, obtaining good agronomic lines from these crosses is a significant challenge. The viney plant type, the small seed size and the extreme pod shattering are all major differences between soybean and wild soybean, and removing these traits from the segregating populations is no small task. All measures of genetic diversity indicate that although phenotypically the wild soybean is much less diverse than the soybean, genotypically it is far more variable. Based on sequence data from 102 random genes, Hyten et al. (2006) demonstrated that a much greater loss in diversity occurred during domestication than in any of the other selection processes that have occurred since domestication. Li and Nelson (2002) compared diversity among soybean and wild soybean accessions from four Chinese provinces using RAPD markers and found that the average genetic distance between the wild soybean and soybean groups was 2 to 3 times greater than the genetic distances among the soybean groups. The complete sequences of a wild soybean accession (Kim et al., 2010) and a soybean cultivar (Schmutz, et al., 2010) provide the opportunity for a very extensive comparison of two genotypes. Kim et al. (2010) identified over 10 000 genes with polymorphisms between the species that could cause differences in protein function. Lam et al. (2010) resequenced 14 *G. max* and 17 *G. soja* accessions to approximately a fivefold genome coverage and reported a higher allelic diversity within *G. soja* than within *G. max*. All the data support the idea that wild soybean can supply not only novel alleles but also unique genes that are not currently available in the domesticated germplasm. Despite the challenges that breeding with wild soybean presents, it is a breeding resource that should not be ignored.

The final category of available genetic diversity is the perennial *Glycine*. There are 26 perennial species. These perennial species are in the soybean tertiary gene pool, which means that fertile progeny can be obtained only with extraordinary means (Harlan and deWet, 1971) and to date, fertile progeny from crosses with soybean have only been produced with one species. Singh et al. (1993) produced the first fertile progeny from crossing soybean with *G. tomentella* ($2n = 78$). This was accomplished by culturing rescued seeds from 19- to 21-day-old pods and developing viable but sterile F_1 plants. Amphidiploid plants were induced by colchicine, and then backcrossed with soybean. In the BC_4, self-fertile $2n = 40$ plants were obtained. No information is available about the fertile progenies developed from this research. Singh and Nelson repeated this research (2015) using a different accession of *G. tomentella* ($2n = 78$) and also made crosses using *G. tomentella* as the female parent and developed self-fertile lines with *G. tomentella* cytoplasm (2014). They attempted hybridizations with accessions from *G. tomentella*, *G. argyrea* and *G. latifolia* but only one *G. tomentella*, PI 441001 ($2n = 78$), parent successfully produced fertile progeny. There is much left to be learned about how perennial *Glycine* can be routinely incorporated into soybean breeding but it has enormous potential for changing cultivar development programmes.

5 Utilizing new genetic diversity

Research on methodologies to expand the genetic base of North American soybean breeding began in the 1960s (Thorne and Fehr, 1970; Hartwig, 1972; Schoener and Fehr, 1979). These projects consistently found larger genetic variance in populations with a greater percentage of exotic germplasm but higher yields from populations with less exotic germplasm. None of these studies concluded that exotic germplasm could be used to increase yield in the short term and none of the research was continued to demonstrate that exotic germplasm could increase yield over the long term.

Breeding methodology studies always involve much smaller populations than actual breeding programmes, so it is difficult for the results of such research to accurately predict the outcomes when expanded to functional proportions. When exotic germplasm was incorporated into breeding programmes, the results were more encouraging using all four classes of exotic germplasm defined earlier. LG90-2550 was released in 1997 (Thompson et al., 1999). It was derived from crossing two experimental lines that each had the pedigree of PI 68658 x Lawrence. Tested in seven environments, LG90-2550 yielded a statistically significant 457 kg/ha more than Lawrence (Thompson and Nelson, 1998b). PI 68658 was introduced from China in 1926, which predated scientific plant breeding. In 2015, Fox et al. confirmed a yield QTL in PI 68658 that consistently increased yield (134 kg/ha), demonstrating that primitive cultivars can be used to increase the yield of current cultivars.

N7001 (Carter et al., 2003), from the cross between the experimental line N77-114 and PI 416937, a cultivar from Japan introduced into the United States in 1977, was released in 2000. In the USDA Southern Uniform Tests, it was lower yielding than Haskell but in state tests in North Carolina it was equivalent to Haskell. N7002 (Carter et al., 2007) was derived from N7001 x Cook and in the USDA Southern Uniform VII Test, it was 224 kg/ha higher yielding than the best check. N8001 (Carter et al., 2008) was released from the same cross and in the USDA Southern Uniform VIII Test, it was 217 kg/ha higher yielding than Cook. Although no statistics were provided for these data, N7002 and N8001 are likely to be superior to the cultivars they were released to replace.

LG04-6000 (Nelson and Johnson, 2012) was released in 2008 and has PI 436682 in its pedigree. PI 436682 is the Chinese cultivar Jilin 15 that was released in 1979 by the Jilin Academy of Agricultural Sciences. LG04-6000 was tested in 22 locations in 2007 and 2008 in the USDA Northern Uniform IV Test. Averaged over both years it was statistically significant 296 kg/ha higher in yield than the best check in the test. More recently, LG11-6760 was tested at 18 locations in the USDA Northern Uniform IV Test in 2014 and 2015 and averaged 396 kg/ha more than the best check. Half of the pedigree of LG11-6760 came from PI 561319A and PI 574477. PI 561319A was introduced from China in 1990. It has a name that indicates that it is a primitive cultivar but two sublines were extracted from the original seed lot, so we cannot be certain about its origin. PI 574477 is Fen dou 31 released in Shanxi province in 1990 (Cui et al., 1999). These examples clearly demonstrate the successful use of released cultivars from other countries to develop cultivars and experimental lines that exceed the yields of the best released cultivars from public soybean breeding.

Although LG90-2550 was significantly higher yielding than its highest-yielding parent, the initial cross between PI 68658 and Lawrence was made in 1979. When it was entered in the USDA Preliminary Uniform III Test in 1996, it yielded only 89% of the best check. In contrast when LG11-6760 was evaluated in the USDA Uniform IV Test in 2014 and 2015, it yielded 110% of the best check. The checks in both cases were the best publicly developed cultivars available at the time. Although these are not definitive data, they present a very strong case that infusing a breeding programme with exotic germplasm can increase the annual rate of gain and in this case that increase is more than 20% over a 20-year period.

The results using wild soybean are not as compelling as those with exotic soybean germplasm but much less effort has been devoted to breeding with this species. Concibido et al. (2003) mapped a yield QTL on chromosome 14 from PI 407305 (*G. soja*) in a BC_2 population. When they attempted to confirm this QTL in several elite soybean backgrounds, the positive effect of the QTL was recorded in only 2 of the 6 populations. These results indicate that this positive allele from wild soybean may already be in the commercially used gene pool. Li et al. (2008) using a BC_2 population with PI 245331 (*G. soja*) as the donor parent also mapped a yield QTL from the wild soybean parent on chromosome 5. Unfortunately, this QTL was not confirmed but it mapped to a location within 1 cM of a QTL from soybean previously mapped by Kabelka et al. (2004).

In 2015, Abdel-Haleem et al. released two germplasm lines derived from the cross of N7103 x PI 366122 (*G. soja*). These lines were tested at 15 locations from 2009 to 2012 and both yielded 88% of the soybean parent, a small-seeded specialty cultivar, and 70% of the best check, N7002. Both lines had very small seeds with 100 seed weights of approximately 6 grams, but were similar to the check cultivars for the other agronomic and seed composition traits measured. The genetic contribution from the wild soybean parent was not reported for these lines. In 2010 Preliminary Uniform IIIB Test, LG07-2309 (IA3023 x LG01-7728) had the same yield as IA3023. LG01-7728 was derived from Williams 82 (2) x PI 479767 (*G. soja*). No genotypic data were collected to determine how much of wild soybean genome remained in this BC_2 line.

Akpertey et al. (2014) tested at 10 locations 93 experimental lines derived from backcrossing with five wild soybean accessions to soybean. Williams 82 was used for the first backcross but other more recent cultivars were used for some of the second backcrosses. The wild soybean parents were PI 65549, PI 479767, PI 483461, PI 507807 and PI 549046 in MGs that ranged from 0 to IV. Of most interest are BC_1 and BC_2 lines with only Williams 82 as the recurrent parent. There were BC_2 lines with each of four wild soybean donor parents that were not significantly different in yield from the recurrent parent. Based on genotypic data using 1536 SNP markers from the Universal Soy Linkage Panel 1.0 (Hyten et al., 2010), the genetic contribution of the wild soybean parents ranged from 6 to 30%. The highest-yielding BC_2 line had 9% of the SNP alleles from the wild soybean parent (PI 479767) and yielded 195 kg/ha more than Williams 82 but that difference was not statistically significant. There were two BC_1 lines that were also equal in yield to Williams 82. One BC_1 line was the result of crossing two BC_1 lines with different donor parents (PI 479767 and PI 549046). It was five days earlier in maturity than Williams 82 and had 56% of the SNP alleles from the wild soybean parents. The other BC_1 line had 24% of the SNP alleles from the wild soybean parent (PI 507807). These results are different from Ertl and Fehr (1985) who found no BC_1 lines that maintained the soybean parent agronomic characteristics. Akpertey et al. (2014) tested 84 BC_2 lines that were derived from four wild soybean donor parents. The average per cent of SNP alleles from the wild soybean parents in the lines within each pedigree was not significantly different from the expected 12.5% with no selection. This was difficult to explain since intense selection was practised through the generations of inbreeding to produce lines with the soybean phenotype.

Swarm (2017) identified two experimental lines from the cross Williams 82 by the wild soybean accession PI 479752 that yielded significantly more than the soybean parent. These lines had phenotypes similar to Williams 82 in most agronomic traits except that both lines had significantly smaller seeds. In this research, intensive phenotypic selection did significantly reduce the wild soybean alleles in the derived lines. Both of the highest-yielding lines had fewer than 20% of the *G. soja* alleles from over 20 000 SNPs identified by using genotyping by sequencing. This is the first example of wild soybean-derived lines that yielded more than the soybean parent and it represents an important milestone in the use of this species in soybean breeding.

There has been the least amount of breeding done with the perennial *Glycine* species but if progress is measured by yield increases over the domestic parent, the results with *G. tomentella* exceed all the other classes of exotic germplasm. To date, fertile progeny have been tested from crosses involving only one accession of *G. tomentella* (PI 441001) and the soybean cultivar Dwight. In five years of testing, we have evaluated 479 backcross-derived lines with *G. max* cytoplasm and 117 backcross-derived lines with *G. tomentella* cytoplasm. Data from replicated tests in 12 environments identified a BC_4 line with the same maturity as Dwight that yielded 242 kg/ha more than Dwight, which was significant (Akpertey, 2015). From the same experiment, a BC_3 line was six days later than Dwight and 477 kg/ha higher yielding (Akpertey, 2015). With preliminary testing in five environments, we identified a BC_2 line with *G. tomentella* cytoplasm that yielded 403 kg/ha more than Dwight (unpublished data, 2016). We do not have sufficient data to determine if there is a cytoplasmic effect on seed yield but the *G. tomentella* cytoplasm does seem to influence the loss of *G. tomentella* chromosomes during backcrossing (Ram Singh, pers. comm., 2016). Of all the lines with *G. tomentella* cytoplasm that we have had in yield test, 71% have fewer than three backcrosses, whereas with the lines with *G. max* cytoplasm, 48% have more than three backcrosses and none fewer than three backcrosses (unpublished data, 2016). This initial research with one accession from one perennial species is enormously promising and provides a strong incentive to pursue expanded wide hybridization research within the genus *Glycine*.

6 Conclusion

In the United States, less than 1% of the available germplasm has been appropriated to create our current cultivars and that figure is applicable to every breeding programme. We still have much to learn about the genetics of seed yield and how we can better select for useful genetic diversity within the wide range of germplasm that is available. Unfortunately, we still do not have the definitive data that can demonstrate a unique contribution from exotic germplasm to the commercially used gene pool that is implicated in increasing yield but research is currently underway that should be able to provide that information. In the meantime, the very high-yielding lines derived from exotic germplasm provide strong evidence that not all the beneficial alleles, or perhaps even genes, are currently captured in our modern cultivars. To provide the best possible future for soybean breeding, it is imperative that every breeding programme be part of the effort to expand the genetic base.

The public sector will certainly do most of the initial breeding with exotic soybean, wild soybean and perennial *Glycine* germplasm, but participation by the commercial sector will be required if the effort is to be successful. As they have already demonstrated, commercial companies can play a critical role in this process by helping to test experimental lines to identify the best genotypes. Since most of the soybean production areas in the United States are planted with cultivars developed by commercial companies, any significant change in the genetic base of the soybean crop will depend on these companies incorporating the improved germplasm developed in the public sector into their new cultivars. Broadening the genetic base is an enormous task that will take years and likely decades. Accomplishing this change will require a commitment by the entire soybean breeding community but the data indicate that the potential rewards are great.

7 References

Abdel-Haleem, H., H. R. Boerma, T. E. Carter, E. Dale Wood and Z. Li. 2015. Registration of G07-6012 and G07-6029 soybean germplasm, which derive 50% pedigree from wild soybean. *J. Plant Regist.* 9: 222–6.

Akpertey, A. 2015. Genetic Introgression from *Glycine tomentella* to Soybean to Increase Seed Yield. Ph.D. diss. University of Illinois at Urbana-Champaign.

Akpertey, A., M. Belaffif, G. L. Graef, M. A. Rouf Mian, J. Grover Shannon, P. B. Cregan, M. E. Hudson, B. W. Diers and R. L. Nelson. 2014. Effects of selective genetic introgression from wild soybean to soybean. *Crop Sci.* 54: 2683–95.

Bernard, R. L., G. A. Juvik, E. E. Hartwig and C. J. Edwards, Jr. 1988. Origins and pedigrees of public soybean varieties in the United States and Canada. *USDA Technical Bulletin No.* 1746. 68p.

Bernard, R. L. and D. A. Lindahl, D. A. 1972. Registration of Williams soybean. *Crop Sci.* 12: 716.

Brown-Guedira, G. L., J. A. Thompson, R. L. Nelson and M. L. Warburton. 2000. Evaluation of genetic diversity of soybean introductions and North American ancestors using RAPD and SSR markers. *Crop Sci.* 40: 815–23.

Carefoot, G. L. and E. R. Sprott. 1967. *Famine on the Wind, Man's Battle Against Plant Disease*. Rand McNally & Company, Chicago, Illinois, USA.

Carter, T. E., Jr., J. W. Burton, D. T. Bowman, Z. Cui, M. R. Villagarcia, A. S. Niewoenher and J. F. Wilder. 2003. Registration of 'N7001' soybean. *Crop Sci.* 43: 1126–7.

Carter, T. E., Jr., J. W. Burton, M. O. Fountain, P. E. Rzewnicki, M. R. Villagarcia and D. T. Bowman. 2007. Registration of 'N7002' soybean. *J. Plant Regist.* 1: 93–4.

Carter, T. E., Jr., J. W. Burton, M. O. Fountain, P. E. Rzewnicki, M. R. Villagarcia and D. T. Bowman. 2008. Registration of 'N8001' soybean. *J. Plant Regist.* 2: 22–3.

Carter, T. E., R. L. Nelson, C. H. Sneller and Z. Cui. 2004. Genetic diversity in soybean. In Boerma, H. R. and Specht, J. E. (Eds) *Soybeans: Improvement, Production, and Uses*. American Society of Agronomy, Crop Science Society of America, Soil Science Society of America, Madison, Wis., Vol. Agronomy; no. 16, pp. 303–416.

Concibido, V. C., B. L. Vallee, P. Mclaird, N. Pineda, J. Meyer, L. Hummel, J. Yang, K. Wu and X. Delannay. 2003. Introgression of a quantitative trait locus for yield from *Glycine soja* into commercial soybean cultivars. *Theor. Appl. Genet.* 106:575–82.

Cook, D. E., T. G. Lee, X. Guo, S. Melito, K. Wang, A. Bayless, J. Wang, T. J. Hughes, D. K. Willis, T. Clemente, B. W. Diers, J. Jiang, M. E. Hudson and A. F. Bent. 2012. Copy number variation of multiple genes at *Rhg1* mediates nematode resistance in soybean. *Science* 338: 1206–9.

Cui, Z., T. E. Carter, Jr., J. Gai, J. Qui and R. L. Nelson. 1999. Origin, description and pedigrees of Chinese soybean cultivars released during 1923–1992. *U.S. Department of Agriculture Technical Bulletin No.1871*.

Dudley, J. W. and R. J. Lambert. 2004. 100 generations of selection for oil and protein in corn. *Plant Breed. Rev.* 24: 79–110.

Ertl, D. S. and W. R. Fehr. 1985. Agronomic performance of soybean genotypes from *Glycine max* x *Glycine soja* crosses. *Crop Sci.* 25: 589–92.

Fehr, W. R., J. B. Bahrenfus, H. L. C. Tachibana and L. C. Card. 1983. Registration of Hardin soybean (Reg. No. 165). *Crop Sci.* 23: 402.

Fox, C. M., T. R. Cary, R. L. Nelson and B. W. Diers. 2015. Confirmation of a Seed Yield QTL in Soybean. *Crop Sci.* 55:1–7.

Gizlice, Z., T. E. Carter, Jr. and J. W. Burton. 1994. Genetic base for North American public soybean cultivars released between 1947 and 1988. *Crop Sci.* 34:1143–51.

Harlan, J. R. and J. M. J. deWet. 1971. Toward a rational classification of cultivated plants. *Taxon.* 20: 509–17.

Hartwig, E. E. 1972. Utilization of soybean germplasm strains in a soybean improvement program. *Crop Sci.* 12:856–9.

Hyten, D. L., I. Y. Choi, Q. Song, J. E. Specht, T. E. Carter, Jr., R. C. Shoemaker, E. Y. Hwang, L. K. Matukumalli and P. B. Cregan. 2010. A high density integrated genetic linkage map of soybean and the development of a 1536 universal soy linkage panel for quantitative trait locus mapping. *Crop Sci.* 50: 960–8.

Hyten, D. L., Q. Song, Y. Zhu, I.-Y. Choi, R. L. Nelson, J. M. Costa, J. E. Specht, R. C. Shoemaker and P. B. Cregan. 2006. Impacts of genetic bottlenecks on soybean genome diversity. *PNAS* 103: 16666–71.

Kabelka, E. A., Diers, B. W., Fehr, W. R., LeRoy, A. R., Baianu, I. C., You, T., Neece, D. J. and R. L. Nelson. 2004. Putative alleles for increased yield from soybean plant introductions. *Crop Sci.* 44: 784–91.

Kim, M. Y., S. Lee, K. Van, T.-H. Kim, S.-C. Jeong, I.-Y. Choi, D.-S. Kim, Y.-S. Lee, D. Park, J. Ma, W.-Y. Kim, B.-C. Kim, S. Park, K.-A. Lee, D. H. Kim, K. H. Kim, J. H. Shin, Y. E. Jang, K. D. Kim, W. X. Liu, T. Chaisan, Y. J. Kang, Y.-H. Lee, K.-H. Kim, J.-K. Moon, J. Schmutz, S. A. Jackson, J. Bhak and S.-H. Lee. 2010. Whole-genome sequencing and intensive analysis of the undomesticated soybean (Glycine soja Sieb. and Zucc.) genome. *PNAS* 107: 22032–7.

Kisha, T. J., B. W. Diers, J. M. Hoyt and C. H. Sneller. 1998. Genetic diversity among soybean plant introductions and North American germplasm. *Crop Sci.* 38:1669–80.

Lam, H.-M., X. Xu, X. Liu, W. Chen, G. Yang, F.-L. Wong, M.-W. Li, W. He, N. Qin, B. Wang, J. Li, M. Jian, J. Wang, G. Shao, J. Wang, S. S.-M. Sun and G. Zhang. 2010. Resequencing of 31 wild and cultivated soybean genomes identifies patterns of genetic diversity and selection. *Nat. Genet.* 42: 1053–61.

Lee, C., M.-S. Choi, H.-T. Kim, H.-T. Yun, B. Lee, Y.-S. Chung, R. W. Kim and H.-K. Choi. 2015a. Soybean [Glycine max (L.) Merrill]: Importance as a crop and pedigree reconstruction of Korean varieties. *Plant Breed. Biotech.* 3: 179–96.

Lee, T. G., I. Kumar, B. W. Diers and M. E. Hudson. 2015b. Evolution and selection of Rhg1, a copy-number variant nematode-resistance locus. *Mol. Ecol.* 24: 1774–91.

Li, D., T. W. Pfeiffer and P. L. Cornelius. 2008. Soybean QTL for yield and yield components associated with *Glycine soja* alleles. *Crop Sci.* 48: 571–81.

Li, Z. and R. L. Nelson. 2002. RAPD marker diversity among soybean and wild soybean accessions from four Chinese provinces. *Crop Sci.* 42: 1737–44.

Li, Z., L. Qiu, J. A. Thompson, M. Welsh and R. L. Nelson. 2001. Molecular genetic analysis of the U.S. and Chinese soybean ancestral lines. *Crop Sci.* 41: 1330–6.

Lorenzen, L. L. and R. C. Shoemaker, R. C. 1996. Genetic relationships within old U.S. soybean cultivar groups. *Crop Sci.* 36: 743–52.

McHale, L. K., W. J. Haun, W. W. Xu, P. B. Bhaskar, J. E. Anderson, D. L. Hyten, D. J. Gerhardt, J. A. Jeddeloh and R. M. Stupar. 2012. Structural variants in the soybean genome localize to clusters of biotic stress-response genes. *Plant Phys.* 159: 1295–308.

Mikel, M. A., B. W. Diers, R. L. Nelson and H. H. Smith. 2010. Genetic diversity and agronomic improvement of North American soybean germplasm. *Crop Sci.* 50: 1219–29.

Miles, M. R., R. D. Frederick and G. L. Hartman. 2006. Evaluation of soybean germplasm for resistance to *Phakopsora pachyrhizi*. *Plant Health Progress*, January: 1–25.

Nelson, R. L. and E. O. C. Johnson. 2012. Registration high-yielding soybean germplasm line LG04-6000. *J. Plant Regist*. 6: 1–4.

Rasmusson, D. C. and R. L. Phillips. 1997. Plant breeding progress and genetic diversity from de novo variation and elevated epistasis. *Crop Sci*. 37: 303–10.

Rincker, K., R. Nelson, J. Specht, D. Sleper, T. Cary, S. Cianzio, S. Casteel, S. Conley, P. Chen, V. Davis, C. Fox, G. Graef, C. Godsey, D. Holshouser, G.-L. Jiang, S. K. Kantartzi, W. Kenworthy, C. L., Rouf Mian, L. McHale, S. Naeve, J. Orf, V. Poysa, W. Schapaugh, G. Shannon, R. Uniatowski, D. Wang and B. Diers. 2014. Genetic improvement of soybean in maturity groups II, III, and IV. *Crop Sci*. 54: 1–14.

Santos J., B. Valliyodan, T. Joshi, S. Khan, Y. Liu, J. Wang, T. Vuong, M. Oliveira, F. Marcelino-Guimarães, D. Xu, H. Nguyen and R. Abdelnoor. 2016. Evaluation of genetic variation among Brazilian soybean cultivars through genome resequencing. *BMC Genomics* 17:110. DOI:10.1186/s12864-016-2431-x.

Schmutz, J., S. B. Cannon, J. Schlueter, J. Ma, T. Mitros, W. Nelson, D. L. Hyten, Q. Song, J. J. Thelen, J. Cheng, D. Xu, U. Hellsten, G. D. May, Y. Yu, T. Sakurai, T. Umezawa, M. K. Bhattacharyya, D. Sandhu, B. Valliyodan, E. Lindquist, M. Peto, D. Grant, S. Shu, D. Goodstein, K. Barry, M. Futrell-Griggs, B. Abernathy, J. Du, Z. Tian, L. Zhu, N. Gill, T. Joshi, M. Libault, A. Sethuraman, X.-C. Zhang, K. Shinozaki, H. T. Nguyen, R. A. Wing, P. Cregan, J. Specht, J. Grimwood, D. Rokhsar, G. Stacey, R. C. Shoemaker and S. A. Jackson. 2010. Genome sequence of the palaeopolyploid soybean. *Nature* 463: 178–83.

Schoener, C. S. and W. R. Fehr. 1979. Utilization of plant introductions in soybean breeding populations. *Crop Sci*. 19:185–8.

Singh, R. J. and G. H. Chung. 2007. Cytogenetics of soybean: Progress and perspectives. *Nucleus* 50: 401–23.

Singh, R. J., K. P. Kollipara and T. Hymowitz. 1993. Backcross (BC_2-BC_4)-derived fertile plants from *Glycine max* and *G. tomentella* intersubgeneric hybrids. *Crop Sci*. 33: 1002–7.

Singh, R. J. and R. L. Nelson. 2014. Methodology for creating alloplasmic soybean lines by using Glycine tomentella as a maternal parent. *Plant Breed*. 133: 624–31.

Singh, R. J. and R. L. Nelson. 2015. Intersubgeneric hybridization between *Glycine max* and *G. tomentella*: production of F 1, amphidiploid, BC_1, BC_2, BC_3, and fertile soybean plants. *Theor. Appl. Genet*. 128: 1117–36.

Sneller, C. H. 1994. Pedigree analysis of elite soybean lines. *Crop Sci*. 34: 1515–22.

Sneller, C. H. 2003. Impact of transgenic genotypes and subdivision on diversity within elite North American soybean. *Crop Sci*. 43: 409–14.

Stefaniak, T. R., D. L. Hyten, V. R. Pantalone, A. Klarer and T. W. Pfeiffer. 2006. Soybean cultivars resulted from more recombination events than unselected lines in the same population. *Crop Sci*. 46: 43–51.

Swarm, S. A. 2017. Overcoming the Domestication Bottleneck in Soybean: Mapping Domestication Traits and Using Wild Soybean to Improve Diversity. Ph.D. diss. University of Illinois at Urbana-Champaign.

Tatum, L. A. 1971. The southern corn leaf blight epidemic. *Science* 171: 1113–16.

Thompson, J. A., P. J. Amdor and R. L. Nelson. 1999. Registration of LG90-2550 and LG91-7350R soybean germplasm. *Crop Sci*. 39: 302–3

Thompson, J. A. and R. L. Nelson. 1998a. Core set of primers to evaluate genetic diversity in soybean. *Crop Sci*. 38: 1356–62.

Thompson, J. A. and R. L. Nelson. 1998b. Utilization of diverse germplasm for soybean yield improvement. *Crop Sci*. 38: 1362–8.

Thompson, J. A., R. L. Nelson and L. O. Vodkin. 1998. Identification of diverse soybean germplasm using RAPD markers. *Crop Sci*. 38: 1348–55.

Thorne, J. C. and W. R. Fehr. 1970. Exotic germplasm for yield improvement in 2-way and 3-way soybean crosses. *Crop Sci*. 10: 677–8.

Walker, D. R., H. R. Boerma, D. V. Phillips, R. W. Schneider, J. B. Buckley, E. R. Shipe, J. D. Mueller, D. B. Weaver, E. J. Sikora, S. H. Moore, G. L. Hartman, M. R. Miles, D. K. Harris, D. L. Wright, J. J. Marois and R. L. Nelson. 2011. Evaluation of USDA soybean germplasm accessions for resistance to soybean rust in the Southern United States. *Crop Sci.* 51:678–93
Zhang, M., Y.-H. Wu, M.-K. Lee, Y.-H. Liu, Y. Rong, T. S. Santos, C. Wu, F. Xie, R. L. Nelson and H.-B. Zhang. 2010. Numbers of genes in the NBS and RLK families vary by more than four-fold within a plant species and are regulated by multiple factors. *Nucleic Acids Res.* 38: 6513–25.
Zhou, X., T. E. Carter, Z. Cui, S. Miyazaki and J. W. Burton. 2000. Genetic base of Japanese soybean cultivars released during 1950 to 1988. *Crop Sci.* 40: 1794–802.

Chapter 3

Advances in conventional soybean breeding techniques

E. E. Large, E. Beche, D. Mutoni and A. Scaboo, University of Missouri, USA

1 Introduction: evolution of the modern soybean

Modern plant breeding is the creative utilization and balance of available tools and techniques to develop the most beneficial crops for human beings (Bernardo, 2002). The primary strategy for improving crops is the selection of one genotype over another and farmers used this technique over the past several millennia. With enough time even the relatively low outcrossing rates of self-pollinating species such as domesticated soybeans (*Glycine max* L. Merrill) and wild soybeans (*G. soja*), <3% and 13%, respectively (Fujita et al., 1997) are enough to generate useful recombination events in the absence of manual cross-pollination. Early plant breeders selected plants using mass selection without modern approaches such as replicate plots or progeny testing and slowly increased genetic gain. The resultant crop improvements were essential to the development of centralized agrarian civilizations around the world. Modern plant breeding still uses the

http://dx.doi.org/10.19103/AS.2017.0034.03

same approach as ancient farmers but provides more options and more efficient means to identify lines with increased breeding value, thereby improving selection efficiency and rapidly achieving genetic gain (Bernardo, 2002).

The first soybean plant breeders were East Asian farmers who domesticated the predecessors of current crops by mass selection. Archaeological and written records indicate that soybean was domesticated in China as early as 4500 years ago (2500 B.C.) and emerged as a food crop in Northeastern China between 1700 and 1100 B.C. Selections from promising wild soybeans may have occurred much earlier given the recent evidence of modern humans in what is now southern China 80–120 000 years ago (Liu et al., 2015). Domesticated soybeans are now a primary source of human protein in China, Japan and Korea. Soybeans (e.g. edamame) and soybean products are the basis of foods such as soy sauce, soy milk, soybean sprouts, tofu, miso, natto, tempeh and many other traditional foods including a type of vodka (Hartman et al., 2011).

Conventional soybean breeding is a relatively recent development in major U.S. and world agriculture programmes outside of East Asia. The earliest documentation of soybeans in the United States was in present-day Savannah, Georgia, in 1765 by Samuel Bowen (Hymowitz, 2008). In 1770, Benjamin Franklin, then ambassador to France, shipped soybean lines from Europe to the United States as a forage and cover crop (Wilson, 2008; Hymowitz, 2008). Soybeans were first evaluated at a land-grant institution in 1879 (New Jersey Agricultural Experiment Station), and by the early 1900s nearly every U.S. agricultural station was growing soybean for use as hay (Hymowitz, 1990). Two major innovations led to the rapid increase in U.S. soybean production: 1) crushing soybeans for oil in 1915 (Hartman et al., 2011) and 2) the discovery heated soy meal could be used as livestock feed in 1917 (Wilson, 2008). A combination of farm management and soybean breeder selections from limited genetic material provided the foundation for cheap readily available high-protein soybean meal by the 1950s leading to a rapid expansion of U.S. livestock and poultry production. The U.S. soybean export industry subsequently began with the opening of the first export office in Japan in 1956 (Hymowitz, 1990; Hartman et al., 2011). Industrial uses of soybean such as biodiesel, soap, resins, adhesives, lubricants, plastics, inks, clothing and cosmetics continue to grow with many more applications in the research pipeline (www.soynewuses.org).

U.S. soybean breeders faced a major challenge by 1990, however, when U.S. soybean production declined in southern states due to lower-than-average yields and competition from more profitable crops. Fortunately, soybeans can be grown in a wide variety of soil and climatic conditions and a refocusing of conventional soybean breeding efforts on developing new seed varieties tolerant of drier conditions with less growing time expanded soybean production into the Northern and Western U.S. (Hymowitz, 1990). The Northern states (e.g. Minnesota, Nebraska, South Dakota and North Dakota), once believed too far north for soybean production, are now some of the top U.S. soybean producing states. Thirty-one U.S. states currently produce soybeans with Nebraska, Illinois and Iowa as the top three soybean producing states, generating 3.9–3.8 metric tons/hectare in 2015 (SoyStats, 2016). The ability of conventional soybean breeding programmes to develop new varieties adapted to novel and emerging environments is crucial for the sustainable cultivation of modern soybeans.

The adaptability and a balance of protein and oil makes soybean the current dominant oil seed in world trade, accounting for 60% of total global oil seed production. The United States led 2015 global soybean production at 106.93 million metric tons (MMT), valued at $19.832 billion dollars, with Brazil a close second (96.50 MMT) and Argentina third (56.80 MMT) (United States Department of Agriculture (USDA), 2016). The remarkable rise of Brazil, which stretches across both subtropical (32°S) and tropical (5°N) zones and did not grow soybeans prior to 1960, highlights the rapid versatility of soybean productivity (Fig. 1). Other soybean countries in decreasing order of production are China, Paraguay, India, Canada, Ukraine, Bolivia, Russia, the European Union and Uruguay (USDA, 2016). In 2015, the global yields in metric tons per hectare were 2.61 tonnes/hectare and the highest average yields from major producer countries such as the United States, Brazil and Argentina were 3.23 tonnes/hectare, 2.92 tonnes/hectare and 2.91 tonnes/hectare, respectively (USDA, 2016). An increasing demand for soybeans and soy products such as oil for human consumption, biodiesel and high-protein animal feed drives the continual increase in production worldwide (Hymowitz, 1990; Hartman et al., 2011; Orf, 2008). Over the last ninety years, conventional U.S. soybean breeding transformed the domestic soybean landscape and provided a leading North American crop for both domestic and export markets (Rincker et al., 2014).

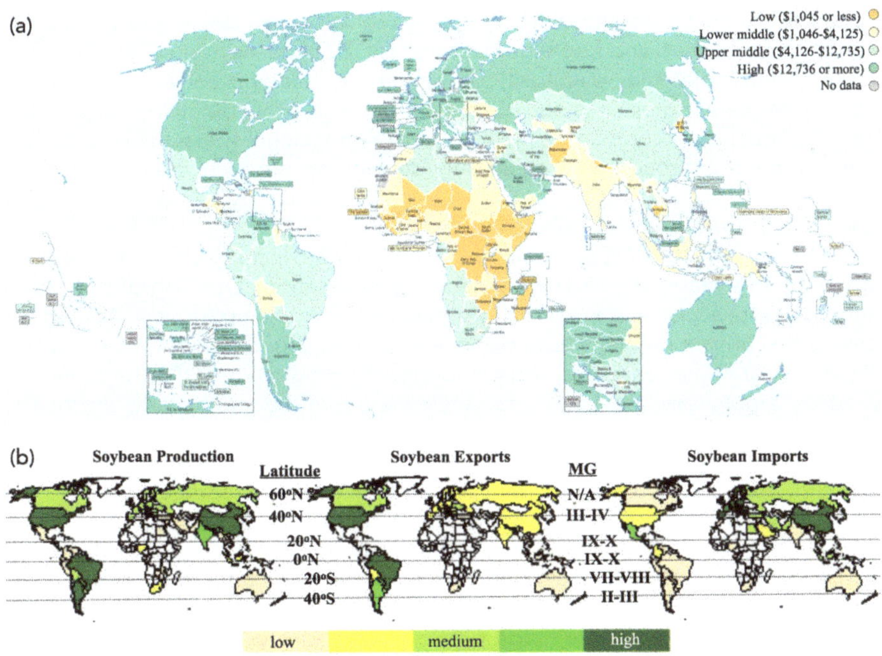

Figure 1 World soybean industry indicators and gross national income levels. (a) Geography of developing and developed countries based on 2015 gross national income (World Bank, 2016). Income levels and coloured circles in the upper right correlate with the distribution of gross national incomes of colour-coded countries. The left box inset shows countries and territories of the Caribbean and the right box inset provides details on select Mediterranean and Eastern European countries. (b) Global soybean production, export and import levels (USDA PSD, 2016) are indicated with colours matching the coloured bar located under the three maps. Latitudinal lines are labelled in intervals of 20° along with common soybean maturity groups (MG; 000 to X) grown at each latitude. Soybeans are not currently grown at 60° N (N/A: not applicable) and the Alaskan production and export levels reflect Continental US soybean data. Tropical regions fall between the Tropic of Cancer (23.5° N) and the Tropic of Capricorn (23.5° S).

2 The rise of genetically engineered soybeans

Conventional soybean breeding programmes are rapidly adapting to the introduction of genetically engineered (GE) soybeans, which were first commercialized in the United States in 1996 (National Academies of Sciences, Engineering, and Medicine, 2016). The new technology was so advantageous to soybean producers that over 80% of the 118 million hectare global soybean crop was GE in 2015 compared to a mere 33 million global hectares of GE soybean in 2001 (National Academies of Sciences, Engineering, and Medicine, 2016). First-generation traits such as herbicide-tolerant (HT) soybean can withstand direct applications of herbicides used to control weeds and provides an efficient means to rapidly select for GE germplasm combined with conventional breeding schemes. The first and most widely used HT soybeans are glyphosate resistant, with glyphosate being the active ingredient in Roundup™ (Greene et al., 2016; National Academies of Sciences, Engineering, and Medicine, 2016). Second-generation varieties can have stacked tolerance to two or more herbicides and modified fatty acid profiles of the soybean seed oil content (e.g. high oleic acid). Third and future generations of GE soybeans using CRISPR-Cas9 and related GE technologies may eventually provide the public and conventional breeders with acceptable breeding options outside the current regulatory definition of 'genetically engineered'. Agriculture companies, such as Dow AgroSciences, are thinking 5, 10, 20 and 30 years down the road and are currently developing products that will be needed 15 years from now, including some of the newest GE technology (Luciano Jaureguy and Julián Lenis, Dow AgroSciences, pers. comm.).

The trickle-down effects of GE technology are not limited to increased farmer productivity. In addition to the material needs of a growing population, Dow AgroSciences take environmental and social sustainability into account. For example, Dow AgroSciences co-sponsors the Environmental Stewardship Award with the National Cattlemen's Beef Association programme and has various outreach programmes. Furthermore, Dow AgroSciences employees are known for their volunteerism, particularly those efforts supporting Science, Technology, Engineering, and Mathematics activities in schools (Luciano Jaureguy and Julián Lenis, Dow AgroSciences, pers. comm.). GE soybeans, therefore, provide opportunities for farmers and breeders in addition to community benefits.

Industry also invested a great deal of money into soybean and other crop science research after the development of GE technology. This has, in turn, facilitated the development of conventional breeding tools such as molecular markers and advanced testing designs. The first high oleic soybeans were made possible using GE technology and new alleles now provide additional options for increasing oleic fatty acid content in soybeans. In the short term, the most efficient means of producing soybeans with favourable fatty acid profiles may be a 'mixed-model' approach using the conventional breeding of existing alleles and relevant GE technologies. In the long term, however, novel GE strategies may be the best option for controlling soybean seed fatty acid composition (Chris Tinius, Bayer CropScience, pers. comm.). The benefits of GE soybeans extend beyond increasing the profitability and productivity of modern soybean cultivation to technological innovations used in conventional soybean breeding programmes.

Producers of conventional soybeans benefit from lower seed costs compared to GE seed and the occasional increased profit margin when non-GE seed is contracted and sold for a premium. Growing HT GE soybeans, on the other hand, allows producers to achieve the same output while lowering weed control costs such as chemicals, chemical applications, mechanical tillage and scouting (Greene et al., 2016; Fernandez-Cornejo & McBride, 2002; National Academies of Sciences, Engineering, and Medicine, 2016). Producers must weigh the benefit of spending more for GE seed but less on crop management strategies. To complicate matters, specific markets for non-GE crops have emerged in response to consumer preferences. Consumers, in turn, are influenced by information and governmental regulations regarding the benefits, risks and ethical considerations of GE crops on the environment and human health. These factors have varied impacts on consumer preferences globally, causing uncertainty about the acceptance of GE crops and their by-products (National Academies of Sciences, Engineering, and Medicine, 2016).

Despite the decline in non-GE soybean breeding in the private sector as a whole, several private companies continue to maintain a strong conventional breeding platform. Maintaining a conventional platform taps into a unique pool of germplasm. Bayer Crop Science currently has ten breeding stations and 15% or more of each breeding station programme consists of non-GE soybean breeding. The 15% level is important because if it drops below 15% say to 10% then it quickly drops to zero. The breeding stations utilize older but improved elite *Glycine max* cultivars, landraces and *G. soja* soybean germplasm. The germplasm diversity emphasis is important for the global sustainability of conventional soybean breeding because 80–88% of soybeans are currently grown in the Americas. These trends may change as other regions begin to invest in the land and infrastructure necessary to produce soybeans on a large scale. It is, therefore, important to have a foundation of sustainable soybean genetic diversity now in order to develop cultivars adapted to the challenges of various global regions in the near future (Chris Tinius, Bayer Crop Science, pers. comm.).

3 Genetic architecture

The year 2016 marked the one hundred and fiftieth anniversary of the publication 'Experiments in Plant Hybridization' by the Austrian monk Gregor Mendel (Mendel, 1866). Mendel used another member of the legume family, the garden pea (*Pisum sativum*), and a quantitative approach to establish the laws of independent assortment, independent segregation and dominance. Mendel focused on seven unlinked traits (i.e. plant height, seed shape, seed colour, pod shape, pod colour, flower location and flower colour) with four of the seven underlying genes now known (Reid and Ross, 2011). The combination of Mendelian genetics with Darwinian evolution led to the modern synthesis and the development of statistical genetics (Fisher, 1930; Haldane, 1932; Wright, 1932). Soybean breeders and geneticists are now armed with sophisticated statistical and molecular tools necessary to evaluate soybean genetic architecture and to predict which genetic combinations will provide the most suitable soybean for a particular environment.

Conventional soybean breeding still relies upon some of the same cross-pollination techniques used to breed peas in Mendel's monastery garden. The soybean cross is the start of three phases for a conventional soybean breeder and begins with two parents selected for one or more beneficial traits. After genetically variable parents are chosen and cross-pollinated, phase two follows with the selection and testing of ideal recombinants between the two parental lines. Finally, during phase three, isogenic lines are released and potentially adopted by farmers within a region of production.

The genetic predictions that go into planning each cross have advanced since the modern synthesis eighty years ago, but the fine-scale manipulation of soybean flowers that unifies two gametes from different plants remains the same. Domesticated soybean (*G. max*) flowers are slightly larger than their wild (*G. soja*) counterparts, but both sets of flowers are still relatively small compared to other angiosperms. Tools such as tweezers and hands-free magnifying glasses are a useful part of a standard soybean crossing kit. Cross-pollination (i.e. hand-pollination) typically takes place for about four hours per day in the early and mid-morning, and an experienced crosser can make about one-hundred cross-pollinations in four to five hours. Therefore, the number of experienced crossers in a particular group should also be considered before planning crosses. Soybean flowers possess self-pollinating complete flowers (i.e. contain both male and female parts). Knowing the location of the male and female structures in addition to the respective developmental periods also helps increase the frequency of successful hybridization and fertilization. Finally, it is important to stage the flower development of designated male and female parents in the crossing block to minimize land use and assure that male and female reproductive organs from different soybean varieties are viable at the same time. Advanced planning of potential male and female gamete donors, knowledge of soybean flower development and anatomy, and the proper use of fine-scale crossing tools increases the chances of useful hybrid germplasm available for the next stage of the breeding cycle.

To manage within-line heterogeneity, pollen is collected from eight to ten different male plants, and no more than five female flowers are fertilized per plant. The most desirable male flowers have the flower petals sticking out above the sepals, which is indicative of a vibrant unopened flower poised to open later the same day. The best female flower petals are just beginning to emerge and are distinguished by sepals with tips even with the petal tips, which is a good indicator the flower will open in about two days, and that the pollen within the flower is not yet viable. Male flowers are collected and placed into a canister containing desiccant. With male flowers in hand, the next step is to emasculate the female flower by removing the ten stamens (Fig. 2c; Fig. 3) and to cross-pollinate the

Figure 2 Soybean flower anatomy. (a) A generic flower with male and female reproductive structures. The anthers and filament of the male stamen and the stigma, style and ovary of the female pistil structure are shown along with a developing ovule. (b) A mature soybean flower with a single standard (i.e. banner) petal, one of two wing petals and a single keel petal. (c) Five soybean petals from panel B are shown with prominent vein patterns. The male reproductive structure (i.e. androecium) is underneath the petal diagrams and exhibits a diadelphous morphology consisting of one set of nine male fused stamens and one free stamen. The diadelphous androecium (i.e. fused stamens) surrounds a single pistil. The stigma, ovary and three developing ovules of the pistil are labelled. The developing ovules will eventually become the soybean seed (not shown).

Figure 3 Emasculation of the flower. (a) A node with three flower buds. (b) Removal of one flower bud and the bract using forceps. (c) Removal of additional flower buds leaves one bud with sepals. (d) Sepals are removed from the flower bud leaving the petals more accessible. (e) A crossing tag is gently placed around the flower bud. (f) The crossing tag is secured around the base of the flower bud. (g) The unfertilized bud with labelled crossing tag. (h) Forceps are used to remove the petals and (i) the diadelphous androecium leaving a single unfertilized pistil with a visible style (white arrow).

chosen female (Fig. 4). It is useful to utilize hands-free magnifying glasses for the first few emasculations and cross-pollinations. If an enlarged stigma is observed in 3–7 days, this is a positive indicator of a successful fertilization.

Simple Mendelian soybean traits can then be used to chart the success of a true cross-pollination, and are an essential part of planning successful breeding designs. For example, purple flowers are dominant while white flowers are recessive and tawny pubescence is dominant while grey pubescence is recessive. Using purple flowers as an indicator of a successful cross-pollination has the added benefit of an associated purple cotyledon trait that can be visualized in the first week after planting. In addition to flower colour and pubescence colour, seed colour, seed coat colour, seed size, pod wall colour, leaf shape, maturity and growth habit (i.e. determinant vs. indeterminate) (Fehr and Hadley, 1980) provide useful visual selectable markers for hybrid population development. Subsequent generations of well-planned crosses can then be validated with visible phenotypes that provide positive indicators of true hybrid populations.

The molecular identity of single genes controlling a variety of qualitative soybean traits are known, some of which have tremendous effects on yield (Table 1). If soybean seed pods are unable to hold their seeds before harvesting (shattering), mechanized harvest and yield is directly affected. The molecular basis of two genes responsible for pod shattering in soybeans has recently been discovered using crosses with wild soybeans (Tian et al., 2010; Ping et al., 2014). In addition to a weedy disposition, small black seeds and small purple flowers, wild soybeans have seed pods that shatter as a possible means of rapid propagation in various environments. The molecular identification of single gene variants

Figure 4 Cross-pollination of the emasculated flower. (a) A labelled canister with desiccant containing male flowers. (b) Mature flower from the canister properly positioned between the thumb and index finger. (c) Removal of the sepal located beneath the keel petals. (d) Forceps are placed on the keel petals with the entire reproductive structure in the centre. (e) The diadelphous androecium is removed from the flower. (f) The anthers of the nine fused stamens are indicated with the white arrow. The free stamen is visible on the opposite side. (g) The anthers of the stamens are placed in contact with the stigma of the pistil. (h) The androecium remains in contact with the stigma and the cross-pollination is complete.

Table 1 Selected single gene traits of agronomic value and known molecular identities

Maturity and flowering	Location	Gene call[a]	Ortholog	References
El	Gm06: 20006928..20007814	Glyma06g23026	Putative legume transcr. factor	Xia et al., 2012; Zhang et al., 2016
E2	GmlO: 44716720..44738268	Glymal0g36600	GIGANTEA	Watanabe et al., 2011
E3	Gml9: 47511095..47520052	Glymal9g41210	Phytochrome A	Watanabe et al., 2009
E4	Gm20: 32087412..32093306	Glyma20g22160	Phytochrome A	Liu et al., 2008
E5	Possible artifact	N/A	N/A	McBlain and Bernard, 1987; Dissanayaka et al., 2016
E6	Unknown	N/A	N/A	Bonato and Velio, 1999
E7	Closely linked (6.0 cM) to El	N/A	N/A	Cober and Voldeng, 2001
E8	Unknown	N/A	N/A	Cober et al., 2010
E9/GmFT2a*	Gml6: 31109991..31115008	Glyma16g26660	Flowering Locus T(FT)	Kong et al., 2010; Zhao et al., 2016
GmFT4*	Gm08: 46606934..46608654	Glyma08g47810	FT	Zhai et al., 2014
GmFT5a*	Gml6: 4115570..4116843	Glymal6g04830	FT	Kong et al., 2010
J	Unknown	N/A	N/A	Ray et al., 1995
Indeterminate/determinate	**Location**	**Gene Call**	**Ortholog**	**References**
Dtl	Gml9: 44979743..44981657	Glymal9g37890	Terminal Flower 1	Liu et al., 2010; Tian et al., 2010
Dt2	Gml8: 59918843..59927288	Glymal8g50910	MADS-Domain Factor	Tian et al., 2010
Shattering	**Location**	**Gene Call**	**Ortholog**	**References**
Pdhl	Gml6: 29601346..29601897	Glymal6g25580	NAC transcr. factors NST1/2	Funatsuki et al., 2014
Shatl-5	Gml6: 1708793..1711532	Glyma 16g02200	NAC transcr. factors NST1/2	Dong et al., 2013

[a] Glyma 1.0/1.1.

* Ten or more known *G. max* FT homologs.

controlling the shattering trait provides the opportunity for soybean breeders to select for domesticated non-shattering variants when crossing with wild soybeans. The selection of a single domesticated gene variant, in turn, directly affects yield and may also allow soybean breeders to harness closely linked loci harbouring additional high-value traits.

Many known soybean genes with large phenotypic effects do not measurably interact with other loci to control a single trait. In contrast, some genes with large phenotypic effects interact with other genes (epistasis) to control soybean traits such as growth habit (i.e. determinate vs. indeterminate). Determinate soybeans have separate vegetative and reproductive stages while indeterminate soybeans have overlapping vegetative and reproductive stages. The determinate growth pattern is well-suited to long and warm growing seasons in the Southern United States and produces a short, bushy plant resistant to lodging and easy to harvest. The indeterminate growth pattern is more suitable for short growing seasons in the Northern U.S. due to the overlap between growth and reproduction, but is more susceptible to lodging. The United States is divided into thirteen latitudinal maturity group (MG) zones, and soybeans are classified according to MG with MG 000 being the earliest and MG X being the latest developing cultivar (Fig. 1b). As a general rule, the Southern United States typically utilizes MG V to VIII varieties with determinate phenotypes while the Northern U.S. utilizes MG 00 to IV varieties with an indeterminate phenotype. On the other hand, Brazil grew soybeans with a determinate growth pattern twenty years ago and now grows predominantly indeterminate soybeans. The Brazilian indeterminate soybeans have several advantages, including decreased lodging. Whether growing conditions or soybean genetics led to a reshaping of Brazilian soybean growth habits remains a future subject of inquiry. The Brazilian example indicates that soybean growth habits are not set in stone and novel stem architecture alleles can rapidly reshape soybean branching landscapes.

The genetics underlying soybean growth habit were known over forty years ago but the molecular identity of the genes was a mystery that unfolded in parallel to the sequencing of the first soybean genome in 2010 (Schmutz et al., 2010). The classic example of soybean epistasis consists of the branching genes *Dt1* and *Dt2* (Bernard, 1972). The genetics of *Dt1* and *Dt2* are slightly more complex than classic Mendelian genetics with the *Dt1* allele incompletely dominant to *dt1*, *Dt2* nearly dominant to *dt2* and *dt1* epistatic to both *Dt2* and *dt2*. All wild soybeans are indeterminate (Tian et al., 2010) but domesticated soybeans can have three types of stem architecture depending on the *Dt1* and Dt2 allelic combinations: determinate, semi-determinate or indeterminate. With the molecular identities of soybean branching genes known, soybean breeders and geneticists can now realize the goal of understanding the molecular basis of a significant epistatic component controlling the genetic architecture of soybean yield.

In addition to soybean reference genomes, modern soybean breeders benefit from other crop genomes. For example, the soybean *Dt2* domestication allele appears to be unique compared to other crop systems. In contrast to soybean, tomato (*Solanum lycopersicum*), chickpea (*Cicer arietinum*) and pigeon pea (*Cajanus cajan*) all have recessive semi-determinacy traits indicating a subtle rewiring of the soybean branching pathway. The recent identification and molecular characterization of *Dt2* as a gain-of-function allele (Ping et al., 2014) leading to semi-determinate soybeans provides evidence that domesticated soybean branching patterns have much more potential for rewiring than previously recognized.

The possibility of increasing soybean yield utilizing novel branching pathway alleles to increase the number of branches per unit of area is illustrated by recent work in tomato using weak alleles of the anti-florigen (*Dt1/SP*; self-pruning) and florigen (*SFT*; single flower truss) pathways. Yield gains of 100% were achieved in tomatoes via the creation of 'bushy' determinate tomatoes using weak alleles of *SP/Dt1* and florigen (Park et al., 2014). Crops with similar growth patterns, such as soybean, may benefit from tweaking similar alleles using the CRISPR-Cas9 system (Belhaj et al., 2013). Future work on the major soybean *SP* homolog (*Dt1*) and the two major soybean SFT homologs (GmFT2a/E9 and GmFT5a) may further enhance the branching architecture of soybeans for increased yield using a combination of novel interacting alleles with large phenotypic effects.

In contrast to qualitative traits, quantitative traits are controlled by multiple genes with smaller effects that can also have measurable interactions. Quantitative traits are the most common genetic trait but dissecting the molecular basis was historically difficult due to a lack of genetic markers. Biochemical markers such as allozymes and isozymes provided initial insights into genome variation but were limited in number compared to DNA-based markers. The advent of restriction fragment length polymorphisms (RFLPs) as genetic markers provided the first-ever genome-wide linkage map for detecting quantitative trait loci (QTL) using recombinant inbred lines (RIL) derived from wild and domesticated tomatoes (Paterson et al., 1988). Some useful markers for tracking DNA fragments have subsequently been developed such as simple sequence repeats (SSRs) and random amplification of polymorphic DNA (RAPD) markers, and several methods are available to detect single nucleotide polymorphisms (SNPs) and other genetic variation with known locations in the genome. Increasingly dense markers eventually led to the cloning of the first causative allele underlying a QTL (Frary et al., 2000) and the first cloning and molecular editing of a causative QTL using CRISPR-Cas9 (Large et al., 2016). Mapping QTLs via linkage uses RILs derived from the crossing of two parents with different genotypes and has considerable statistical power to detect causative alleles but suffers from a lack of population diversity and low mapping resolution. Near-isogenic lines must often be constructed to refine QTL resolution and then the QTL must be tested for consistency in multiple genetic backgrounds. The growth in genome-wide markers and genome sequences lead to the development of genome-wide association study (GWAS) mapping approaches. With GWAS a given phenotype can be mapped using the inferred recombination events derived from large sets of genome markers in linkage disequilibrium. GWAS false positives are reduced by using Bayesian statistical tools to account for population structure (Risch and Merikangas, 1996; Pritchard and Rosenberg, 1999; Pritchard et al., 2000). The first plant GWAS QTL, a candidate flowering time allele, was subsequently mapped with this approach using 92 diverse maize inbred lines without any crosses (Thornsberry et al., 2001). In contrast to linkage analysis, GWAS approaches have high resolution and greater genetic diversity but require more markers and have lower power to detect causative alleles. The ability of researchers to map QTL using linkage or association has rapidly accelerated in multiple species along with increasing computing power, more sophisticated algorithms, and increasingly inexpensive and accurate genome sequence information.

The most important soybean quantitative trait is seed yield, but only a few confirmed QTLs are known (Table 2). In developed countries, where mature plants are harvested by mechanical means, yield is controlled by the number of pods per plant, the number of seeds per pod, the seed size, the seed weight, lodging and pod shattering resistance, although these traits are highly susceptible to biotic and abiotic environmental stress. Out of 19 confirmed soybean QTLs for various traits (Grant et al., 2010), there is one confirmed yield QTL on chromosome 5 with some environmental sensitivity

Table 2 Confirmed soybean QTL[a]

QTL name	Chr	LG	LE	RE
cqSeed weight-006	1	D1a	17.52	25.28
cqSeed weight-010	1	D1a	17.52	24.13
cqSeed oil-014	2	D1b	114.49	118.94
cqSeed oil-011	2	D1b	117.48	124.73
cqSeed protein-010	3	N	15.85	22.65
cqSeed protein-004	3	N	31.85	35.30
cqSeed oil-005	3	N	35.77	39.34
cqSeed phytate-001	3	N	73.98	75.98
cqSeed oil-001	4	C1	89.72	91.72
cqSeed yield-002	5	A1	29.93	31.93
cqSeed protein-011	5	A1	93.23	96.73
cqSeed oil-008	5	A1	96.72	102.05
cqSeed oil-012	5	A1	96.73	102.05
cqSeed oil-015	5	A1	96.73	102.05
cqSeed protein-014	5	A1	96.73	102.05
cqSeed weight-013	6	C2	22.54	40.29
cqSeed protein-005	6	C2	33.47	36.91
cqSeed oil-006	6	C2	34.30	34.70
cqSeed oil-009	6	C2	34.30	34.71
cqSeed protein-015	6	C2	34.30	45.75
cqSeed oil-016	6	C2	34.34	51.84
cqSeed protein-007	6	C2	34.34	40.29
cqSeed weight-008	6	C2	34.34	51.84
cqSeed weight-011	6	C2	34.34	51.84
cqSeed protein-012	6	C2	113.50	128.22
cqSeed protein-006	7	M	18.87	46.65
cqSeed protein-009	7	M	43.74	46.65
cqSCN-002	8	A2	47.00	49.00
cqSeed protein-013	8	A2	51.50	53.91
cqSeed protein-016	8	A2	51.50	53.99
cqSeed oil-002	12	H	85.80	87.80
cqSeed protein-008	15	E	9.99	10.15
cqSeed oil-007	15	E	16.13	16.78

(*Continued*)

QTL name	Chr	LG	LE	RE
cqSeed oil-010	15	E	16.13	16.78
cqSeed protein-001	15	E	29.89	31.89
cqSeed oil-013	15	E	38.78	41.82
cqSCN-006	15	E	39.07	45.08
cqSCN-003	16	J	67.79	78.57
cqSDS-001	17	D2	76.69	85.15
cqSCN-005	17	D2	87.01	89.01
cqSCN-001	18	G	3.70	5.70
cqSeed weight-002	18	G	44.40	46.40
cqSeed weight-001	18	G	52.42	54.42
cqSCN-007	18	G	75.64	94.83
cqSeed oil-003	19	L	30.58	34.54
cqSeed weight-009	19	L	68.35	70.36
cqSeed weight-012	19	L	68.35	70.36
cqSeed weight-005	19	L	68.36	70.36
cqSeed weight-007	19	L	68.36	70.36
cqSeed phytate-002	19	L	70.44	72.44
cqR8 Full maturity-001	20	I	35.40	37.40
cqSeed oil-004	20	I	35.40	37.40
cqSeed protein-003	20	I	35.40	37.40
cqSeed weight-003	20	I	35.40	37.40
cqSeed yield-001	20	I	35.40	37.40

Abbreviations: Chr = Chromosome; LG = Linkage Group; LE = Left End of QTL; RE = Right End of QTL.
[a] David Grant (pers. comm.); see also: http://soybase.org/resources/QTL.php

(Fox et al., 2015) and one confirmed yield QTL on chromosome 18 associated with lower seed protein (Nichols et al., 2006). Additional confirmed yield QTL may soon be found using nested association mapping (NAM) populations. Both linkage and association mapping are currently being implemented in NAM populations using a high-yielding hub parent (IA3032) crossed with forty other parents (Diers et al., 2015). The forty parental lines consist of eight lines known to have high yields in drought, fifteen lines with diverse genetic ancestry and seventeen high-yielding lines. 140 F5 RILs were developed for a total of 5600 RILs, and both parental and RIL populations will be genotyped. The identification of confirmed yield QTL may provide useful markers for introgressing yield traits into different soybean backgrounds. Markers associated with yield and consistent in stressful environments, such as drought, will be particularly useful for conventional soybean breeders.

Additional QTLs controlling yield have been identified via recurrent backcrosses with wild soybeans (Concibo et al., 2003; Wang et al., 2004), although only two of six genetic backgrounds provided increased yield in one study (Concibo et al., 2003). Molecular markers for the Concibo et al. (2003) QTL were used to rapidly select for progeny containing the QTL using a technique known as marker-assisted selection (MAS). Markers can be used to either select for one parent's genotype or against another genotype, and multiple genomic regions can be selected for simultaneously. Furthermore, MAS becomes genomic selection (GS) when many markers are utilized with phenotypic data for a set of individuals as a training set to develop a genomic prediction model to select individuals who are genotyped but not phenotyped (Meuwissen et al., 2001; Lorenz et al., 2011). The benefits of GS over traditional phenotypic selection continue to improve with enhanced models.

Efforts to advance genetic gain achieved via conventional soybean breeding are still rooted in the breeder's equation ($R = h^2S$). Small useful advances in genetic gain have been made with molecular markers using newer models, but the application of markers to tease out G x E interactions is still in its infancy and somewhat reminiscent of RFLP technology thirty years ago. More advances will undoubtedly be made with increased access to DNA sequence information and more sophisticated algorithms, but there is still a lot of sand in the gears of the machinery needed to tease out G x E interactions. Modern conventional soybean breeders have a wealth of technology and updated models at their fingertips, but basic information can still be worked out by a soybean breeder walking through fields running older algorithms in his or her head (Chris Tinius, Bayer Crop Science, pers. comm.).

Depending on the goals of the breeder and the starting material, a variety of selection methods exist to develop suitable germplasm for testing in multiple environments. Parents are carefully selected and, following the initial cross, several options are available that have been evaluated with soybeans and other self-pollinated crops. Six commonly used inbreeding and selection methods include pedigree selection, the bulk method, single seed descent, the early generation testing procedure, backcrossing and recurrent selection. Selection methods can be implemented in both developed and developing countries using various strategies (Fig. 5). The time for variety development typically follows 6–10 year timeline depending on the goals and resources of the breeding programme: Cross (0.5 years), Selfing (1–2 years), Selection (1 year), Yield Evaluation (2–4 years), Product Evaluation (1 year) and Seed Multiplication (1–2 years). Selection schemes can be modified according to the breeder's needs and years can be adjusted to accommodate winter nursery and target area availability.

In addition to qualitative single gene traits, complex multi-genic traits and gene interactions, conventional soybean breeders must be keenly aware of environmental interactions. The variable phenotypic effects of the environment on genetically similar plants were well known even in ancient times. For example, the Annals of Master Yan (third century BC; 晏子春秋; pinyin: yànzi chūnqiū) references an orange tree growing north of the Huai river bearing bitter fruit, although the same species of orange tree growing south of the Huai river bears sweet fruit (南橘北枳; pinyin: nán jú běi zhǐ; 'An orange tree produces oranges in the south but bitter fruits in the north'). In this ancient story, two different adjacent environments are sufficient to change the plant phenotype enough to make it appear to be an entirely different species. A more succinct modern version is expressed as 'A character measured in two different environments is to be regarded not as one character but as two' (Falconer and Mackay, 1996). Both the ancient and modern expressions of the effect environments can have on plant phenotypic plasticity is also an important consideration for soybean breeders.

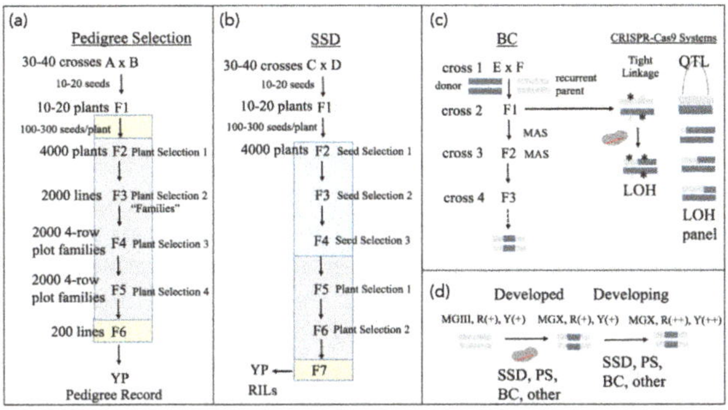

Figure 5 Soybean Breeding for Developing Countries. (a) An example of the Pedigree Selection method using an early maturing variety 'A' with a desirable trait crossed to a late maturing variety 'B' adapted to a target area in the tropics. Ten to twenty seeds are derived from crosses and planted 12" to 20" apart in rows. The time of flowering of F1 plants in the tropics will be more similar to the early variety (Shanmugasundaram, 1981) so visible selection for maturity is limited. Plants are advanced to the F2 generation with seed spacing 4" apart in rows given 100–300 seeds/plant in summer tropical conditions (Hinson et al., 1982). A medium-sized population of 4000 plants is then assessed for the target trait, and late maturing lines are selected for advancement to the F3 generation. Individual lines are harvested and 60–75 viable seeds are planted in a single 10' row. In this example, 2000 lines remain following selection planted in 2000 rows. At the F3 stage, four plants are selected from each row, harvested and planted in adjacent rows in the next generation. The F4 plants now consist of 2000 four-row plots and each plot is a single F2-derived 'family'. The best families and the best row from each family are selected and the four plants are selected from this row for advancement to the F5 generation. Two hundred lines are selected from the F5 generation with no two rows coming from a single family. These rows are bulk harvested for yield trials in the F6 generation. Knowledge of the pedigree helps map genetic variation associated with a desirable trait. (b) An example of the single seed descent method is provided. Selection does not take place until the F4 generation so lines can be advanced to the next generation in a non-target environment and without strict field spacing requirements. F1 seeds are planted in conditions suitable to produce seeds, bulk harvested, and planted. A single pod is selected from the middle of the F2 plant, and a single seed is selected from the pod. The single seeds are bulked and planted, and a single seed is picked again from the F3 plant population and planted. A single seed is then picked from the F4 plants, bulked, and seeds are planted 4" apart in rows in the target environment. F5 plants are selected for target traits, and individual plants are harvested. The seeds from a single F5 plant are planted in a 10' progeny row and F6 lines with desirable target traits are identified and selected. The progeny row is then harvested for yield testing. recombinant inbred lines (RILs) provide useful mapping tools. (c) The backcross method is used to introgress a desirable gene or genes from a donor to a recurrent parent. F1 and subsequent generations are crossed back into the recurrent parent until the gene or genes of interest reach homozygosity. Alternatively, MAS, including biomarkers, can be used to detect the gene or gene products at an earlier stage. The products of a cross can be refined using the CRISPR-Cas9 system to map QTL or to investigate tightly linked loci using loss of heterozygosity approaches (Sadhu et al., 2016). (d) Pre-breeding efforts in developing countries using a combination of genomic engineering and common selection methods provides rapid line development. In this example, diverse early maturing varieties (MGIII) with desirable biotic or abiotic resistance (R) and yield (Y) traits are rapidly advanced to MGX lines with the R and Y traits intact. The new lines are provided to developing countries in MGX zones providing local breeders opportunities to focus on breeding for enhanced R and Y traits. *Key*. Light yellow box: indicates strict field conditions in the absence of selection. Light grey box: indicates selection that occurs with strict field and target environment conditions. Light blue box: indicates selection in the absence of strict field or environment conditions. YP: Yield plots with similar conditions to local soybean producers.

Soybean breeders design experiments knowing soybeans grown in the same small plot can have dramatically different growth patterns caused by genotype x environment (G x E) interactions. For example, optimized yield testing of soybeans utilizes uniform genetic buffers (filler) around field blocks to protect against environmental effects. Yield plots can be double rows, four rows, eight rows or larger. Four rows provide an ideal balance between having enough seed and mitigating environmental effects. A four-row yield plot contains two inner data rows and two outer border rows that mitigate the environmental perturbations. Within one field a single experimental soybean genotype has replicates, and within a single mega-environment multiple fields are tested in an attempt to reproduce the average farmer's field within the mega-environment. Soybean breeders must, therefore, incorporate G x E interaction considerations into plot, field and experimental designs, particularly at the more advanced stages of the soybean breeding cycle (see Section 6).

A growing knowledge of the genetic architecture of various soybean traits provides a roadmap for future selection schemes. Unique and similar suites of 'domestication genes' are becoming more evident in crop systems (Fuller et al., 2014; Meyer and Purugganan, 2013). Some crop system studies indicate that domestication alleles provide buffers against epistatic and G x E interactions (Doust et al., 2014). These revelations provide the conventional soybean breeder with tangible tools to select the best soybean varieties using the latest technology and to reduce the time from the first cross to the first seeds being planted in the farmer's field.

4 Germplasm diversity

Before GE soybeans, soybean breeders would exchange seed packets freely. Now, however, it is very difficult to acquire and test soybean germplasm material because of intellectual property associated with soybean varieties and traits. For example, if one wants to grow a test with a public breeder as a collaborator, and experimental lines come from different public breeding programmes, several MTAs will be necessary. The legal hurdles associated with acquiring soybean germplasm with increased diversity are even more challenging. Therefore, legally binding contracts are useful for protecting the unauthorized use of soybean germplasm in the short term but inhibit soybean germplasm diversity in the long run (Luciano Jaureguy and Julián Lenis, Dow AgroSciences, pers. comm.).

Dow AgroSciences has a global breeding programme, and the material developed for one geography can sometimes be successfully utilized in different parts of the world. For example, some material developed for the U.S. Midwest can be tested for adaptation in Argentina if the germplasm is freely available and does not have any associated intellectual property protection. Most of the publicly available germplasm that fits in this category is MG III or later (Luciano Jaureguy and Julián Lenis, Dow AgroSciences, pers. comm.).

Dow AgroSciences soybean breeding has used and continues to make use of public soybean germplasm diversity. For example, Dr. Randall Nelson and other public breeders have successfully collaborated for many years with Dow AgroSciences and provided diverse soybean material at various stages of variety development for Dow AgroSciences to utilize in its breeding programme. Dow AgroSciences has also found useful publicly available varieties by consulting with other public breeders and mining publicly available databases. The largest public soybean germplasm database is accessible via the Germplasm Resources Information Network and provides information on native traits of a variety of conventional soybean material while top-performing lines can be gauged from soybean cyst nematode (SCN) regional tests and some uniform tests. Some of the more useful USDA and university breeding material is high-yielding with advantageous disease-related traits (Luciano Jaureguy and Julián Lenis, Dow AgroSciences, pers. comm.).

In terms of sustainability, genetic diversity is essential for soybean breeders to continue to increase soybean yield and to combat a myriad of abiotic and biotic stresses. All public and private U.S. soybean varieties are derived from East Asian plant introductions, so a complete understanding of the domestication history is important for quantitating the level of genetic diversity in current and past soybean collections. Current molecular evidence suggests that the domesticated soybean (*G. max*) shares a recent common ancestor with the wild soybean (*G. soja*). The world collection of elite domesticated soybean cultivars can, therefore, be traced to one or many domestication events several thousand years ago from a *G. soja*-like ancestor. The subsequent myriad of semi-domesticated landraces would eventually become the genetic basis for modern elite soybean cultivars. The genetic diversity of northern and southern U.S. soybean breeding programmes is narrow (Gizlice et al., 1994). The gradual development of modern soybeans via breeding and selection led to genetic bottlenecks and a substantial decrease in genetic diversity over time (Hyten et al., 2006). As sequencing and genotyping technologies become more accessible, it is important for modern conventional soybean breeders to continually evaluate and maintain genetic diversity in their breeding programmes.

Efforts to increase soybean sustainability led to the development of manageable soybean germplasm collections known as core or mini-core collections. A core collection should capture about 70% of the genetic diversity of an entire collection using only 10% of the collection with an upward limit of 3000 species (Brown, 1989). The concept of a more manageable mini-core collection was subsequently introduced utilizing a phenotypic clustering matrix to capture 10% of the diversity of a chickpea core collection (Upadhyaya and Ortiz, 2001). Whole-genome resequencing and affordable genotyping platforms now provide more accurate methods of evaluating the genetic diversity of various soybean collections to develop core and mini-core collections, although detailed phenotyping platforms lag behind.

Soybeans were domesticated in ancient China and the largest and most diverse soybean collections are found there, consisting of more than 23 000 *G. max* accessions and more than 7000 *G. soja* accessions (Qiu et al., 2013). Several core and mini-core collections have been developed in China, Japan and Korea, although the Chinese core collection remains the largest and most well-studied wild and domesticated soybean collection (Qiu et al., 2013). The ancestral Asian hot spots of soybean domestication are quickly disappearing so these soybean collections represent the irreplaceable legacy of thousands of years of soybean domestication efforts.

Public core and mini-core collections also exist outside of Asia. A soybean core collection was developed in Brazil (Priolli et al., 2013) and the largest soybean collection outside of China is the USDA collection consisting of 18 480 *G. max* accessions, 1168 *G. soja* accessions and 919 perennial *Glycine* species accessions (Oliveira et al., 2010; Song et al., 2015). USDA-based core collections have been established for *G. max* (Oliveira et al., 2010) and perennial *Glycine* species (Brown et al., 1987). The diversity captured by the USDA soybean collection is continually maintained, utilized, evaluated and annotated and has recently been assessed using a SoySNP50K chip (Song et al., 2013, 2015). Storage and maintenance of diverse soybean collections is vital to sustainability but the value of the genetic diversity is often not realized without several years of commitment from conventional soybean breeders.

There is a tremendous need for public soybean breeder support in order to create sustainable soybean breeding programmes with more genetically diverse germplasm. Soybean varieties developed by public breeders have more genetic diversity than varieties developed by private breeders due to the limited number of elite lines used to quickly bring a product to market. The number of small private seed companies is in decline, and almost 50% of the world soybean seed market is currently controlled by three companies (Monsanto, DuPont and Syngenta). The increasing consolidation of private soybean breeding programmes revolves around a limited gene pool of elite cultivars, which compounds the lack of genetic diversity in private soybean collections. Private breeding programmes utilize both GE and conventional breeding material but the range of material released is limited due to licensing and patent issues. Private companies do, however, utilize varieties developed from public programmes to increase the diversity of private collections. The minimal amount of soybean genetic diversity generated by commercial enterprises places the burden of improving genetic sustainability on public soybean breeding programmes in partnership with private companies.

U.S. public soybean breeder/geneticist positions are limited to eighteen people at the university level and seven people at the USDA – Agricultural Research Service (ARS) level (Fig. 6). In addition to public and private breeding collaborations, each breeder has a unique programme focus but all public programmes tend to emphasize conventional breeding approaches leading to high-yielding, abiotic and biotic stress-resistant soybeans that generate high-quality seeds with a variety of characteristics (e.g. seed size, seed colour, hilum colour and protein/fat/carb composition). Within the public sector only ten states have soybean breeding programmes focused on developing soybean varieties for specialized use in food products (nimame, tempeh, tofu, natto, soy milk, miso, soy sauce, soybean sprouts and edamame). Furthermore, public soybean breeders are responsible for educating the next generation of public and private sector breeders and affiliated personnel. Therefore, a limited number of public U.S. soybean breeders provide educated and trained personnel, conventional breeding techniques and specialized soybean germplasm to both the public and private soybean sectors.

The efficiency of a modern conventional soybean breeding programme is improved with knowledge of the genome sequences used for breeding. The recent publication of the Williams 82 reference genome, which represents a historical Northern U.S. germplasm source (PI 518671; MG III), was a major milestone for soybean research (Schmutz et al., 2010). The level of sequence variation such as SNPs, indels, presence-absence variation, copy-number variants and structural rearrangements (e.g. translocations and inversions) was assessed in one anchored and assembled reference genome, while approximately 15% of the estimated 1.115 Gb genome remains in flux. Continuous annotations provide further insight into soybean genome organization and a second reference genome representing a significant Southern United States germplasm source (Lee – PI 548656; MG VI) has been sequenced in collaboration with private companies, with a release date scheduled before 2020. The genetic diversity of DNA-containing organelles such as chloroplasts and mitochondria continues to be a subject of study in addition to rhizosphere-related organisms (e.g. rhizobia and mycorrhizal fungi). Furthermore, hundreds of *G. soja* genomes have been resequenced and at least one *G. soja* pan-genome has been sequenced. Seven years after the release of the Williams 82 soybean genome (Schmutz et al., 2010), Williams 82 remains the gold standard reference genome for both *G. soja* and *G. max* resequencing efforts.

The soybean genome is unique amongst other published legume genomes (adzuki, common bean, pigeon pea, medicago, chickpea and *Lotus*) in having both the highest repeat content (58.74%) and a recent genome duplication event (~13 Myr ago) not found

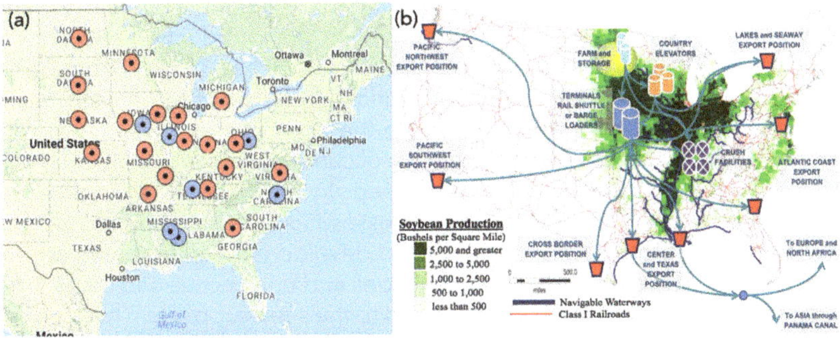

Figure 6 Distribution of US public soybean breeders in relation to US soybean production and distribution. (a) The United States currently has eighteen university soybean breeders (red circles: AR, GA, 2x IA, 2x IL, IN, KS, KY, MI, MN, 2x MO, NE, ND, OH, SD, TN and VA) and seven USDA/ARS soybean breeders (blue circles: IA, IL, 2x MS, NC, OH and TN). (b) Soybeans are transported via rail (red lines), barges (dark blue lines) and roads (not shown) to loading terminals (e.g. rail shuttle or barge loader), country grain elevators, storage facilities and crush facilities. The major soybean production centres are shown in dark green (>5000 bushels/sq.mi.) in addition to less productive areas shown in lighter shades of green. Source: United Soybean Board, Field to Market (2012).

in other legumes (Yang et al., 2015). The recent (~13 Myr ago) duplication event started with a proto-soybean containing ~30 000 genes that duplicated to generate ~60 000 genes. The current soybean genome has thrown out the paralogs of 15 166 genes and has retained 31 264 genes with paralogs for a total of 46 430 predicted genes (Schumtz et al., 2010). In addition, a common legume duplication event occurred ~58 Myr ago (Schmutz et al., 2010; Yang et al., 2015) followed by a rapid purging and subsequent stabilization of many gene duplicates. Therefore, some traits, such as oleic and linolenic acid composition, are controlled by genes sharing a common ancestor. Furthermore, most of the repeat sequences occur in low recombination heterochromatin regions near the centromeres with the remaining euchromatin regions containing most of the gene content and most recombination events (Schumtz et al., 2010). Efforts to provide soybean breeders and geneticists with affordable and accurate sequence information have recently begun to focus on the euchromatin regions, while sequence alignment algorithms suitable for evaluating a high frequency of soybean gene duplicates and repeat sequences lag behind. The ability of soybean breeders and geneticists to locate genes and gene regions responsible for desirable traits will continue to be facilitated by comparative genomic research utilizing wild and domesticated soybean and legume genome sequence information.

5 Seed certification and mechanization for plant breeding

5.1 Seed certification

At the end of the day, the phenotype most important to soybean producers is seed yield with seed quality also influencing potential profits. The potential yield and quality of seed derived from a given cultivar can be gauged via seed certification. Before seed certification programmes were first introduced in the United States in 1915, most farmers depended on saved seed from previous harvests. This changed with the introduction and expansion of seed certification programmes leading to more farmers purchasing seed from commercial traders instead of using saved seed. Seed certification is now the officially recognized system for maintaining soybean varietal purity on the open market.

Certification programmes led to a rise in the role of commercial seed markets by providing farmers with quality assurances of purchased seed. In the beginning, commercial seed suppliers were mainly small family-owned businesses lacking the resources necessary to invest in research and development. Research and development of improved varieties was almost exclusively conducted by land-grant colleges and universities and state agricultural experimental stations who would pass on the seed to the seed suppliers. The seed suppliers would then multiply and sell these seeds (Fernandez-Cornejo, 2004). Depending on the region, large seed firms now directly market and distribute their end product to regional, national and international markets using complex distribution networks including farmer dealers, farmer associations, company salespeople, private wholesalers and retailers.

The seed certification programme maintains and makes available to the public high-quality seeds and propagation materials of genetically distinct crop varieties. Certified seed is produced by farmers and seed producers starting with pedigreed planting stock and followed by careful quality control, multiple field inspections during the growing season and seed inspections following harvest to ensure varietal line purity. There are four classes of certified seed: Breeders Seed, Foundation Seed, Registered Seed and Certified Seed. Breeders Seed is commercially unavailable seed produced directly by the soybean breeder or representative organization in amounts ranging from a few pounds to several hundred bushels. Foundation Seed is produced from the Breeders Seed under the direction of an approved agronomist to maintain genetic identity and purity. The Foundation Seed is used to produce either Registered Seed or Certified Seed using the same high standards as Foundation Seed, although the Registered Seed step is sometimes omitted to increase the speed of varietal release. Once the soybean seeds pass both field inspections and laboratory analyses they are considered Certified Seed. Blue tags attached to the Certified Seed bags will contain information such as the name of the soybean variety, the percentage of germination, the percentage of inert matter and the percentage of pure seed (MCIA, 2016).

In the United States, seed certification is a state responsibility. Each state has an agency designated to certify the seed that is also empowered by the state seed law. Several state certification programmes are administered by the state departments of agriculture, while a few states have certification programmes administered by the Cooperative Extension Service. Most state certification programmes are administered by grower-controlled crop improvement associations. These state agencies maintain a close association with university personnel who may serve on their board of directors. Regardless of how the programme is organized, they are generally non-profit programmes and must generate funds to cover salaries, and overhead and operating expenses (Fernandez-Cornejo, 2004).

Plant breeding is the foundation of the modern seed industry and both public and private soybean programmes place a great deal of resources, planning and effort into developing new soybean varieties. A cultivar's success in the market depends on its improved traits which are a direct result of the research and development effort to create them. Large-scale research and development is expensive and limited to land-grant colleges and universities, a relatively small number of large companies, and Federal Government agencies such as the USDA – ARS. Prior to 1970 the USDA and State Agriculture Experiment

Stations were responsible for soybean variety development. The 1970 U.S. Plant Variety Protection Act (PVPA) helped spur investment into soybean cultivar development and led to the increased adoption of soybean varieties developed by private breeders. The PVPA provides legal protection of sexually produced seeds, including soybean, for 20 years if the soybean variety has three properties: Distinctiveness, Uniformity and Stability. A plant utility patent has been an option for soybean variety protection since 1985 and provides protection for 17 years for soybean varieties possessing Novelty and Non-obviousness. Soybean breeders create a unique and marketable product by developing seeds with farmer-preferred traits such as high yield, resistance to disease and pests, and adaptation to regional agro-climatic conditions. The high research and development costs associated with variety development means private sector soybean varieties must be commercially viable and highly competitive, while both public and private sector soybean varieties must be protected by intellectual property rights.

5.2 Mechanization of planting and harvesting

The mechanization of plant breeding in the late 1980s was responsible for increasing the number of experimental lines tested by 5–10 times, which significantly increased the size of breeding programmes (Janick, 2008). It is desirable to plant the field using mechanical planters; however, hand planting is still a viable alternative in small breeding programmes and to overcome technical limitations in large breeding programmes. Cone planters are the most common plot planters available and have fairly simple operating systems. As the planter moves forward the seed is poured over the cone, the cone revolves and the seed passes into a hole that leads through a tube or disk into the soil. A range of plot lengths and seeding rates can be achieved by adjusting the gears controlling the speed of cone rotation. Planter efficiency and accuracy has improved with the implementation of GPS technology. For example, ALMACO®, an agriculture research equipment company, provides a cable-free SkyTrip GPS field planting system that can be coupled with other auto-steer GPS systems. The SkyTrip GPS system produces and utilizes digital field maps for accurately planning and planting hundreds of different soybean lines in one hectare.

Automated cone planters are also available. For example, the ALMACO PowerPlant Planting System utilizes a magazine preloaded with the seeds for each plot and row. With automated cone planters, breeders can plant different genotypes in each plot or row and thereby increase the planting efficiency and accuracy by bypassing manual seed packet dumping into the cone. When an even more precise seed distribution is required precision planters can be used. The most common precision planter is the vacuum planter, which applies suction to the distribution plate in order to allow only a single seed to be sucked into each hole position and placed uniformly in the seed furrow.

To harvest the plots a variety of small plot combines are available that will cut, thresh and clean the seed samples. Modern plot combines can measure the plot weight and moisture and upload the data into a tablet and/or computer. ALMACO recently released a new experimental plot combine with a rotary threshing system that can decrease seed damage and plot loss and increase plot cleanliness and harvest efficiency. However, in small breeding programmes or when the programme cannot afford the equipment, a cheaper option is using a large plot thresher with hand harvesting.

Another important step in experimental plot preparation is packaging and labelling. To minimize errors, breeders must take many precautions before starting the packaging process: 1 – Label all the experimental lines and envelopes with unique entry numbers,

2 – Check if the entry numbers between experimental lines and envelopes match and 3 – Print the entry list to use as your guideline during the packaging. In general, the packaging process is laborious and tedious. However, plant breeders are adopting automated systems to make the process more precise and faster. Envelope and label printers currently print all the information needed to recognize experimental lines such as plot number, name of the line, name of the experiment, year, location and other useful information. In addition, many breeding programmes now use barcodes on their labels to implement a digital tracking system, thereby providing the means to rapidly scan labels on envelopes packed with seeds for shipping, planting or laboratory experiments.

Seed counters are another useful piece of equipment used during the packaging process, especially with different seed weights and planting density experiments. Standard seed counters count seeds and distribute a desired number of seeds into separate envelopes but still require manual intervention to switch out envelopes between sets. An automated packaging system is now available with cartridges instead of envelopes. The cartridges are stored in magazines and each cartridge has a unique barcode. The automated packaging system increases the quality and accuracy of the process and saves time in the preparation of experiments.

6 Minimizing phenotypic variation due to non-genetic factors

As the amount of arable land decreases in developed countries and the human population continues to increase, higher soybean yields are needed from less land. Furthermore, the gradual warming of the planet results in variable regional climate changes that may or may not affect traditional testing locations and target regions. In order to meet the challenges ahead it is imperative for farmers and breeders to continually assess growing conditions (e.g. air, water, soil and pests) in the target region and testing locations. The number of crosses, number of years, number of replicates and number of locations needed to evaluate a specific number of genotypes within a given field and/or mega-environment continues to be a subject of investigation (Hanson and Brim, 1963; Sprague and Federer, 1951; Yan et al., 2011). Calculations for optimal test parameters must include the amount of soybean seeds needed to carry out the test, the amount of land, the time and labour of various personnel, and the trial costs. Furthermore, more or less locations, years or replicates may be necessary depending on the trait in question, with soybean yield being the ultimate deciding factor. The cost of adding one replicate or one location to mitigate the experimental error variance or the genotype-by-location interaction variance must be carefully weighed.

Use of the correct field design is important in all stages of a breeding programme, and will impact the capacity to identify the most suitable phenotypic output generated by a given genotype in a specific environment. Experimental field designs are based on the concept of replication, randomization and control of field variation (Hinkelman and Kempthorne, 2006). The main focus of experimental designs is minimizing the variance associated with error, thereby increasing the accuracy and repeatability of variance associated with other sources of variation in field experiments. Replication and randomization of the genotypes in the field allows an accurate estimate of the experimental error, as well as estimates of the variance components associated with genotype, replication, location and their respective interactions. Several experimental designs are utilized to evaluate experimental soybean lines, such as randomized complete block (RCB), incomplete block (lattice and alpha-lattice) and augmented designs (Casler, 2015; Federer et al., 2001; Jones et al., 2015). Moreover, field designs can range from unreplicated trials at one location to replicated trials including many locations over many years. By minimizing variance, soybean breeders are able to make selections of the best performing populations and lines for advancement through the breeding programme.

The most common designs utilized in breeding programmes are RCB and augmented designs. The RCB is very simple and flexible and has been used in intermediate and final tests in a breeding programme, when fewer genotypes are tested (Fig. 7). In this case the field is divided into blocks, maintaining the homogeneity within each block, and each block contains at least one replicate of each genotype and check (i.e. a well-established cultivar serving as a control for various traits). The augmented block design has some advantages when the land, amount of seed, and resources are limited, which is often the case in the early generations of a soybean breeding programme. In this case the augmented design has replicated checks and unreplicated entries within an environment (Federer, 1956). An augmented design can increase the number of genotypes and environments tested, since the number of replications per environment is reduced (Federer and Crossa, 2012). In some cases, two replicates can be included to obtain a better estimate for the genotypic effects, thus increasing the seed requirements, land area and operational costs. The availability of several experimental field design options provides soybean breeders with choices for minimizing phenotypic variance using a limited amount of resources (e.g. seed, land and labour). This, in turn, maximizes the quality of soybean lines selected from experimental trials in a particular environment.

Soybean farmers and breeders desire and require the best soybean varieties adapted to their region. Field testing is essential to gauging the performance of new soybean varieties compared to established varieties. Field testing takes place in multiple locations (i.e. environments) with soil and weather conditions similar to the intended commercial production environment and often taking place on actual production soybean farms. Soybean testing generates information about the genotypes grown in each environment that determines how well the genotype performs in each environment. The field test data includes the identity of the target environment, the identity of the test location and a variety of traits including seed yield. Public testing utilizes seed from private seed companies, growers, breeders and agricultural experiment stations. The seed quality (i.e. purity and germination rate) is carefully assessed prior to planting to ensure optimal variety performance and yield, while maturity, lodging and height are typically assessed during the trials. The data is usually provided as a genotype-by-trait two-way table for each location that includes trait summary statistics (e.g. least significant difference, mean and standard error), a genotype-by-location two-way table for each trait containing absolute values, a genotype-by-location two-way table for each trait containing values relative to the check or trial mean, genotypic means across all locations within the target environment, and occasionally genotypic means from recent past years. Testing is therefore essential to ensure useful and accurate data for farmers and breeders to advance the best lines forward for a particular environment.

A useful measure of the effectiveness of a trial is the heritability of a trait. The heritability of a trait can be deduced by partitioning the genetic (additive: V_A; dominance: V_D; and epistatic: V_I), phenotypic (V_P) and environmental (V_E) variance into individual components with the phenotypic variance being equal to the sum of the genetic and environmental

Figure 7 Soybean randomized complete block (RCB) design yield test using lines derived from a *G. max* x *G. soja* cross. (a) *G. max* (MG III) mature flower (white scale bar = 7 mm). (b) *G. soja* (MG I) mature flower (white scale bar = 7 mm). (c) A grid showing locations of relevant field plots including field checks (i.e. controls). (d) Four-row *G. soja* (MG I) yield plot corresponding to the small white rectangle with the dotted line in panel 6E. The picture was taken 84 days after harvest using an 8 megapixel (MP) camera. (e) No-till field experiment with two tests (MGIII and MGIV) using RCB design with two replicates of each genotype. The MGIII test is located in the right block and the MGIV test is located in the left block. Each four-row plot is twelve feet long with a four-foot alley between plots. Six hundred seeds were planted in each plot with thirty inches between rows. The field picture was acquired 95 days after planting using a single image at 300 ft with a 14 MP camera. A 16 ft blue field truck is pictured in the lower right-hand corner for scale. (f) Resolution of the box with the black dotted line from panel 6E with 14 MP camera at 300 ft. The single rectangle with the white dotted line shows the two data rows while the two rectangles with black dotted lines show the border rows. (g) Resolution of the box with black dotted line from panel 6E with 14 MP camera at 200 ft. (h) Resolution of box with black dotted line from panel 6E with 14 MP camera at 100 ft. Panel 6E key: dark grey = border rows (i.e. filler); light grey = filler used for experimental duplicates (not enough seed); yellow = MGIII check; red = MGIV check; dark green = *G. soja* (MG0) parent; light green = *G. soja* (MG I) parent; blue = *G. max* (MG III) parent. The MGIV test consists of light blue blocks on the right and the MGIII test contains tan blocks on the left.

variance components. The ratio of V_A to the total phenotypic variance of a trait provides a measure of the narrow sense heritability (h^2) (Falconer and Mackay, 1996). This value represents the fraction of the total quantitative genetic variance of a trait transmitted from parent to progeny within a population that can be used for short-term selection in the absence of any DNA sequence information. If one location within a particular year is taken as a single environment, growing soybean lines with multiple replicates at several locations provides a reasonable estimate of heritability. Various approaches optimize the number of replicates and locations to achieve a predetermined heritability threshold for a particular trait (e.g. yield) within complex or simple mega-environments (Yan, 2014). The increased ability to monitor microclimates, including growing conditions, within a complex mega-environment provides public soybean breeders the opportunity to continuously assess and improve the effectiveness of various trials.

The consistency and reliability of seed testing is also heavily dependent on the coordination of testing authorities. Testing authorities must ensure that standard seed testing guidelines are met at public testing locations. Testing authorities should also coordinate with farmers and breeders to continually evaluate testing locations and the target environment using available data. Even small amounts of genetic gain are useful for soybean breeders, and proper coordination of soybean breeders, farmers and testing authorities ensures testing locations are well-suited for the target environment.

Choosing a precise breeding target environment is complex and defined by the needs of target market segments. The end goals of assessing G x E are to identify genotypes that have outstanding performance in the testing environments as well as to single out those with broad versus narrow adaptability. Therefore, the breeding target needs to meet the needs of specific market segments and the soybean breeding material should capture that variability. Site selection is based on the size of the location and the homogeneity of the location. Geographic information systems mapping is very useful in deciphering the homogeneity of a region by identifying low spots, high spots and the slope of the site. Variable soil type is another concern. Experimental design is determined by working closely with quantitative geneticists and an information technology group. Later in the season, quantitative geneticists provide tools that can evaluate data from different locations and estimate the value of a plot based on spatial variability analysis. The information technology group, in turn, provides visualization and analysis tools that help understand the performance of lines in the geography they are tested and help make decisions for variety placement. Target region consideration is also evaluated by a commercial group, agronomists, and a product development group. Therefore, a large multidisciplinary team, a great deal of data and many decisions are involved in the variety development and release process (Luciano Jaureguy and Julián Lenis, Dow AgroSciences, pers. comm.).

7 Field and laboratory phenotyping

Traditional phenotyping approaches are more time consuming and labour intensive, compared to high-throughput phenotyping (HTP) platforms, and inefficiently measure a limited set of phenotypes (Furbank and Tester, 2011). Much progress has been made in wheat and corn using HTP at Bayer Crop Science, and a large group is examining the potential for HTP in soybean. Drones, conveyor belts and various sensing tools are being tested in an effort to efficiently phenotype soybeans, although GS is currently a much bigger tool. HTP and GS should eventually go hand-in-hand to select the most suitable soybeans but more people are currently working on genomic prediction and selection (Chris Tinius, Bayer Crop Science, pers. comm.).

The approach of the private sector to phenotyping is currently different in terms of marker development and usage. While the public sector may identify a few markers associated with a phenotype and publish them, the private sector uses many published and unpublished markers to reduce the amount of phenotyping. The affordability and speed of marker development provides the opportunity to increase selection intensity using genomic prediction algorithms. This provides a fast and cheap pipeline to bypass expensive and laborious phenotyping using GSs (Luciano Jaureguy and Julián Lenis, Dow AgroSciences, pers. comm.).

Phenotyping is the most expensive part of a breeding programme and efforts are underway to reduce it as much as possible. HTP of complex traits promises to revolutionize the breeding process by improving the precision of selection and by allowing for more material to be evaluated. Advances in field phenotyping make screening large populations and experimental lines possible with considerably lower cost and less time. The use of HTP platforms is a non-invasive and non-destructive method of characterizing plant functions during the growing season. These methods could improve selection for targeted traits and allow plant breeders to select at earlier stages. For example, drone and sensor-based technology (Table 3) might prove useful by reducing the time and cost associated with phenotyping disease plots and taking maturity notes (Luciano Jaureguy and Julián Lenis, Dow AgroSciences, pers. comm.).

Population growth and climate change are on the minds of Bayer management, the management of other companies and various influential public sector organizations such as the FOA (Food and Agriculture Organization of the United Nations). To meet the growing challenges of sustainable crop breeding, Bayer CropScience recently established Hypercare farms in the United States, Canada, Germany and France. Hypercare Farms optimize crop yield potential using the latest in HTP and genetic analyses to determine plant fitness in response to biological and chemical applications in real field environments. The effectiveness of various digital monitoring systems, chemical products or biological products is then passed on to farmers who can use less resources and land to produce the same amount of crops, including soybeans (Chris Tinius, Bayer Crop Science, pers. comm.).

One of the key advantages of HTP is the ability to rapidly phenotype several traits in parallel. Three major data sets are currently phenotyped using HTP: 1) remote sensing and imaging in the field and in the lab, 2) laboratory analyses of seed and plant tissue samples and 3) near-infrared reflectance (NIR) spectroscopy of seed and plant tissue samples (White et al., 2012). NIR can capture physical and chemical characteristics of soybean seed samples and generates reliable mathematical model and sample-dependent predictions of seed composition traits such as protein, oil, sugars, fatty acids, amino acids, moisture and starch (Hacisalihoglu et al., 2010). Wet lab chemistry and nucleic acid extraction can, in turn, provide accurate seed composition data and increasingly inexpensive information about the DNA, RNA or epigenetic profiles of laboratory samples. Remote sensing and imaging of field experiments is a novel technique in the HTP toolkit that provides reliable unbiased phenotypic data with greater speed and precision than expensive manual labour approaches, while lab imaging of specimens (e.g. seed and leaf images) combined with feature extraction software provides useful quantitative phenotypic data. HTP data combined with genomic data and environmental data will help refine our understanding of the realized phenotype derived from soybean genetic architecture in a given micro-environment.

Remote sensing and imaging techniques are based on the information provided by visible/near-infrared (VIS-NIR) radiation reflected and far-infrared radiation emitted by the crop (Araus and Cairns, 2014). Canopy reflectance properties allow for the estimation of a range of traits such as canopy temperature, water status, pigment content and photosynthetic activity. Furthermore, canopy reflectance can provide additional benefits such as screening for yield potential, biotic stresses (pests and diseases), and abiotic stresses

Table 3 Sensors and applications in HTP studies

Sensor type	Photosynthesis	Chlorophyll	Water stress	Visible properties	Growth	Greenness	Nutrient deficiency	Disease/defects	Plant stress	Crop response	Plant height	Canopy density	Plant/canopy volume	Canopy temperature
	Traits measured													
Fluorescent (visible and near-infrared)	×	×	×											
Digital camera (RGB)				×	×	×	×	×						×
Multispectral/colour-infrared camera			×				×	×						×
Hyperspectral camera									×					
Thermal sensor/camera			×					×						×
Visible-near infrared spectrometer						×		×	×	×				
3D camera											×	×		
LIDAR (Light Detection and Ranging)											×		×	
SONAR (Sound Navigation and Ranging)													×	

Modified from Sankaran et al., 2015.

(drought tolerance, extreme temperature, salinity, nutrients) (Munns et al., 2010; Weber et al., 2012; Araus and Cairns, 2014; Sankaran et al., 2015). Keep et al. (2016) recently found an association between grain yield and low canopy temperature and chlorophyll content in a set of cultivars released between 1920 and 2010, while Christenson et al. (2014) found significant genetic differences for spectral reflectance in soybean cultivars. The same trend was found in maize, where 40% of variation in grain yield was explained by spectral reflectance (Weber et al., 2012). These traits are relatively straightforward to measure using remote sensing technology and can be applied in breeding programmes on a large scale to increase the speed and accuracy of selection of superior progeny.

A great deal of progress has been made using camera-based approaches installed in ground vehicles similar to a tractor (Deery et al., 2014) despite the challenge of applying remote sensing and imaging of traits on a field scale. White and Conley (2013) proposed a low-cost cart for field phenotyping. The basic cart consists of a rectangular frame welded onto two bicycle frames, weighs 40 kg and a single person can push the cart through the field. Crain et al. (2016) created the Phenocart using a similar approach as used by White and Conley (2013). The Phenocart weighs 12 kg and is more portable between fields. Both systems are inexpensive to build, highly manoeuvrable in the field

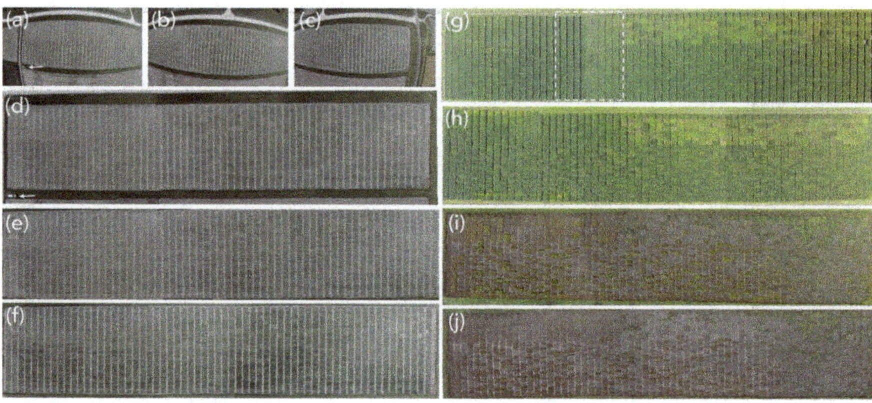

Figure 8 Soybean field experiment development over a single season. (a) Uncorrected fisheye image of the left side of the field in panel D acquired at 300 ft. A white arrow points to a 20-ft-long field truck for scale. (b) Uncorrected fisheye image of the middle of the field in panel D acquired at 300 ft. (c) Uncorrected fisheye image of the right side of the field in panel D acquired at 300 ft. (d) Fisheye corrected images merged from panels A–C and cropped. A white arrow points to a 20-ft-long field truck for scale. (e) Cropped image from panel D for comparison to panels F–J. The image was acquired 40 days after planting. The field contains multiple experiments with different designs and variable numbers of rows per plot. Each row is twelve feet long with a four-foot alley between ranges and thirty inches of spacing between rows. (f) Merged fisheye corrected image of field 50 days after planting. (g) Merged fisheye corrected image of field 82 days after planting. (h) Merged fisheye corrected image of field 103 days after planting. (i) Merged fisheye corrected image of field 137 days after planting. (j) Merged fisheye corrected image of field 146 days after planting. Automated lens correction and photomerge were used to correct fisheye lens distortion and to merge images, respectively, in panels D–J. No adjustments were made to the white balance of any images. In the case of panel G, a white balance distortion occurs between merged images and is outlined with a rectangular white dotted line.

and could substantially increase the amount of phenotypic data a crop breeder obtains per season. The type of sensors/cameras to be used is adjustable to accommodate the needs of different crops.

There is some progress using unmanned aerial vehicle (UAV) approaches since the use of UAVs (i.e. drones) is less expensive and more popular, and images larger areas faster (0.4 ha/10 sec) than stationary approaches (White et al., 2012; Liebisch et al., 2015). Furthermore, on 29 August 2016 new Federal Aviation Administration (FAA) regulations (Part 107, Final Rule for Operation and Certification of Small Unmanned Aircraft Systems) paved the way for small unmanned aircraft systems to become part of the U.S. airspace. The new FAA regulations provide opportunities to test and develop more robust aerial sensing-based crop phenotyping methods (Fig. 8) and to significantly accelerate HTP in plant breeding. For example, Yu et al. (2016) recently used multispectral data collected from a UAV to estimate yield and soybean maturity, and achieved over 93% accuracy in classifying maturity date. UAV-HTP platforms represent a significant advantage for soybean breeders, particularly for acquiring maturity notes, which change quickly and need to be taken every week in potentially thousands of experimental plots.

Field conditions may occasionally prevent ground vehicles from being deployed while weight limitations prevent aerial approaches from carrying as much imaging and sensing equipment as ground-based vehicles. A combination of the two complementary techniques is now a viable option, particularly with the decreasing affordability and accessibility of imaging, sensing and drone equipment. Low resource breeding programmes now have the opportunity to accurately phenotype their experiments with unprecedented accuracy and speed. Collaborations with computer scientists and engineers can help streamline the HTP pipeline and maximize the amount of data captured by imaging and sensing equipment. The potential time and money saved and HTP data gained is now limited solely by the imagination and initiative of conventional soybean breeding personnel.

8 Breeding to reduce losses from biotic and abiotic stresses

There are a wide variety of known and unknown diseases and pests that affect soybean and many of these are unique to a specific location. To increase yield, breeders focus on strategies to eliminate crop losses by breeding for biotic stresses. Major public breeding efforts focus on economically important fungal, viral and bacterial pathogens; nematodes; pests; and weeds that can have an undesirable impact on soybean yield. For example, the most economically important viral pathogens are bean pod mottle virus, bean yellow mosaic virus and soybean mosaic virus (Tolin & Lacy, 2004) while the most economically important nematode is SCN (*Heterodera glycines*) (Niblack et al., 2004). Soybean pests and pathogens infect and attack all parts of the plant, and the degree of damage will depend on the pathogen, plant development stage, tissues under attack, disease severity, environmental conditions favouring disease spread, host plant susceptibility and plant stress level (Hartman et al., 2011). Strategies to increase yield can involve a combination of pesticide or nematicide applications, cultural and seed sanitation techniques, and deployment of genetically resistance varieties (Hartman et al., 2011; Grau et al., 2004). Control of these pathogens relies heavily on cultural practices and genetic resistance.

Fungal pathogens of soybean vary based on frequency of infection and associated yield loss (Palmer et al., 2004). Fungal pathogens of particular note are sudden death syndrome (SDS), caused by *Fusarium solani* f.sp. *glycines*; Asian soybean rust (ASR), caused by *Phakopsora pachyrhizi*; stem and root rot, caused by *Phytophthora sojae*; and white mould, caused by *Sclerotinia sclerotiorum*. Major resistance genes have been deployed against *P. pachyrhizi* and *P. sojae* (Tyler, 2008) but resistance conferred by single or multiple resistance (R) genes has steadily eroded via genetic changes in pathogen populations (Tyler, 2008). A solution may lie in breeding for quantitative resistance to fungal pathogens.

ASR is one of the deadlier soybean diseases and causes up to 80% losses under the right environmental conditions (Rosa et al., 2015). The active agent of ASR, biotrophic fungus *P. pachyrhizi*, causes premature defoliation of soybeans. The fungus was initially discovered in Southeast Asia in 1914 but has since then spread to Africa in 1975 and the Americas in the last decade (Schneider et al., 2005; Yorinori et al., 2005). ASR favours warm and humid regions, and rust epidemics have been limited to the southern states of the United States. In the absence of fungicide management, U.S. yield losses attributable to ASR of up to 33% have been reported (Mueller et al., 2008). ASR has thrived in Brazil, however, causing more than $10 billion in crop losses since the first epidemic outbreak (Yorinori et al., 2005). Soybean breeders and geneticists continue to discover new candidate genes and germplasm sources to improve soybean resistance to the pathogen (Kim et al., 2012; Yamanaka et al., 2016; Kawashima et al., 2016).

Fungicide applications are currently the most effective control for ASR in Brazil. However, the high number of applications needed to increase soybean yield is expensive (Godoy et al., 2015). Furthermore, ASR has recently become insensitive to some fungicide classes (Godoy et al., 2012). Six dominant resistant genes (*Rpp* 1–6 genes) have been discovered (McLean and Byth, 1980; Bromfield and Hartwig, 1980; Hartwig and Bromfield, 1983; Hartwig, 1986; Garcia et al., 2008; Li et al., 2012) and three recessive genes have been identified (Calvo et al., 2008; Pierozzi et al., 2008). Pyramiding the R genes in a single genotype is a desirable strategy for conferring a more durable and broad-spectrum (multiple *P. pachyrhizi* isolates) resistance in soybean. Yamanaka et al. (2015) observed highly resistant phenotypes in the pyramided lines (*Rpp4* + *Rpp5*, *Rpp2* + *Rpp3* + *Rpp4* and *Rpp2* + *Rpp4* + *Rpp5*) with almost no sporulation compared to the original sources carrying single *Rpp* genes.

SDS of soybean, caused by *Fusarium virguliforme*, is also an important fungal disease. The only effective management available is using tolerant or resistant cultivars complemented by cultural practices. However, there is currently no highly effective and durable SDS-resistant germplasm available. The genetic architecture of SDS resistance is complex and governed by many QTLs (Iqbal et al., 2001; Kazi et al., 2008). In addition, the pathogen/disease is highly influenced by environmental conditions (de Farias Neto et al., 2006), which makes evaluation of resistance performance challenging in soybean breeding programmes. For example, wet and cool conditions early in the season favour the initial infection of soybean roots by *F. virguliforme* leading to a great variability in the evaluation of SDS resistance. Bao et al. (2015) used association mapping and GS techniques and suggested that both methods could be used in the genetic evaluation of SDS-associated root lesion severity. However, a more precise dissection of the genetic architecture of SDS resistance is still necessary and improvements in the phenotypic screening of SDS resistance will remain a challenge in current and future SDS breeding programmes.

Plant parasitic nematodes are also of concern to soybean breeders. Root-knot nematodes in the *Meloidogyne* genus occur in many important soybean growing areas of the world and resistance to *M. hapla* was developed in the Southern United States prior to introduction into Brazil. The *M. hapla*-resistant soybeans were susceptible to Brazilian *M. javanica*, which led to the development of high-yielding Brazilian cultivars resistant to both species of root-knot nematode (Kiihl et al., 1981). In the same way, SCN resistance has received major attention by plant breeders. Two major genes (*Rhg1* and *Rhg4*) and many minor QTLs have been identified for SCN resistance (Concibido et al., 2004; Vuong et al., 2010; Jiao et al., 2015). Since many genes are involved in SCN resistance and the genes can have different numbers of copies (Cook et al., 2012), breeding for SCN resistance is complex.

The reduction in soybean yield due to pests is estimated to be in the range of 26–29% globally (Oerke, 2006) and the amount of pesticides used in agriculture increased seven times in the last forty years. However, pesticides and nematicides are not always successful in reducing crop losses (Tilman et al., 2001). A sustainable and environmentally friendly approach to control agricultural pests includes screening and introducing soybean alleles for pest and disease-resistant genotypes into high-yielding varieties adapted to a specific region. A novel emerging option includes the control or elimination of regional soybean nematode and insect pests without repeated chemical applications using CRISPR-Cas9 mutagenic chain reaction approaches (Gantz et al., 2015).

Another focus for breeders is developing cultivars that perform better in stressful environments. Breeding for adaptation to abiotic stress such as drought, heat, salt tolerance, soil mineral deficiency and soil toxicity has increased soybean yield gains. Maximum yields are rarely realized but can be improved with tolerance to stressful environments. Abiotic constraints affecting soybean production include weather-related phenomena, soil nutrient availability, salinity, and response to photoperiod (Hartman et al., 2011).

Since the discovery of photoperiodism nearly a century ago (Garner and Allard, 1920), much is now known about the relationship between soybean flowering and day length. Soybeans are short-day annual plants that flower when the photoperiod is sufficiently short and soybean breeders must develop cultivars specific to the day length of the latitude of the target environment. Soybeans with a critical day length photoperiod of fourteen hours, for example, will initiate flowering when the day is shorter than fourteen hours but will continue vegetative growth if the day length is longer. If the variety is adapted to a temperate latitude with a shorter growing season and is grown in a latitude closer to the equator, flowering will occur early but at the expense of vegetative growth. This results in reductions in leaf area, plant height, node number and yield. If the variety is moved farther north, flowering may occur too late in the season and reproduction will be interrupted by the autumn frost. In temperate areas the soybean is planted after the soil has warmed in the spring but when the day is sufficiently long to encourage vegetative growth. When the photoperiod reaches the critical length of a particular variety flowering occurs. As the photoperiod continues to decrease, soybean reproduction and maturation ensue. Optimal yields in temperate areas are achieved with full season varieties that require a planting and maturity cycle ranging from 140 to 180 days. Several of the genes controlling maturity have been cloned (Table 1). Ongoing work to understand the genetics of maturity alleles may eventually provide targets for rapid breeding of MGs using MAS.

Farming practices can reduce the impact of some abiotic factors but cannot for others like flooding, frost and drought. Soybeans submerged for extended periods in water have reduced survival rates, leading to significant yield losses, while a killing frost can occur either after emergence or before plants fully mature at the end of the season (Palmer et al.,

2004). Drought is the primary cause of yield loss in arid climates (Hartman et al., 2011) and causes flowers and pods to abort leading to fewer pods, fewer seeds smaller seeds per pod, and between 24 and 88% reduced soybean yields (Eck et al., 1987; Frederick et al., 2001). Irrigation can prevent losses due to seasonal drought; however, water availability and irrigation installation and maintenance are major limiting factors.

Drought tolerance in soybeans is related to many different factors, such as nitrogen (N) fixation (Sinclair et al., 2007), the root system (Fenta et al., 2014) and slow wilting canopy trait (Sinclair et al., 2008; King et al., 2009). Drought-tolerant landraces and PIs have been identified as donor material for improving drought tolerance in soybean cultivars (Carter et al., 2016; Pathan et al., 2014) and drought tolerance QTLs have been identified (Mian et al., 1998; Specht et al., 2001; Abdel-Haleem et al., 2012). For example, two germplasm lines with improved yield and N fixation under drought stress (R01-416F and R01-581F) are available (Chen et al., 2007). Soybean breeding for drought-tolerant germplasm continues to be a major effort with benefits to sustainable soybean production in both developing and developed countries.

The slow canopy-wilting trait is unique because it is used both as a morphological marker for breeding drought tolerance into soybeans and for introgression into high-yielding germplasm. For example, Carter et al. (2016) recently released a soybean cultivar (USDA-N8002) with sustained N fixation during drought stress and slow or delayed canopy wilting. USDA-N8002 responds to vapour pressure deficits with a water-conserving transpiration response. The cultivar yielded 5–7% higher than the checks across different yield trials in five years of evaluation. Charlson et al. (2009) also found four QTLs representing 47% of the phenotypic variation in the slow canopy-wilting trait in a biparental population across three environments. A recent meta-QTL analysis of multiple populations (Hwang et al., 2016) identified six meta-QTLs that appear to be stable across years. The stability of these QTLs provides potential candidates for MAS of the slow canopy-wilting trait in the near future.

Nutrient availability and soil pH also contribute to soybean production potential. Soybeans require a proper supply of soil nutrients and annual soil testing is recommended to optimize yield. High soil salinity sensitivity is another concern, resulting in leaf chlorosis, poor root development, reduced plant vigour and eventually depressed yield (Hartman et al., 2011). Fertilizers can be applied to low nutrient fields while other sustainable practices like proper crop rotation, soil amendments and tillage treatments contribute to peak sustainable soybean production.

Breeders should be prepared for the impact of global climate change on agriculture, which will likely decrease yields and make gains more elusive (Searchinger et al., 2013). Current climate variation explains 32–39% of soybean, maize, rice and wheat global yield variability (Ray et al., 2015). Increased levels of carbon dioxide (CO_2) and ozone (O_3) could increase or decrease the incidence of some diseases and could possibly affect photosynthesis. For example, increased CO_2 may increase the photosynthetic activity of crops, although some evidence suggests the combined effects of a raise in CO_2 and temperature does not lead to significant improvements in photosynthesis compared to an increase in CO_2 alone (Ruiz-Vera et al., 2013). Temperature and rainfall extremes, which are hallmarks of global climate change, will also have direct and indirect effects on crop productivity (Hartman et al., 2011). For example, crop models predict high temperatures (>30°C per day) will reduce soybean yield up to 6% in the United States even under rainfed conditions (Schauberger et al., 2017). To meet projected crop needs by increasing production without expanding the annual area harvested, crop yields would need to grow an average of 32% more from 2006 to 2050 than crop yields from 1962 to 2006 (Searchinger et al., 2014). Although substantial potential remains for yield increases, boosting yields at an even more rapid rate going forward presents a unique challenge to soybean breeders, farmers and geneticists.

9 Management and analysis of phenotyping data

There is an increasing focus in the life sciences on the management and processing of big data. This is particularly important for soybean breeding data covering multiple years with many locations. Accurate records can be combined with meteorological data and genome sequencing data to refine field observations and to predict soybean variety performance. Furthermore, the increasing concern of the effects of climate change on soybean crop yield can be more accurately quantified with reliable multi-year digital records.

Efficient statistical evaluation of phenotyping data requires standardized data storage to ensure long-term data availability and accessibility. Platforms that can handle phenotypic and genotypic data are essential for plant breeding. Traditionally, plant breeders used conventional methods of data recording and management in individual formats on paper without backup copies. Currently, most data is digital and stored and analysed with information management programmes such as Microsoft® Excel. Furthermore, individual plant breeder preferences dictate file storage structure with data often spread across several files. These data management practices make access to and evaluation of the data slow and limited to individual breeders.

The use of database software to collect and manage phenotypic and genotypic data has grown in recent years. Field data is now easily collected in the field using mobile computing devices such as tablets, and several database management options are available for soybean breeders. Database software should have several features such as 1) database flexibility with regard to new developments in genomics or breeding methods, 2) the database should be accessible by a single user or with a client server for many users and 3) the database should be compatible with commonly used statistical data analysis software (Heckenberger et al., 2008). AGROBASE® (Agronomix, Inc., Winnipeg, Canada) is one popular database storage and management software in public plant breeding programmes. Breeding programmes can be managed in many steps, such as the design and randomization of fields and experiments, statistical analyses, the management of seed inventory, label printing, barcoding and the generation of detailed reports. The Plant Research Information Sharing Manager (PRISM®) (Central Software Solutions, Inc.) is another option with many of the same applications and functions as AGROBASE. The breeding management system (BMS), which is part of the integrated breeding platform, is also becoming more common in international breeding efforts. BMS provides MAS support functions along with several overlapping PRISM and AGROBASE functions (The IBP Breeding Management System Version 3.0.8, 2015). In addition to publicly available proprietary software, large private breeding companies maintain their own proprietary software for data storage and breeding programme management.

Data analysis is an important step in plant breeding to make objective and correct decisions about the genotypes tested. Plant breeders used several statistical methods to group genotypes based on phenotypic data and make accurate predictions about the performance of breeding material. Several statistical software packages are available for plant breeders. Statistical Analysis Software (SAS®) (www.sas.com) is the most commonly used, albeit expensive, software. Another option is the R (www.r-project.org), which is a widely used free software environment for statistical computing and graphics with open source code. R contains several plant science research packages, such as plant breeding (http://plantbreeding.r-forge.r-project.org/), that can be used for the analysis and visualization of data from plant breeding and genetics experiments. Additional functionalities of the package include: analysing plant breeding specialized designs, estimation of variance components, heritability, genetic correlation, genotype x environment interaction analysis, selection and selection index, and molecular breeding tools. Options also exist for linkage and association mapping of QTLs. For accurate and thorough data analyses SAS is still the industry standard and includes excellent technical support while R is gaining in popularity and has the added benefit of being free.

10 Soybean cultivation in developing countries

The global population is projected to increase to nine billion by 2050 with most of the growth taking place in developing countries (Searchinger et al., 2014; Ronald, 2011). Concerns at the forefront for breeders are how to increase crop production without expanding the annual area harvested while anticipating the effects of a changing climate. The rate at which soybean yield increases across the world would need to double to meet the projected population demands (Rincker et al., 2014). Over the last eighty years, U.S. soybean breeders have increased soybean seed yields by 29 kg per hectare per year (Rincker et al., 2014). An analysis of on-farm yield gains with genetic gains estimates two-thirds of on-farm yield gains resulting from increased genetic gains and one-third from agronomic improvements (Rincker et al., 2014). Although substantial potential remains for yield increases, boosting yields at a more rapid rate going forward is a tall order. The green revolution's successes were due to the world's farmers adopting improved seeds varieties, fertilizer and doubling the irrigated area. Less water is now available to expand irrigation and no comparable transformative technologies are readily available (Searchinger et al., 2014). Millions of lives may depend upon the extent to which plant breeders can increase genetic gain to contend with a growing global population, changing climate and shrinking environmental resources (Ronald, 2011).

The amount of arable farm land is decreasing in developed countries but increasing in developing countries. There is increasing efficiency in irrigation and irrigated land, but some regions of the world will face deficits in renewable water supplies. Climate change is slowly redefining traditional mega-environments used to develop soybean varieties, and a combination of globalization and climate change is leading to new invasive soybean pests (e.g. plant, animal and microbes). There is also an alarming trend upward in the amount of herbicide-resistant weeds as a result of herbicide-resistant crops. Seedbed preparation is another important consideration for farmers with minimal or no mechanical equipment and farmer's experience with local soils is important for determining the best method. Conventional tilling, minimum tillage, hand labour methods, cover crops, intercropping and local farming practices should all be taken into consideration in addition to soil fertility problems. These challenges make filling the gap between potential soybean yield and farm yield more difficult in the near future.

Soybean rhizobia are typically not found in fields where soybeans have not been grown, including farmland in developing countries. Soybeans are desirable crops partly because of the large amount of protein produced. The element nitrogen (N) is a fundamental building block of proteins, and soybeans require large amounts of N either directly from the soil as inorganic N or via symbiotic nitrogen fixation with one of several species of rhizobia. Supplementing the soil with N is expensive and occasionally interferes with the symbiosis between rhizobia and soybeans leading to reduced nodule size and numbers. Therefore, it is impractical for developing countries to produce soybeans without nitrogen-fixing bacteria and it is important to inoculate soils with the best available inoculants and methods prior to planting soybeans.

Some developing regions in Africa, Asia and Latin America that could potentially utilize soybeans are closer to the equator than major soybean producing regions, with the exception of northern Brazil. Breeding for tropical and subtropical regions presents its own set of unique challenges. To date, the International Institute of Tropical Agriculture (ITTA) programme has spearheaded research into developing soybeans for these unique geographic regions. Current efforts to produce soybeans in tropical regions are affected by seeds with low viability, low yield, poor nodulation and pods that shatter easily upon maturity. Seed harvesting, handling and storage should minimize damage to the seed, which may reduce seed viability. High temperature and high moisture should be avoided. Seeds should be tested for germination days before the desired planting date and special care should be used to ensure proper planting depth, soil temperature and plant spacing to provide optimal germination and to maximize the absorption of sunlight.

The possibility exists to improve seed germination rate and seed fungal infection caused by storage in less than ideal conditions using improved genetics and smaller seeds. The expansion of soybeans into West Africa was, in part, caused by a lack of soybean varieties that produce high-quality seeds (Mercer-Quarshie and Nsowah, 1975). Highly productive U.S. varieties introduced into central India were similarly rejected based on poor seed quality (Hinson et al., 1982). Smaller seeds have several advantages over larger seeds for farmers in developing countries. For example, several pounds of small seeds can generate a larger population and are less susceptible to mechanical damage during harvest. Furthermore, smaller seeds survive storage conditions better than large seeds (Singh, 1976; Minor and Paschal, 1982) and can be bred for larger more resilient seeds over time.

As new varieties are developed, farmers must be informed about variety performance and how they may obtain new lines. Well-developed seed systems are necessary resources for farmers in low, medium and high resource areas. For example, U.S. farmers have extension and outreach resources such as state- and federal-funded land-grant universities, cooperative extension systems and agricultural experiment stations. These systems represent advancements that evolved in the United States to make university knowledge and research more accessible to the general public (Smith, 2007). The primary task of these systems is 1) to identify public problems for researchers, 2) to facilitate public education as a means of addressing the problems and 3) to acquire political support from extension programme beneficiaries for continued funding of research and education (Smith, 2007). These extension and outreach systems are operated by public sector extension services, private seed companies and non-government organizations (NGOs) or directly by farmer's organizations. Malfunctioning of either private/public sector extension services affects the dissemination of improved varieties to farmers in low and medium resource areas (FAO, 2011).

Most developing countries do not have a well-documented seed policy or flexible legislation to facilitate growth of the seed sector. This is crucial since a seed policy serves as a guiding principle by which seed systems, and entire seed value chains, can be developed. National seed polices and regulatory frameworks are vital to the success of a nation's seed sector. Key gaps of seed systems to be addressed are policy and regulatory gaps, variety development, seed production, seed quality and assurance, seed conditioning and storage, seed distribution, seed marketing, and seed technology education (FAO, 2011).

Currently, approximately two billion people are employed in agriculture or agriculture-related activities globally, many of whom are poor (Searchinger et al., 2014). Agriculture, and particularly conventional plant breeding, can act as driver for poverty eradication by providing economic activities for the poor. Plant breeders can improve the profitability of farming by developing higher yielding varieties better suited to farmer's environments. The improved economic condition of farmers, in turn, contributes to the growth and structure of communities. Furthermore, in many developing countries women are predominately involved in agriculture. Research shows raising women's income through the use of improved varieties and agricultural inputs increases food security and reduces poverty (Searchinger et al., 2014). To address these issues, in 2013 the United States Agency for International Development (USAID) funded an international development programme, the Soybean Innovation Lab (SIL), to support the soybean value chain in sub-Saharan African Feed the Future countries.

11 Case study: the USAID SIL

Currently, there is a major global effort to focus national and international research into coordinated projects that span expertise, countries, institutions and research centres. In 2013, the USAID created the SIL, a Feed the Future Innovation Lab for Soybean Value Chain Research, at the University of Illinois. The mission of SIL is 'to provide researchers, extensionists, the private sector, NGO and funders operating across the entire value chain the critical information needed for successful soybean development'. The members of SIL consist of a multi-institutional and international collaboration of stakeholders, scientists, NGOs and USAID personnel.

The SIL implements research in four key pillars (Fig. 9) that are the foundation for developing and supporting sustainable soybean production in sub-Saharan Africa. Those four pillars include 1) genetic improvement, 2) crop productivity and quality, 3) nutrition and 4) value chains and socio-economic research. The four pillars support ten managed research areas that have inclusive research objectives revolving around capacity building issues facing the tropical soybean value chain:

Pillar 1 (Genetic Improvement): Increase the capacity and improve the practices of the soybean breeding programmes in target countries by improving variety performance and providing highly trained soybean breeders.

Pillar 2 (Crop Productivity and Quality): Make research information on soybean production and varietal testing available to provide clients with best management practices for seed storage and environmental stewardship.

Pillar 3 (Nutrition): Improve and establish utilization of soybean for livestock and poultry nutrition by introducing poultry nutrition centres. This pillar also addresses research devoted to using soybean for human nutrition by providing information for soybean processing, product development and recipe design.

Pillar 4 (Value Chains and Socio-economic Research): Evaluate the role of gender in the adoption of soybean as a sustainable crop by defining socio-economic principles for

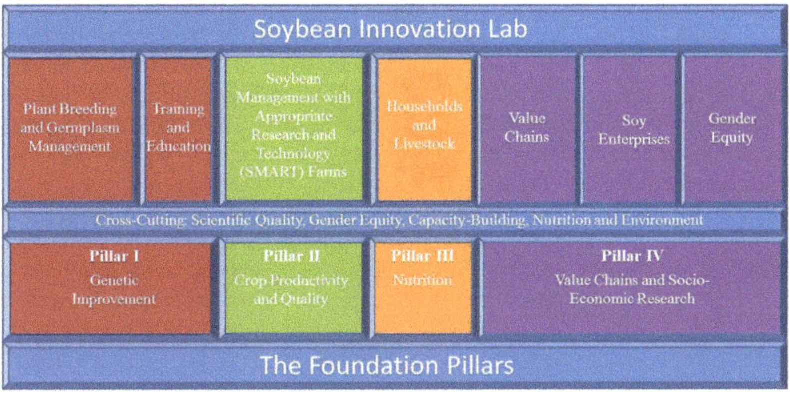

Figure 9 Soybean Innovation Lab foundation pillars for maximizing collaboration amongst the consortium members. See text and the SIL website (www.soybeaninnovationlab.illinois.com) for details.

adoption and by estimating and quantifying the impact throughout the soybean value chain.

The four research pillars and their associated objectives support ten managed research area(s) (MRA) with specific activities. Research Pillar 1 supports three MRAs centred around the genetic improvement of soybean. The Plant Breeding and Germplasm MRA is focused on increasing the productivity of new and current soybean breeding programmes by education, training and the introduction of elite germplasm from the United States and Brazil. The Grain and Seed Quality MRA activities include promotion of improved seed management practices through evaluation of threshing equipment and techniques, seed storage and seed quality, while also developing low-processing soybean varieties for human nutrition. The Plant Breeder Education MRA is establishing the first Master's Degree programme in sub-Saharan Africa and developing internships in the United States for plant breeding students.

Research Pillar 2 supports two MRAs for crop production and quality research. The Production and Agronomy MRA has established and operates Soybean Management and Appropriate Research and Technology farms, which help guide clients on best management practices for soybean production. The Environmental Impacts MRA is concerned with the environmental impact of soybean production and devotes research efforts to develop recommendations for best practices in reducing environmental impact due to soybean cultivation and research.

Research Pillar 3 supports two MRAs focused on nutrition for both humans and livestock. The Utilization for Human Nutrition MRA aims to increase soy in household diets by researching small-scale technologies such as SoyCow (soymilk) and VitaGoat and to determine how this research impacts the continued adoption of soybean in the household. The Utilization for Livestock Nutrition MRA is developing a Poultry Nutrition Center in the region for sustained poultry research and to inform producers of best management practices through demonstrations and hands-on training.

Research Pillar 4 supports three MRAs for value chains and socio-economic research. The Gender Impacts MRA is conducting assessments of gender empowerment by identifying issues for women utilizing soybean for their household diet as well as for income. The Economic Impacts MRA is assessing socio-economic parameters that control adoption of soybean by measuring the economic impact throughout the soybean value chain, particularly for small and medium holders. The Seed Systems MRA will provide information to clients regarding seed storage, seed germination, seed quality and seed supply.

Currently, the SIL programme is implementing its research efforts in five Feed the Future countries in Africa including Ghana, Mozambique, Ethiopia, Zambia and Malawi and shall span five years (2013–2018). The SIL has active collaborations with researchers at the Ethiopian Institute for Agricultural Research, the Jimma Agricultural Research Center, the Savanna Agricultural Research Institute, the IITA, the Syngenta Foundation for Sustainable Agriculture and many others. Through research and education centred on capacity building, the SIL is currently one of the largest soybean research efforts in Africa and will certainly have a large impact on the sustainability of the soybean value chain in Africa.

12 Summary and future trends

12.1 Summary

The rise of soybean from an exotic twentieth-century Western world forage crop to a global multi-billion dollar industry in the twenty-first century is a true Cinderella story. The genetic potential for yield gain across many environments continues to improve with greater access to well-characterized germplasm and improved plant breeding methodologies and technology. The list of known alleles and confirmed QTLs responsible for agronomic traits that benefit farmers in both the developed and developing worlds continues to grow (Table 1; Table 2). Better algorithms, faster computers and increasingly inexpensive genome-wide molecular markers are improving selection speed and reducing the amount of land and resources needed for variety development. Automation of planting, harvesting and phenotyping promises to provide higher yielding soybeans in multiple environments.

Developing countries face many challenges but may soon realize the benefits of sustainable conventional soybean breeding with continued outreach efforts. Soybeans are a novel source of human and livestock food in developing nations, and educational efforts are necessary to promote and sustain soybean production. Furthermore, new equipment is needed to process soybeans in addition to the integration of novel farm management techniques that harness the long-term benefits of improved soil health from a nitrogen-fixing legume. Community building efforts through workshops, scholarships and joint research projects provide the opportunity for people from developing and developed nations to learn from one another about the unique soybean production requirements of particular areas. Public and private funding for recruiting, educating and training the next century of soybean breeders from communities around the world is essential to making these efforts possible and ensuring sustainable growth of the soybean Cinderella crop well into the next century.

12.2 Future trends in sustainable conventional soybean breeding

For developing countries, new core collections of adaptable germplasm suitable for the needs of specific region need to be developed. The molecular identities of several maturity genes, determinacy genes and shattering genes are known and can be used to develop diverse germplasm suitable for farmers in developing countries. In particular, developing regions may benefit from a balance of specialized food-grade soybean and production-grade soybean varieties, which allows developing country farmers to play an active role in selecting varieties suitable for integration into their region and culture. The speed and efficiency with which these efforts move forward are dependent upon complex factors but modern conventional soybean breeding combined with cutting-edge GE techniques may provide developing countries with beneficial soybean cultivars in the years ahead.

The ability to rapidly and accurately introduce foreign genes, modify existing genes, and precisely control recombination events is now possible in low-budget soybean breeding programmes using novel genome-editing techniques (e.g. CRISPR-Cas9). The public perception of what makes a GE soybean 'genetically engineered' (i.e. genetically modified organism; GMO) currently lags behind the rapid pace of recent technological advances. In the long term, however, public perception may become increasingly favourable for various sub-categories of GE soybeans in parallel to the greater awareness of various medical and societal benefits provided by efficient genome engineering (Cyranoski, 2016).

In contrast to the promise of future GE efforts, many potentially interesting loss-of-function alleles have already been discovered but lie hidden in public databases. Hundreds of soybean genomes have been resequenced and deposited in public online databases, although more coordinated efforts are needed to identify and categorize existing loss-of-function alleles, and to deposit these results in a comprehensive database. A more centralized and regularly updated public online database will benefit soybean breeders searching for causative alleles under QTL peaks and basic researchers interested in finding conserved gene functions between plant model organisms.

The most recent Animal and Plant Health Inspection Service (APHIS) proposal (Docket No. APHIS-2015-0057) provides a glimpse of the future of GE soybean production with the first update since 1987 on the 'Importation, Interstate Movement, and Environmental Release of Certain Genetically Engineered Organisms'. The current framework of the APHIS proposal merges traditional conventional soybean breeding with the latest advances in genome engineering. Under the new guidelines, GE soybeans that could have been generated via traditional breeding or classic mutagenesis will not be regulated. Deregulation of this subset of GE soybeans could be the game changer; future conventional soybean breeders need to keep up with a growing population, limited resources and variable global climates.

Soybeans are in a unique genomic engineering position from both a basic and applied research standpoint. Efficient genome editing or induced recombination hot spots are now possible using DNA-based soybean transformation events with CRISPR-Cas9 transgene constructs (Sadhu et al., 2016), while CRISPR-Cas9 gene drives can generate homozygous alleles in a single generation (Gonen et al., 2017). Future efforts may utilize transgene-free approaches such as recombinant CRISPR-Cas9 loaded with a gRNA and introduced into soybean protoplasts generated from the dicot morphogenic equivalent of *Wus* and *Bbm* (Lowe et al., 2016). The deletion or inactivation of multiple gene duplicates provides the opportunity to evaluate the genetic redundancy of soybean gene duplicate families missed by classic reverse and forward genetic screens. The time and resources saved, combined with the knowledge gained from nascent GE soybean breeding techniques, provide incredible opportunities for sustainable conventional soybean breeding in the years ahead.

Bayer CropScience is a forward-thinking company that currently has a large team working on GE technologies. As the resolution and public acceptance of GE technologies continue to increase, plant breeding will change. What it means to be a plant breeder is currently a moving target and the terms 'plant engineer' or 'soybean engineer' may be equally applicable. On the other hand, the foundations of plant breeding, such as identifying fundamental differences between two cultivars, will remain the same. The next big ideas leading to increasingly better genomic editing resolution may come from the public sector (e.g. university-based research) while others may come from small companies. A grad student may be working on an enzyme or technique right now that could revolutionize the field. It is up to the public and policy makers to ensure that recent and future scientific advances are being used to make soybean breeding, and cultivation, sustainable (Chris Tinius, Bayer Crop Science, pers. comm.).

13 Where to look for further information

United Soybean Board

- http://unitedsoybean.org/

Bayer Crop Science

- http://www.bayer.com/en/sustainability.aspx
- https://www.cropscience.bayer.com/en/stories/2016/hyper-care-for-healthy-crops
- http://www.globalhealthaction.co.uk/world-hunger/innovation-is-key-for-food-security

Dow AgroSciences

- http://www.dowagro.com/en-us/sustainability
- http://www.dow.com/en-us/science-and-sustainability/
- http://www.dow.com/en-us/science-and-sustainability/2025-sustainability-goals

United States Agency for International Development (USAID)

- https://www.usaid.gov/

Soybean Innovation Lab (SIL)

- http://soybeaninnovationlab.illinois.edu/

International Institute of Tropical Agriculture (IITA) programme

- http://www.iita.org/home
- International Soybean Program (INTSOY)
- http://nsrl.illinois.edu/nsrl-intsoy-short-course

Germplasm Resources Information Network

- http://www.ars-grin.gov/

SoyBase

- http://soybase.org/

Phytozome

- https://phytozome.jgi.doe.gov/pz/portal.html#!info?alias=Org_Gmax

Soybean Knowledge Base

- http://soykb.org/
- SoyStats 2016
- http://soystats.com/

14 References

Abdel-Haleem, H., Carter Jr., T. E., Purcell, L. C., King, C. A., Ries, L. L., Chen, P., Schapaugh Jr., W., Sinclair, T. R. and Boerma, H. R., 2012. Mapping of quantitative trait loci for canopy-wilting trait in soybean (*Glycine max* L. Merr). *Theoretical and Applied Genetics*, 125(5), pp. 837–46.

Araus, J. L. and Cairns, J. E., 2014. Field high-throughput phenotyping: the new crop breeding frontier. *Trends in Plant Science*, 19(1), pp. 52–61.

Bao, Y., Kurle, J. E., Anderson, G. and Young, N. D., 2015. Association mapping and genomic prediction for resistance to sudden death syndrome in early maturing soybean germplasm. *Molecular Breeding*, 35(6), pp. 1–14.

Belhaj, K., Chaparro-Garcia, A., Kamoun, S. and Nekrasov, V., 2013. Plant genome editing made easy: targeted mutagenesis in model and crop plants using the CRISPR/Cas system. *Plant Methods*, 9(1), p. 1.

Bernard, R. L., 1972. Two genes affecting stem termination in soybeans. *Crop Science*, 12(2), pp. 235–9.

Bernardo, R., 2002. *Essentials of Plant Breeding*, Stemma Press, Woodbury, MN.

Bonato, E. R. and Vello, N. A., 1999. E6, a dominant gene conditioning early flowering and maturity in soybeans. *Genetics and Molecular Biology*, 22(2), pp. 229–232.

Bromfield, K. R. and Hartwig, E. E., 1980. Resistance to soybean rust and mode of inheritance. *Crop Science*, 20(2), pp. 254–5.

Brown, A. H., Grace, J. P. and Speer, S. S., 1987. Designation of a core collection of perennial Glycine. *Soybean Genetics Newsletter-United States*, Agricultural Research Service (USA), 14: 59–70.

Brown, A. H. D., 1989. Core collections: a practical approach to genetic resources management. *Genome*, 31(2), pp. 818–24.

Calvo, É. S., Kiihl, R. A., Garcia, A., Harada, A. and Hiromoto, D. M., 2008. Two major recessive soybean genes conferring soybean rust resistance. *Crop Science*, 48(4), pp. 1350–4.

Carter, T. E., Todd, M. and Gillen, A. M., 2016. Registration of 'USDA-N8002' soybean cultivar with high yield and abiotic stress resistance traits. *Journal of Plant Registrations*, 10, 238–45.

Casler, M. D., 2015. Fundamentals of experimental design: Guidelines for designing successful experiments. *Agronomy Journal*, 107(2), pp. 692–705.

Charlson, D. V., Bhatnagar, S., King, C. A., Ray, J. D., Sneller, C. H., Carter Jr., T. E. and Purcell, L. C., 2009. Polygenic inheritance of canopy wilting in soybean [Glycine max (L) Merr]. *Theoretical and Applied Genetics* 119, pp. 587–94.

Chen, P., Sneller, C. H., Purcell, L. C., Sinclair, T. R., King, C. A. and Ishibashi, T., 2007. Registration of soybean germplasm lines R01-416F and R01-581F for improved yield and nitrogen fixation under ray stress. *Journal of Plant Registrations*, 1, pp. 166–7.

Christenson, B. S., Schapaugh, W. T., An, N., Price, K. P. and Fritz, A. K., 2014. Characterizing changes in soybean spectral response curves with breeding advancements. *Crop Science*, 54(4), pp. 1585–97.

Cober, E. R. and Voldeng, H. D., 2001. A new soybean maturity and photoperiod-sensitivity locus linked to E1 and T. *Crop Science*, 41(3), pp. 698–701.

Cober, E. R., Molnar, S. J., Charette, M. and Voldeng, H. D., 2010. A new locus for early maturity in soybean. *Crop Science*, 50(2), pp. 524–7.

Concibido, V., La Vallee, B., Mclaird, P., Pineda, N., Meyer, J., Hummel, L., Yang, J., Wu, K. and Delannay, X., 2003. Introgression of a quantitative trait locus for yield from *Glycine soja* into commercial soybean cultivars. *Theoretical and Applied Genetics*, 106(4), pp. 575–82.

Concibido, V. C., Diers, B. W. and Arelli, P. R., 2004. A decade of QTL mapping for cyst nematode resistance in soybean. *Crop Science*, 44(4), pp. 1121–31.

Cook, D. E., Lee, T. G., Guo, X., Melito, S., Wang, K., Bayless, A. M., Wang, J., Hughes, T. J., Willis, D. K., Clemente, T. E. and Diers, B. W., 2012. Copy number variation of multiple genes at *Rhg1* mediates nematode resistance in soybean. *Science*, 338(6111), pp. 1206–9.

Crain, J. L., Wei, Y., Barker, J., Thompson, S. M., Alderman, P. D., Reynolds, M., Zhang, N. and Poland, J., 2016. Development and deployment of a portable field phenotyping platform. *Crop Science*, 56(3), pp. 965–75.

Cyranoski, D., 2016. CRISPR gene-editing tested in a person for the first time. *Nature News*, 539(7630), p. 479.

de Farias Neto, A. L., Hartman, G. L., Pedersen, W. L., Li, S., Bollero, G. A. and Diers, B. W., 2006. Irrigation and inoculation treatments that increase the severity of soybean sudden death syndrome in the field. *Crop Science*, 46(6), pp. 2547–54.

Deery, D., Jimenez-Berni, J., Jones, H., Sirault, X. and Furbank, R., 2014. Proximal remote sensing buggies and potential applications for field-based phenotyping. *Agronomy*, 4(3), pp. 349–79.

Diers, B., Specht, J., Hyten, D., Cregan, P., Nelson, R., and Beavis, B., 2015. SoyNAM project update. *Soybean Breeders Workshop*. 16–18 February, St. Louis, MO.

Dissanayaka, A., Rodriguez, T. O., Di, S., Yan, F., Githiri, S. M., Rojas Rodas, F., Abe, J. and Takahashi, R., 2016. Quantitative trait locus mapping of soybean maturity gene E5. *Breeding Science*, 66(3), 407–15.

Dong, Y., Yang, X., Liu, J., Wang, B. H., Liu, B. L. and Wang, Y. Z., 2014. Pod shattering resistance associated with domestication is mediated by a NAC gene in soybean. *Nature Communications*, 5, p. 3352. https://doi.org/10.1038/ncomms4352

Doust, A. N., Lukens, L., Olsen, K. M., Mauro-Herrera, M., Meyer, A. and Rogers, K., 2014. Beyond the single gene: How epistasis and gene-by-environment effects influence crop domestication. *Proceedings of the National Academy of Sciences*, 111(17), pp. 6178–83.

Eck, H. V., Mathers, A. C. and Musick, J. T., 1987. Plant water stress at various growth stages and growth and yield of soybeans. *Field Crops Research*, 17(1), pp. 1–16.

Falconer, D. S. and Mackay, T. F. C., 1996. *Introduction to Quantitative Genetics*. Longman Group Ltd., Essex, UK.

FAO, 2011 . *Strengthening Seed Systems: Gap Analysis of the Seed Sector*. Commission on Genetic Resources for Food and Agriculture (CGRFA). Thirteenth Regular Session, Rome, 18–22 July 2011. Food and Agriculture Organization (FAO) of the United Nations, Rome, Italy.

Federer, W. T. and Crossa, J., 2012. I. 4 screening experimental designs for quantitative trait loci, association mapping, genotype-by environment interaction, and other investigations. *Front. Physiol.*, 3: 1–8.

Federer, W. T., Reynolds, M. and Crossa, J., 2001. Combining results from augmented designs over sites. *Agronomy Journal*, 93(2), pp. 389–95.

Federer, W. T., 1956. Augmented (or hoonuiaku) designs. *Hawaii Plant Rec*, 55, pp. 191–208.

Fehr, W. R. and Hadley, H. H., 1980. *Hybridization of Crop Plants*. American Society of Agronomy and Crop Science Society of America, Madison, WI.

Fenta, B. A., Beebe, S. E., Kunert, K. J., Burridge, J. D., Barlow, K. M., Lynch, J. P. and Foyer, C. H., 2014. Field phenotyping of soybean roots for drought stress tolerance. *Agronomy*, 4(3), pp. 418–35.

Fernandez-Cornejo, J. and McBride, W. D., 2002. *Adoption of Bioengineered Crops*. USDA Economic Research Service, Washington, DC, AER810, pp. 1–67.

Fernandez-Cornejo, J., 2004. *The Seed Industry in US agriculture: An Exploration of Data and Information on Crop seed Markets, Regulation, Industry Structure, and Research and Development* (No. 33671). United States Department of Agriculture, Economic Research Service, Washington, DC

Fisher, R. A., 1930. *The Genetical Theory of Natural Selection*. Clarendon Press, Oxford.

Fox, C. M., Cary, T. R., Nelson, R. L. and Diers, B. W., 2015. Confirmation of a seed yield QTL in soybean. *Crop Science*, 55(3), pp. 992–8.

Frary, A., Nesbitt, T. C., Frary, A., Grandillo, S., Van Der Knaap, E., Cong, B., Liu, J., Meller, J., Elber, R., Alpert, K. B. and Tanksley, S. D., 2000. fw2. 2: a quantitative trait locus key to the evolution of tomato fruit size. *Science*, 289(5476), pp. 85–8.

Frederick, J. R., Camp, C. R. and Bauer, P. J., 2001. Drought-stress effects on branch and mainstem seed yield and yield components of determinate soybean. *Crop science*, 41(3), pp. 759–63.

Fujita, R., Ohara, M., Okazaki, K. and Shimamoto, Y., 1997. The extent of natural cross-pollination in wild soybean (Glycine soja). *Journal of Heredity*, 88(2), pp. 124–8.
Fuller, D. Q., Denham, T., Arroyo-Kalin, M., Lucas, L., Stevens, C. J., Qin, L., Allaby, R. G. and Purugganan, M. D., 2014. Convergent evolution and parallelism in plant domestication revealed by an expanding archaeological record. *Proceedings of the National Academy of Sciences*, 111(17), pp. 6147–52.
Funatsuki, H., Suzuki, M., Hirose, A., Inaba, H., Yamada, T., Hajika, M., Komatsu, K., Katayama, T., Sayama, T., Ishimoto, M. and Fujino, K., 2014. Molecular basis of a shattering resistance boosting global dissemination of soybean. *Proceedings of the National Academy of Sciences*, 111(50), pp. 17797–802.
Furbank, R. T. and Tester, M., 2011. Phenomics–technologies to relieve the phenotyping bottleneck. *Trends in Plant Science*, 16(12), pp. 635–44.
Gantz, V. M., Jasinskiene, N., Tatarenkova, O., Fazekas, A., Macias, V. M., Bier, E. and James, A. A., 2015. Highly efficient Cas9-mediated gene drive for population modification of the malaria vector mosquito Anopheles stephensi. *Proceedings of the National Academy of Sciences*, 112(49), pp. E6736–43.
Garcia, A., Calvo, É. S., de Souza Kiihl, R. A., Harada, A., Hiromoto, D. M. and Vieira, L. G. E., 2008. Molecular mapping of soybean rust (*Phakopsora pachyrhizi*) resistance genes: discovery of a novel locus and alleles. *Theoretical and Applied Genetics*, 117(4), pp. 545–53.
Garner, W. W. and Allard, H. A., 1920. Effect of the relative length of day and night and other factors of the environment on growth and reproduction in plants 1. *Monthly Weather Review*, 48(7), pp. 415–15.
Gizlice, Z., Carter, T. E. and Burton, J. W., 1994. Genetic base for North American public soybean cultivars released between 1947 and 1988. *Crop Science*, 34(5), pp. 1143–51.
Godoy, C. V., 2012. Risk and management of fungicide resistance in the Asian soybean rust fungus *Phakopsora pachyrhizi*. In Thind, T. S. (Ed.), *Fungicide Resistance in Crop Protection: Risk and Management*, CAB International, Wallingford, UK, pp. 87–95.
Godoy, C. V., Bueno, A. D. F. and Gazziero, D. L. P., 2015. Brazilian Soybean Pest Management and Threats to its Sustainability. *Outlooks on Pest Management*, 26(3), pp. 113–17.
Gonen, S., Jenko, J., Gorjanc, G., Mileham, A. J., Whitelaw, C. B. A. and Hickey, J. M., 2017. Potential of gene drives with genome editing to increase genetic gain in livestock breeding programs. *Genetics Selection Evolution*, 49(1), p. 3.
Grau, C. R., Dorrance, A. E., Bond, J. and Russin, J. S., 2004. Fungal diseases. In Boerma, H. R. and Specht, J. E., (Eds), *Soybeans: Improvement, Production, and Uses*, 3rd ed., pp. 679–763 Agronomy Monograph No. 16, American Society of Agronomy, Crop Science Society of America, Soil Science Society of America, Madison, WI.
Grant, D., Nelson, R. T., Cannon, S. B. and Shoemaker, R. C., 2010. SoyBase, the USDA-ARS soybean genetics and genomics database. *Nucleic Acids Research*, 38(suppl_1), pp. D843–D846.
Greene, C., Wechsler, S. J., Adalja, A. and Hanson, J., 2016. *Economic Issues in the Coexistence of Organic, Genetically Engineered (GE), and Non-GE crops*. Economic Information Bulletin, Vol. 149, (Washington, DC: USDA, 2016).
Hacisalihoglu, G., Larbi, B. and Settles, A. M., 2010. Near-infrared reflectance spectroscopy predicts protein, starch, and seed weight in intact seeds of common bean (*Phaseolus vulgaris* L.). *Journal of Agricultural and Food Chemistry*, 58(2), pp. 702–6.
Haldane, J. B. S., 1932. *The Causes of Evolution*. Princeton University Press, Princeton, NJ.
Hanson, W. D. and Brim, C. A., 1963. Optimum allocation of test material for two-stage testing with an application to evaluation of soybean lines. *Crop Science*, 3(1), pp. 43–9.
Hartman, G. L., West, E. D. and Herman, T. K., 2011. Crops that feed the World 2. Soybean – worldwide production, use, and constraints caused by pathogens and pests. *Food Security*, 3(1), pp. 5–17.
Hartwig, E. E. and Bromfield, K. R., 1983. Relationships among three genes conferring specific resistance to rust in soybeans. *Crop Science*, 23(2), pp. 237–9.
Hartwig, E. E., 1986. Identification of a fourth major gene conferring resistance to soybean rust. *Crop Science*, 26(6), pp. 1135–6.

Heckenberger, M., Maurer, H. P., Melchinger, A. E. and Frisch, M., 2008. The Plabsoft database: a comprehensive database management system for integrating phenotypic and genomic data in academic and commercial plant breeding programs. *Euphytica*, 161(1–2), pp. 173–9.

Hinson, K., Hartwig, E. E. and Minor, H. C., 1982. *Soybean Production in the Tropics*. Revised by H. C. Minor. FAO Plant Production and Protection Papers (FAO). no. 4 (Rev. 1).

Hwang, S., King, C. A., Chen, P., Ray, J. D., Cregan, P. B., Carter, T. E., Li, Z., Abdel-Haleem, H., Matson, K. W., Schapaugh, W. and Purcell, L. C. 2016. Meta-analysis to refine map position and reduce confidence intervals for delayed-canopy-wilting QTLs in soybean. *Molecular Breeding*, 36, p. 91.

Hymowitz, T., 1990. Soybeans: The success story. In *Advances in New Crops. Proceedings of the First National Symposium 'New Crops: Research, Development, Economics'*, Indianapolis, Indiana, USA, 23-26 October 1988, pp. 159–63. Timber Press, Portland, OR.

Hymowitz, T., 2008. The history of the soybean. In Johnson, L., White, P. J., and Galloway, R. (Eds), *Soybeans: Chemistry, Production, Processing, and Utilization*, pp. 1–32. American Oil Chemists' Society Press, Urbana, IL.

Hyten, D. L., Song, Q., Zhu, Y., Choi, I. Y., Nelson, R. L., Costa, J. M., Specht, J. E., Shoemaker, R. C. and Cregan, P. B., 2006. Impacts of genetic bottlenecks on soybean genome diversity. *Proceedings of the National Academy of Sciences*, 103(45), pp. 16666–71.

IBP Breeding Management System Version 3.0.8, 2015. *The Integrated Breeding Platform*. https://www.integratedbreeding.net/breeding-management-system.

Iqbal, M. J., Meksem, K., Njiti, V. N., Kassem, M. A. and Lightfoot, D. A., 2001. Microsatellite markers identify three additional quantitative trait loci for resistance to soybean sudden-death syndrome (SDS) in Essex × Forrest RILs. *Theoretical and Applied Genetics*, 102(2–3), pp. 187–92.

Janick, J., 2008. *Plant Breeding Reviews* (Vol. 30). John Wiley & Sons, Hoboken, NJ.

Jiao, Y., Vuong, T. D., Liu, Y., Meinhardt, C., Liu, Y., Joshi, T., Cregan, P. B., Xu, D., Shannon, J. G. and Nguyen, H. T., 2015. Identification and evaluation of quantitative trait loci underlying resistance to multiple HG types of soybean cyst nematode in soybean PI 437655. *Theoretical and Applied Genetics*, 128(1), pp. 15–23.

Jones, M., Woodward, R. and Stoller, J., 2015. Increasing precision in agronomic field trials using Latin square designs. *Agronomy Journal*, 107(1), pp. 20–4.

Kawashima, C. G., Guimarães, G. A., Nogueira, S. R., MacLean, D., Cook, D. R., Steuernagel, B., Baek, J., Bouyioukos, C., do VA Melo, B., Tristão, G. and de Oliveira, J. C., 2016. A pigeonpea gene confers resistance to Asian soybean rust in soybean. *Nature Biotechnology*, 34: 661–65.

Kazi, S., Shultz, J., Afzal, J., Johnson, J., Njiti, V. N. and Lightfoot, D. A., 2008. Separate loci underlie resistance to root infection and leaf scorch during soybean sudden death syndrome. *Theoretical and Applied Genetics*, 116(7), pp. 967–77.

Keep, N. R., Schapaugh, W. T., Prasad, P. V. V. and Boyer, J. E., 2016. Changes in physiological traits in soybean with breeding advancements. *Crop Science*, 56(1), pp. 122–31.

Kiihl, R. A. S., Leones, A. A., Campelo, G. J. A., Bays, I. A., Gomes, E. R., Monteiro, P. M. F. O., and Miranda, M. A. C. 1981. Soja: Cultivar Tropical – Descricao. II *Seminario Nacional de Pesquisa de Soja*, Brasilia, Brasil, p. 147.

Kim, K. S., Unfried, J. R., Hyten, D. L., Frederick, R. D., Hartman, G. L., Nelson, R. L., Song, Q. and Diers, B. W., 2012. Molecular mapping of soybean rust resistance in soybean accession PI 561356 and SNP haplotype analysis of the *Rpp1* region in diverse germplasm. *Theoretical and Applied Genetics*, 125(6), pp. 1339–52.

King, C. A., Purcell, L. C. and Brye, K. R., 2009. Differential wilting among soybean genotypes in response to water deficit. *Crop science*, 49(1), pp. 290–8.

Kong, F., Liu, B., Xia, Z., Sato, S., Kim, B. M., Watanabe, S., Yamada, T., Tabata, S., Kanazawa, A., Harada, K. and Abe, J., 2010. Two coordinately regulated homologs of FLOWERING LOCUS T are involved in the control of photoperiodic flowering in soybean. *Plant physiology*, 154(3), pp. 1220–31.

Large, E. E., Xu, W., Zhao, Y., Brady, S. C., Long, L., Butcher, R. A., Andersen, E. C. and McGrath, P. T., 2016. Selection on a Subunit of the NURF Chromatin Remodeler Modifies Life

History Traits in a Domesticated Strain of *Caenorhabditis elegans*. *PLoS Genetics*, 12(7), p. e1006219.
Li, S., Smith, J. R., Ray, J. D. and Frederick, R. D., 2012. Identification of a new soybean rust resistance gene in PI 567102B. *Theoretical and Applied Genetics*, 125(1), pp. 133–42.
Liebisch, F., Kirchgessner, N., Schneider, D., Walter, A. and Hund, A., 2015. Remote, aerial phenotyping of maize traits with a mobile multi-sensor approach. *Plant Methods*, 11(1), p. 1.
Liu, B., Kanazawa, A., Matsumura, H., Takahashi, R., Harada, K. and Abe, J., 2008. Genetic redundancy in soybean photoresponses associated with duplication of the phytochrome A gene. *Genetics*, 180(2), pp. 995–1007. https://doi.org/10.1534/genetics.108.092742
Liu, B., Watanabe, S., Uchiyama, T., Kong, F., Kanazawa, A., Xia, Z., Nagamatsu, A., Arai, M., Yamada, T., Kitamura, K. and Masuta, C., 2010. The soybean stem growth habit gene Dt1 is an ortholog of Arabidopsis TERMINAL FLOWER1. *Plant Physiology*, 153(1), pp. 198–210.
Liu, W., Martinon-Torres, M., Cai, Y. J., Xing, S., Tong, H. W., Pei, S. W., Sier, M. J., Wu, X. H., Edwards, R. L., Cheng, H. and Li, Y. Y., 2015. The earliest unequivocally modern humans in southern China. *Nature*, 526(7575), pp. 696–9.
Lorenz, A. J., Chao, S., Asoro, F. G., Heffner, E. L., Hayashi, T., Iwata, H., Smith, K. P., Sorrells, M. E. and Jannink, J. L., 2011. 2 Genomic selection in plant breeding: Knowledge and prospects. *Advances in Agronomy*, 110, p. 77.
McBlain, B. A. and Bernard, R. L., 1987. A new gene affecting the time of flowering and maturity in soybeans. *Journal of Heredity*, 78(3), pp. 160–2.
MCIA, 2016. Seed Certification. Missouri Crop Improvement Association. Available at: http://www.moseed.org/SeedCertification.htm [Accessed 18 September 2016].
McLean, R. J. and Byth, D. E., 1980. Inheritance of resistance to rust (Phakopsora pachyrhizi) in soybeans. *Crop and Pasture Science*, 31(5), pp. 951–6.
Meuwissen, T. H., Hayes, B. J. and Goddard, M. E., 2001. Prediction of total genetic value using genome-wide dense marker maps. *Genetics*, 157(4), p. 1819.
Mendel, G. 1866. 'Versuche über Pflanzen-Hybriden' [Experiments Concerning Plant Hybrids]. *Verhandlungen des naturforschenden Vereines in Brünn* [Proceedings of the Natural History Society of Brünn] IV(1865): 3–47. Available online: www.mendelweb.org/Mendel.html (accessed on 20 July 2016).
Mercer-Quarshie, H. and Nsowah, G. F., 1975. Soyabean in Ghana. *INTSOY Ser Int Soybean Program.*
Meyer, R. S. and Purugganan, M. D., 2013. Evolution of crop species: genetics of domestication and diversification. *Nature Reviews Genetics*, 14(12), pp. 840–52.
Mian, M. A. R., Ashley, D. A. and Boerma, H. R., 1998. An additional QTL for water use efficiency in soybean. *Crop Science*, 38(2), pp. 390–3.
Minor, H. C. and Paschal, E. H., 1982. Variation in storability of soybeans under simulated tropical conditions. *Seed Science and Technology (Netherlands)*, 10, pp. 1–9
Mueller, J. D., Koenning, S. R., Kemerait Jr., R. C. and Phipps, P. M. (Eds). 2008. *Soybean Rust Management in the Mid-Atlantic States*. Clemson University Extension Service, Clemson, SC, p. 51.
Munns, R., James, R. A., Sirault, X. R., Furbank, R. T. and Jones, H. G., 2010. New phenotyping methods for screening wheat and barley for beneficial responses to water deficit. *Journal of Experimental Botany*, 61, pp. 3499–507.
National Academies of Sciences, Engineering, and Medicine, 2016. *Genetically Engineered Crops: Experiences and Prospects*. The National Academies Press, Washington, DC.
Niblack, T. L., Tylka, G. L. and Riggs, R. D., 2004. Nematode pathogens of soybean. In *Soybeans: Improvement, Production, and Uses*, 3rd ed., pp. 821–51. Agronomy Monograph No. 16, American Society of Agronomy, Crop Science Society of America, Soil Science Society of America, Madison, WI.
Nichols, D. M., Glover, K. D., Carlson, S. R., Specht, J. E. and Diers, B. W., 2006. Fine mapping of a seed protein QTL on soybean linkage group I and its correlated effects on agronomic traits. *Crop Science*, 46(2), pp. 834–9.
Oerke, E. C., 2006. Crop losses to pests. *The Journal of Agricultural Science*, 144(01), pp. 31–43.

Oliveira, M. F., Nelson, R. L., Geraldi, I. O., Cruz, C. D. and de Toledo, J. F. F., 2010. Establishing a soybean germplasm core collection. *Field Crops Research*, 119(2), pp. 277–89.

Orf, J. H., 2008. Breeding, genetics, and production of soybeans. In Johnson, L., White, P. J., and Galloway, R. (Eds) *Soybeans: Chemistry, Production, Processing, and Utilization*, pp. 33–66, AOCS Press, Urbana, IL.

Palmer, R. G., Pfeiffer, T. W., Buss, G. R. and Kilen, T. C., 2004. Qualitative genetics. In *Soybeans: Improvement, Production, and Uses*, 3rd ed., pp. 137–233. Agronomy Monograph No. 16, American Society of Agronomy, Crop Science Society of America, Soil Science Society of America, Madison, WI.

Park, S. J., Jiang, K., Tal, L., Yichie, Y., Gar, O., Zamir, D., Eshed, Y. and Lippman, Z. B., 2014. Optimization of crop productivity in tomato using induced mutations in the florigen pathway. *Nature Genetics*, 46(12), pp. 1337–42.

Paterson, A. H., Lander, E. S., Hewitt, J. D., Peterson, S., Lincoln, S. E. and Tanksley, S. D., 1988. Resolution of quantitative traits into Mendelian factors by using a complete linkage map of restriction fragment length polymorphisms. *Nature*, 335(6192), p. 721.

Pathan, S. M., Lee, J. D., Sleper, D. A., Fritschi, F. B., Sharp, R. E., Carter, T. E., Nelson, R. L., King, C. A., Schapaugh, W. T., Ellersieck, M. R. and Nguyen, H. T., 2014. Two soybean plant introductions display slow leaf wilting and reduced yield loss under drought. *Journal of Agronomy and Crop Science*, 200(3), pp. 231–6.

Pierozzi, P. H. B., Ribeiro, A. S., Moreira, J. U. V., Laperuta, L. D. C., Rachid, B. F., Lima, W. F., Arias, C. A. A., Oliveira, M. F. D. and Toledo, J. F. F. D., 2008. New soybean (*Glycine max Fabales, Fabaceae*) sources of qualitative genetic resistance to Asian soybean rust caused by *Phakopsora pachyrhizi* (*Uredinales, Phakopsoraceae*). *Genetics and Molecular Biology*, 31(2), pp. 505–11.

Ping, J., Liu, Y., Sun, L., Zhao, M., Li, Y., She, M., Sui, Y., Lin, F., Liu, X., Tang, Z., Nguyen, H., Tian, Z., Qiu, L., Nelson, R. L., Clemente, T. E., Specht, J. E., and Ma, J., 2014. *Dt2* is a gain-of-function MADS-domain factor gene that specifies semideterminacy in soybean. *The Plant Cell*, 26(7), pp. 2831–42.

Priolli, R. H. G., Wysmierski, P. T., Cunha, C. P. D., Pinheiro, J. B. and Vello, N. A., 2013. Genetic structure and a selected core set of Brazilian soybean cultivars. *Genetics and Molecular Biology*, 36(3), pp. 382–90.

Pritchard, J. K. and Rosenberg, N. A., 1999. Use of unlinked genetic markers to detect population stratification in association studies. *The American Journal of Human Genetics*, 65(1), pp. 220–8.

Pritchard, J. K., Stephens, M. and Donnelly, P., 2000. Inference of population structure using multilocus genotype data. *Genetics*, 155(2), pp. 945–59.

Qiu, L. J., Xing, L. L., Guo, Y., Wang, J., Jackson, S. A. and Chang, R. Z., 2013. A platform for soybean molecular breeding: the utilization of core collections for food security. *Plant Molecular Biology*, 83(1–2), pp. 41–50.

Ray, D. K., Gerber, J. S., MacDonald, G. K., West, P. C. 2015. Climate variation explains a third of global crop yield variability. *Nature Communications*, 6, p. 5989.

Ray, J. D., Hinson, K., Mankono, J. and Malo, M. F., 1995. Genetic control of a long-juvenile trait in soybean. *Crop Science*, 35(4), pp. 1001–6.

Reid, J. B. and Ross, J. J., 2011. Mendel's genes: Toward a full molecular characterization. *Genetics*, 189(1), pp. 3–10.

Rincker, K., Nelson, R., Specht, J., Sleper, D., Cary, T., Cianzio, S. R., Casteel, S., Conley, S., Chen, P., Davis, V. and Fox, C., 2014. Genetic improvement of US soybean in maturity groups II, III, and IV. *Crop Science*, 54(4), pp. 1419–32.

Risch, N. and Merikangas, K., 1996. The future of genetic studies of complex human diseases. *Science*, 273(5281), pp. 1516–17.

Ronald, P., 2011. Plant genetics, sustainable agriculture and global food security. *Genetics*, 188(1), pp. 11–20.

Rosa, C. R. E., Spehar, C. R. and Liu, J. Q., 2015. Asian Soybean Rust Resistance: An Overview. *Journal of Plant Pathology and Microbiology*, 6(307), p. 2.

Ruiz-Vera, U. M., Siebers, M., Gray, S. B., Drag, D. W., Rosenthal, D. M., Kimball, B. A., Ort, D. R. and Bernacchi, C. J., 2013. Global warming can negate the expected CO2 stimulation in photosynthesis and productivity for soybean grown in the Midwestern United States. *Plant Physiology*, 162, 410–23.

Sankaran, S., Khot, L. R., Espinoza, C. Z., Jarolmasjed, S., Sathuvalli, V. R., Vandemark, G. J., Miklas, P. N., Carter, A. H., Pumphrey, M. O., Knowles, N. R. and Pavek, M. J., 2015. Low-altitude, high-resolution aerial imaging systems for row and field crop phenotyping: A review. *European Journal of Agronomy*, 70, pp. 112–23.

Sadhu, M. J., Bloom, J. S., Day, L. and Kruglyak, L., 2016. CRISPR-directed mitotic recombination enables genetic mapping without crosses. *Science*, 352(6289), pp. 1113–16.

Schauberger, B., Archontoulis, S., Arneth, A., Balkovic, J., Ciais, P., Deryng, D., Elliott, J., Folberth, C., Khabarov, N., Müller, C., Pugh, T. A. M., Rolinski, S., Schaphoff, S., Schmid, E., Wang, X., Schlenker, W., Frieler, K. 2017. Consistent negative response of US crops to high temperatures in observations and crop models. *Nature Communications*, 8, p. 13931.

Schmutz, J., Cannon, S. B., Schlueter, J., Ma, J., Mitros, T., Nelson, W., Hyten, D. L., Song, Q., Thelen, J. J., Cheng, J. and Xu, D., 2010. Genome sequence of the palaeopolyploid soybean. *Nature*, 463(7278), pp. 178–83.

Schneider, R. W., Hollier, C. A., Whitam, H. K., Palm, M. E., McKemy, J. M., Hernandez, J. R., Levy, L. and DeVries-Paterson, R., 2005. First report of soybean rust caused by *Phakopsora pachyrhizi* in the continental United States. *Plant Disease*, 89(7), pp. 774.

Searchinger, T., Hanson, C., Ranganathan, J., Lipinski, B., Waite, R., Winterbottom, R., Dinshaw, A., Heimlich, R., Boval, M., Chemineau, P. and Dumas, P., 2014. Creating a sustainable food future. A menu of solutions to sustainably feed more than 9 billion people by 2050. World resources report 2013–14: interim findings.

Shanmugasundaram, S. 1981. *Varietal Difference and Genetic Behavior for the Photoperiodic Responses in Soybeans*. Bull. Inst. Trop. Agr. Kyushu Univ., Japan 4: 1–61.

Sinclair, T. R., Purcell, L. C., King, C. A., Sneller, C. H., Chen, P. and Vadez, V., 2007. Drought tolerance and yield increase of soybean resulting from improved symbiotic N2 fixation. *Field Crops Research*, 101(1), pp. 68–71.

Sinclair, T. R., Zwieniecki, M. A. and Holbrook, N. M., 2008. Low leaf hydraulic conductance associated with drought tolerance in soybean. *Physiologia Plantarum*, 132(4), pp. 446–51.

Singh, B. B., 1976. Breeding soybean varieties for the tropics, pp. 11–16. In R. M. Goodman (Ed.) *Expanding the Use of Soybeans*. INTSOY Series International Soybean Program. No. 10. Univ. of Illinois, Urbana-Champaign., USA.

Smith, E. G., 2007. The evolution of extension education in the United States: a Texas perspective. In *World Cotton Research Conference-4*, Lubbock, Texas, USA, 10–14 September 2007. International Cotton Advisory Committee (ICAC), Washington, DC.

Song, Q., Hyten, D. L., Jia, G., Quigley, C. V., Fickus, E. W., Nelson, R. L. and Cregan, P. B., 2013. Development and evaluation of SoySNP50K, a high-density genotyping array for soybean. *PLoS One*, 8(1), p. e54985.

Song, Q., Hyten, D. L., Jia, G., Quigley, C. V., Fickus, E. W., Nelson, R. L. and Cregan, P. B., 2015. Fingerprinting soybean germplasm and its utility in genomic research. *G3: Genes| Genomes| Genetics*, 5(10), pp. 1999–2006.

Soystats, 2016. U.S. Yield & Production: Yield by State. Available at: http://soystats.com/u-s-yield-production-yield-by-state/.

Specht, J. E., Chase, K., Macrander, M., Graef, G. L., Chung, J., Markwell, J. P., Germann, M., Orf, J. H. and Lark, K. G., 2001. Soybean response to water. *Crop Science*, 41(2), pp. 493–509.

Sprague, G. F. and Federer, W. T., 1951. A comparison of variance components in corn yield trials: II. Error, year x variety, location x variety, and variety components. *Agron J.* 43, pp. 535–41.

Thornsberry, J. M., Goodman, M. M., Doebley, J., Kresovich, S., Nielsen, D. and Buckler, E. S., 2001. Dwarf8 polymorphisms associate with variation in flowering time. *Nature Genetics*, 28(3), pp. 286–9.

Tian, Z., Wang, X., Lee, R., Li, Y., Specht, J. E., Nelson, R. L., McClean, P. E., Qiu, L. and Ma, J., 2010. Artificial selection for determinate growth habit in soybean. *Proceedings of the National Academy of Sciences*, 107(19), pp. 8563–8.

Tilman, D., Fargione, J., Wolff, B., D'Antonio, C., Dobson, A., Howarth, R., Schindler, D., Schlesinger, W. H., Simberloff, D. and Swackhamer, D., 2001. Forecasting agriculturally driven global environmental change. *Science*, 292(5515), pp. 281–4.

Tolin, S. A. and Lacy, G. H., 2004. Viral, bacterial, and phytoplasmal diseases of soybean. In *Soybeans: Improvement, Production, and Uses*, 3rd ed., pp. 765–819. Agronomy Monograph No. 16, American Society of Agronomy, Crop Science Society of America, Soil Science Society of America, Madison, WI.

Tyler, B. M., 2008. Genomics of fungal-and oomycete-soybean interactions. In Stacey, G. (Ed.) *Genetics and Genomics of Soybean*, pp. 243–67. Springer, New York, NY.

Upadhyaya, H. D. and Ortiz, R., 2001. A mini core subset for capturing diversity and promoting utilization of chickpea genetic resources in crop improvement. *Theoretical and Applied Genetics*, 102(8), pp. 1292–8.

USDA, 2016. Crop Production 2015 Summary. (January).

Vuong, T. D., Sleper, D. A., Shannon, J. G. and Nguyen, H. T., 2010. Novel quantitative trait loci for broad-based resistance to soybean cyst nematode (*Heterodera glycines Ichinohe*) in soybean PI 567516C. *Theoretical and applied genetics*, 121(7), pp. 1253–66.

Wang, D., Graef, G. L., Procopiuk, A. M. and Diers, B. W., 2004. Identification of putative QTL that underlie yield in interspecific soybean backcross populations. *Theoretical and Applied Genetics*, 108(3), pp. 458–67.

Watanabe, S., Hideshima, R., Xia, Z., Tsubokura, Y., Sato, S., Nakamoto, Y., Yamanaka, N., Takahashi, R., Ishimoto, M., Anai, T. and Tabata, S., 2009. Map-based cloning of the gene associated with the soybean maturity locus E3. *Genetics*, 182(4), pp. 1251–62.

Watanabe, S., Xia, Z., Hideshima, R., Tsubokura, Y., Sato, S., Yamanaka, N., Takahashi, R., Anai, T., Tabata, S., Kitamura, K. and Harada, K., 2011. A map-based cloning strategy employing a residual heterozygous line reveals that the GIGANTEA gene is involved in soybean maturity and flowering. *Genetics*, 188(2), pp. 395–407.

Weber, V. S., Araus, J. L., Cairns, J. E., Sanchez, C., Melchinger, A. E. and Orsini, E., 2012. Prediction of grain yield using reflectance spectra of canopy and leaves in maize plants grown under different water regimes. *Field Crops Research*, 128, pp. 82–90.

White, J. W. and Conley, M. M., 2013. A flexible, low-cost cart for proximal sensing. *Crop Science*, 53(4), pp.1646–9.

White, J. W., Andrade-Sanchez, P., Gore, M. A., Bronson, K. F., Coffelt, T. A., Conley, M. M., Feldmann, K. A., French, A. N., Heun, J. T., Hunsaker, D. J. and Jenks, M. A., 2012. Field-based phenomics for plant genetics research. *Field Crops Research*, 133, pp. 101–12.

Wilson, R. F., 2008. Soybean: market driven research needs. In Stacey, G. (Ed.) *Genetics and Genomics of Soybean*, pp. 3–15. Springer, New York, NY.

Wright, S. G., 1932. Evolution in Mendelian populations. *Genetics*, 16(2), pp. 97–159.

Xia, Z., Watanabe, S., Yamada, T., Tsubokura, Y., Nakashima, H., Zhai, H., Anai, T., Sato, S., Yamazaki, T., Lü, S. and Wu, H., 2012. Positional cloning and characterization reveal the molecular basis for soybean maturity locus E1 that regulates photoperiodic flowering. *Proceedings of the National Academy of Sciences*, 109(32), pp. E2155–64.

Yamanaka, N., Morishita, M., Mori, T., Lemos, N. G., Hossain, M. M., Akamatsu, H., Kato, M. and Yamaoka, Y., 2015. Multiple *Rpp*-gene pyramiding confers resistance to Asian soybean rust isolates that are virulent on each of the pyramided genes. *Tropical Plant Pathology*, 40(5), pp. 283–90.

Yamanaka, N., Morishita, M., Mori, T., Muraki, Y., Hasegawa, M., Hossain, M., Yamaoka, Y. and Kato, M., 2016. The locus for resistance to Asian soybean rust in PI 587855. *Plant Breeding*, 135(5), pp. 621–6.

Yan, W., 2014. *Crop Variety Trials: Data Management and Analysis*. John Wiley & Sons, Hoboken, NJ.

Yan, W., Pageau, D., Frégeau-Reid, J. and Durand, J., 2011. Assessing the representativeness and repeatability of test locations for genotype evaluation. *Crop Science*, 51(4), pp. 1603–10.

Yang, K., Tian, Z., Chen, C., Luo, L., Zhao, B., Wang, Z., Yu, L., Li, Y., Sun, Y., Li, W. and Chen, Y., 2015. Genome sequencing of adzuki bean (*Vigna angularis*) provides insight into high starch and low fat accumulation and domestication. *Proceedings of the National Academy of Sciences*, 112(43), pp. 13213–18.

Yorinori, J. T., Paiva, W. M., Frederick, R. D., Costamilan, L. M., Bertagnolli, P. F., Hartman, G. E., Godoy, C. V. and Nunes Jr., J., 2005. Epidemics of soybean rust (*Phakopsora pachyrhizi*) in Brazil and Paraguay from 2001 to 2003. *Plant Disease*, 89(6), pp. 675–7.

Yu, N., Li, L., Schmitz, N., Tian, L. F., Greenberg, J. A., Diers, B. W. 2016. Development of methods to improve soybean yield estimation and predict plant maturity with an unmanned aerial vehicle based platform. *Remote Sensing of Environment*, 187, pp. 91–101.

Zhai, H., Lü, S., Liang, S., Wu, H., Zhang, X., Liu, B., Kong, F., Yuan, X., Li, J. and Xia, Z., 2014. GmFT4, a homolog of FLOWERING LOCUS T, is positively regulated by E1 and functions as a flowering repressor in soybean. *PLoS One*, 9(2), p. e89030.

Zhang, X., Zhai, H., Wang, Y., Tian, X., Zhang, Y., Wu, H., Lü, S., Yang, G., Li, Y., Wang, L. and Hu, B., 2016. Functional conservation and diversification of the soybean maturity gene E1 and its homologs in legumes. *Scientific Reports*, 6: 29548.

Zhao, C., Takeshima, R., Zhu, J., Xu, M., Sato, M., Watanabe, S., Kanazawa, A., Liu, B., Kong, F., Yamada, T. and Abe, J., 2016. A recessive allele for delayed flowering at the soybean maturity locus E9 is a leaky allele of FT2a, a FLOWERING LOCUS T ortholog. *BMC Plant Biology*, 16(1), p. 1.

Chapter 4

Mapping the soybean genome

Xiaobo Wang, Anhui Agricultural University, China; and Lijuan Qiu, Chinese Academy of Agricultural Sciences, China

1 Introduction

Soybean [*Glycine max* (L.) Merr.] is one of the most important grain legumes (Li et al., 2014). It is an important source of plant protein and oil, and is also a natural source of nutraceutical compounds such as isoflavones and lunasin (de Lumen, 2005; Mateos-Aparicio, 2008). The development of new cultivars has played an important role in improving the crop. As soybean is a strict self-pollination crop, and the natural outcrossing rate is less than 1%, its genetic mapping has been relatively underdeveloped compared to other crops (Nakayama and Yamaguchi, 2002), which has limited the molecular breeding of modern soybean. The soybean community has made significant progress in mapping the soybean genome. With the appearance of DNA molecular markers, which have greatly facilitated genetic mapping, it has been possible to build up a density linkage map of soybean. In many research fields, genetic mapping has played an important role from marker-assisted selection (MAS) in plant improvement to map-based cloning (MBC) in molecular genetic studies. The development of a precise, high-density linkage map has thus made an important contribution to soybean research. In recent years, many essential agronomic and quality traits have been studied by developing a genetic linkage map and identifying quantitative trait loci (QTLs) or genes to improve yield and nutritional quality, as well as biotic and abiotic stress tolerances (Concibido

http://dx.doi.org/10.19103/AS.2017.0034.04

et al., 2004; Panthee et al., 2006; Tuyen et al., 2010; Cardinal et al., 2014; Kato et al., 2014). Additionally, the next-generation sequencing (NGS) technology provides the capacity for parallel sequencing of genomes and development of a sequencing-based high-throughput genotyping method that combines the advantages of ultra-high-density marker coverage, while eliminating the likelihood of overlooking double crossovers (Huang et al., 2009; Xu et al., 2010). The technology ensures high mapping accuracy and resolution together with more comparable genome and genetic maps among the mapping populations (Huang et al., 2009).

In this chapter, we summarize and describe recent advances in mapping the soybean genome.

2 Mapping the soybean genome: development of genome maps

Soybean genome mapping started with conventional markers; it accelerated following the development of molecular markers in the early 1990s, and a number of genetic maps have been constructed since then (Keim et al., 1990; Gore et al., 2002; Hyten et al., 2010; Gutierrez-Gonzalez et al., 2011). With these genetic maps, more than a thousand QTLs associated with essential traits have been identified (Soybase, http://www.soybase.org). Meanwhile, high-throughput sequencing technology provides the ability to develop massive single-nucleotide polymorphism (SNP) markers and whole-genome sequencing (Varshney et al., 2009; China soybean data center: www.soybean-resources.cn).

Conventional mapping based on morphological, pigment, isozyme and other markers in soybean was slow. There were only seven linkage groups (LGs) (Bernard and Weiss, 1973) until 1973, reaching 13 (Palmer et al., 1985) in 1985. Although there were 250 morphological and isozyme markers, most of them had not been mapped on the genome before the molecular map was completed for soybean. Only 57 of these markers were located on 17 linkage maps with a length of 420 cM (Palmer and Kilen, 1987). Until 1999, there were still only 68 markers located on 20 LGs and an average of 3.4 loci on each linkage map (Palmer et al., 1998). Subsequently, Cregan et al. (1999) integrated the MS (A81-356022 × PI468916), CH (Clark × Harosoy) and MN (Minsoy × Noir 1) of three genetic linkage maps in a population by 606 simple sequence repeat (SSR) markers, which was the first public genetic map containing 20 LGs corresponding to chromosomes in soybean. Later, Song et al. (2004) integrated the genetic map spanning 2523.6 cM of Kosambi's map distance across 20 LGs that contained 1849 markers, including 1015 SSRs, 709 restriction fragment length polymorphisms (RFLPs), 73 randomly amplified polymorphic DNAs (RAPDs), 24 classical traits, 6 amplified fragment length polymorphism (AFLPs), 10 isozymes and 12 others increased by 420 newly developed SSRs and MA (Minsoy × Archer) and Atlas AN (Archer × Noir 1) in two populations. Xia et al. (2007) encrypted the genetic linkage map that was constructed by Misuzudaizu (*G. max*) and Moshidou Gong with 503 (*Glycine gracilis*) hybrid offspring F_2 populations using RFLP, SSR, AFLP and STS markers (Yamanaka et al., 2000, 2001) giving a total map length of 3080 cM in 20 LGs. Thus, an accurate and more saturated genetic linkage map of soybean is essential for studies on modern soybean genomics, enabling identification of subtle or new trait loci including QTL, MBC and physical map construction or even whole-genome sequencing.

2.1 Mapping genome by RFLP markers

The RFLP molecular markers, the first generation of DNA markers, were widely used in gene mapping, diversity and evolutionary studies because of their reliable repeatability and co-dominance (Botstein et al., 1980). Apuya et al. (1988) constructed the first global RFLP map using the F_2 population obtained from a hybrid of two cultivars 'Minsoy' and 'Noir 1', which included four LGs and 11 RFLP markers on them. Subsequently, Keim et al. (1990) recorded the second molecular linkage map constructed with AFLP markers of soybean. This map consisted of 26 genetic LGs containing a total of 150 RFLP loci and was based on an F_2 population derived from an inter-specific cross of *G. max* (A81-356022) × *G. soja* (PI468916). Lark et al. (1993) used 132 RFLPs and morphological and isozyme markers to construct a soybean genetic map comprising 31 LGs. Shoemaker and Specht (1995) mapped 110 RFLPs and 8 RAPDs and 7 pigmentation, 6 morphological and 7 isozyme markers in an F_2-population derived from a mating of isolines from the important soybean cultivars 'Clark' and 'Harosoy'. These early genetic maps were primarily based on RFLP markers.

Zhang et al. (1997) constructed the first molecular genetic mapping of soybean in China using a hybridization of cultivated Changnong 4 and semi-wild Xinmin 6 of soybean, using a completed LG analysis of 84 F_2 populations based on RFLP markers. A total of 20 LGs were constructed, with a total distribution of 71 markers, including 63 RFLP and 8 RAPD markers; the average distance between markers was 20.4 cM, covering a length of 1446.8 cM. Then, Liu et al. (2000) constructed the second molecular genetic mapping of soybean in China using 88-individual recombinant inbred line (RIL) populations derived from a cross between cultivated Changnong 4 and semi-wild Xinmin 6. This map contains 22 LGs and 240 markers, including 100 RFLPs, 33 SSRs, 62 RAPDs, 42 AFLPs, 1 sequence-characterized amplified region and 2 morphological markers, covering approximately 3713.5 cM of the soybean genome.

Although RFLP was widely used in various fields of biology, studies by Cregan et al. (1999) showed that RFLP markers have two major shortcomings: first, only rarely have more than two alleles been identified at RFLP loci in soybean. As these two alleles generally have asymmetric frequencies, for example, $p > 0.9$, $q < 0.1$ (Keim et al., 1989, 1992), the likelihood that any two genotypes will be polymorphic at a particular RFLP locus is relatively low. This is particularly true when both parents of the mapping population come from adapted soybean germplasm pools (Apuya et al., 1988; Lark et al., 1993). A second factor that complicates the use of RFLP markers in soybean is the detection of multiple DNA fragments (i.e. multiple loci) with most probes. This may be the result of the tetraploid origin of soybean (Hymowitz and Singh, 1987). Due to the lack of polymorphism of RFLP loci in soybean and/or the complexity of multiple DNA banding patterns detected with most RFLP probes, SSRs were proposed for map development (Akkaya et al., 1992), thus launching a new era in soybean genome mapping.

2.2 Mapping genome by SSR markers

Most SSRs are single-locus markers and many SSR loci are multi-allelic. These characteristics make SSRs an ideal marker system not only for understanding the genetic diversity of germplasms and creating genetic maps, but also as an unambiguous means of defining LG homology across soybean mapping populations (Song et al., 2004; Qiu et al., 2013). The SSRs are widely distributed throughout the soybean genome. The first report of the polymorphism and heritability of SSR markers in plants was in soybean (Akkaya et al., 1992). Subsequently, 606 soybean SSR loci were developed and mapped in three mapping populations to create the first version of a soybean-integrated genetic linkage map (Cregan et al., 1999). Five years later, Song et al. (2004) produced the second version of the integrated genetic linkage map, which was based on five mapping populations and contained a total of 1015 SSR loci.

In 2001, Wu et al. developed a high-density genetic map constructed with RILs population (NJRIKY) from a cross of two cultivated soybeans, Kefeng 1 and Nannong 1138-2, using four markers (RFLP, SSR, RAPD and AFLP). This map consists of 24 LGs including 792 genetic markers and covers 2320.7 cM with an average genetic distance of 2.9 cM. The results indicated that SSR markers with higher polymorphisms between two parents could be used as anchored markers to integrate different maps constructed with different genetic markers due to their stability and simplicity. Wan et al. (2005) reported that 227 SSRs were mapped in soybean mapping populations: 'Jindou 23 × ZDD 2315', an F_{10} RIL comprising 474 lines using a total of 441 SSR markers from Cregan's soybean public linkage map published in 1999. In contrast to the soybean genetic linkage map published by Song et al. in 2004, this soybean genetic linkage map was highly comparable in the grouping and sequencing of the mapped SSR markers, allowing it to become a very effective bridge with the international soybean map. Thereafter, Gong et al. (2006) integrated the soybean genetic linkage map using the mapping strategy of anchoring SSR markers with two F_2 mapping populations based on the results of Wan et al. (2005). This integrated map covers 1951.8 cM of the soybean genome with 355 markers on 20 non-gap LGs, where the length of the LG varied from 40.8 to 184.4 cM with an average genetic distance between adjacent markers of about 5.48 cM of Kosambi. This integrated map identifies more closely with the soybean consensus map. In 2009, Zhou et al. constructed individual genetic linkage maps containing 560, 223, 195 and 133 markers using four RIL populations, 'Ke feng1 × Nan nong 1138-2', 'Nan nong 87-23 × NG 94-156', 'Su 88-M 21 × Xin yi xiao hei dou' and 'Wan 82-178 × Tong shan bo pi huang dou jia', derived from crosses between distinct elite cultivars of soybean, respectively. The result contained 795 markers spanning 2772.9 cM of the soybean genome, distributed over 20 LGs, with the length of LGs varying from 77.1 to 224.7 cM, the marker number from 24 to 69 and an average marker distance of 3.49 cM. This map shows good coincidence except with an additional six SSR markers located on different LGs and five new SSR markers.

2.3 Mapping genome by high-throughput genotyping markers

In recent years, single-DNA base differences between homologous DNA fragments, plus small insertions and deletions (Indels) collectively referred to as SNPs, were shown to be the most abundant source of DNA polymorphisms in soybean (Zhu et al., 2003; Hyten et al., 2006). The SNPs are the markers of choice for many researchers due to their abundance and high-throughput methods available for their multiplex analysis (Song et al., 2016). The SNP markers have only recently been available to researchers in soybean with the release of the third version of the consensus genetic linkage map that added 1,141 SNP markers to the map (Hyten et al., 2010). In 2010, a high-density integrated genetic linkage map of soybean and a 1,536 universal soy linkage panel for QTL mapping were published (Hyten et al., 2010). Li et al. (2010) subsequently reported that they genotyped a sample of 303 accessions of domesticated soybean (*G. max*) and its wild progenitor *G. soja* with 99 microsatellite markers and 554 SNP markers. The results showed that the SSR loci averaged 21.5 alleles per locus and the overall Nei's gene diversity was 0.77 and SNPs had substantially lower genetic diversity (0.35) than SSRs. A unique feature of SNP markers is the modest cost per data point and speed of data acquisition (Fan et al., 2006). Additionally, the development of next-generation high-throughput sequencing technology provides the capability for developing massive SNP markers (Varshney et al., 2009).

The Soybean Consensus Map 4.0 facilitated the anchoring of 95.6% of the whole soybean genome sequence, but its marker density was only sufficient to properly orient 66% of the sequence scaffolds. The discovery and genetic mapping of more SNP markers were needed to anchor and orient the remaining genome sequence. The NGS and high-throughput genotyping were combined to obtain a much higher resolution genetic map that could be used to anchor and orient most of the remaining sequences and help validate the integrity of the existing scaffold builds. Hyten et al. (2010) reported that a total of 7108–25 047 predicted SNPs were discovered by the Illumina sequence-by-synthesis method on the clonal single-molecule array (CSMA) platform. Those SNPs were then used to create a high-resolution genetic map that assisted in the assembly of scaffolds from the 8x whole-genome shotgun sequences into pseudomolecules corresponding to chromosomes of the organism and would help address the challenges faced in assembling and orienting the remainder of the soybean genome (Hyten et al., 2010).

Lee et al. (2015) reported that a high-density genetic map was constructed using 2 545 SNPs by the Infinium BARCSoySNP6K BeadChip array. The total length of this genetic map was 2 346 cM and made up of 2 213 unique SNP loci with 86–162 unique loci per chromosome. The average marker interval ranged from 0.9 to 1.3 cM with an overall mean of 1.1 cM. This high-density genetic map shows the utility of the BARCSoySNP6K BeadChip array and will lead to a better understanding of the genetic architecture of complex traits and identification of SNPs tightly linked to QTL for many important soybean traits. Song et al. (2016) constructed two high-density genetic linkage maps based on 21 478 SNP loci mapped in the Williams 82 × *G. soja* (Sieb. & Zucc.) PI479752 and Essex × Williams 82 populations. They found that there were 37 regions or single markers where the marker order in the two populations was in agreement, but was not consistent with the physical position in the Glyma1.01 build. Map data were also used to identify false joins in the Glyma1.01 assembly, and the corresponding scaffolds were broken and re-assembled in the new assembly, Wm82.a2.v1. The genomic positions of the commonly used markers contained in the BARCSOYSSR_1.0 database and the SoySNP50K BeadChip were updated based on the Wm82.a2.v1 assembly. Moreover, the SoySNP50K Illumina BeadChip containing 52 041 SNPs was also used to genotype two natural soybean populations, 342 landraces and 1062 improved lines. A total of 125 candidate selection regions were identified within the modern soybean genome, which is meaningful from the perspective of evolutionary biology and for revealing the genetic architecture of agronomic traits (Wen et al., 2015).

3 New mapping techniques: specific-locus amplified fragment sequencing (SLAF-seq) mapping, whole-genome re-sequencing mapping and comparative genome mapping

Large-scale genotyping plays an important role in genetic association studies. It has provided new opportunities for gene discovery, especially when combined with high-throughput sequencing technologies (Sun et al., 2013). Although QTL mapping is an efficient way to reveal the genetic architecture underlying complex quantitative traits, the low density of molecular markers in genetic maps has limited the efficiency and accuracy of QTL mapping (Li et al., 2014). SLAF-seq is a recently developed high-resolution strategy for large-scale de novo discovery and genotyping of SNPs based on NGS (Sun et al., 2013; Li et al., 2014; Qi et al., 2014). The SLAF-seq technology has several distinguishing characteristics:

1. Deep sequencing to ensure genotyping accuracy;
2. Reduced representation strategy to reduce sequencing costs;
3. Pre-designed reduced representation scheme to optimize marker efficiency; and
4. Double barcode system for large populations (Sun et al., 2013).

Recently, Sun et al. (2013) developed SLAF-seq as a high-resolution strategy for large-scale de novo SNP discovery and genotyping; this approach was successfully used to create a genetic map for common carp (*Cyprinus carpio* L.); it was the highest density genetic map yet for any organism without a reference genome sequence. Sun et al. (2013) also tested the efficiency of SLAF-seq on rice and soybean data. Both sets of results showed strong consistency between predicted and practical SLAFs and considerable genotyping accuracy. An F_{21} population of 147 RILs derived from a cross between 'Charleston' and 'Dongnong 594' was constructed based on SLAF-seq, including 5308 markers on 20 LGs and was 2 655.68 cM in length, with an average distance of 0.5 cM between adjacent markers (Qi et al., 2014). Moreover, SLAF-seq was employed to construct a high-density soybean genetic map using RILs (Luheidou2 × Nanhuizao, F5:8) (Li et al., 2014). The results showed that the average coverage for each SLAF marker was 11.20-fold for the female parent, 12.51-fold for the male parent and an average of 3.98-fold for individual RILs. The final map consisted of 5 785 SLAFs on 20 LGs and spanned 2 255.18 cM in genome size with an average distance of 0.43 cM between adjacent markers. This result will facilitate the identification of genes and QTLs underlying essential agronomic traits in soybean. The major novel QTL for isoflavone content is useful not only for further study on the genetic basis of isoflavone accumulation, but also for MAS in soybean breeding in the future.

3.1 Whole-genome re-sequencing mapping

The QTL mapping and cloning using DNA markers are useful tools to unlock the genetic basis of complex phenotypic variations. Moreover, the application of molecular markers developed from the QTL analysis enhances breeding efficiency by enabling MAS for particular agronomic traits. However, traditional approaches for the QTL analysis require significant time and manpower because of the low efficiency of polymorphic marker development in constructing high-resolution maps and the difficulties in identifying suitable candidate genes. Whole-genome sequencing, using next-generation technology, can overcome the limitations of QTL analysis. The NGS technology coupled with a growing number of genome sequences provides an opportunity to re-design genotyping strategies, enabling more effective genetic mapping and genome analysis (Huang et al., 2009). A sequencing-based genotyping method uses SNPs identified by whole-genome sequencing of a mapping population (Huang et al., 2009). This approach could substantially reduce the amount of time and effort required to genotype a mapping population. Zhang et al. (2011) have also reviewed soybean nuclear gene codon composition, usage patterns and determining factors using full-length transcript sequencing.

Lam et al. (2010) produced a large-scale analysis of the patterns of genome-wide genetic variation in soybeans using re-sequencing. They re-sequenced a total of 17 wild and 14 cultivated soybean genomes to an average of approximately ×5 depth and >90% coverage using the Illumina platform. Comparing the patterns of genetic variations between wild and cultivated soybeans, they identified higher allelic diversity in wild soybeans. Their study results suggest that marker-assisted breeding of soybean will be less challenging than MBC. They also identified a set of 205 614 tag SNPs that may be useful for QTL mapping and associated studies. These data will provide a valuable resource for the analysis of wild soybeans and facilitate future breeding and quantitative trait analysis.

Li et al. (2013) re-sequenced 25 diverse soybean accessions from the China National GeneBank (http://www.cgris.net/#), which represent three distinct gene pools:

- The pre-domesticated annual wild progenitor species (*G. soja*);
- Domesticated local landraces (*G. max*); and
- Modern elite cultivars (*G. max*).

To achieve a more comprehensive analysis, they integrated these re-sequenced data with the re-sequenced data previously generated from 14 wild and 17 cultivated soybean genomes (Lam et al., 2010). As a result, a total of 5 102 244 SNPs and 707 969 indels were identified. The results found that artificial selection during domestication led to more pronounced reduction in the genetic diversity of soybean than the switch from landraces to elite cultivars. The genomic resources and information provided by this study would help reveal underlying agronomically important traits in genes/loci (China soybean data center: www.soybean-resources.cn). The availability of the soybean genome sequence (Schmutz et al., 2010) and high-throughput sequencing technologies provide an unprecedented opportunity to track the evolutionary history of the domesticated soybean and dissect the genetic bases for soybean domestication and varietal diversification.

Zhou et al. (2015) re-sequenced 302 soybean accessions to more than 11 × depth and analysed genomic variation dynamics during soybean domestication, improvement and local breeding. The result revealed the associations between ten selected regions and nine domestication or improvement traits and identified 13 previously uncharacterized loci for agronomic traits including oil content, plant height and pubescence form using new sequences. Meanwhile, 96 correlates with reported oil QTLs and 21 containing fatty acid biosynthesis genes were identified by combining with previous QTL information. Several key genes or loci associated with the main agronomic traits of soybean were detected, including *Dt1*, *T*, *W1* and *I* locus, *E1* and *Ln* gene and one copy number variation (CNV). Moreover, some traits and loci were also identified related to geographical regions, demonstrating soybean populations are structured geographically.

Based on the above, whole-genome re-sequencing has been widely used in genotyping, domestication, genetic improvement, genetic diversity and selection, discovery of new genes, SNPs, indels and other structural variation. Moreover, whole-genome re-sequence data have also substantially enhanced the efficiency of polymorphic marker development for QTL, fine mapping and the identification of possible candidate genes in soybean.

3.2 Comparative genome mapping

Comparative genomics is an important discipline with applications in evolutionary, genetic and genome rearrangement studies (Swidan et al., 2006; Chang et al., 2014). Comparative genome mapping was pioneered by Tanksley et al. (1988) and similar works (Bonierbalem et al., 1988) were done using tomato RFLP probes to map the tomato, chilli pepper and potato genomes. This approach allows the determination of the extent and nature of chromosomal rearrangements between cross-incompatible species and opens up comparisons between distantly related genomes, which are not amenable to analysis by traditional cytogenetic techniques (Whitkus et al., 1992). Comparative genome mapping highlights similarities and differences in the genome organization of species and can lead to concrete applications in the domains of genome evolution and quantitative genetics (Nadeau and Sanko, 1998; Paterson et al., 2000; Innes et al., 2012). As a result of genetic processes such as duplication, deletion, insertion, inversion or translocation of genes, comparative mapping helps us to achieve a better understanding of chromosomal evolution (Ahn and Tanksley 1993; Paterson et al., 2000; Ma et al., 2005; Chang et al., 2014).

Comparative genome mapping has been widely used to study soybean, including the *bacterial artificial chromosome* (*BAC*) contigs (Yan et al., 2004), *Rsv4* (Hwang et al., 2006), *Dt1* (Tian et al., 2010; Zhou et al., 2015), *CDF3*, *VRN1*, *SVP*, *AP3* and *PIF3* genes, paralogue-rich genes associated with flowering (Jung et al., 2012), *VQ* motif-containing genes (Wang et al., 2014), *SWEET* gene family (Patil et al., 2015), *W1*, *I* and *T* locus, *E1* and *Ln* gene (Zhou et al., 2015). Wang et al. (2014) discovered a *VQ* gene family in soybean using a comparative analysis, and based on that investigated genetic diversity in different types of soybean populations. Genetic variation and domestication analysis in cultivated and wild soybeans revealed there were selected site(s) in *GmVQ6*, *GmVQ7*, *GmVQ10*, *GmVQ26* and *GmVQ61*, suggesting these genes have undergone artificial selection during soybean domestication. The results also suggested that soybean VQ-motif containing genes may act as key regulators in soybean growth, development and nitrogen metabolism. Similarly, Zhou et al. (2016) identified the leucine-rich repeat receptor-like protein kinase genes in soybean and gene analysis that might involve evolutionary history. In summary, comparative genomic analysis is an effective method for the identification and functional characterization of new genes and their family using the soybean genome.

Comparative genomic analysis not only determines the similarities between species, but also the characteristics of soybean. Tian et al. (2010) showed that the soybean *Dt1* gene is the functional homolog (designated as *GmTfl1*) of the *Arabidopsis* terminal flower. They identified 1 (*TFL1*) a regulatory gene encoding a signalling protein of shoot meristems by a comparative genomics approach and confirmed *GmTfl1* complemented the functions of *TFL1* in *Arabidopsis*. However, the *GmTfl1* homolog, despite its more recent divergence from *GmTfl1* than from *Arabidopsis TFL1*, appears to be sub- or neo-functionalized, as revealed by the differential expression of the two genes at multiple plant developmental stages and by allelic analysis at both loci. Liu et al. (2015) tested 1120 Chinese soybean varieties. They found that the dominant GmTfl1 allele was prevalent in accessions from the Northern ecoregion of China; the recessive allele Gmtfl1 was more common in the Southern ecoregion and the proportions of GmTfl1 and recessive alleles were 40.1% and 59.9%, respectively, in the Huang-Huai ecoregion. The distributions of GmTfl1 and Gmtfl1 have shown no large changes in nearly 60 years of breeding, but the frequency of the recessive genotype Gmtfl1 has shown a rising trend in the last 20 years. This study provides a theoretical foundation for breeding new soybean varieties for different ecoregions. Patil et al. (2015) reported that the soybean (*G. max*) *SWEET* (*MtN3_saliva*) gene families, a recently identified group editing efflux transporters, play an indispensable role in sugar efflux, phloem loading, plant–pathogen interaction and reproductive tissue development by analysing their phylogeny, gene structure, domain architecture, expression profiles and natural genetic variation. The results indicated that the *SWEET* gene family has a role as a key component in reproductive tissue development, nutrient unloading and pathogen resistance, and manipulating *SWEET* expression in specific tissues (phloem sap, pedicel and developing seeds) could increase yield by enhancing sugar delivery to developing seeds.

4 Case study: comparing wild and cultivated soybean varieties

Annual wild soybean (*G. soja* Sieb. & Zucc.) is the closest relative and antecedent of cultivated soybean [*G. max* (L.) Merr.], but their gene repertoire remains largely unexplored (Li et al., 2014). As some technologies (such as re-sequencing) are limited in terms of capturing many types of structural variations, the de novo construction of a pan-genome for a species, consisting of a core genome shared among individuals and individual-specific or partially shared dispensable genome, is necessary to capture the majority of genetic diversity within a species (Tettelin et al., 2005). To fully characterize the genomic content and molecular evolutionary history of *G. soja*, the undomesticated gene pool of soybean, soybean pan-genome, was constructed by de novo sequencing and assembling seven *G. soja* accessions.

A computational pipeline was designed, which takes advantage of the sequencing reads as well as assembled genomes to catalogue variations including SNPs, indels, CNVs and presence–absence variations (PAV) and identify differences between *G. soja* genomes and GmaxW82 (Li et al., 2014). The results showed that the assembly-based method enabled the identification of more SNPs than by re-sequencing alone. Compared to GmaxW82, about 0.50–0.77 million indels in *G. soja* were detected. Although many indels were trinucleotides, 2 989–4 181 resulted in frameshifts. For example, three guanines of *Glyma02g25230* were deleted at positions 695, 707 and 714 in all seven *G. soja* genomes, resulting in five amino acid changes in one of the HEAT-repeat motifs. Thus, it was concluded that *Glyma02g25230* is a candidate gene that may be responsible for the change from the twining growth habit found in *G. soja* to erect growth found in *G. max*. Compared to GmaxW82, in *G. soja* accessions 1 179 had CNV loss, 726 had CNV gain and 73 had both CNV loss and gain (Li et al., 2014). Gene ontology (GO) analysis indicated that these genes are evolutionarily labile and may be involved in adaptation to local environments and/or interaction with pathogens. The *G. soja*-specific PAV (2.3–3.9 Mbp) is less than the 8.3 Mbp reported for a single previously re-sequenced *G. soja* genome, and the total length of *G. max*-specific PAV (1.86 Mbp) was also less than that of *G. soja*–specific PAV. The GC content for these regions (32.9%) was not significantly different from the whole-genome average (34.1%). These results indicated that the variation between *G. soja* and GmaxW82 was distributed unevenly throughout the genome.

There were many variations in agronomic traits between *G. soja* and *G. max*, such as flowering time, oil and fatty acid content, organ size, stems, plant height and adaptation to biotic or abiotic stresses (Liu et al., 2007; Kim et al., 2010; Li et al., 2014; Qi et al., 2014). The R-genes with NBS domains mediate effector-triggered immunity acting as detectors for pathogen virulence proteins (Innes et al., 2008). The 11 lineage-specific R-gene domain architectures found in *G. soja* accessions were part of the dispensable genome, possibly reflecting adaptation to biotic stresses. By comparing the *G. soja* genomes and GmaxW82 flowering time-related genes with large-effect, variations were found, including light receptor *PHYA*, floral integrator *FLOWERING LOCUS T* (*FT*) and meristem identity protein *LEAFY* (*LFY*) genes (Li et al., 2014). Thus, it was speculated that the gene controlling flowering time was probably involved in adaptation to new climatic regions. Our study results also found that the oil and fatty acid contents of soybean have been intensely selected so that *G. max* produces almost twice as much oil as *G. soja*. In addition, the cultivated soybean has larger sized organs, thicker stems and is shorter in height compared to *G. soja*. Inter-genomic comparisons identified lineage-specific genes and genes with CNV or large-effect mutations, some of which show evidence of positive selection and may contribute to a variation in agronomic traits such as biotic resistance, seed composition, flowering and maturity time, organ size and final biomass (Li et al., 2014). This will facilitate the harnessing of untapped genetic diversity from wild soybean enhancing elite cultivars.

5 Case study: identifying salinity tolerance in soybean

Based on the above study results, Guan et al. (2014) identified the *GmSALT3* (*G. max* salt tolerance-associated gene on chromosome 3) in a soybean population derived from commercial cultivars Tiefeng 8 and 85-140 using MBC. A combination of de novo sequencing data and re-sequencing data was also used to identify this gene, and rapid gain-of-function tests found the protective effects of *GmCHX1* (*GmSALT3*) towards salt stress (Qi et al., 2014). A total of nine haplotypes including two salt-tolerant haplotypes and seven salt-sensitive haplotypes were identified for *GmSALT3* by sequencing 31 soybean landraces and 22 wild soybeans (*G. soja*). Among which, haplotype 1 (H1, found in Tiefeng 8) was found to be strongly associated with salt tolerance and was likely to be the ancestral allele by analysing the distribution of haplotypes among 172 Chinese soybean landraces and 57 wild soybeans. Based on the results analysis, it was speculated that coupled with the low genetic diversity of the salt-tolerant alleles, *GmSALT3* is under strong selection pressure when required, but can be lost when this pressure is removed resulting in a decrease in the salt tolerance of the plant (Guan et al., 2014). These results indicated that the identification of *GmSALT3* by a comparative analysis of different genome mapping, which encodes a cation/proton exchanger family member, was likely to have significance for soybean and other crop-breeding programmes (Guan et al., 2014, Qi et al., 2014).

6 Summary

Soybean [*G. max* (L.) Merr.] is one of the most important grain legumes (Li et al., 2014) and is also an important source of plant protein and oil. In many research fields, genetic mapping has played an important role from MAS in plant improvement to MBC in molecular genetic studies. Therefore, the development of a high-density and precise genetic map will have great significance for further study of the important economic crop of soybeans, including classic morphological markers, isozyme markers and a large number of high-degree DNA molecular markers.

Soybean genome mapping based on molecular markers started in the early 1990s (Keim et al., 1990). In recent years, the development of a genetic map extended conventional mapping (including RFLP, SSR and AFLP markers) SNP, SALF-seq, whole-genome re-sequencing mapping and comparative genome mapping. These genetic mapping technologies have made important contributions to the study of growth development and important related agronomic traits in soybean, which effectively promoted the healthy global development of soybean. Currently, the development of high-throughput sequencing technology combined with developed genetic stocks such as the mutants library (Li et al., 2017) and the development of an integrated phenotype and genome database (www.soybean-resources.cn) provides new opportunities and directions for related research on soybean in the future.

7 Future trends

From the perspective of gene discovery and breeding practice, the existing genetic markers and technologies are far from meeting the demands of soybean genome research. This is not only from the point of view of technological development, but also from the application requirements of soybean products. The development of the soybean genome and comparative genome mapping need to be structured and refined. On the technical side, single-molecule sequencing (SMS) or single-molecule real-time (SMRT) sequencing, which can read the base sequence directly from individual strands of DNA or RNA present in a sample of interest will truly reflect the nature of the soybean genome.

7.1 SMS or SMRT sequencing

This sequencing technique is based on successive enzymatic degradation of fluorescently labelled single-DNA molecules and the detection and identification of the released monomer molecules according to their sequential order in a microstructured channel (Dörrey et al., 1997). There are currently two principal approaches:

1. DNA polymerase-based single-molecule DNA sequencing, and
2. direct inspection of individual DNA strands (Korlach and Turner, 2010).

The main technique characteristics of SMS include:

1. Increasing intrinsic sequencing speeds;
2. Increasing the length of DNA sequencing reads;
3. Avoiding processing steps prior to sequencing;
4. Resolving heterogeneous DNA mixtures; and
5. Reading DNA base modifications as part of sequencing (Korlach and Turner, 2010).

Thus, SMS is an attractive approach due to its simplicity and the lack of cloning or amplification in sample preparation (Pushkarev et al., 2009). Previous reports of SMS have established the principle (Greenleaf and Block, 2006; Harris et al., 2008; Eid et al., 2006; Lipson et al., 2009; Pushkarev et al., 2009; Hart et al., 2010; Koren et al., 2012), and their sequencing throughput has not been competitive with alternative methods.

In 2008, SMS had already been used successfully to sequence the genome of the M13 virus (Harris et al., 2008). Subsequently, Lipson et al. (2009) demonstrated the quantification of the yeast transcriptome using the SMS digital gene expression (smsDGE) technique. They believe that smsDGE is optimized for mRNA quantification rather than full transcriptome sequencing. The effectiveness of counting by smsDGE is driven by the fact that only a single read is generated from each cDNA molecule, thereby maintaining a faithful representation of transcript distribution (Lipson et al., 2009). Pushkarev et al. (2009) described the use of single-molecule methods to sequence an individual human genome. The results demonstrated that SMS enabled the analysis of human genomic information without the need for cloning, amplification or ligation. This milestone should allow the widespread application of genome sequencing to many aspects of genetics and human health, including personal genomics.

After that, SMS technology has been widely applied to the field of plant research. A 72× sequencing coverage of the *Oropetium* genome was completed using SMRT sequencing (Van Buren et al., 2015). The resulting sequence had a read N50 length of over 16 kb, and there was a 10× coverage of reads over 20 kb in length. The assembly contains 650 contigs spanning 99% (244 Mb) of the estimated 245 Mb genome size with a contig N50 length of 2.4 Mb. The final assembly consists of 625 contigs after removing the complete chloroplast genome, mitochondria-derived contigs and contaminants. The *Oropetium* genome demonstrates the utility of SMRT sequencing for assembling high-quality plant and other eukaryotic genomes and serves as a valuable resource for the plant comparative genomics community. Dong et al. (2015) investigated the transcripts in common wheat, developing caryopses using SMRT sequencing technology and assessed the resultant data, thus improving common wheat genome annotation and grain transcriptome research. As a result, a total of 91 881 high-quality full-length non-chimeric reads were identified and mapped to 16 188 chromosomal loci, corresponding to 13 162 known genes and 3 026 new genes not previously annotated. These data results facilitated the identification of 6 030 genes differentially regulated during caryopsis development and full-length transcripts for 72 transcribed gluten gene members that are important for the end-use quality control of common wheat. Booher et al. (2015) revealed the dynamic structure and complex transcription activator-like (*TAL*) effector gene relationships by SMRT sequencing in *Xanthomonas oryzae* genomes. The results from their study paved the way for population-level studies to inform resistance breeding, improve biotechnology and probe *TAL* effector evolution.

The SMS technology has been described as 'third-generation' or 'next–next generation' sequencing technology (Gupta, 2008) Also, it will be much faster and cheaper so that researchers in the near future should be able to pursue new scientific enquiries. The SMS technology will either be the dominant DNA sequencing method or co-exist with other technologies in the future (Gupta, 2008). The SMS or SMRT sequencing has been widely used in wheat (Dong et al., 2015) *X. oryzae* (Booher et al., 2015) and grass *Oropetium thomaeum* (VanBuren et al., 2015). The limitations of sequencing technology and the large number of repeat sequences affected the assembly of the soybean genome. Therefore, there are significant changes from the first to the second version of the soybean reference genome. The new technologies will provide a unique view of genome biology through direct sequencing of cellular and extracellular nucleic acids in an unbiased manner for the soybean genome and provide both quantitative and sequence information. Therefore, we hope that SMS or SMRT sequencing will also be able to reflect the true nature of the soybean genome in the future.

7.2 Comparative genome mapping

Comparative genome mapping is an effective way to rapidly transfer the research results of the plant to soybeans. In a narrow sense, the comparative genome is a comparison between cultivated and wild species, while, in the broad sense, it is a comparison between *soja* subgenus and other species. The genome development of plants and even crops is not balanced; however, the comparison of different genome species was completed by comparative genomic methods which help bridge the gap between basic and applied research. This approach will play an important role in understanding the mechanisms of genes and genomes, the relationship between gene structure and function and how DNA changes lead to biodiversity, etc. (Li et al., 2000; Qiu et al., 2011). On the other hand, comparative genome sequencing has also revealed that plant genome evolution has involved numerous rearrangements, including translocations and Indels (Gaut, 2002; Hall et al., 2002). Moreover, comparative genome mapping allowed the genome organization of cryptic polyploids to be fully elucidated. In summary, comparative genomics has provided a method of new gene discovery, helping to identify more allelic variations in soybean and thus play a vital role in the research and development of the soybean genome.

7.3 Construction of fine pan-genome of cultivated and wild soybean mapping

From the construction of seven wild soybean pan-genomes, there are significant variations between the genomes of different wild soybean and the genetic variation among individuals. This indicated that the reference genome of Williams82 could not fully reflect the characteristics of the soybean genome. Therefore, improvement in the assembly and expansion of pan-genome for wild and cultivated soybean should be given far more attention, for which more SNP, INDEL, structural variants and other information could be located on the soybean genome. The extensive use of soybean mini core collection (MCC) showed that this population is highly representative of soybean genetic diversity (Wang et al., 2006; Qiu et al., 2009; Tian et al., 2010), and selecting the appropriate sample from an extension of MCC will undoubtedly have a multiple effect on the construction of the soybean pan-genome.

The completion of the soybean pan-genome will greatly accelerate the speed of gene discovery and cloning. However, soybean is a strictly self-pollinated crop with a large linkage disequilibrium and an ancient tetraploid with 75% duplication of genes, which is difficult for map-based gene cloning. Therefore, it is important not only to place emphasis on mapping soybean pan-genome at the DNA level, but also on metabolic levels of RNA and protein. The construction of integrated mapping of the soybean genome by omics will greatly promote high-throughput mapping genes, quickly discovering favourite traits and assisting genomic selection from research (Ma et al., 2016) to practice breeding new soybean cultivars.

8 Where to look for further information

The following books and websites provide related current knowledge or information on genetics and breeding of soybean:

Books:

Recent Advances in Molecular Marker Systems for Soybean Breeding editor by Wenbin Li and David A. Lightfoot (Scientific Press, 2012, Beijing, China).

Genetics and Genomics of soybean editor by Gary Stacey (Springer Science+Business Media, 223 Spring Street LLC, New York, NY 10013, USA).

Single-Molecule Electronics: An Introduction to Synthesis, Measurement and Theory by Manabu Kiguchi (Springer Science+Business Media, 2016 Spring Street LLC, New York, NY 10013, USA).

Websites:

The *National Center for Biotechnology Information* (NCBI, www.ncbi.nlm.nih.gov/pubmed) enables you to search for literature on related genetic modification of soybean and genome-editing technology.

The *Phytozome* (https://phytozome.jgi.doe.gov/pz/portal.html) website and database provides you with access to gene families, individual genes, diversity and expression data of soybean.

China Soybean Data Platform (www.soybean-resources.cn). This website integrates Chinese soybean genetic resources information, phenotype data, molecular markers (SSR, SNP), resequencing and pan genome data for soybean breeders and researchers.

The *SoyBase and the Soybean Breeder's Toolbox* (SoyBase, http://www.soybase.org/), website integrates genetics and molecular biology for soybean researchers.

PLANT GENOME DUPLICATION DATABASE (PGDD, http://translate.google.cn.sixxs.org/) is a public database to identify and catalogue plant genes in terms of intra-genome or cross-genome syntonic relationships. Current efforts focus on flowering plants with available whole-genome sequences (preferably assembled pseudomolecules with ordered gene models).

USDA GRIN database (http://www.ars-grin.gov/). United States Department of Agriculture (USDA) Genetic Resources Information Network (GRIN) Database, USA.

Chinese CGRIS database (http://www.cgris.net/#), Chinese Crop Germplasm Information System, Institute of Crop Sciences, Chinese Academy of Agricultural Sciences, Beijing, China.

9 References

Ahn, S. and Tanksley, S. D. (1993) Comparative linkage maps of the rice and maize genomes. *Proc. Natl. Acad. Sci. USA*, 90: 7980–84.

Akkaya, M. S., Bhagwat, A. A. and Cregan P. B. (1992) Length polymorphisms of simple sequence repeat DNA in soybean. *Genetics*, 132: 1131–9.

Apuya, N. R. (1988) Restriction fragment length polymorphism as genetic marker in soybean. *Theor. Appl. Genet.*, 75: 889–901.

Bernard, R. and Weiss, M. U. (1973) Qualitative genetics, pp. 112–54. In: B. E. Caldwell (Ed.), *Soybeans: Improvement, Production, and Uses*. ASA, Madison, WI.

Bonierbalem M. W., Plaisted R. L. and Tanksley S. D. (1988) RFLP maps based on a common set of clones reveal modes of chromosomal evolution in potato and tomato. *Genetics*, 120: 1095–103.

Booher, N. J., Carpenter, S. C., Sebra, R. P., Wang, L, Salzberg, S. L., Leach, J. E. and Bogdanove A. J. (2015) Single molecule real-time sequencing of Xanthomonas oryzae genomes reveals a dynamic structure and complex TAL (transcription activator-like) effector gene relationships. *Microb. Genom.*, 1(4). doi:10.1099/mgen.0.000032.

Botstein, D., White, R. L., Skolnick, M. and Davis R. W. (1980) Construction of a genetic linkage map in man using restriction fragment length polymorphisms. *Am. J. Hum. Genet.*, 32(3): 314–31.

Cardinal, A. J., Whetten, R., Wang, S., Auclair, J., Hyten, D., Cregan, P., Bachlava, E., Gillman, J., Ramirez, M., Dewey, R., Upchurch, G., Miranda, L. and Burton J. W. (2014) Mapping the low palmitate fap1 mutation and validation of its effects in soybean oil and agronomic traits in three soybean populations. *Theor. Appl. Genet.*, 127(1): 97–111.

Chang, S., Thurber, C. S., Brown, P. J., Hartman, G. L., Lambert, K. N. and Domier, L. L. (2014) Comparative mapping of the wild perennial Glycine latifolia and soybean (G. max) reveals extensive chromosome rearrangements in the genus Glycine. *PLoS One*, 9(6): e99427.

Concibido, V. C., Diers, B. W. and Arelli, P. R. (2004) A decade of QTL mapping for cyst nematode resistance in soybean. *Crop Sci.*, 44(4): 1121–31.

Cregan, P. B., Jarvik, T., Bush, A. L., Shoemaker, R. C., Lark, K. G., Kahler, A. L., Kaya, N., Vantoai, T. T., Lohnes, D. G., Chung, J. and Specht, J. E. (1999) An integrated genetic linkage map of the soybean genome. *Crop Sci.*, 39(5): 1464–90.

de Lumen B. O. (2005) Lunasin: A cancer-preventive soy peptide. *Nutr. Rev.*, 63(1): 16–21.

Dong, L. L., Liu, H. F., Zhang, J. C., Yang, S. J., Kong, G. Y., Chu, J. S.C, Chen, N. S. and Wang, D. W. (2015) Single- molecule real-time transcript sequencing facilitates common wheat genome annotation and grain transcriptome research. *BMC Genomics*, 16: 1039.

Dörrey, K., Brakmann, S., Brinkmeier, M., Han, K. T., Riebeseel, K., Schwille, P., Stephan, J., Wetzel, T., Lapczyna, M., Stuke, M., Bader, R., Hinz, M., Seliger, H., Holm, J., Eigen, M. and Rigler R. (1997) Techniques for single molecule sequencing. *Bioimaging*, 5: 139–52.

Eid, J., Fehr, A., Gray, J., Luong, K., Lyle, J., Otto, G., Peluso, P., Rank, D., Baybayan, P., Bettman, B., Bibillo, A., Bjornson, K., Chaudhuri, B., Christians, F., Cicero, R., Clark, S., Dalal, R., Dewinter, A., Dixon, J., Foquet, M., Gaertner, A., Hardenbol, P., Heiner, C., Hester, K., Holden, D., Kearns, G., Kong, X., Kuse, R., Lacroix, Y., Fan, J. B., Chee, M. S. and Gunderson K. L. (2006) Highly parallel genomic assays. *Nat. Rev. Genet.*, 7: 632–44.

Fan, J.B., Chee, M.S., and Gunderson, K. L. (2006). Highly parallel genomic assays. *Nat. Rev. Genet.* 7:632–44.

Gaut, B. S. (2002) Evolutionary dynamics of grass genomes. *New Phytol.*, 154: 15–28.

Gong, P. T., Mu, J. G., Zhao, J. R., Wang, X. L., Bai, Y. N. and Fang X. J. (2006) An integrated soybean genetic linkage map comprising 315 SSRs and 40 AFLPs. *Mol. Plant Breed.*, 4(3): 309–16.

Gore, M. A., Hayes, A. J., Jeong, S. C., Yue, Y. G., Buss, G. R. and Maroof, S. (2002) Mapping tightly linked genes controlling potyvirus infection at the Rsv1 and Rpv1 region in soybean. *Genome*, 45(3): 592–9.

Greenleaf, W. J. and Block, S. M. (2006) Single-molecule, motion-based DNA sequencing using RNA poly-merase. *Science*, 313(5788): 801.

Guan, R. X., Qu, Y., Guo, Y., Yu, L. L., Liu, Y, Jiang, J. H., Chen, J. G., Ren, Y. L., Liu, G. Y., Tian, L, Jin, L. G., Liu, Z. X., Hong, H. L., Chang, R. Z., Gilliham, M. and Qiu, L. J. (2014) Salinity tolerance in soybean is modulated by natural variation in GmSALT3. *Plant J.*, 80: 937–50.

Gupta, P. K. (2008) Single-molecule DNA sequencing technologies for future genomics research. *Trends Biotechnol.*, 26(11): 602–11.

Gutierrez-Gonzalez, J. J., Vuong, T. D., Zhong, R., Yu, O., Lee, J. D., Shannon, G., Ellersieck, M., Nguyen, H. T. and Sleper D. A. (2011) Major locus and other novel additive and epistatic loci involved in modulation of isoflavone concentration in soybean seeds. *Theor. Appl. Genet.*, 123(8): 1375–85.

Hall, A. E., Fiebig, A. and Preuss, D. (2002) Beyond the Arabidopsis genome: opportunities for comparative genomics. *Plant Physiol.*, 129: 1439–47.

Harris, T. D., Buzby, P. R., Babcock, H., Beer, E., Bowers, J., Braslavsky, I., Causey, M., Colonell, J., Dimeo, J., Efcavitch, J. W., Giladi, E., Gill, J., Healy, J., Jarosz, M., Lapen, D., Moulton, K., Quake, S. R., Steinmann, K., Thayer, E., Tyurina, A., Ward, R., Weiss, H. and Xie, Z. (2008) Single-molecule DNA sequencing of a viral genome. *Science*, 320(5872), 106–9.

Hart, C., Lipson, D., Ozsolak, F., Raz, T., Steinmann, K., Thompson, J. and Milos, P. M. (2010) Single molecule sequencing: sequence methods to enable accurate quantitation. *Meth. Enzymol.*, 472: 407–30.

Huang, X. H., Feng, Q., Qian, Q., Zhao, Q., Wang, L., Wang, A. H., Guan, J. P., Fan, D. L., Weng, Q. J., Huang, T. J., Dong, G. J., Sang, T. and Han, B. (2009) High-throughput genotyping by whole-genome resequencing. *Genome Res.*, 19(6): 1068–76.

Hwang, T. Y., Moon, J. K., Yu, S., Yang, K., Mohankumar, S., Yu, Y. H., Lee, Y. H., Kim, H. S., Kim, H. M., Maroof, M. A. and Jeong S. C. (2006) Application of comparative genomics in developing molecular markers tightly linked to the virus resistance gene Rsv4 in soybean. *Genome*, 49(4): 380–8.

Hymowitz, T. and Singh, R. J. (1987) Taxomony and speciation, pp. 23–48. In: R. J. Wilcox (Ed.) *Soybeans: Improvement, Production, and Uses*. 2nd ed. Agron. Monogr. 16. ASA, CSSA, SSSA, Madison, WI.

Hyten, D. L., Choi, I. Y., Song, Q. J., Specht, J. E., Carter Jr., T. E., Shoemaker, R. C., Hwang, E. Y., Matukumalli, L. K. and Cregan, P. B. (2010) A high density integrated genetic linkage map of soybean and the development of a 1536 universal soy linkage panel for quantitative trait locus mapping. *Crop Sci.*, 50(3): 960–8.

Hyten, D. L., Song, Q. J., Zhu, Y. L., Choi, I. Y., Nelson, R. L., Costa, J. M., Specht, J. E. and Shoemaker R. C., Cregan P. B. (2006) Impacts of genetic bottlenecks on soybean genome diversity. *Proc. Natl. Acad. Sci.*, 103: 16666–71.

Hyten, D. L., Cannon, S. B., Song, Q., Weeks, N., Fickus, E. W., Shoemaker, R. C., Specht, J. E., Farmer, A. D., May, G. D. and Cregan, P. B. (2010). High-throughput SNP discovery through deep resequencing of a reduced representation library to anchor and orient scaffolds in the soybean whole genome sequence. *BMC Genomics*, 11: 38.

Innes, R. W., Ameline-Torregrosa, C., Ashfield, T., Cannon, E., Cannon, S. B., Chacko, B., Chen, N. W., Couloux, A., Dalwani, A., Denny, R., Deshpande, S., Egan, A. N., Glover, N., Hans, C. S., Howell, S., Ilut, D., Jackson, S., Lai, H., Mammadov, J., Del Campo, S. M., Metcalf, M., Nguyen, A., O'Bleness, M., Pfeil, B. E., Podicheti, R., Ratnaparkhe, M. B., Samain, S., Sanders, I., Segurens, B., Sevignac, M., Sherman-Broyles, S., Thareau, V., Tucker, D. M., Walling, J., Wawrzynski, A., Yi, J., Doyle, J. J., Geffroy, V., Roe, B. A., Maroof, M. A. and Young, N. D. (2008) Differential accumulation of retroelements and diversification of NB-LRR disease resistance genes in duplicated regions following polyploidy in the ancestor of soybean. *Plant Physiol.* 148: 1740–59.

Jung, C. H., Wong, C. E., Singh, M. B. and Bhalla, P. L. 2012. Comparative genomic analysis of soybean flowering genes. *Plos One*, 7(6): e38250

Kato, S., Sayama, T., Fujii, K., Yumoto, S., Kono, Y., Hwang, T. Y., Kikuchi, A., Takada, Y., Tanaka, Y., Shiraiwa, T. and Ishimoto, M. (2014) A major and stable QTL associated with seed weight in soybean across multiple environments and genetic backgrounds. *Theor. Appl. Genet.*, 127(6): 1365–74.

Keim, P., Beavis, W., Schupp, J. and Freestone, R. (1992) Evaluation of soybean RFLP marker diversity in adapted germplasm. *Theor. Appl. Genet.*, 85: 205–12.

Keim, P., Diers, B. W., Olson, T. C. and Shoemaker R. C. (1990) RFLP mapping in soybean: association between marker loci and variation in quantitative traits. *Genetics*, 126: 735–42.

Keim, P, Shoemaker, R. C. and Palmer, R. G. (1989) Restriction fraglength polymorphism diversity in soybean. *Theor. Appl. Genet.*, 77: 786–92.

Kim, M. Y., Lee, S., Van, K., Kim, T. H., Jeong, S. C., Choi, I. Y., Kim, D. S., Lee, Y. S., Park, D., Ma, J., Kim, W. Y., Kim, B. C., Park, S., Lee, K. A., Kim, D. H., Kim, K. H., Shin, J. H., Jang, Y. E., Kim, K. D., Liu, W. X., Chaisan, T., Kang, Y. J., Lee, Y. H., Kim, K. H., Moon, J. K., Schmutz, J., Jackson, S. A., Bhak, J. and Lee S. H. (2010) Whole-genome sequencing and intensive analysis of the undomesticated soybean (Glycine soja Sieb. and Zucc.) genome. *Proc. Natl. Acad. Sci. USA*, 107, 22032–7

Koren, S., Schatz, M. C., Walenz, B. P., Martin, J., Howard, J. T., Ganapathy, G., Wang, Z., Rasko, D. A., McCombie, W. R., Jarvis, E. D. and Adam M. P. (2012) Hybrid error correction and de novo assembly of single molecule sequencing reads. *Nat. Biotechnol.*, 30(7): 693–700.

Korlach, J. and Turner, S. W. (2010) Single-molecule sequencing. *Springer Berlin Heidelberg*, 472(10): 2344–7.

Lam, H. M., Xu, X., Liu, X., Chen, W. B., Yang, G. H., Wong, F. L., Li, M. W., He, W. M., Qin, N., Wang, B., Li, J., Jian, M., Wang, J., Shao, G. H., Wang, J., Sun, S. S. M. and Zhang, G. Y. (2010) Re-sequencing of 31 wild and cultivated soybean genomes identifies patterns of genetic diversity and selection. *Nat. Genet.*, 42, 1053–9.

Lark, K. G., Weisemann, J. M., Matthews, B. F., Palmer, R., Chase, K. and Macalma, T. (1993) A genetic map of soybean (Glycine max L.) using an intraspecific cross of two cultivars: 'Minsoy' and 'Noir 1'. *Theor. Appl. Genet.*, 86: 901–6.

Lee, S., Freewalt, K., McHale, L., Song, Q., Taehwan, J., Michel, A., Dorrance, A. E., Mian, R. M. (2015) A high-resolution genetic linkage map of soybean based on 357 recombinant inbred lines genotyped with BARCSoySNP6K. *Mol. Breed.*, 35: 58.

Li, B., Tian, L., Zhang, J. Y., Huang, L., Han, F. X., Yan, S. R., Wang, L. Z., Zheng, H. K. and Sun J. M. (2014) Construction of a high-density genetic map based on large-scale markers developed by specific length amplified fragment sequencing (SLAF-seq) and its application to QTL analysis for isoflavone content in *Glycine* max. *BMC Genomics*, 15: 1086.

Li, Y., Wang, T. Y. and Jia, J. Z. (2000) Research advances in plant comparative genomics. *Biotechnol. Info.*, 5: 11–14.

Li, Y. H., Li, W., Zhang, C., Yang, L., Chang, R. Z., Gaut, B. S. and Qiu L. J. (2010) Genetic diversity in domesticated soybean (Glycine max) and its wild progenitor (Glycine soja) for simple sequence repeat and single- nucleotide polymorphism loci. *New Phytol.*, 188: 242–53.

Li, Y. H., Zhao, S. C., Ma, J. X., Li, D., Yan, L., Li, J., Qi, X. T., Guo, X. S., Zhang, L., He, W. M., Chang, R. Z., Liang, Q. S., Guo, Y., Ye, C., Wang, X. B., Tao, Y., Guan, R. X., Wang, J. Y., Liu, Y. L., Jin, L. G., Zhang, X. Q., Liu, Z. X., Zhang, L. J., Chen, J., Wang, K. J., Nielsen, R., Li, R. Q., Chen, P. Y., Li, W. B., Reif, J. C., Purugganan, M., Wang, J., Zhang, M. C., Wang, J., Qiu, L. J. (2013) Molecular footprints of domestication and improvement in soybean revealed by whole genome re-sequencing. *BMC Genom.*, 14: 579. doi:10.1186/1471-2164-14-579.

Li, Y. H., Zhou, G., Ma, J., Jiang, W., Jin, L. G., Zhang, Z., Guo, Y., Zhang, J., Sui, Y., Zheng, L., Zhang, S. S., Zuo, Q., Shi, X. H., Li, Y. F., Zhang, W. K., Hu, Y., Kong, G., Hong, H. L., Tan, B., Song, J., Liu, Z. X., Wang, Y., Ruan, H., Yeung, C. K., Liu, J., Wang, H., Zhang, L. J., Guan, R. X., Wang, K. J., Li, W. B., Chen, S. Y., Chang, R. Z., Jiang, Z., Jackson, S. A., Li, R. and Qiu, L. J. (2014) De novo assembly of soybean wild relatives for pan-genome analysis of diversity and agronomic traits. *Nat. Biotechnol.*, 32, 1045–52.

Li, Z. F., Jiang, L. X., Ma, Y. S., Wei, Z. Y., Hong, H. L., Liu, Z. X., Lei, J., Liu, Y., Guan, R. X., Guo, Y., Jin, L. G., Zhang, L. J., Li, Y. H., Ren, Y. L., He, W., Liu, M., Phyu Sin Htwe, N. M., Liu, L., Guo, B. F., Song, J., Tan, B., Liu, G. F., Li, M. Q., Zhang, X. L., Liu, B., Shi, X. H., Han, S. N., Hua, S. N., Zhou, F. L., Yu, L. L., Li, Y. F., Wang, S., Wang, J., Chang, R. Z. and Qiu, L. J. (2017). Development and utilization of a new chemically-induced soybean library with a high mutation density. *J. Integ. Plant Biol.*, 59(1): 60–74.

Lipson, D., Raz, T., Kieu, A., Jones, D. R., Giladi, E., Thayer, E., Thompson, J. F., Letovsky, S., Milos, P. and Causey, M. (2009) Quantification of the yeast transcriptome by single-molecule sequencing. *Nat. Biotechnol.*, 27(7): 652–8.

Liu, B. H., Fujita, T, Yan, Z. H., Sakamoto, S., Xu, D. H. and Abe, J. (2007) QTL mapping of domestication-related traits in soybean (Glycine max). *Ann. Bot.*, 100: 1027–38.

Liu, F., Zhuang, B. C., Zhang, J. S. and Chen S. Y. (2000) Construction and analysis of soybean genetic map. *Acta Genetica Sinica*, 27(11): 1018–26.

Liu, G., Zhao, L., Averitt, B. J., Liu, Y., Zhang, B., Chang, R., Ma, Y. S., Luan, X. Y., Guan, R. X. and Qiu, L. (2015) Geographical distribution of GmTfl1 alleles in Chinese soybean varieties. *Crop J.*, 3(5): 371–8.

Ma, J., SanMiguel, P., Lai, J., Messing, J. and Bennetzen J. L. (2005) DNA rearrangement in orthologous Orp regions of the maize, rice and sorghum genomes. *Genetics*, 170: 1209–20.

Ma, Y. S., Reif, J., Jiang, Y., Wen, Z. X., Wang, D. C., Liu, Z. X., Guo, Y., Wei, S. H., Wang, S. M., Yang, C. M., Wang, H. C., Yang, C. Y., Lu, W. G., Xu, R., Zhou, R., Wang, R. Z., Sun, Z. D., Chen, H. Z., Zhang, W. H., Wu, J., Hu, G. H., Liu, C. Y., Luan, X. Y., Fu, Y. S., Guo, T., Han, T. F., Zhang, M. C., Sun, B. C., Zhang, L., Chen, W. Y., Wu, C. X., Sun, S, Yuan, B. J., Xinan Zhou, X. A., Han, D. Z., Yan, H. R., Li, W. B. and Qiu L. J. (2016), Potential of marker selection to increase prediction accuracy of genomic selection in soybean (Glycine max. L.). *Mol. Breed.* 36(113): 1–10.

Mateos-Aparicio, I., Redondo Cuenca, A., Villanueva-Suarez, M. J. and Zapata-Revilla, M. A. (2008) Soybean, a promising health source. *Nutr. Hosp.*, 23(4): 305–12.

Nadeau, J. H., Sanko, V. D. (1998) Counting on comparative maps. *Trends Genet.*, 14: 495–501.

Nakayama, Y. and Yamaguchi, H. (2002) Natural hybridization in wild soybean (Glycine max ssp.soja) by pollen flow from cultivated soybean (Glycine max ssp.max) in a designed population. *Weed Biol. Manage.*, 2: 25–30.

Palmer, R. G. and Kilen, T. C. (1987) Qualitative Genetics and Cyto-genetics. In *Soybean: Improvement, Production, and Use*. American Inc., Madison, pp. 135–209.

Palmer, R. G. (1985) *World Soybean Research Conference* Ⅲ, pp. 337–44.

Panthee, D. R., Pantalone, V. R., Sams, C. E., Saxton, A. M., West, D. R., Orf, J. H. and Killam, A. S. (2006) Quantitative trait loci controlling sulfur containing amino acids, methionine and cysteine, in soybean seeds. *Theor. Appl. Genet.*, 112(3): 546–53.

Paterson, A. H., Bowers, J. E., Burow, M. D., Draye, X., Elsik, C. G., Jiang, C. X., Katsar, C. S., Lan, T. H., Lin, Y. R., Ming, R. and Wright, R. J. (2000) Comparative genomics of plant chromosomes. *Plant Cell*, 12: 1523–39.

Patil, G., Valliyodan, B., Deshmukh, R., Prince, S., Nicander, B., Zhao, M. Z., Sonah, H., Song, L., Lin, L., Chaudhary, J., Liu, Y., Joshi, T., Xu, D. and Nguyen, H. T. (2015) Soybean (Glycine max) SWEET gene family: Insights through comparative genomics, transcriptome profiling and whole-genome re-sequence analysis. *BMC Genomics*, 16: 520.

Pushkarev, D., Neff, N. F. and Quake S. R. (2009) Single-molecule sequencing of an individual human genome. *Nat Biotechnol.*, 27(9): 847–50.

Qiu, L. J., Guo, Y., Li, Y., Wang, X. B., Zhou, G. A., Liu, Z. X., Zhou, S. R., Li, X. H., Ma, Y. Z., Wang J. K. and Wan J. M. (2011) Novel gene discovery of crops in China: status, challenging, and perspective. *Acta Agronomica Sinica*, 37(1): 1–17.

Qiu, L. J., Li, Y. H., Guan, R. X., Liu, Z. X., Wang, L. X. and Chang R. Z. (2009) Establishment, representative testing and research progress of soybean core collection and mini core collection. *Acta Agronomica Sinica*, 35(4): 571–9.

Qiu, L. J., Xing, L. L., Guo, Y., Wang, J., Jackson, S. A. and Chang R. Z. (2013) A platform for soybean molecular breeding: the utilization of core collections for food security. *Plant Mol. Biol.*, 83(1–2): 41–50.

Qi, X., Li, M. W., Xie, M., Liu, X., Ni, M., Shao, G., Song, C., Yim, A. K., Tao, Y., Wong, F. L., Isobe, S., Wong, C. F., Wong, K. S., Xu, C, Li, C., Wang, Y., Guan, R., Sun, F., Fan, G., Xiao, Z., Zhou, F., Phang, T. H., Liu, X., Tong, S. W., Chan, T. F., Yiu, S. M., Tabata, S., Wang, J., Xu, X. and Lam, H. M. (2014). Identification of a novel salt tolerance gene in wild soybean by whole-genome sequencing. *Nat. Commun.*, 5: 4340.

Qi, Z. M., Huang, L., Zhu, R. S., Xin, D. W., Liu, C. Y., Han, X., Jiang, H. W., Hong, W. G., Hu, G. H., Zheng, H. K. and Chen, Q. S. (2014) A high-density genetic map for soybean based on specific length amplified fragment sequencing. *PloS One*, 9(8): e104871.

Schmutz, J., Cannon, S. B., Schlueter, J., Ma, J., Mitros, T., Nelson, W. and et al. (2010) Genome sequence of the palaeopolyploid soybean. *Nature*, 463(7278): 178–83.

Shoemaker, R. C. and Specht J. E. (1995) Integration of the soybean molecular and classical genetic linkage groups. *Crop Sci.*, 35: 436–46.

Song, J., Liu, Z. X., Hong, H. L., Ma, Y. S., Tian, L, Li, X. X., Li, Y. Y., Guan, R. X., Guo, Y. and Qiu, L. J. (2016) Identification and validation of loci governing seed coat color by combining association mapping and bulk segregation analysis in soybean. *PLoS ONE*, 11(7): e0159064.

Song, Q. J., Marek, L. F., Shoemaker, R. C., Lark, K. G., Concibido, V. C., Delannay, X., Specht, J. E. and Cregan, P. B. (2004) A new integrated genetic linkage map of the soybean. *Theor. Appl. Genet.*, 109(1): 122–8.

Song, Q. J., Jenkins, J., Jia, G. F., Hyten, D. L., Pantalone, V., Jackson, S. A., Schmutz, J. and Cregan, P. B. (2016) Construction of high resolution genetic linkage maps to improve the soybean genome sequence assembly Glyma1.01. *BMC Genomics*, 17: 33.

Sun, X. W., Liu, D. Y., Zhang, X. F., Li, W. B., Liu, H., Hong, W. G., Jiang, C. B., Guan, N., Ma, C. X., Zeng, H. P., Xu, C. H., Song, J., Huang, L., Wang, C. M., Shi, J. J., Wang, R., Zheng, X. H., Lu, C. Y., Wang, X. W. and Zheng H. K. (2013) SLAF-seq: An efficient method of large-scale De novo SNP discovery and genotyping using high-throughput sequencing. *PLoS One*, 8(3): e58700.

Swidan, F., Rocha, E. P. C., Shmoish, M. and Pinter R. Y. (2006) An integrative method for accurate comparative genome mapping. *PLoS Comput Biol.*, 2(8): e75.
Tanksley, S. D., Bernatzky, R., Lapitan, N. L. and Prince, J. P. (1988) Conservation of gene repertoire but not gene order in pepper and tomato. *Proc. Natl. Acad. Sci. USA*, 85: 6419–23.
Tettelin, H., Masignani, V., Cieslewicz, M. J., Donati, C., Medini, D., Ward, N. L., Angiuoli, S. V., Crabtree, J., Jones, A. L., Durkin, A. S., Deboy, R. T., Davidsen, T. M., Mora, M., Scarselli, M., Margarit y Ros, I., Peterson, J. D., Hauser, C. R., Sundaram, J. P., Nelson, W. C., Madupu, R., Brinkac, L. M., Dodson, R. J., Rosovitz, M. J., Sullivan, S. A., Daugherty, S. C., Haft, D. H., Selengut, J., Gwinn, M. L., Zhou, L., Zafar, N., Khouri, H., Radune, D., Dimitrov, G., Watkins, K., O'Connor, K. J., Smith, S., Utterback, T. R., White, O., Rubens, C. E., Grandi, G., Madoff, L. C., Kasper, D. L., Telford, J. L., Wessels, M. R., Rappuoli, R. and Fraser, C. M. (2005) Genome analysis of multiple pathogenic isolates of Streptococcus agalactiae: Implications for the microbial 'pan-genome'. *Proc. Natl. Acad. Sci. USA* 102(39): 13950–5.
Tian, Z. X., Wang, X. B., Lee, R., Li, Y. H., Spechtd, J. E., Nelson, R. L., McClean, P. E., Qiu, L. J. and Ma, J. X. 2010 Artificial selection for determinate growth habit in soybean. *Proc. Natl. Acad. Sci. USA*, 107(19): 8563–8.
Tuyen, D. D., Lal, S. K. and Xu, D. H. (2010) Identification of a major QTL allele from wild soybean (Glycine soja Sieb. & Zucc.) for increasing alkaline salt tolerance in soybean. *Theor. Appl. Genet.*, 121(2): 229–36.
VanBuren, R., Bryant, D., Edger, P. P., Tang, H., Burgess, D., Challabathula, D., Spittle, K., Hall, R., Gu, J., Lyons, E., Freeling, M., Bartels, D., Ten Hallers, B., Hastie, A., Michael, T. P. and Mockler T. C. (2015) Single molecule sequencing of the desiccation tolerant grass Oropetium thomaeum. *Nature*, 527(7579): 508–11.
VanBuren, R., Bryant, D., Edger, P. P., Tang, H. B., Burgess, D., Challabathula, D., Spittle, K., Hall, R., Gu, J., Lyons, E., Freeling, M., Bartels, D., Hallers, B. T., Hastie, A., Michael, T. P. and Mockler, T. C. (2015) Single-molecule sequencing of the desiccation tolerant grass Oropetium thomaeum. *Nature*, 527(7579): 508–11.
Varshney, R. K., Nayak, S. N., May, G. D. and Jackson S. A. (2009) Next-generation sequencing technologies and their implications for crop genetics and breeding. *Trends Biotech.*, 27(9): 522–30.
Wan, Y. S., Wang, Z, Xiao, Y. H., Lv, B. and Fang X. J. (2005) A soybean genetic linkage map comprising of 227 SSR loci in a soybean RIL population. *Mol. Plant Breed.*, 3(1): 15–20.
Wang, L. X., Guan, Y., Guan, R. X., Li, Y. H., Ma, Y. S., Dong, Z. M., Liu, X, Zhang, H. Y., Zhang, Y. Q., Liu, Z. X., Chang, R. Z., Xu, H. M., Li, L. H., Lin, F. Y., Luan, W. J., Ya, N. Z., Ning, X. C., Zhu, L, Cui, Y. H., Piao, R. H., Liu, Y, Chen, P. Y. and Qiu, L. J. (2006) Establishment of Chinese soybean (Glycine max) core collections with agronomic traits and SSR markers. *Euphytica*, 151(2): 215–23.
Wang, X. B., Zhang, H. W., Sun, G. L., Jin, Y. and Qiu L. J. (2014) Identification of active VQ motif-containing genes and the expression patterns under low nitrogen treatment in soybean. *Gene*, 543: 237–43.
Wen, Z. X., Boyse, J. F., Song, Q. J., Cregan, P. B. and Wang D. C. (2015) Genomic consequences of selection and genome-wide association mapping in soybean. *BMC Genomics*, 16: 671.
Whitkus, R., Doebley, J. and Lee, M. (1992) Comparative genome mapping of sorghum and maize. *Genetics*, 132: 1119–30.
Wu, X. L., He, C. Y., Wang, Y. J., Zhang, Z. Y., DongFang, Y., Zhang, J. S., Chen, S. Y. and Gai, J. Y. (2001) Construction and analysis of a genetic linkage map of soybean. *Acta Genetica Sinica*, 28(11): 1051–61.
Xia, Z. J., Tsubokura, Y., Hoshi, M., Hanawa, M., Yano, C., Okamura, K., Ahmed, T. A., Anai, T., Watanabe, S., Hayashi, M., Kawai, T., Hossain, K. G., Masaki, H., Asai, K., Yamanaka, N., Kubo, N., Kadowaki, K., Nagamura, Y., Yano, M., Sasaki, T. and Harada, K. (2007) An integrated high-density linkage map of soybean with RFLP, SSR, STS, and AFLP markers using a single F_2 population. *DNA Res.*, 14(6): 257–69.
Xu, J. J., Zhao, Q, Du, P. N., Xu, C. W., Wang, B. H., Feng, Q., Liu, Q. Q., Tang, S. Z., Gu, M. H., Han, B. and Liang, G. H. (2010) Developing high throughput genotyped chromosome segment

substitution lines based on population whole-genome re-sequencing in rice (Oryza sativa L.). *BMC Genomics*, 11: 656.

Yamanaka, N., Nagamura, Y. and Tsubokura, Y. (2000) Quantitative trait locus analysis of flowering time in soybean using a RFLP linkage map. *Breed. Sci.*, 50: 109–15.

Yamanaka, N., Ninomiya, S. and Hoshi, M. (2001) An informative linkage map of soybean reveals QTLs for flowering time, leaf let morphology and regions of segregation distortion. *DNA Res.*, 8: 61–72.27.

Yan, H. H., Mudge, J., Kim, D. J., Shoemaker, R. C., Cook, D. R. and Young, N. D. (2004) Comparative physical mapping reveals features of microsynteny between Glycine max, Medicago truncatula, and Arabidopsis thaliana. *Genome*, 47 (1): 141–55.

Zhang, D. S., Dong, W., Hui, D. W., Chen, S. Y. and Zhuang B. C. (1997) Construction of molecular marker genetic linkage map using the F2 population of the hybrid between cultivated and semi-wild in soybean. *Chinese Sci. Bull.*, 42(12): 1326–30.

Zhang, L., Jin L. G., Luo. L., Wang. Y. P., Dong. Z. M., Sun. S. H., Qiu. L. J. (2011) Analysis of nuclear gene codon bias on soybean genome and transcriptome. *Acta Agronomica Sinica*, 37(6): 965–74

Zhou, B., Xing, H., Chen, S. Y. and Gai J. Y. (2009) An integrated genetic linkage map of soybean and its application. *Soybean Sci.*, 28(4): 557–65.

Zhou, F. L., Guo, Y. and Qiu, L. J. (2016) Genome-wide identification and evolutionary analysis of leucine-rich repeat receptor-like protein kinase genes in soybean. *BMC Plant Biol.*, 16(58): 1–13(3.631).

Zhou, Z. K., Jiang, Y., Wang, Z., Gou, Z. H., Lyu, J., Li, W. Y., Yu, Y. J., Shu, L. P., Zhao, Y. J., Ma, Y. M., Fang, C., Shen, Y. T., Liu, T. F., Li, C. C., Li, Q, Wu, M., Wang, M., Wu, Y. S., Dong, Y., Wan, W. T., Wang, X., Ding, Z. L. and Gao, X. D., Xiang, H., Zhu, B. G., Lee, S. H., Wang, W. and Tian Z. X. (2015) Resequencing 302 wild and cultivated accessions identifies gene related to domestication and improvement in soybean. *Nat Biotechnol.*, 33(4): 408–14.

Zhu, Y. L., Song, Q. J., Hyten, D. L., Van Tassell, C. P., Matukumalli, L. K., Grimm, D. R., Hyatt, S. M., Fickus, E. W., Young, N. D. and Cregan P. B. (2003) Single-nucleotide polymorphisms in soybean. *Genetics*, 163: 1123–34.

Chapter 5

Advances in marker-assisted breeding of soybean

T. Vuong, University of Missouri, USA; and D. Walker, USDA-ARS and University of Illinois, USA

1 Introduction

For centuries, plant breeding has had a significant impact on the improvement of crop productivity and quality worldwide. With extensive knowledge and practical experience in plant genetics, plant breeders have developed effective breeding methodologies and have made tremendous progress in phenotypic selection of superior genotypes. However, breeders using these methods also often encounter difficulties related to phenotyping procedures, which can be time-consuming, expensive, unreliable for traits with a low heritability and subject to genotype × environment (G × E) interactions (Francia et al., 2005). In recent decades, advances in molecular genetics and genomics have revolutionized breeding approaches for a number of crops, and DNA marker technologies offer great promise for increasing the precision and rate of progress in plant breeding through marker-assisted selection (MAS). The recent development of next-generation sequencing (NGS) technologies, coupled with reduced sequencing costs, facilitates the simultaneous discovery of large numbers of single nucleotide polymorphism (SNP) markers, even in crop species that are not major commodities. Many SNPs are located near or within genes or gene structural variants, revealing allelic variations in genes that affect traits of interest (Collard and Mackill, 2008). Such NGS-based DNA marker technologies, in combination with robust high-throughput genotyping assays, offer a wide

http://dx.doi.org/10.19103/AS.2017.0034.05

range of applications in modern molecular breeding. For example, analyses of genetic diversity can be conducted using whole-genome sequencing (WGS) data. Quantitative trait loci (QTL) can be identified using a genome-wide association study (GWAS) as an alternative mapping approach. Superior genotypes can be efficiently selected using accurate genomic-enabled prediction (GP) models for conducting genome-wide selection (GS) (Bhat et al., 2016; Crossa et al., 2017; Duhnen et al., 2017; Meuwissen et al., 2001).

Breeders have made considerable improvements in a number of traits that have allowed soybean to become and remain one of the most important agricultural commodities internationally. Until the mid-1990s, this progress was based almost entirely on phenotypic selection and evaluations that had various disadvantages. For example, the phenotypic evaluations of seed quality traits could only be conducted after harvest of large volumes of seeds to select a few lines with desired seed composition traits. One of the biggest advantages of MAS over phenotypic selection is that it can be implemented at the seedling, or even seed, stage to select plants with genes that affect traits which could not be selected phenotypically until at or after full maturity.

Although many studies had already used molecular markers to detect and map genes or QTL associated with phenotypic variations, and the tremendous potential for MAS to supplement phenotypic selection was recognized, the cost of conducting MAS with the markers available at that time was prohibitive. Moreover, genotyping assays were time-consuming, and probe-based assays to detect DNA polymorphisms, such as restriction fragment length polymorphisms (RFLPs), required relatively large amounts of purified DNA, which further increased the cost and time required per data point. The invention of reliable polymerase-chain-reaction (PCR)-based DNA markers, such as simple sequence repeats (SSRs), SNPs, represented a major technological breakthrough that made large-scale MAS an economically viable alternative to phenotype-based selection for many complex traits. In recent years, with a large number of DNA markers developed through genotyping-by-sequencing (GBS) (Elshire et al., 2011), GP and GS can be efficiently conducted in soybean to select for quantitative traits, like seed yield, using markers that are not necessarily linked to known QTL, but are statistically associated with improvements in the target trait within a defined breeding population (Jarquín et al., 2014; Sebastian et al., 2010).

The objectives of this chapter are to summarize different disciplines of modern molecular soybean breeding, including the development of DNA marker technologies, marker detection assays and genotyping platforms along with practical achievements using MAS strategy to select superior soybean genotypes with high yields, excellent seed quality and resistance or tolerance to biotic and abiotic stresses. In addition, future perspectives for applications of NGS-based breeding for the improvement of soybean will also be discussed.

2 Molecular marker types

The discovery of genetic variation in genomic DNA sequences in various crop plants led to the development of several types of molecular markers that have been used for MAS and mapping loci associated with variation in important traits. These marker classes include RFLPs (Botstein et al., 1980), random amplified polymorphic DNAs (RAPDs) (Welsh and McClelland, 1990; Williams et al., 1990), sequence characterized amplified regions (SCARs), DNA amplification fingerprinting (DAF) markers (Caetano-Anolles et al., 1992), sequence tagged sites (STSs), amplified fragment length polymorphisms (AFLPs) (Vos et al., 1995), cleaved amplified polymorphic sequences (CAPS) and SSRs or microsatellites (Litt and Luty, 1989). Recently, advances in genome sequencing technology have led to the development of large numbers of SNP markers, which detect differences in single nucleotides or the presences of DNA deletions/insertions (Collins et al., 1998; Lindblad-Toh et al., 2000).

Although the marker types mentioned above have been useful for mapping genes and QTL associated with important traits, several of them have some disadvantages that limit their practical applications for MAS, such as the hybridization-based markers (i.e. RFLPs) and some classes of PCR-based markers (e.g. RAPDs, DAFs, SCARs, STSs, CAPs and AFLPs). In contrast, SSR and SNP markers have been popular for genetic analyses and molecular breeding applications due to their practical advantages relative to other markers.

2.1 Simple sequence repeats

SSR markers detect length variations in tandemly repeated units of short nucleotide motifs that are 1–6 base pairs (bp) in length. Nucleotide motifs, such as di-(CA), tri-(AAT) and tetra-nucleotide (GATA) repeats, are widely distributed throughout the genomes of plants and animals (Tautzl and Renz, 1984). Markers are developed for SSRs by screening either sequences in databases or libraries of clones, and PCR primers are designed to anneal to conserved sequences that flank the repeats. If no sequence information is available, SSRs can also be developed using other methods (Xu, 2010). The recent availability of WGS information for the *Glycine max* soybean genome has facilitated the development of new SSR markers providing a valuable resource to the soybean research community (Song et al., 2010).

SSR loci are relatively abundant and are highly polymorphic and can be used to distinguish multiple alleles within eukaryotic genomes (Cho et al., 2000; Cregan et al., 1999; Eujayl et al., 2002; Thiel et al., 2003). These unique characteristics make SSR markers useful for DNA fingerprinting, genetic mapping, genetic diversity analysis, population genetics and MAS (Vuong et al., 2007). However, the SSR marker system also has drawbacks, such as a high cost of development and a need for prior knowledge of the conversed flanking sequences to design locus-specific PCR markers. Moreover, it is often difficult to distinguish alleles that differ in length by only two or three bases (Cregan et al., 1994). The development of a fluorescence-based genotyping system made it possible to multiplex SSR amplicons labelled with different fluorescent dyes for semi-automated detection using either polyacrylamide gel electrophoresis (PAGE) or capillary electrophoresis DNA sequencer systems.

In soybean, a small set of SSR markers were first developed in the 1990s using cultivated (*G. max*) and wild soybean (*G. soja*) genotypes (Akkaya et al., 1992; Morgante and Olivieri, 1993). Since then, many efforts have been made to study allelic specificity and to develop additional SSR markers. High levels of allelic variation with up to 26 alleles at one locus were founded in a group of 91 cultivated and five wild soybean genotypes (Cregan et al., 1994; Maughan et al., 1995; Morgante et al., 1994; Rongwen et al., 1995). Subsequently, more than 600 SSR markers were developed and mapped in an integrated molecular linkage map of the soybean genome (Cregan et al., 1999). Song et al. (2004) developed PCR primer sets for more than 1000 SSR loci, which were genetically mapped to 20 molecular linkage groups (LGs) corresponding to the 20 soybean chromosomes of the soybean genome (https://www.soybase.org).

Recently, advances in NGS technologies have enabled soybean researchers to successfully sequence the entire soybean genome (Schmutz et al., 2010). The WGS information has allowed genome-wide identification of SSR markers *in silico* and the development of locus-specific SSR markers. Using the soybean WGS database (http://www.phytozome. net/index.php; verified 10 September 2017), Song et al. (2010) investigated the abundance of SSR motifs and successfully developed over 33 000 new SSR markers that were available in the BARCSOYSSR_1.0 database (https://www.soybase.org). Several of these SSR markers were found to be located within genomic regions or genes associated with traits of interest and have been used for molecular breeding (Jun et al., 2012b; Rosso et al., 2011; Sun et al., 2014; Wang et al., 2014).

2.2 Single nucleotide polymorphisms

SNP markers detect differences in individual nucleotide bases (C, G, T and A) between homologous DNA sequences or polymorphisms resulting from small insertions or deletions (Indels). Edwards et al. (2007) indicated that SNP frequencies vary among different crop plants with estimated SNP frequencies typically in the range of one every 100–300 nucleotides. One of the major merits of SNPs is their abundance and density in plant genomes compared to SSRs. Another major advantage of SNP markers is that they are amenable to automated detection technologies, making them the most ideal class of marker for high-throughput genotyping. One disadvantage of SNPs relative to SSRs is that SNP polymorphisms typically involve only two alternative nucleotide bases, so haplotypes consisting of multiple linked SNPs may be necessary to tag certain alleles at a locus of interest. Depending on their physical locations in the genomes, SNPs may occur in non-coding or coding regions of genes. The SNPs in coding regions can be synonymous or non-synonymous. Synonymous SNPs do not alter the protein sequence, due to degeneracy of the genetic code, whereas non-synonymous SNPs change the amino acid sequence of a protein and can be of two types: missense and nonsense (https://www.ncbi.nlm.nih.gov).

For SNP discovery, either a global or a targeted approach can be used (Kwok and Chen, 2003). For global or random SNP discovery, an early approach was to scan for alternations in restriction sites in the human genome (Botstein et al., 1980). Following the completion of a physical map of the human genome in the mid-1990s, large numbers of SNPs were developed using STSs from mapped SSRs (Kwok et al., 1996; Wang et al., 1998). Remarkable progress in NGS technologies has provided much more efficient approaches for discovering SNPs on a large scale, as well as enabling high-throughput genotyping at a low cost. For GBS, Elshire et al. (2011) used specific restriction enzymes to reduce the complexity of the maize and barley genomes, from which multiple DNA samples were genotyped to identify hundreds of thousands of SNPs for genetic analysis. With the availability of the *G. max* reference genome sequence information, it has become feasible to use the WGS method (https://www.illumina.com) in order to economically produce several million SNPs for studies of wild relative species, exotic germplasm and landraces, as well as elite cultivars (Li et al., 2013; Valliyodan et al., 2016).

In the targeted or regional approach to SNP discovery, SNPs in a genomic region of interest are screened and validated. PCR amplicons, varying from 300 to 600 bp in length, were originally sequenced using the conventional Sanger sequencing method (Sanger and Coulson, 1975). Polymorphism scanning methods were later developed, facilitating the identification of SNPs and mutations in DNA fragments (Kwok and Chen, 2003). With the development of better DNA polymerases and sequencing chemistries, direct DNA sequencing has become highly competitive relative to other methods because it yields complete sequence information for DNA samples (Kwok and Chen, 2003).

Zhu et al. (2003) studied nucleotide diversity in coding and noncoding regions of the soybean genome and reported θ values of 0.00053 and 0.00111 for these regions, respectively. This value indicated that SNP frequency in the soybean genome was much lower than the similar estimates of *Arabidopsis* (θ = 0.0017; Olsen et al., 2002) and maize (θ = 0096; Tenaillon et al., 2001) and that the SNP discovery should focus on noncoding perigenic regions, where greater sequence polymorphism is present. This comparatively low SNP density is offset somewhat, however, by the relatively high linkage disequilibrium (LD) in the *G. max* genome (Hyten et al., 2007a). Using publicly available EST libraries (www.ncbi.nlm.nih.gov/), Van et al. (2005) successfully developed several SNP markers in coding and noncoding regions in a diverse set of soybean accessions from South Korea and the United States.

Researchers, using the regional SNP detection approach, have successfully identified SNPs associated with traits of interest, such as grassy-beany flavours, affected by the lipoxygenase-2 gene (*Lx2*) (Kim et al., 2004), soybean mosaic virus resistance, conditioned by the *Rsv1* and *Rsv3* genes (Moon et al., 2009), soybean cyst nematode (SCN) resistance conditioned by genes at the *Rhg1* and *Rhg4* loci (Cook et al., 2012, 2014; Liu et al., 2012), high oleic acid content controlled by the *fad2-1a* and *fad2-1b* genes (Pham et al., 2010), and salt tolerance provided by the *GmCHX1* gene (Patil et al., 2016; Qi et al., 2014). For example, from a previous study using the re-sequencing approach to develop a set of 384 SNPs, Hyten et al., (2008, 2010) developed 2500 additional SNP markers, from which a subset of 1536 was selected to create a 'universal soybean linkage panel' (USLP 1.0) for high-throughput genotyping assays using the Illumina GoldenGate (Illumina, Inc., San Diego, CA, USA). This USLP 1.0 array has been then widely used to map several traits (Bao et al., 2014; Gutierrez-Gonzalez et al., 2011; Jiao et al., 2015a; Nguyen et al., 2012; Vuong et al., 2010, 2011; Wu et al., 2010). Using advanced Illumina NGS technology (http://www.illumina.com/), Song et al. (2013) developed a set of over 50 000 SNPs for the *G. max* genome and created the SoySNP50K iSelect BeadChip. This sequence-based SNP chip has provided a powerful tool for characterizing soybean genetic diversity and LD mapping (Bandillo et al., 2015; Dhanapal et al., 2015; Hwang et al., 2014; Vuong et al., 2015). Using WGS technologies, different laboratories have successfully developed millions of SNP markers from both cultivated (*G. max*) and wild (*G. soja*) soybean genomes (Valliyodan et al., 2016; Li et al., 2013, 2014; Zhou et al., 2015; Lam et al., 2010).

3 Marker assays and genotyping platforms for marker-assisted selection

3.1 SSR marker genotyping

3.1.1 Gel electrophoresis

To detect size polymorphisms between SSR alleles, PCR amplicons can be stained and separated using different gel-based systems, such as agarose gel electrophoresis (AGE), PAGE and capillary gel electrophoresis (CGE) (Cregan et al., 1994).

As the most common and easily accessible method, AGE with 3.0–3.5% of MetaPhor agarose has been used in a wide range of molecular analysis for past decades. Despite some disadvantages, such as low resolution and low throughput, ethidium bromide (EtBr) staining was popularly utilized to visualize the amplification products in AGE. It also raises a great concern because EtBr is a known potent mutagen. Recently, many commercial products, such as EZ-Vision™ (AMRESCO LLC, https://www.amresco-inc.com) and Nancy-520 (Sigma-Aldrich, https://www.sigmaaldrich.com), were adapted as a non-toxic, safer alternative to EtBr with comparable results.

In many cases, when the difference in fragment sizes of SSRs is too small and the products cannot be effectively resolved by AGE, PAGE has been used as an alternative. Although some technical challenges may be encountered, the PAGE method has many advantages over AGE with respect to throughput capacity and ability to separate amplicons of similar sizes (Barril and Nates, 2012). Besides EtBr staining, the DNA fragments can be visualized using silver staining technique to improve gel images and less risk of staining artefacts (Bassam and Gresshoff, 2007).

To improve the separation speed and resolution of allele sizes, CGE instruments were developed (Akkaya et al., 1992; Marino et al., 1995). This system is capable of separating amplicons that differ in size by a little as 2 bp and a capillary can be reused multiple times (Marino et al., 1995). Wang et al. (2003) developed a relatively inexpensive and simple non-denaturing PAGE system in a Mega-Gel High Throughput Vertical Unit (CBS Scientific Co., Del Mar, CA, USA). This system was successfully used for earlier genetic analyses in both soybean and wheat (Wang et al., 2001) and was shown to be inexpensive, have a relatively high-throughput and be a good alternative to existing detection systems. However, this system necessitates a further optimization with non-EtBr staining technique to ease concerns of carcinogen that is associated with the use of EtBr.

3.1.2 Fluorescence dye-labelling for detection of SSR amplicons

For this system, four fluorescent dyes corresponding to the T, A, C and G dideoxy terminators were used to label primers in PCR amplification. The resulting single-stranded fluorescently labelled DNA products were then detected using a DNA sequencing instrument (Applied Biosystems Inc., Forest, CA, USA), in which an argon ion laser beam scans across the lanes of a sequencing gel (Cregan et al., 1994). With ABI sequencing instruments, like the DNA Prism 3100 and the DNA Sequencer 3730, PCR amplicons with different allele sizes and fluorescence could be multiplexed and analysed (Fig. 1), which significantly increased the throughput. This technology has widely been used in soybean research for a number of years (Nguyen et al., 2012; Vuong et al., 2010, 2011, 2016; Wu et al., 2009; Ziegle et al., 1992). Although fluorescently labelled SSRs have been of limited value for large MAS programmes, they were very useful for pyramiding resistance genes or for determining which plants were homozygous for a selected gene and which were heterozygous.

3.2 SNP marker genotyping

3.2.1 LightCycler® 480 system

To detect allelic variation of SNPs, the PCR products can be labelled and detected using different SNP genotyping systems, such as TaqMan (Holland et al., 1991), SimpleProbes, melting curves (https://usdiagnostics.roche.com) and Kompetitive allele-specific PCR

(a)

(b)

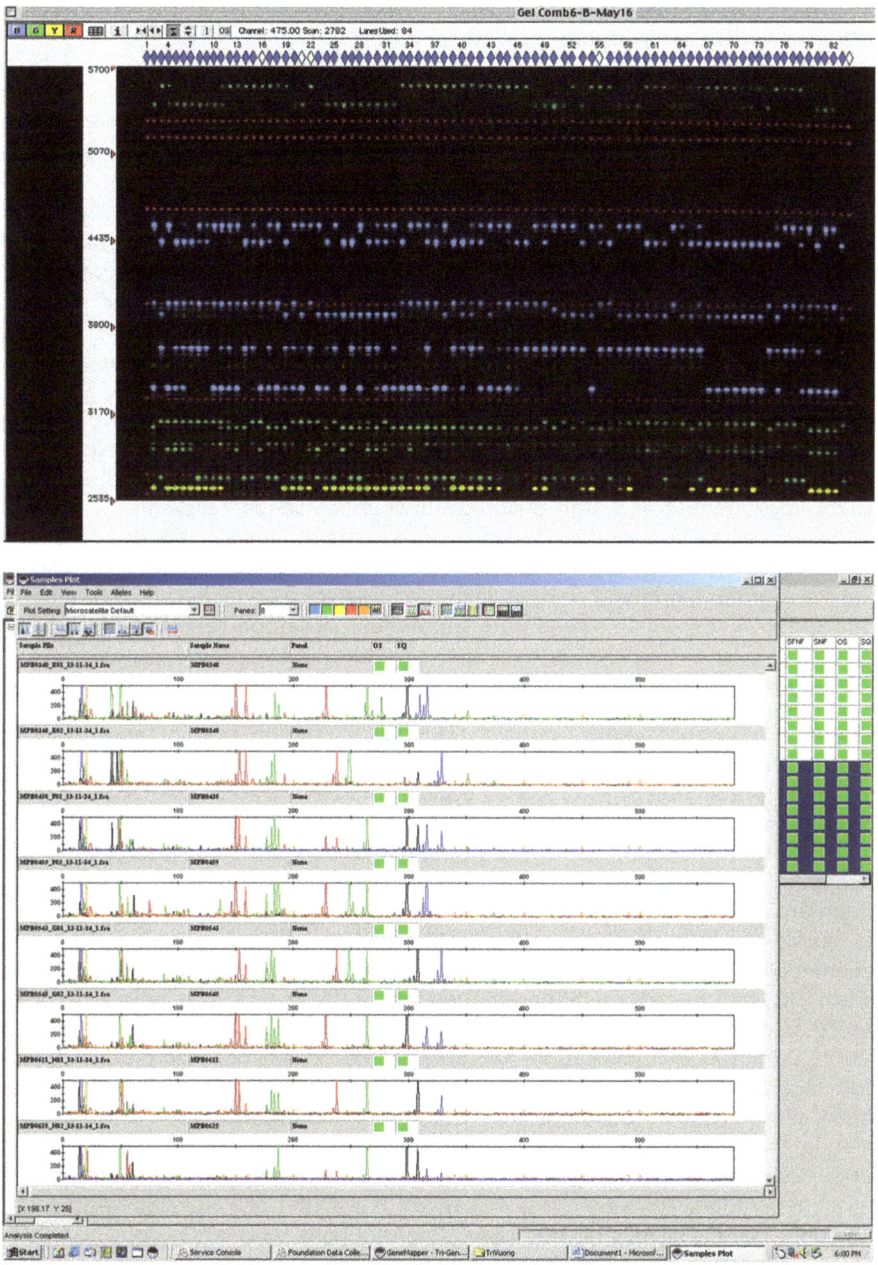

Figure 1 Gel images show the resulting PCR products of fluorescence-dye-labelled simple sequence repeat (SSR) markers multiplexed and separated during electrophoresis based on fragment sizes. Marker polymorphisms are easily detected using either the ABI DNA Prism 3100 (Panel a) or the ABI DNA Sequencer 3730 system (Panel b).

(KASP) (He et al., 2014a). The LightCycler® 480 (Roche Diagnostics, Indianapolis, IN, USA) has proved to be versatile for detecting SNP products generated by these SNP genotyping assays (Fig. 2).

The LightCycler® 480 instrument is a high-performance, versatile, medium-throughput PCR platform designed for gene expression and SNP genotyping based on melting curves, endpoint genotyping and high-resolution melting analysis. It can be used for a wide range of probe formats (HybProbes, SimpleProbes and hydrolysis probes) or fluorescent dyes (SYBR Green and KASP). Moreover, this instrument is also flexible in switching PCR plates and thermal block cyclers between 96-well and 384-well formats to increase throughput level and significantly improves efficiency in MAS implementation. The state-of-the-art optical system and user-friendly data analysis software of the LightCycler® 480 system make it a sensitive and reliable platform to support a wide range of SNP genotyping activities that are normally required in many molecular breeding laboratories of the public sector (http://www.roche-applied-science.com/).

In soybean, the LightCycler® 480 system has commonly been used for SNP genotyping assays to conduct MAS or MABC in numerous traits of interest. Using gene-specific SNPs, Lenis et al. (2010) and Pham et al. (2010) successfully developed SimpleProbes detection assays for lipoxygenase and high oleic acid in soybean seeds, respectively. Rosso et al. (2011) developed KASP assays to detect sucrose content. Similarly, Shi et al. (2015b) and Kadam et al. (2016) developed many KASP assays for SCN. Patil et al. (2016) used WGS SNP information for haplotype analysis of the salt tolerance gene, from which gene-based markers and KASP detection assays were successfully developed and used for MAS.

3.2.2 Fluorescence resonance energy transfer

Fluorescence resonance energy transfer (FRET) is a physical phenomenon that occurs between two dye molecules, resulting from the transfer of energy from an excited donor dye to an acceptor molecule. Oligonucleotide with DNA backbones labelled with one or several fluorophores can be used as DNA-based FRET probes and can be detected in a FRET system (Didenko, 2001). For years, the FRET system has been widely utilized in research to visualize molecular interaction between nucleic acids, which has led to the development of SNP detection assays for plant breeding.

The TaqMan technology involves two specific primers that amplify target SNP alleles in genomic DNA (gDNA) and two probes that are labelled with different fluorophores, 6-FAM or VIC, attached to the 5′ end a quencher at the 3′ end. The TaqMan probes rely on the 5′–3′ exonuclease activity of the enzyme *Taq*-polymerase and fluorophore-base detection. During PCR amplification, *Taq*-polymerase cleaves only the probes specifically hybridized to the target gDNA, resulting in the fluorescence of the fluorophore after it is liberated, allowing the detection of the reporter dye fluorescence.

Recently, the KASP assay has been developed as a novel SNP genotyping chemistry (He et al., 2014a). It is based upon competitive allele-specific oligo extension and FRET for signal generation. The assay uses two types of mix: the SNP-specific KASP assay mix and the universal KASP master mix. The former contains three assay-specific unlabelled oligos: two allelic-specific F primers and one common R primer. Each of the F primers harbours a unique tail sequence that corresponds with a universal FRET cassette. After the completion of PCR amplification, the fluorescence intensities of labelled PCR products are quantified using endpoint genotyping analysis.

(a)

(b)

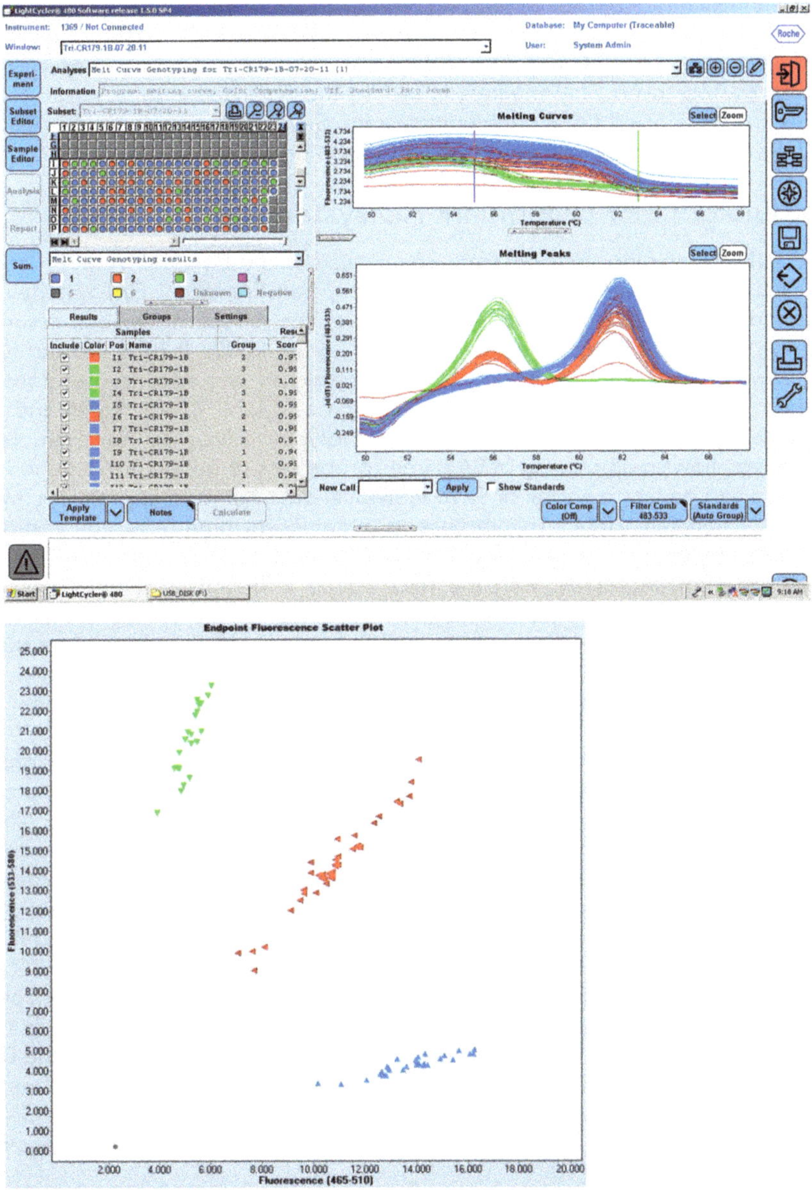

Figure 2 Graphs show the resulting PCR products of labelled single nucleotide polymorphism (SNP) markers detected using two different detection methods in LightCycler® 480 system. (Panel a): with SimpleProbe assays, allelic polymorphisms of the marker are detected by the differentiation of typical melting curve peaks. (Panel b): with KASP assays, allelic polymorphisms are identified by cluster classification of genotypes using the endpoint analysis.

In soybean, Shi et al. (2015b) developed high-throughput and functional SNP detection assays for oleic and linolenic acids in soybean using the FRET system Tecan M1000 Pro Infinite Reader (Tecan Group Ltd., Morrisville, NC, USA). Using a similar approach, King et al. (2016b) used SNP markers linked to the *Rpp* genes of soybean rust (SBR) resistance and developed KASP genotyping assays. Several near-isogenic lines (NILs) were successfully developed with pyramided genes for SBR resistance.

3.2.3 Fluidigm EP1 or Biomark HD system

A novel SNP genotyping method that uses a microfluidic Dynamic Array™ Integrated Fluidic Circuit (IFC) system (Fluidigm Corporation, San Francisco, CA) was developed. This method has been demonstrated to be simple, cost-effective, flexible, accurate and suitable for high-throughput multiplexing (Wang et al., 2009). It provides the flexibility of a microwell plate and the density of a microarray in one easy-to-use, consumable IFC system. Different dynamic array formats, such as 24.24, 48.48, 96.96 and 192.24, can be specifically selected for different SNP genotyping purposes. For example, 96.96 dynamic array allows multiplexing of 96 primer/probe sets to genotype 96 genomic DNA samples and can generate a total of 9216 individual data points in approx. 3 h (Maughan et al., 2011; Thomson, 2014). This represents a significant time reduction compared to other genotyping systems, such as the Illumina GoldenGate or Sequenom iPLEX (https://www.fluidigm.com). Moreover, with the microfluidic architecture of the dynamic array IFC coupled with high-throughput multiplexing, DNA samples, reagents and primer-probes can be combined in a final volume of 5.0–6.0 μL for PCR, resulting in a significant reduction in the cost per data point. This genotyping method has also provided high-quality genotypic data with an excellent 'call' rate, >99.5%, and an accuracy, >99.8% (Wang et al., 2009).

In addition to SNP Type™ and TaqMan assays customized by the Fluidigm Corporation (https://www.fluidigm.com), KASP genotyping assays have been adapted and successfully implemented in the Dynamic Array™ IFC system (Fig. 3; Maughan et al., 2012; Thomson, 2014). This showed the remarkable advantages that KASP genotyping assays have over other SNP genotyping systems in pre-breeding discovery as well as SNP employment for crop improvement (Maughan et al., 2011, 2012; Thomson, 2014). In soybean, although TaqMan and KASP detection assays have widely been employed in breeding, there has been little known about their implementations in the Dynamic Array™ IFC system. Recently, a subset of over 400 informative SNPs, which were derived from the SoySNP6K Illumina Infinium BeadChip (Illumina, Inc., San Diego, CA) and were genetically mapped throughout the soybean genome, were selected to develop KASP assays for the Dynamic Array™ FTC system. The assay validations were performed in the Soybean Genetics and Genomics Laboratory at the University of Missouri in an effort to generate a 384-SNP panel (Nguyen Lab, unpublished data). It will provide a useful resource for genomic-assisted breeding (GAB) in soybean.

3.2.4 Genotyping-by-sequencing

Advances in NGS technologies have led to the development of GBS, an advanced SNP genotyping approach with flexibility and cost reduction. It has made the GBS applications feasible and affordable for plant genomics research and crop breeding. This genotyping method can simultaneously identify several thousands to millions of SNPs distributed

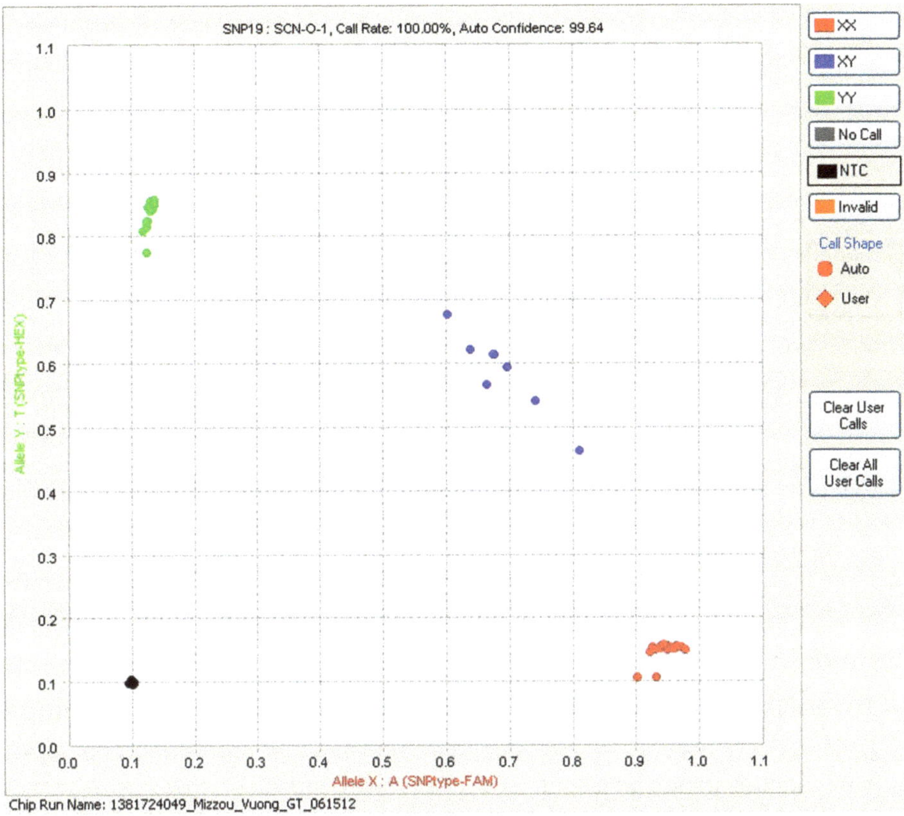

Figure 3 A graph shows the cluster classification of the resulting PCR products of labelled single nucleotide polymorphism (SNP) markers detected using SNP Type™ genotyping assays in the Fluidigm Dynamic Array™ IFC system. Allelic polymorphisms of the marker are effectively detected by the endpoint analysis.

across the entire genome, which enables effective detection of genomic regions or genes associated with traits of interest (Poland and Rife, 2012). Compared to other methods, GBS is an ideal platform for the development of markers either for single genes or for whole-genome profiling because there is no requirement for prior knowledge of genomic sequence data for the species. It allows breeders to study genomic diversity, discover DNA markers, genotype large breeding populations and perform GS on a large scale in plant breeding programmes (He et al., 2014b).

Although GBS offers a novel high-throughput SNP genotyping approach that enhances the efficiency and capacity of crop breeding, it also has potential drawbacks and challenges. For most users, the main challenges of using GBS are the requirement for efficient and accurate computational processing, and variant and genotype calling of large amounts of sequence information generated from sequencing libraries (Torkamaneh et al., 2017). Numerous bioinformatics pipelines have been developed (Bradbury et al., 2007; Glaubitz et al., 2014; Lu et al., 2013; Nielsen et al., 2011); however, all need to undergo several similar steps that represent specific challenges and ambiguities. To overcome such drawbacks, Torkamaneh et al. (2017) successfully developed Fast-GBS, a new bioinformatics analysis pipeline that was demonstrated to possess advantages such as high efficiency and reliability for SNP calling, accurate genotyping, modest computing resources requirement and ease of use. For soybean, Fast-GBS analysis had an estimated genotyping accuracy of 98.7% in comparison with different SNP genotyping methods. These results suggested that Fast-GBS can be an efficient and powerful alternative for GBS-derived SNP genotyping, and that it can accelerate GAB for soybean improvement (Torkamaneh et al., 2017).

The GBS approach has been utilized in several soybean studies to develop SNPs for various applications. Jarquín et al. (2014) developed more than 16 000 SNPs that were used for GP in a breeding population and concluded that GBS holds good potential to expedite genetic gain for yield. Similarly, Bastien et al. (2014) generated more than 7800 SNPs that were used for GWAS of *Sclerotinia sclerotiorum* resistance and identified and confirmed major QTL for this disease. After that, Sonah et al. (2015) generated over 47 000 SNPs utilized for a GWAS of agronomic traits and showed that the integration of GBS and GWAS can be a robust approach complementary to traditional genetic linkage mapping of complex traits. More recently, around 3000 polymorphic, high-quality SNPs that were obtained through GBS were used for mapping of QTL associated with stearic acid content in soybean seeds, resulting in the identification and characterization of minor and major effect QTL that could be selected for improving this trait (Heim and Gillman, 2017).

3.3 Other developments crucial to high-throughput genotyping

While advances in marker types and technologies have been tremendously important for high-throughput genotyping and MAS in soybean and other crops, simultaneous advances in methods for DNA extraction and bioinformatics have also been crucial components of the success. Seed companies developed 'seed chippers' capable of slicing off small pieces of soybean seeds without damaging the embryos. These devices can sample tissue from 96 or more seeds simultaneously, depositing the seed tissue into plates that can then be used for fully automated DNA extraction by robots. Plates and samples can be identified using bar codes, thus making it possible to handle hundreds of thousands of DNA samples in a central genotyping facility. The ability to genotype seeds makes it possible to discard those seeds that lack the target genes or QTL prior to planting them, thus saving the costs associated with evaluating entire populations of plants in the field or greenhouse. Alternative technologies and products have also been invented, which expedite and simplify the collection of DNA from plants in the field and greenhouse. For example, Whatman FTA™ cards (GE Healthcare Bio-Sciences Corp., Piscataway, NJ, USA) can be pressed against leaves to capture DNA released from the crushed tissue.

With the sharp increase in the amount of marker data collected, it is essential to have computer hardware and software capable of storing and analysing the data, particularly for GS and other applications that require marker coverage of the entire genome. It is also essential to have personnel who have the skill and experience needed to analyse and interpret the DNA marker data in a breeding programme. Modern soybean breeding in the private sector, and in many public breeding programmes, has thus become a team effort, often involving specialized personnel 'compartmentalized' in geographically distant locations, often with a central genotyping facility and numerous research stations where crosses are made, populations advanced and tissue samples collected for DNA extraction.

4 Marker-assisted selection in soybean breeding

Molecular breeding has greatly accelerated the development of new soybean varieties with desirable agronomic traits. Some of the traits for which MAS has been used in soybean breeding programmes are briefly discussed below to highlight some of the advantages and successful applications of MAS in soybean breeding. Since the main sources of information are published in literature, the examples chosen are not necessarily major MAS objectives in the private sector.

4.1 Soybean nematodes

Soybean plants can be parasitized by several nematode species, including the (SCN, *Heterodera glycines*), root-knot nematodes (RKN, *Meloidogyne* spp.) and the reniform nematode (RN, *Rotylenchulus reniformis*). Wrather et al. (2010) estimated that, in the top eight soybean-producing countries, SCN caused yield losses of approximately 7.19 million metric tonnes in 2006, and an additional 2.69 metric tonnes were lost to RKN and other nematode parasites. Many efforts have been made to breed resistant cultivars, which have been shown to be the most effective means to reduce losses to these pests. For decades, advances in molecular genetics and marker technologies have made it possible to efficiently identify and characterize DNA markers associated with genomic locations underlying resistance to nematodes.

Although many RFLP, AFLP and RAPD markers were associated with SCN resistance QTL (Boutin et al., 1992; Chang et al., 1997; Concibido et al., 1994; Kabelka et al., 2005; Meksem et al., 2001), these markers were mainly used for genetic mapping and were not employed for molecular breeding. In contrast, several SSR markers located close to the major SCN resistance loci on chromosome (Chr.) 18 (*Rhg1* locus) and Chr. 8 (*Rhg4* locus) (Cregan et al., 1999; Guo et al., 2006) have been used for MAS. Approximately 20 000 soybean breeding lines were genotyped with these SSR markers for breeding programmes of the University of Missouri (Vuong et al., 2007, 2013).

In addition to the DNA markers associated with the *Rhg1* and *Rhg4* loci, several SSR and SNP markers associated with two novel QTL for broad-based resistance to multiple SCN races (PA1, -2, -3, -5, -14, and LY1) were first identified and mapped to Chrs. 10 and 18 (Vuong et al., 2010). Among these, three SSR markers close to the QTL were effectively utilized for MAS to develop new SCN-resistant soybean germplasm, including the lines JTN-5203, JTN-4307, JTN-5303 and JTN-5503 (Arelli et al., 2007, 2015, 2017).

Recently, candidate SCN resistance genes from the *Rhg1* and *Rhg4* loci were successfully cloned from PI 88788 (Cook et al., 2012) and 'Forrest' (Liu et al., 2012), respectively. *Rhg1* was found to actually be a complex locus with three different genes contributing to SCN resistance, and copy number variation was identified as a major factor determining different levels of resistance among several sources of resistance. With allelic variation of the cloned genes in combination with genomic resources, Shi et al. (2015b) developed three KASP genotyping assays for two SNPs derived from the gene *Glyma18g02590* at the *Rgh1* locus and one SNP marker derived from the gene *Glyma08g11490* at the *Rhg4* locus. In a separate effort, Kadam et al. (2016) also developed and validated a subset of SNP markers for the *Rhg1* and *Rhg4* loci along with KASP assays. Those functional markers, coupled with robust genotyping assays, not only provided genotype information for selection of SCN resistance from Peking and PI 88788, but also were able to differentiate between the genomic regions of Peking and PI 88788. For many years, MAS has been routinely performed in the Nguyen Laboratory at the University of Missouri and in the Li Laboratory at the University of Georgia using these gene-based SNP markers to select superior genotypes with SCN resistance.

Efforts have also been made to study and characterize genetic control of RKN resistance. Four RKN species, *Meloidogyne incognita* (Kofoid & White) Chitwood (Mi), *M. arenaria* (Neal) (Ma), *M. javanica* (Treub) Chitwood (Mj) and *M. hapla* (Chitwood) (Mh) commonly parasitize soybean and caused 95% of the incidence of RKN damages in important crop plants worldwide (Doucet and Pinochet, 1992; Hussey and Janssen, 2002). Among these, the Mi and Ma species cause the most trouble to soybean, tobacco, peanut and cotton growers in the southern United States (Boerma and Hussey, 1992). Many additional sources of resistance to RKN species have been reported (Harris et al., 2003; Luzzi et al., 1994; Yates et al., 2009; Xu et al., 2013), from which SSR and SNP markers associated with QTL for RKN resistance were detected and mapped (Ha et al., 2004; Jiao et al., 2015b; Li et al., 2001; Xu et al., 2013). Ha et al. (2007) developed two SNPs that have been highly effective in MAS for Mi nematode resistance from PI 96354. Subsequently, four candidate genes were identified at the *Rmi1* locus on Chr. 10 (Pham et al., 2013), from which a subset of functional SNPs and KASP genotyping assays were developed for large-scale MAS. From a different resistance source, PI 438489B (Vuong et al., 2011), a subset of candidate-gene-based SNPs and KASP genotyping assays were developed for RKN resistance QTL on Chr. 10 and were validated in breeding materials (Nguyen Lab, unpublished data). These should enable MAS for RKN resistance to be performed with a high-throughput genotyping platform.

For RN, more than 20 sources of resistance have been reported (Davis et al., 1996; Kim et al., 2016; Robbins et al., 1994) and QTL for RN resistance have been mapped (Ha et al., 2007; Jiao et al., 2015b). Even though several SSRs and SNPs were associated with RN resistance genomic regions, it is not known whether they have been used for MAS. PI 438489B is resistant to both SCN and RN species, and preliminary genetic analysis showed that significant QTL for RN resistance are mapped close to the *Rhg1* SCN resistance locus on Chr. 18 (Klepadlo et al., 2017). It is thus possible that the genes at the *Rhg1* locus contribute to resistance to both SCN and RN, or that different genes are closely linked. Hence, the functional markers and robust genotyping assays developed for MAS of SCN resistance will also be useful in MAS for RN resistance.

4.2 Disease: soybean rust

Soybean rust (SBR, *Phakopsora pachyrhizi* Syd. & Syd.) is one of the most economically important and destructive foliar diseases of soybean worldwide and causes yield losses up to 80% under environmental conditions favourable to this fungal pathogen (Hartman et al., 2005). Although it has not become a major disease in the Midwestern United States, SBR has occurred each year in southern states and was observed as far north as Ontario, Canada, in 2007 (sbr.ipmpipe.org/cgi-bin/sbr/public.cgi). To control this disease, applications of fungicides have been widely utilized, but these can be expensive and may have negative impacts on the environment. Thus, the development of resistant soybean genotypes has been recommended as an effective step in long-term SBR management (Bonde et al., 2006; Hartman et al., 2005; Miles et al., 2006; Paul and Hartman, 2009; Ray et al., 2009).

SBR resistance is conferred by resistance genes at several loci in the soybean genome, including *Rpp1* (McLean and Byth, 1980), *Rpp2* and *Rpp3* (Hartwig and Bromfield, 1983), *Rpp4* (Hartwig, 1986), alleles of *Rpp5* and *rpp5* (Garcia et al., 2008), *Rpp6* (Li et al., 2012) and *Rpp7* (Childs et al., 2017). Individual genes provide resistance to some fungal pathotypes, but not others. The rust pathogen exhibits considerable variation in virulence among isolates and field populations, and certain strains have evolved the ability to overcome single-gene resistance that previously had been effective (Hartman et al., 2005). Thus, there has been a need to identify additional SBR resistance sources with novel genes and different resistance mechanisms for future development of resistant varieties (Walker et al., 2011, 2014). Because of the pathotype variation within and among *P. pachyrhizi* populations, pyramiding two or more *Rpp* genes in the same cultivar could enhance the breadth and durability of resistance to SBR (Yamanaka et al., 2013, 2015). More comprehensive information on *Rpp* genes/alleles and sources of SBR resistance can be found in other chapters in this volume.

Hyten et al. (2007b) mapped the *Rpp1* locus to Chr. 18, and Chakraborty et al. (2009) mapped a different resistance allele and designated *Rpp1-b*, in the same region. Subsequently, Hossain et al. (2015) confirmed the *Rpp1* locus on Chr. 18, which is flanked by the same SSR markers in a different germplasm accession. Silva et al. (2008) identified the *Rpp2* and *Rpp4* loci and mapped them to Chrs. 16 and 18, respectively. The *Rpp3* locus from PI 462312 was mapped by Hyten et al. (2009) to Chr. 6 within an interval that overlaps the genomic region in 'Hyuuga' as previously reported by Monteros et al. (2007). Hossain et al. (2015) subsequently confirmed the *Rpp3* locus on Chr. 6 in PI 416764. Garcia et al. (2008) mapped the *Rpp5* locus to Chr. 3 from PI 200456, PI 471904 and PI 200526. Li et al. (2012) mapped the *Rpp6*, gene from PI 567102B to Chr. 18. Recently, an allele at the *Rpp6* locus was detected and mapped on Chr. 18 in PI 567068A (King et al., 2016a), and Liu et al. (2016) found that PI 567104B also carries a resistance gene at the *Rpp6* locus. Childs et al. (2017) discovered a resistance gene (*Rpp7*) at a seventh independent locus in PI 605823. Some germplasm accessions carry two SBR resistance genes. Kendrick et al. (2011) found that 'Hyuuga' has a resistance gene at the *Rpp5* locus as well as at the *Rpp3* locus. Recently, Vuong et al. (2016) discovered that PI 635999 carries resistance genes at both the *Rpp3* and *Rpp4* loci on Chrs. 6 and 18, respectively, and reported that several SNP markers were associated with these two loci.

For developing SBR-resistant soybean breeding lines with resistance genes/alleles derived from known resistance sources, Diers et al. (2014) utilized SSR markers in an MABC programme to select plants carrying the *Rpp1*, *Rpp1-b*, *Rpp*?(Hyuuga) and *Rpp5* genes with the genetic backgrounds of LD10-30052 and LD09-16057 (MG II), or LD10-14205 and LD10-13091 (MG IV). The backcross lines were not significantly different from their recurrent parents for agronomic traits, including seed yield. These lines should be useful to soybean breeders in developing SBR-resistant cultivars. Similarly, using known resistance sources of *Rpp1*, *Rpp2*, *Rpp3* and *Rpp4* genes, King et al. (2016a) used KASP genotyping assays with SNP markers to perform MABC breeding and successfully developed four NILs, G00-3213*Rpp1*, G00-3213*Rpp2*, G00-3213*Rpp3* and G00-3213*Rpp4*, all in MG VII. These NILs exhibited a similar level of resistance to a bulk local isolate of *P. pachyrhizi* and can be useful for gene pyramiding to develop new SBR-resistant varieties. Studying the impact of multiple *Rpp* genes on SBR resistance, Yamanaka et al. (2013) and Yamanaka et al. (2015) developed and evaluated several pyramided lines carrying multiple *Rpp* genes. They found that soybean lines with two or three *Rpp* gene combinations exhibited higher SBR resistance than the original resistance sources with single *Rpp* genes.

Various analysis approaches, such as SNP haplotype analysis (Kim et al., 2012), fine-mapping (Yu et al., 2015) and genomic characterization (Vuong et al., 2016), have been used to identify several SNPs tightly linked to the *Rpp1* to *Rpp4* loci in other sources of SBR resistance. The linked SNPs have been used to develop robust KASP genotyping assays, which have been successfully evaluated and incorporated into the soybean breeding programmes supported by the United States Agency for International Development (USAID), whose one objective is to develop high-yielding and rust-resistant varieties for different soybean production areas in sub-Saharan Africa. The progress from this international collaboration can be monitored at the homepage of the Soybean Innovation Laboratory at the University of Illinois (http://soybeaninnovationlab.illinois.edu).

4.3 Insect: soybean aphid

Soybean aphid (SA, *Aphis glycines* Matsumura) is an invasive insect pest that originated from Asia (Wu et al., 2004). In the United States, SA was first discovered in the Midwest in 2000 (Hartman et al., 2001). Since then, this pest has rapidly spread throughout major soybean production areas in North America, including Canada (Ragsdale et al., 2011), and it poses a significant threat to soybean productivity. This insect pest can cause yield losses of up to 40% by direct stylet-feeding, which removes nutrients and water from the host plants and can eventually result in premature death under heavy infestation (Wu et al., 2004). Also, it was reported that SA can transmit plant viruses to soybean and several other crops (Hartman et al., 2001; Iwaki et al., 1980).

To control this pest, various approaches, such as biological control and insecticide applications, have been individually or integratively implemented. Insecticide applications are commonly used to manage this pest; however, restraining aphid populations below damaging levels and reducing the possibility of SA populations developing resistance to commonly used insecticides can be difficult (Gao et al., 1993; Wu et al., 2004). Thus, host plant genetic resistance would be a valuable alternative means to manage this pest (Chandrasena et al., 2015; Hill et al., 2004; Mian et al., 2008a; Zhang et al., 2010).

Genetic studies have identified several single dominant or recessive genes conferring aphid resistance in different soybean germplasm accessions. Hill et al. (2006a) and Hill et al. (2006b) first identified a dominant gene, *Rag1*, in cultivar Dowling and a second dominant gene, *Rag*, in cultivar Jackson. These resistance genes were then genetically mapped to the same region on Chr. 7 (LG M) (Li et al., 2007). Mian et al. (2008a,b) identified another dominant resistance gene, *Rag2*, in PI 243540. *Rag2* was mapped to Chr. 13 (LG F), and resistance alleles at this locus were subsequently found in PI 200538 (Hill et al., 2009) and PI 587732 (Kim et al., 2014). Recessive genes conferring an antibiosis type of resistance have also been identified. Among these, *rag1c* and *rag4* were mapped to Chrs. 7 and 13, respectively, in PI 567541B (Mensah et al., 2008; Zhang et al., 2009), while *rag1b* was mapped to Chr. 7 (Mensah et al., 2008). From other aphid-resistant PI lines, Zhang et al. (2010, 2013) and Bales et al. (2013) detected and mapped *Rag3*, *Rag3b* and *rag3* genes to a genomic region on Chr. 16. Using a BARC-SNP-based high-density genetic map, Jun et al. (2012a) mapped another dominant resistance gene, named *Rhg5*, to Chr. 13 of PI 567301B. Although it confers antixenosis-type resistance, *Rag5* is located very closely to the previously identified *Rag2* gene, which conditions antibiosis resistance (Hill et al., 2009;

Mian et al., 2008b), so this locus was subsequently referred to as the *Rag2/Rhg5* locus. Zhang et al., (2017b) detected two major QTL significantly associated with aphid antibiosis resistance from a wild soybean (*G. soja*) accession called 85-32 (Yang et al., 2004). One QTL was mapped to Chr. 8, named *Rag6*. A second QTL named *Rag3c* was mapped to Chr. 16. More comprehensive information on *Rag* genes/alleles and for genetic characterization of resistance in different germplasm accessions can be found in other chapters in this volume.

In an effort to refine the location of the *Rag1* locus on Chr. 7, Kim et al. (2010a) used SNP markers previously reported by Kaczorowski et al. (2008) and developed additional SNP markers to narrow the location of the locus to a 115-kb interval. After that, Kim et al. (2010b) fine-mapped the *Rag2* locus (Chr. 13) to a 54-kb interval. Following earlier primary QTL analysis from the *G. soja* accession 85-32, Zhang et al. (2017a) utilized Illumina Infinium BeadChip- or WGS-based SNPs to fine-map the two novel SA resistance loci to smaller intervals of 49 kb for the *Rag6* locus (Chr. 8) and 150 kb for the *Rag3c* locus (Chr. 16), resulting in the identification of candidate genes along with associated SNP markers. DNA markers closely linked to known *Rag* genes can be used for MAS and gene pyramiding to develop aphid-resistant breeding lines and cultivars.

The durability of SA resistance is dependent upon not only the nature of the resistance genes, but also aphid biotypes and the environmental and physiological conditions impacting aphid growth and reproduction. Kim et al. (2008) and Alt and Ryan-Mahmutagic (2013) discovered new SA biotypes in North America, which were able to colonize breeding lines carrying *Rag1*, *Rag2* or the combination of both genes. By using SSRs and SNPs in robust KASP detection assays, Chandrasena et al. (2015) successfully stacked various combinations of the recessive genes *rag1b*, *rag1c*, *rag3* and *rag4* in lines that were later tested in greenhouse and field evaluations. The authors reported that lines with the *rag3* or *rag1c* genes alone or in a combination had significant levels of resistance. In contrast, lines with the *rag4* or *rag1b* genes had a lower level of resistance, and the *rag4* + *rag1b* combination was the least effective against SA. These results indicated that pyramiding multiple *Rag* genes could be an effective strategy for developing soybean varieties with more durable SA resistance (Hesler et al., 2012; Wiarda et al., 2012).

4.4 Seed oil: fatty acids

Soybean oil is used in many food and non-food products and accounts for nearly 60% of the world's oil seed production (http://www.soystats.com). Five fatty acids, palmitic acid (12%), stearic acid (4%), oleic acid (23%), linoleic acid (53%) and linolenic acid (8%), make up nearly the entire oil portion of a soybean seed. Oil made from typical commodity-grade soybean is low in saturated fatty acids and high in unsaturated fatty acids compared to many other fats and oils (Fehr, 2007). However, it has been shown that the high concentrations of linoleic acid and linolenic acid are responsible for the low oxidative stability and frying stability of soybean oil. Oxidation of these polyunsaturated fatty acids results in rancidity, a rapid decrease in optimum flavour, and reduced shelf lives for manufactured food products (Warner and Fehr, 2008). Moreover, the partial hydrogenation process used to reduce the polyunsaturated fatty acid content in soybean oil used to make margarine and some other food products can result in a 10–40% *trans* fat content. Such *trans* fats are unhealthy for human consumption because of their association with a higher risk of obesity and cardiovascular problems. To overcome the shortcomings of soybean oil, numerous efforts have been made using traditional breeding and/or genetic engineering approaches to produce soybean varieties with elevated oleic acid and reduced linolenic acid.

Genetic studies have identified several genes involved in fatty acid biosynthetic pathways, including several FAD2 and FAD3 genes located on Chrs. 10 and 20, respectively, in developing soybean seeds. These two genes are responsible for the conversion of oleic acid to linoleic acid, and for a reduction in linolenic acid (Dierking and Bilyeu, 2009). Both genetic engineering and candidate genes have been utilized in the development of new soybean cultivars with an oil content that is over 80% oleic acid (Buhr et al., 2002; Pham et al., 2010). Transgenic high-oleic acid soybean lines were successfully produced using a ribozyme-terminated antisense approach to down-regulate *FAD2-1* gene transcript levels (Buhr et al., 2002). However, the production and importation of these soybean varieties has faced regulatory issues and social acceptance obstacles, particularly in Europe and Asia (Pham et al., 2011). Alternatively, a candidate-gene-based approach involved searching for soybean germplasm with natural genetic variations in the FAD2-1A and FAD2-1B genes. Allelic variation analysis identified several SNPs in these two genes in various soybean mutants or germplasm, such as M23 (Takagi and Rahman, 1996), 17D (Dierking and Bilyeu, 2009) and PI 603452 (Pham et al., 2011) for FAD2-1A, and in PI 283327 (Pham et al., 2010) for FAD2-1B.

Linolenic acid levels in soybean seeds are associated with at least three independent genetic loci *fan1* located on Chr.14 (Rennie and Tanner, 1991), *fan2* on Chr. 18 (Fehr et al., 1992) and *fan3* on Chr. 2 (Fehr and Hammond, 2000). Multiple mutant alleles at the *fan* locus have been reported in many soybean genotypes (Pham et al., 2014). In an effort to discover new sources of low linolenic acid, a mutagenesis approach was used in some breeding programmes, resulting in several mutant lines with a significant reduction in linolenic acid content. It has been shown, however, that natural mutations at the *fan* locus in these accessions were either allelic or identical to the original *fan* allele in line C1640 (Rennie and Tanner, 1989; Rennie et al., 1988). Since then, efforts in soybean breeding have been continued to develop soybean oil with lower linolenic content. Remarkably, Ross et al. (2000) successfully developed line A29 with very low linolenic acid (1%) by combining three independent mutant alleles, including *fan1* from line A5, *fan2* from A23 and *fan3* from A89-144003.

In soybean seeds, three microsomal ω-3 fatty acid desaturate genes, *GmFAD3A*, *GmFAD3B* and *GmFAD3C*, control the conversion of linoleic acid precursors into linolenic acid. Earlier studies have analysed different low linolenic sources and identified many allelic mutations in these three FAD3 genes (Anai et al., 2005; Bilyeu et al., 2003, 2006). For example, for the GmFAD3A gene, a deletion was found in line A5, while a point mutation was detected in a mutant line C1640. Genetic analysis has shown that these genes contribute unequally, but additively to the linolenic acid content (Bilyeu et al., 2005). Thus, combinations of the mutant alleles from these sources, breeding lines with a significantly reduced linolenic acid content, varying from 3% to 1%, can be successfully developed (Bilyeu et al., 2011).

The remarkable discoveries of allelic mutations in the FAD2 and FAD3 genes led to the identification of 'perfect' functional DNA markers, from which various types of marker detection assays, such as TaqMan (Pham et al., 2014), SimpleProbes (Pham et al., 2010) and KASP (Patil et al., 2017; Shi et al., 2015a), have been successfully developed for different genotyping platforms. These have been effectively utilized for MAS in soybean breeding programmes. For example, Pham et al. (2010) successfully developed several conventional soybean breeding lines with over 80% of oleic acid and environmental stability by combining mutant alleles of the FAD2-1A and FAD2-1B genes along with perfect gene-based SNP markers.

La et al. (2014) evaluated over ten F4:7 breeding populations derived from backcrossing of mutant FAD2-1A and -1B alleles and elite soybean lines over six growing environments. They concluded that the combinations of a mutant allele of FAD2-1A from line 17D and mutant allele of FAD2-1B from PI 283327 produced soybean lines with high oleic acid (>79%) and comparable seed yields across environments. The results indicated that the high oleic trait had no detrimental effects on the yield of the lines, and that it had a positive effect on both seed oil and protein content. The study suggested that, by combining appropriate mutant FAD2 alleles that have no yield drag, soybean breeders were able to develop high oleic soybean with higher protein and high seed yield.

With support from the United Soybean Board, a new molecular breeding programme was established using an MABC strategy to develop soybean lines with high oleic (>75%) and low linolenic (<2.5%) acids for the food industry market (https://unitedsoybean.org). Several backcrossing (BC) cycles were conducted to introduce novel mutant alleles of the FAD2 and FAD3 genes into high-yielding soybean lines derived from different US soybean breeding programmes covering maturity groups (MGs) 0 to VIII. Robust and efficient SNP genotyping assays for MAS have been conducted in the Nguyen Laboratory at the University of Missouri and in the Li Laboratory at the University of Georgia. Segregating BC populations have been generated for progeny selection, after which yield tests will be conducted to select lines with the potential to be released as specialty soybean varieties in a few years. Further information can be found at the United Soybean Board website (https://unitedsoybean.org).

4.5 Abiotic stress: salinity

In soybean fields, salinity can cause a 20–40% reduction in seed yield (Chinnusamy et al., 2005; Papiernik et al., 2005). Yield losses are due to poor seedling germination, severely affected vegetative growth and development and poor seed quality (Blanco et al., 2007; Phang et al., 2008). Salt stress also reduces the efficiency of symbiotic nitrogen fixation as a result of lowered numbers and biomass of root nodules (Elsheikh and Wood, 1995; Singleton and Bohlool, 1984). It has been predicted that about 50% of the arable land in the world could be affected by salinity by the year 2050 because of ineffective fertilizer practices, salt-water intrusion and the use of poor-quality irrigation water (Blumwald and Grover, 2006).

For decades, genetic studies have been conducted and breeding efforts have been made to identify and characterize soybean germplasm with tolerance to salinity that could be used in breeding programmes (Abel, 1969; Lee et al., 2004, 2008; Lenis et al., 2011). An exotic accession, S-100 (https://npgsweb.ars-grin.gov), was found to have a genomic region on Chr. 3 that was significantly associated with salt tolerance (Lee et al., 2004). This line has been used in breeding programmes, leading to the development of many salt-tolerant soybean varieties, such as 'Lee', 'Forrest' and 'Wright' (Lee et al., 2004). Using a population derived from a cross of 'Williams 82' and 'Fiskeby III', Do et al. (2017b) confirmed the same major salt tolerance locus on Chr. 3, with a favourable allele inherited from Fiskeby III. They also detected a second salt tolerance locus on Chr. 13 with a favourable allele inherited from Williams 82.

In addition to cultivated soybean (*G. max*), Lenis et al. (2011) evaluated other *Glycine* species, including wild relative and perennial species and identified many new sources of salt tolerance, such as PI 468916 (*G. soja*), PI 441008 (*G. tomentella*) and PI 595792 (*G. argyrea*). Lee et al. (2009) and Ha et al. (2013) also detected and confirmed the existence of QTL for salt tolerance at the same previously reported locus on Chr. 3 in a wild relative soybean, PI 483463.

Taking advantages of NGS technology by combining *de novo* sequencing of a wild soybean accession called W05, and the construction of GBS-based genetic map in a RIL population, Qi et al. (2014) successfully identified a novel ion transporter gene, *GmCHX1 (Glyma03g32900)*, as the primary gene responsible for salt tolerance in that population. Patil et al. (2016) used high-quality (17× coverage) WGS information from a diverse panel of 106 soybean accessions to perform a GWAS and haplotype analysis. They identified three major structural variants (SV), SV-1 (salt sensitivity), SV-2 (salt tolerance) and SV-3 (salt sensitivity), in addition to allelic variation in the promoter and genic regions of the *GmCHX1* gene. They discovered six gene-associated SNPs, M1 to M6, from which robust KASP genotyping assays were developed and validated in a subset of germplasm and a mapping population. It has been demonstrated that these SNPs can be used to differentiate three major SVs representative of the three groups (Patil et al., 2016). For instance, SNP markers M2 and M3 were able to differentiate the SV-1 and SV-2 variants from SV-3, while SNP marker M5 was able to differentiate the SV-1 variant from SV-2 and VS-3. In subsequent studies, Do et al. (2017a) incorporated three functional SNPs, M1, M4 and M6, into genetic analysis and also a GWAS. They successfully mapped the gene *GmCHX1* at the same locus on Chr. 3 in Fiskeby III and other recently identified salt-tolerant germplasm.

The results of gene cloning and haplotype analysis have enhanced our understanding of the genetic basis and the functionality of salt tolerance gene in soybean. The gene-based SNPs facilitate the identification of new salt tolerance sources that possibly carry a novel genomic locus different from the *GmCHX1* gene based on the marker-trait association analysis (Do et al., 2017a). Robust and cost-effective KASP genotyping assays have been efficiently utilized for MAS in the soybean breeding programme at the University of Missouri, leading to the development of several high-yielding elite lines carrying the salt tolerance gene, such as S11-9618, S12-4755 and S13-2390 (J. G. Shannon, personal communication).

5 Genomic selection

Advances in DNA marker technologies have enabled plant breeders to conduct MAS effectively for traits that have been challenging or expensive to select for using phenotypic evaluations (Fallen et al., 2015; Reynab and Sneller, 2001). The use of MAS to improve complex traits like seed protein content, yield or abiotic stress tolerance has been a challenge. Most reported QTL associated with these traits have minor phenotypic effects or environmental instability, and some appear to be important in some genetic backgrounds, but not others. Thus, the introgression of quantitative traits into elite backgrounds using MAS or MABC approaches has generally encountered difficulties (Lorenz et al., 2011).

Unlike traditional forms of MAS or MABC, GS involves the use of genotype data from all polymorphic DNA markers distributed across the whole genome. These data are then used to calculate genomic estimated breeding values (GEBVs) for complex traits, allowing selection to be effectively implemented (Lorenz et al., 2011). GS was developed from a concept of GP first proposed by Meuwissen et al. (2001), in which a 'training population' (or reference population) with available genomic and phenotypic data is used to develop statistical prediction models. These models can be then applied to a 'validation set' of selected individuals that have been genotyped, but not phenotypically evaluated. Over the years, several prediction models have been developed and tested (Heslot et al., 2012); however, the accuracy of a prediction model depends on a number of factors, such as the LD, the training population size, the heritability of given traits and marker density (Lin et al., 2014). As demonstrated in GP studies, larger training populations, higher trait heritability and higher marker density generally lead to better prediction accuracy of genomic breeding values in different crop plants (Lin et al., 2014; Lorenz et al., 2011).

As great progress in NGS technologies facilitated WGS, as well as a GBS strategy, GS has become a popular approach to select for quantitative traits in modern breeding programmes. GS holds great promise for plant breeding because it can theoretically achieve multiple cycles of selection in the amount of time required to complete one cycle using phenotypic selection approaches (Bernardo and Yu, 2007; Heffner et al., 2009; Lipka et al., 2015). For many traits, phenotyping has become a greater impediment to breeding progress than genotyping. GS has also been proven to be useful for traits with low heritability, or traits for which phenotyping is expensive (Endelman et al., 2014; Heslot et al., 2015). Even though the availability of the high density of SNP markers made GS possible and affordable in plant breeding, some aspects of NGS-derived SNPs also need to be practically considered. For example, SNP markers generated by GBS tend to have higher missing rates, and the use of a single reference genome can result in ascertainment bias in the markers (Lipka et al., 2015). Moreover, as mentioned earlier, lack of expertise in bioinformatics and efficient and accurate computational processing are main challenges for plant breeders. Recent development of bead-array- or chip-derived SNP markers enables plant breeders to overcome the challenges of GBS strategy; however, cost-effectiveness as using the GoldenGate arrays or Illumina Infinium BeadChip (Illumina, Inc., San Diego, CA, USA) in molecular breeding programmes is also a practical challenge that needs be considered.

In soybean, many studies have analysed prospects of GS method for different quantitative traits. Researchers from Pioneer were among the first breeders to use a form of GS, which they referred to as context-specific MAS, for soybean breeding (Sebastian et al., 2010). Using that method, they observed average yield gains of 3.1% in reselected lines, with a maximum gain of 5.8%. Jarquín et al. (2014) developed GBS-derived SNP markers to evaluate the accuracy of prediction models for grain yield, plant height and maturity in a set of 301 soybean experimental lines. The authors concluded that standard additive G-BLUP models can be used on unfiltered, imputed GBS data without a loss in accuracy, and that GP had a good potential to expedite genetic gain. Studying resistance to SCN and sudden death syndrome, Bao et al. (2014, 2015) assessed GS accuracy using 282 representative breeding lines from the University of Minnesota soybean breeding programme and concluded that GS using the full 1536 SNP marker set produced average prediction accuracy for SCN resistance significantly more accurate than traditional MAS. Zhang et al. (2016) compared the prediction accuracies of GS and MAS for seed weight using over 300 soybean germplasm accessions, and reported that the performance of GS was better than that of MAS. Studying predictive properties of GS for six quantitative traits using 5K SNP BeadChips to genotype a soybean nested-association mapping (NAM) population, Xavier et al. (2016) concluded that training population size was the main limiting factor for prediction accuracy. Using the BARC SoySNP6K BeadChip developed by Song et al. (2014) to genotype a population of 235 soybean varieties and to assess prediction accuracy of GS for plant height and yield, Ma et al. (2016) showed that different factors, such as marker density, marker preselection procedure and haplotype block analysis, had different impacts on prediction accuracy for each trait. Duhnen et al. (2017) compared different prediction models of GS for yield and protein content in a breeding population and showed that the prediction accuracy was improved by including epistasis in the BLUP model.

The above analyses and recent studies of GS in soybean breeding suggest that GS offers an excellent opportunity to increase genetic gain of complex traits per unit time and cost. The availability of the Williams 82 reference genome sequence (Schmutz et al., 2010), as well as new reference genomes of *G. max* and *G. soja* germplasm accessions (Nguyen, pers. comm.), will ease ascertainment bias for the high-throughput marker generation and will facilitate the implementation of GS. NGS-based SNP genotyping has increased the prediction accuracy of GEBV over other established marker platforms; however, to obtain the maximum benefits from GS, a good combination of advanced marker genotyping and high-throughput phenotyping (HTP) methods is required (Bhat et al., 2016). Moreover, the continuous improvement of statistical approaches for more accuracy, in conjunction with significant cost reduction for marker genotyping and computational efficiency, will greatly facilitate GS applications in soybean breeding.

6 Conclusion and future trends

Remarkable advances in NGS technologies during recent decades have made DNA sequencing approaches possible and affordable due to significant cost reductions and high-throughput improvements. The availability of a reference genome for *G. max* facilitates resequencing efforts (Schmutz et al., 2010). The genome sequence information, coupled with efficient bioinformatics tools, has enabled soybean scientists to identify large numbers of SNPs distributed throughout the genome, including some potentially located in coding regions where causal genes are associated with traits of interest. With the powerful capacity to generate genome-wide marker data coupled with robust, cost-effective and high-throughput genotyping chemistries, the GBS strategy has made GP possible as a powerful approach to effectively predict desirable genotypes for subsequent selection using a GS approach. Furthermore, WGS information also enables soybean scientists to conduct genome-wide haplotype analysis to identify structural variants of target genes, from which the breeders are able to select the most appropriate allelic combinations of the corresponding loci for a subsequent genome-wide breeding strategy.

In addition to effective and robust marker genotyping technologies, successful GAB also requires accurate and HTP platforms. With a HTP platform, large breeding populations or a number of germplasm accessions grown under given treatment conditions can be efficiently assessed and utilized for early phenotypic prediction and subsequent GS (Cabrera-Bosquet et al., 2012). Although HTP platforms have been adapted for breeding programmes in cotton, corn and rice, they seem to be less prevalent in soybean breeding, particularly in public breeding programmes. Compared to the innovations of DNA marker discovery and genotyping methods, the ability to rapidly and accurately measure plant performance in greenhouse or field environments remains a limiting factor in modern plant breeding.

The NGS revolution, integrated with various effective genome-wide approaches, has paved the way for the advent of next-generation breeding (NGB). It is anticipated that in the post-genomic era, NGB should be able to produce new crop plants, not only for a few selected traits, but also for most loci in the genome. Moreover, the knowledge of haplotypes of favourable alleles can be utilized for crossing or backcrossing to effectively pyramid not only for QTL, but also for specific allelic variants present in diverse germplasm for desirable genotypes (Barabaschi et al., 2016). Despite the fact that genetic resources are still essential for breeding programmes, the applications of NGS technologies, bioinformatics and HTP tools continue revolutionizing the NGB through the concept of breeding-by-design.

6.1 Patent issues

Since the 1990s, there has been a frenzy to patent the use of MAS to select for a wide range of desirable traits in soybean, primarily in the private sector. As a result, there are technically many potential legal obstacles to the use of MAS in soybean breeding programmes. The less stringent requirements for patent approvals relative to peer-reviewed publication have permitted the issue of patents covering claims that are not always supported by data or persuasive evidence that the patent applicants actually achieved, or are capable of carrying out all of the 'inventions' covered by the patents, such as introgression of loci from perennial *Glycine* species. Previous publication of research results in a journal does not prevent other individuals from filing a patent claim on using MAS to select for the discoveries reported. Since the genomic locations of many of the QTL covered by MAS patents are typically described proprietary DNA markers developed by the patent applicants, it can be challenging for a patent reviewer to determine whether the loci described are the same as ones that have already been reported in the literature or in presentations given at scientific meetings.

The 'gold rush' to patent the use of MAS in soybean breeding apparently began with an application filed by Pioneer Hi-Bred International, Inc. in January 1994 to restrict the use of nucleic acid markers to introgress SCN resistance into elite soybean germplasm, and US Patent 5,491,081 was issued in February 1996 (Webb, 1996). Since then, dozens of other patent applications pertaining to the use of MAS and/or the introgression of 'loci' associated with useful traits inherited from specific germplasm accessions in the USDA Soybean Germplasm Collection have been filed and approved. An example of these is US Patent 9,091.681, issued to Baley et al. (2015) in July 2015, the fifth in a series of patents with the broad title 'Method to identify resistance quantitative trait loci in soybean and compositions thereof'. This patent essentially restricts the unapproved use of MAS to select soybean plants that carry the *Rpp1* gene from PI 200492 of the USDA Soybean Germplasm Collection, though this gene is referred to as 'ASR resistance locus 1' in the patent. PI 200492 was reported to be resistant to SBR in Taiwan in 1974, and the single dominant gene conditioning resistance was later named *Rpp1* by Hartwig and Bromfield (1983). The *Rpp1* locus was mapped to Chr. 18 (LG G) through a collaboration between USDA-ARS and Monsanto (Hyten et al., 2007b). Despite the long and well-documented history of research that led to the mapping of the *Rpp1* locus, scientists representing Monsanto were able to obtain a patent (Anonymous) on the use of MAS to introgress the re-named 'ARS resistance locus 1' into breeding lines. MAS for virtually any trait of economic importance in soybean and its wild relatives has been the subject of patents and patent applications, many of which substitute surrogate names for published gene names approved by the Soybean Genetics Committed after rigorous investigation.

It is unclear whether, and under which circumstances, patents restricting the use of MAS and MABC in public sector soybean breeding programmes would be enforced by the patent holders, but in any case the potential for MAS to expedite soybean breeding progress is, and will continue to be hindered by legal concerns about patents. While such patents are unlikely to impede investigative research activities, they seem likely to reduce the value and availability of promising breeding lines developed with the use of MAS to pyramid genes and QTL, or to select for traits with low heritability.

7 Acknowledgements

We acknowledge funding support from the United Soybean Board towards soybean genetics and genomics research and breeding programme at the University of Missouri, Columbia, MO, 65311. The USDA's Agricultural Research Service is also acknowledged for its support. We would also like to thank Dr. Anne E. Dorrance, the Ohio State University, and Dr. Dechun Wang, the Michigan State University, for their constructive discussion on marker-assisted selection for soybean diseases and aphids.

8 Where to look for further information

For further information related to advances in molecular marker genotyping technology and the applications of MAS, the books by Xu (2010) and Varshney and Tuberosa (2007) are recommended. In addition, Lorenz et al. (2011) and Poland et al. (2012) have provided a comprehensive overview of knowledge and prospects of different genotyping-by-sequencing approaches that can be effectively applied in genomic-assisted breeding in crops, including soybean.

9 References

Abel, G. H. (1969). Inheritance of the capacity for chloride inclusion and chloride exclusion by soybeans. *Crop Science* 9, 697–8.

Akkaya, M. S., Bhagwat, A. A. and Cregan, P. B. (1992). Length polymorphism of single sequence repeat DNA in soybean. *Genetics* 132(4), 1131–9.

Alt, J. and Ryan-Mahmutagic, M. (2013). Soybean aphid biotype 4 identified. *Crop Science* 53(4), 1491–5.

Anai, T., Yamada, T., Kinoshita, T., Rahman, S. M. and Takagi, Y. (2005). Identification of corresponding genes for three low-alpha-linolenic acid mutants and elucidation of their contribution to fatty acid biosynthesis in soybean seed. *Plant Science* 168(6), 1615–23.

Arelli, P. R., Pantalone, V. R., Allen, F. L. and Mengistu, A. (2007). Registration of soybean germplasm JTN-5303. *Journal of Plant Registrations* 1, 69–70.

Arelli, P. R., Pantalone, V. R., Allen, F., Mengistu, A. and Fritz, L. A. (2015). Registration of JTN-5203 soybean germplasm with resistance to multiple cyst nematode populations. *Journal of Plant Registrations* 9, 108–14.

Arelli, P. R., Shannon, J. G., Mengistu, A., Gillen, A. M. and Fritz, L. A. (2017). Registration of conventional soybean germplasm JTN-4307 with resistance to nematodes and fungal diseases. *Journal of Plant Registrations* 11(2), 192–9.

Bales, C., Zhang, G. R., Liu, M. H., Mensah, C., Gu, C. H., Song, Q. J., Hyten, D., Cregan, P. and Wang, D. C. (2013). Mapping soybean aphid resistance genes in PI 567598B. *Theoretical and Applied Genetics* 126(8), 2081–91.

Baley, G. J., George, J., Butruille, D. V., Eathington, S. R., Haverdink, M. D., Kruger, W. M., Ledeaux, J. R., Concibido, V. C., Narvel, J., Pitkin, J. W., Tamulonis, J. and Xie, C. (2015). Method to identify disease resistant quantitative trait loci in soybean and compositions thereof, Monsanto Technology LLC, St. Louis, MO, US Patent No. 9,091,681.

Bandillo, N., Jarquín, D., Song, Q. J., Nelson, R., Cregan, P., Specht, J. and Lorenz, A. (2015). A population structure and genome-wide association analysis on the USDA Soybean Germplasm Collection. *The Plant Genome* 8(3), 1–13.

Bao, Y., Vuong, T., Meinhardt, C., Tiffin, P., Denny, R., Chen, S. Y., Nguyen, H. T., Orf, J. H. and Young, N. D. (2014). Potential of association mapping and genomic selection to explore PI 88788 derived soybean cyst nematode resistance. *The Plant Genome* 7(3), 1–13.

Bao, Y., Kurle, J. E., Anderson, G. and Young, N. D. (2015). Association mapping and genomic prediction for resistance to sudden death syndrome in early maturing soybean germplasm. *Molecular Breeding* 35, 128–42.

Barabaschi, D., Tondelli, A., Desiderio, F., Volante, A., Vaccino, P., Vale, G. and Cattivelli, L. (2016). Next generation breeding. *Plant Science* 242, 3–13.

Barril, P. and Nates, S. (2012). Introduction to agarose and polyacrylamide gel electrophoresis matrices with respect to their detection sensitivities. In: Magdeldin S. (Ed.), *Gel Electrophoresis - Principles and Basics*, InTech. DOI:10.5772/38573. Available from: https://www.intechopen.com/books/gel-electrophoresis-principles-and-basics/introduction-to-agarose-and-polyacrylamide-gel-electrophoresis-matrices-with-respect-to-their-detect (Accessed 12 November 2017).

Bassam, B. J. and Gresshoff, P. M. (2007). Silver staining DNA in polyacrylamide gels. *Nature Protocols* 2(11), 2649–54.

Bastien, M., Sonah, H. and Belzile, F. (2014). Genome wide association mapping of *Sclerotinia sclerotiorum* resistance in soybean with a genotyping-by-sequencing approach. *The Plant Genome* 7(1), 1–13.

Bernardo, R. and Yu, J. (2007). Prospects for genomewide selection for quantitative traits in maize. *Crop Science* 47, 1082–90.

Bhat, J. A., Ali, S., Salgotra, R. K., Mir, Z. A., Dutta, S., Jadon, V., Tyagi, A., Mushtaq, M., Jain, N., Singh, P. K., Singh, G. P. and Prabhu, K. V. (2016). Genomic selection in the era of next generation sequencing for complex traits in plant breeding. *Frontiers in Genetics* 7, 221. DOI:10.3389/fgene.2016.00221. Available from: https://www.ncbi.nlm.nih.gov/pmc/articles/PMC5186759/.

Bilyeu, K. D., Palavalli, L., Sleper, D. A. and Beuselinck, P. R. (2003). Three microsomal omega-3 fatty-acid desaturase genes contribute to soybean linolenic acid levels. *Crop Science* 43, 1833–8.

Bilyeu, K., Palavalli, L., Sleper, D. and Beuselinck, P. (2005). Mutations in soybean microsomal omega-3 fatty acid desaturase genes reduce linolenic acid concentration in soybean seeds. *Crop Science* 45(5), 1830–6.

Bilyeu, K., Palavalli, L., Sleper, D. A. and Beuselinck, P. (2006). Molecular genetic resources for development of 1% linolenic acid soybeans. *Crop Science* 46(5), 1913–8.

Bilyeu, K., Gillman, J. D. and Leroy, A. R. (2011). Novel FAD3 mutant allele combinations produce soybeans containing 1% linolenic acid in the seed oil. *Crop Science* 51(1), 259–64.

Blanco, F. F., Folegatti, M. V., Gheyi, H. R. and Fernandes, P. D. (2007). Emergence and growth of corn and soybean under saline stress. *Scientia Agricola* 64, 451–9.

Blumwald, E. and Grover, A. (2006). Salt tolerance. In: Halford, N. G. (Ed.), *Plant Biotechnology: Current and Future Applications of Genetically Modified Crops*, John Wiley & Sons, Ltd., Chichester, UK, pp. 206–24. DOI:10.1002/0470021837.ch11. Available from: http://onlinelibrary.wiley.com/doi/10.1002/0470021837.ch11/summary.

Boerma, H. R. and Hussey, R. S. (1992). Breeding plants for resistance to nematodes. *Journal of Nematology* 24, 242–52.

Bonde, M. R., Nester, S. E., Austin, C. N., Stone, C. L. and Frederick, R. D. (2006). Evaluation of virulence of *Phakopsora pachyrhizi* and *P. meibomiae* isolates. *Plant Disease* 90, 708–16.

Botstein, D., White, R. L., Skolnick, M. and Davis, R. W. (1980). Construction of a genetic linkage map in man using restriction fragment length polymorphisms. *American Journal of Human Genetics* 32(3), 314–31.

Boutin, S., Ansari, H., Concibido, V., Denny, R., Orf, J. and Young, N. D. (1992). RFLP analysis of cyst nematode resistance in soybean. *Soybean Genetics Newsletter* 19, 123–7.

Bradbury, P. J., Zhang, Z., Kroon, D. E., Casstevens, T. M., Ramdoss, Y. and Buckler, E. S. (2007). TASSEL: Software for association mapping of complex traits in diverse samples. *Bioinformatics* 23(19), 2633–5.

Buhr, T., Sato, S., Ebrahim, F., Xing, A., Zhou, Y., Mathiesen, M., Schweiger, B., Kinney, A., Staswick, P. and Tom, C. (2002). Ribozyme termination of RNA transcripts down-regulate seed fatty acid genes in transgenic soybean. *Plant Journal* 30(2), 155–63.

Cabrera-Bosquet, L., Crossa, J., Von Zitzewitz, J., Serret, M. D. and Araus, J. L. (2012). High-throughput phenotyping and genomic selection: The frontiers of crop breeding converge. *Journal of Integrative Plant Biology* 54(5), 312–20.

Caetano-Anolles, G., Bassam, B. J. and Gresshoff, P. M. (1992). Primer-template interactions during DNA amplification fingerprinting with single arbitrary oligonucleotides. *Molecular and General Genetics* 235(2–3), 157–65.

Chakraborty, N., Curley, J. C., Frederick, R. D., Hyten, D. L., Nelson, R. L., Hartman, G. L. and Diers, B. W. (2009). Mapping and confirmation of a new allele at *Rpp1* from soybean PI 594538A conferring RB lesion–type resistance to soybean rust. *Crop Science* 49, 783–90.

Chandrasena, D., Wang, Y., Bales, C., Yuan, J. Z., Gu, C. H. and Wang, D. C. (2015). Pyramiding *rag3, rag1b, rag4,* and *rag1c* aphid-resistant genes in soybean germplasm. *Crop Science* 55(5), 2108–15.

Chang, S. J. C., Doubler, T. W., Kilo, V. Y., Abu-Thredeih, J., Prabhu, R., Freire, V., Suttner, R., Klein, J., Schmidt, M. E., Gibson, P. T. and Lightfoot, D. A. (1997). Association of loci underlying field resistance to soybean sudden death syndrome (SDS) and cyst nematode (SCN) race 3. *Crop Science* 37, 965–71.

Childs, S. P., King, Z. R., Walker, D. R., Harris, D. K., Pedley, K. F., Buck, J. W., Boerma, H. R. and Li, Z. (2017). Discovery of a seventh Rpp soybean rust resistance locus in soybean accession PI 605823. *Theoretical and Applied Genetics* 131(1), 27–41. Online, DOI:10.1007/s00122-017-2983-4. Available from: https://link.springer.com/article/10.1007%2Fs00122-017-2983-4.

Chinnusamy, V., Jagendorf, A. and Zhu, J. K. (2005). Understanding and improving salt tolerance in plants. *Crop Science* 45, 437–48.

Cho, Y. G., Ishii, T., Temnykh, S., Chen, X., Lipovich, L., McCouch, S. R., Park, W. D., Ayres, N. and Cartinhour, S. (2000). Diversity of microsatellites derived from genomic libraries and GenBank sequences in rice (*Oryza sativa* L.). *Theoretical and Applied Genetics* 100(5), 713–22.

Collard, B. C. and Mackill, D. J. (2008). Marker-assisted selection: An approach for precision plant breeding in the twenty-first century. *Philosophical Transactions of the Royal Society B: Biological Sciences* 363(1491), 557–72.

Collins, F. S., Brooks, L. D. and Chakravarti, A. (1998). A DNA polymorphism discovery resource for research on human genetic variation. *Genome Research* 8(12), 1229–31.

Concibido, V. C., Denny, R. L., Boutin, S. R., Hautea, R., Orf, J. H. and Young, N. D. (1994). DNA marker analysis of loci underlying resistance to soybean cyst nematode (*Heterodera glycines* Ichinohe). *Crop Science* 34, 240–6.

Cook, D. E., Lee, T. G., Guo, X., Melito, S., Wang, K., Bayless, A. M., Wang, J., Hughes, T. J., Willis, D. K., Clemente, T. E., Diers, B. W., Jiang, J., Hudson, M. E. and Bent, A. F. (2012). Copy number variation of multiple genes at *Rhg1* mediates nematode resistance in soybean. *Science* 338(6111), 1206–9.

Cook, D. E., Bayless, A. M., Wang, K., Guo, X., Song, Q., Jiang, J. and Bent, A. F. (2014). Distinct copy number, coding sequence, and locus methylation patterns underlie *Rhg1*-mediated soybean resistance to soybean cyst nematode. *Plant Physiology* 165(2), 630–47.

Cregan, P. B., Bhagwat, A. A., Akkaya, M. S. and Rongwen, J. (1994). Microsatellite fingerprinting and mapping of soybean. *Methods in Molecular and Cellular Biology* 5, 49–61.

Cregan, P. B., Mudge, J., Fickus, E. W., Danesh, D., Denny, R. and Young, N. D. (1999). Two simple sequence repeat markers to select for soybean cyst nematode resistance conditioned by the rhg1 locus. *Theoretical and Applied Genetics* 99(5), 811–18.

Crossa, J., Pérez-Rodríguez, P., Cuevas, J., Montesinos-López, O., Jarquín, D., De Los Campos, G., Burgueño, J., González-Camacho, J. M., Pérez-Elizalde, S., Beyene, Y., Dreisigacker, S., Singh, R., Zhang, X., Gowda, M., Roorkiwal, M., Rutkoski, J. and Varshney, R. K. (2017). Genomic selection in plant breeding: Methods, models, and perspectives. *Trends in Plant Science* 22(11), 961–75.

Davis, E. L., Koenning, S. R., Burton, J. W. and Barker, K. R. (1996). Greenhouse evaluation of selected soybean germplasm for resistance to North Carolina populations of *Heterodera glycines, Rotylenchulus reniformis*, and *Meloidogyne Species. Journal of Nematology* 28(4S), 590–8.

Dhanapal, A. P., Ray, J. D., Singh, S. K., Hoyos-Villegas, V., Smith, J. R., Purcell, L. C., Andy King, C., Cregan, P. B., Song, Q. and Fritschi, F. B. (2015). Genome-wide association study (GWAS) of carbon isotope ratio (delta13C) in diverse soybean [*Glycine max* (L.) Merr.] genotypes. *Theoretical and Applied Genetics* 128(1), 73–91.

Didenko, V. V. (2001). DNA probes using fluorescence resonance energy transfer (FRET): Designs and applications. *Biotechniques* 31(5), 1106–21.

Dierking, E. C. and Bilyeu, K. D. (2009). New sources of soybean seed meal and oil composition traits identified through TILLING. *BMC Plant Biology* 9, 89–100.

Diers, B. W., Kim, K. S., Frederick, R. D., Hartman, G. L., Unfried, J., Schultz, S. and Cary, T. (2014). Registration of eight soybean germplasm lines resistant to soybean rust. *Journal of Plant Registrations* 8, 96–101.

Do, T. D., Vuong, T. D., Dunn, D., Clubbs, M., Dang, N. D. N., Chen, P., Nguyen, H. T. and Shannon, J. G. (2017a). Identification of new loci for salt tolerance in soybean by high-resolution genome-wide association mapping (*in preparation*).

Do, T. D., Vuong, T. D., Dunn, D., Smothers, S., Patil, G., Yungbluth, D. C., Chen, P., Scaboo, A., Xu, D., Carter, T. E., Nguyen, H. T. and Shannon, J. G. (2017b). Mapping and confirmation of loci for salt tolerance in a novel soybean germplasm, Fiskeby III. *Theoretical and Applied Genetics* (*in Press*). DOI:https://doi.org/10.1007/s001.

Doucet, M. E. and Pinochet, J. (1992). Occurrence of Meloidogyne spp. in Argentina. *Supplement Journal Nematology* 24, 765–70.

Duhnen, A., Gras, A., Teyssèdre, S., Romestant, M., Claustres, B., Daydé, J. and Mangin, B. (2017). Genomic selection for yield and seed protein content in soybean: A study of breeding program data and assessment of prediction accuracy. *Crop Science* 57, 1325–37.

Edwards, D., Forster, J. W., Chagné, D. and Batley, J. (2007). What are SNPs? In: Oraguzie, N. C., Rikkerink, E. H. A., Gardiner, S. E. and Silva, H. N. D. (Eds), *Association Mapping in Plants*, Springer, New York, NY, pp. 41–52. DOI:10.1007/978-0-387-36011-9_3.

Elsheikh, E. A. E. and Wood, M. (1995). Nodulation and N_2 fixation by soybean inoculated with salt-tolerant Rhizobia or salt-sensitive Bradyrhizobia in saline soil. *Soil Biology and Biochemistry* 27, 657–61.

Elshire, R. J., Glaubitz, J. C., Sun, Q., Poland, J. A., Kawamoto, K., Buckler, E. S. and Mitchell, S. E. (2011). A robust, simple genotyping-by-sequencing (GBS) approach for high diversity species. *PLoS One* 6(5), e19379. Available from: https://doi.org/10.1371/journal.pone.0019379.

Endelman, J. B., Atlin, G. N., Beyene, Y., Semagn, K., Zhang, X., Sorrells, M. E. and Jannink, J. L. (2014). Optimal design of preliminary yield trials with genome-wide markers. *Crop Science* 54, 48–59.

Eujayl, I., Sorrells, M. E., Baum, M. and Wolters, P. (2002). Isolation of EST-derived mircosatellite markers for genotyping the A and B genomes of wheat. *Theoretical and Applied Genetics* 104(2–3), 399–407.

Fallen, B. D., Allen, F. L., Kopsell, D. A., Saxton, A. M., McHale, L., Shannon, J. G., Kantartzi, S. K., Cardinal, A. J., Cregan, P. B., Hyten, D. L. and Pantalone, V. R. (2015). Selective genotyping for marker-assisted selection strategies for soybean yield improvement. *Plant Genetics, Genomics, and Biotechnology* 2(1), 95–119.

Fehr, W. R. (2007). Breeding for modified fatty acid composition in soybean. *Crop Science* 47(Supplement_3), S-72–S-87.

Fehr, W. R. and Hammond, E. G. (2000). Reduced linolenic acid production in soybeans, US Patent No. 6,133,509.

Fehr, W. R., Welke, G. A., Hammond, E. G., Duvick, D. N. and Cianzio, S. R. (1992). Inheritance of reduced linolenic acid content in soybean genotypes A16 and A17. *Crop Science* 32, 903–6.

Francia, E., Tacconi, G., Crosatti, C., Barabaschi, D., Bulgarelli, D., Dall'aglio, E. and Vale, G. (2005). Marker assisted selection in crop plants. *Plant Cell, Tissue and Organ Culture* 82(3), 317–42.

Gao, J. F., Zhang, G. X., Qin, Y. C., Yu, K., Li, M. H. and Qin, Y. (1993). Field survey on overwintering syrphids in Changbai Mountain and artificial protection of overwintering syrphids. *Journal of Biological Control* 19, 142–3.

Garcia, A., Calvo, E. S., De Souza Kiihl, R. A., Harada, A., Hiromoto, D. M. and Vierira, L. G. E. (2008). Molecular mapping of soybean rust (*Phakopsora pachyrhizi*) resistance genes: discovery of a novel locus and alleles. *Theoretical and Applied Genetics* 117(4), 545–53.

Glaubitz, J. C., Casstevens, T. M., Lu, F., Harriman, J., Elshire, R. J., Sun, Q. and Buckler, E. S. (2014). TASSEL-GBS: A high capacity genotyping by sequencing analysis pipeline. *PLoS One* 9(2), e90346. DOI:https://doi.org/10.1371/journal.pone.0090346. Available from: http://journals.plos.org/plosone/article?id=10.1371/journal.pone.0090346.

Guo, B., Sleper, D. A., Nguyen, H. T., Arelli, P. R. and Shannon, J. G. (2006). Quantitative trait loci underlying resistance to three soybean cyst nematode populations in soybean PI 404198A. *Crop Science* 46, 224–33.

Gutierrez-Gonzalez, J. J., Vuong, T. D., Zhong, R., Yu, O., Lee, J. D., Shannon, G., Ellersieck, M., Nguyen, H. T. and Sleper, D. A. (2011). Major locus and other novel additive and epistatic loci involved in modulation of isoflavone concentration in soybean seeds. *Theoretical and Applied Genetics* 123(8), 1375–85.

Ha, B. K., Bennett, J. B., Hussey, R. S., Finnerty, S. L. and Boerma, H. R. (2004). Pedigree analysis of a major QTL conditioning soybean resistance to southern root-knot nematode. *Crop Science* 44, 758–63.

Ha, B. K., Hussey, R. S. and Boerma, H. R. (2007). Development of SNP assays for marker-assisted selection of two southern root-knot nematode resistance QTL in soybean. *Crop Science* 47(2), S73–S82.

Ha, B. K., Vuong, T. D., Velusamy, V., Nguyen, H. T., Shannon, J. G. and Lee, J. D. (2013). Genetic mapping of quantitative trait loci conditioning salt tolerance in wild soybean (*Glycine soja*) PI 483463. *Euphytica* 193(1), 79–88.

Harris, D. K., Boerma, H. R., Hussey, R. S. and Finnerty, S. L. (2003). Additional sources of soybean germplasm resistant to two species of rootknot nematode. *Crop Science* 43, 1848–51.

Hartman, G. L., Domier, L. L., Wax, L. M., Helm, C. G., Onstad, D. W., Shaw, J. T., Solter, L. F., Voegtlin, D. J., D'Arcy, C. J., Gray, M. E. and Steffey, K. L. (2001). Occurrence and distribution of Aphis glycines on soybeans in Illinois in 2000 and its potential control. *Plant Health Progress*. Online, DOI:10.1094/PHP-2001-0205-01-HN. Available from: https://naldc.nal.usda.gov/download/11854/.

Hartman, G. L., Miles, M. R. and Frederick, R. D. (2005). Breeding for resistance to soybean rust. *Plant Disease* 89, 664–6.

Hartwig, E. E. (1986). Identification of a fourth major gene conferring resistance to soybean rust. *Crop Science* 26, 1135–6.

Hartwig, E. E. and Bromfield, K. R. (1983). Relationships among three genes conferring specific resistance to rust in soybeans. *Crop Science* 23, 237–9.

He, C., Holme, J. and Anthony, J. (2014a). SNP genotyping: The KASP Assay. In: Fleury, D. and Whitford, R. (Eds), *Crop Breeding – Methods and Protocols*, Springer Science, New York, NY, pp. 75–86.

He, J., Zhao, X., Laroche, A., Lu, Z. X., Liu, H. and Li, Z. (2014b). Genotyping-by-sequencing (GBS), an ultimate marker-assisted selection (MAS) tool to accelerate plant breeding. *Frontiers in Plant Science* 5, 484.

Heffner, E. L., Sorrells, M. E. and Jannink, J. L. (2009). Genomic selection for crop improvement. *Crop Science* 49, 1–12.

Heim, C. B. and Gillman, J. D. (2017). Genotyping-by-sequencing based investigation of the genetic architecture responsible for a ~sevenfold increase in soybean seed stearic acid. *G3 Genes Genomes Genetics* 7(1), 299–308.

Hesler, L. S., Prischmann, D. A. and Dashiell, K. E. (2012). Field and laboratory evaluations of soybean lines against soybean aphid (Hemiptera: Aphididae). *Journal of Economic Entomology* 105(2), 608–15.

Heslot, N., Yang, H. P., Sorrells, M. E. and Jannink, J. L. (2012). Genomic selection in plant breeding: A comparison of models. *Crop Science* 52, 146–60.

Heslot, N., Jannink, J. L. and Sorrells, M. E. (2015). Perspectives for genomic selection applications and research in plants. *Crop Science* 55, 1–12.

Hill, C. B., Li, Y. and Hartman, G. L. (2004). Resistance to the soybean aphid in soybean germplasm. *Crop Science* 44(1), 98–106.

Hill, C. B., Li, Y. and Hartman, G. L. (2006a). A single dominant gene for resistance to the soybean aphid in the soybean cultivar Dowling. *Crop Science* 46 (4), 1601–5.

Hill, C. B., Li, Y. and Hartman, G. L. (2006b). Soybean aphid resistance in soybean Jackson is controlled by a single dominant gene. *Crop Science* 46(4), 1606–8.

Hill, C. B., Kim, K. S., Crull, L., Diers, B. W. and Hartman, G. L. (2009). Inheritance of resistance to the soybean aphid in soybean PI 200538. *Crop Science* 49(4), 1193–200.

Holland, P. M., Abramson, R. D., Watson, R. and Gelfand, D. H. (1991). Detection of specific polymerase chain reaction product by utilizing the 5′----3′ exonuclease activity of *Thermus aquaticus* DNA polymerase. *Proceedings of the National Academy of Sciences of the USA* 88 (16), 7276–80.

Hossain, M. M., Akamatsu, H., Morishita, M., Mori, T., Yamaoka, Y., Suenaga, K., Soares, R. M., Bogado, A. N., Ivancovich, A. J. G. and Yamanaka, N. (2015). Molecular mapping of Asian soybean rust resistance in soybean landraces PI 594767A, PI 587905 and PI 416764. *Plant Pathology* 64, 147–56.

Hussey, R. S. and Janssen, G. J. W. (2002). Root knot nematodes: *Meloidogyne species*. In: Starr, J. L., Cook, R. and Bridge, J. (Eds), *Plant Resistance to Parasitic Nematodes*. CAB International, New York, NY, pp. 43–70.

Hwang, E. Y., Song, Q., Jia, G., Specht, J. E., Hyten, D. L., Costa, J. and Cregan, P. B. (2014). A genome-wide association study of seed protein and oil content in soybean. *BMC Genomics* 15, 1.

Hyten, D. L., Choi, I. Y., Song, Q., Shoemaker, R. C., Nelson, R. L., Costa, J. M., Specht, J. E. and Cregan, P. B. (2007a). Highly variable patterns of linkage disequilibrium in multiple soybean populations. *Genetics* 175(4), 1937–44.

Hyten, D. L., Hartman, G. L., Nelson, R. L., Frederick, R. D., Concibido, V. C., Narvel, J. M. and Cregan, P. B. (2007b). Map location of the *Rpp1* locus that confers resistance to soybean rust in soybean. *Crop Science* 47(2), 837–40.

Hyten, D. L., Song, Q., Choi, I. Y., Yoon, M. S., Specht, J. E., Matukumalli, L. K., Nelson, R. L., Shoemaker, R. C., Young, N. D. and Cregan, P. B. (2008). High-throughput genotyping with the GoldenGate assay in the complex genome of soybean. *Theoretical and Applied Genetics* 116(7), 945–52.

Hyten, D. L., Smith, J. R., Frederick, R. D., Tucker, M. L., Song, Q. and Cregan, P. B. (2009). Bulked segregant analysis using the GoldenGate assay to locate the *Rpp3* locus that confers resistance to soybean rust in soybean. *Crop Science* 49, 265–71.

Hyten, D. L., Cannon, S. B., Song, Q., Weeks, N., Fickus, E. W., Shoemaker, R. C., Specht, J. E., Farmer, A. D., May, G. D. and Cregan, P. B. (2010). High-throughput SNP discovery through deep resequencing of a reduced representation library to anchor and orient scaffolds in the soybean whole genome sequence. *BMC Genomics* 11, 38.

Iwaki, M., Roechan, M., Hibino, H., Tochihara, H. and Tantera, D. M. (1980). A persistent aphidborne virus of soybean, Indonesian soybean dwarf virus. *Plant Disease* 64(11), 1027–30.

Jarquín, D., Kocak, K., Posadas, L., Hyma, K., Jedlicka, J., Graef, G. and Lorenz, A. (2014). Genotyping by sequencing for genomic prediction in a soybean breeding population. *BMC Genomics* 15, 740.

Jiao, Y. Q., Vuong, T. D., Liu, Y., Meinhardt, C., Liu, Y., Joshi, T., Cregan, P. B., Xu, D., Shannon, J. G. and Nguyen, H. T. (2015a). Identification and evaluation of quantitative trait loci underlying resistance to multiple HG types of soybean cyst nematode in soybean PI 437655. *Theoretical and Applied Genetics* 128(1), 15–23.

Jiao, Y. Q., Vuong, T. D., Liu, Y., Li, Z. L., Noe, J., Robbins, R. T., Joshi, T., Xu, D., Shannon, J. G. and Nguyen, H. T. (2015b). Identification of quantitative trait loci underlying resistance to southern root-knot and reniform nematodes in soybean accession PI 567516C. *Molecular Breeding* 35(6), 131.

Jun, T. H., Mian, M. A. R., Freewalt, K., Mittapalli, O. and Michel, A. P. (2012a). Development of genic-SSRs markers from soybean aphid sequences generated by high-throughput sequencing of cDNA library. *Journal of Applied Entomology* 136(8), 614–25.

Jun, T. H., Mian, R. M. A., Kang, S. T. and Michel, A. P. (2012b). Genetic mapping of the powdery mildew resistance gene in soybean PI 567301B. *Theoretical and Applied Genetics* 125(6), 1159–68.

Kabelka, E. A., Carlson, S. R. and Diers, B. W. (2005). Localization of two loci that confer resistance to soybean cyst nematode from *Glycine soja* PI 468916. *Crop Science* 45(6), 2473–81.

Kaczorowski, K. A., Kim, K. S., Diers, B. W. and Hudson, M. E. (2008). Microarray-based genetic mapping using soybean near-isogenic lines and generation of SNP markers in the *Rag1* aphid-resistance interval. *Plant Genome* 1(2), 89–98.

Kadam, S., Vuong, T. D., Qiu, D., Meinhardt, C. G., Song, L., Deshmukh, R., Patil, G., Wan, J. R., Valliyodan, B., Scaboo, A. M., Shannon, J. G. and Nguyen, H. T. (2016). Genomic-assisted phylogenetic analysis and marker development for next generation soybean cyst nematode resistance breeding. *Plant Science* 242, 342–50.

Kendrick, M. D., Harris, D. K., Ha, B. K., Hyten, D. L., Cregan, P. B., Frederick, R. D., Boerma, H. R. and Pedley, K. F. (2011). Identification of a second Asian soybean rust resistance gene in Hyuuga soybean. *Phytopathology* 101, 535–43.

Kim, E., Muga, S. J. and Fischer, S. M. (2004). Identification and characterization of a phorbol ester-responsive element in the murine 8S-lipoxygenase gene. *Journal of Biological Chemistry* 279(12), 11188–97.

Kim, C. S., Schaible, G., Garrett, L., Lubowski, R. and Lee, D. (2008). Economic impacts of the US soybean aphid infestation: A multi-regional competitive dynamic analysis. *Agricultural and Resource Economics Review* 37(2), 227–42.

Kim, K. S., Bellendir, S., Hudson, K. A., Hill, C. B., Hartman, G. L., Hyten, D. L., Hudson, M. E. and Diers, B. W. (2010a). Fine mapping the soybean aphid resistance gene *Rag1* in soybean. *Theoretical and Applied Genetics* 120(5), 1063–71.

Kim, K. S., Hill, C. B., Hartman, G. L., Hyten, D. L., Hudson, M. E. and Diers, B. W. (2010b). Fine mapping of the soybean aphid-resistance gene *Rag2* in soybean PI 200538. *Theoretical and Applied Genetics* 121(3), 599–610.

Kim, K. S., Unfried, J. R., Hyten, D. L., Frederick, R. D., Hartman, G. L., Nelson, R. L., Song, Q. and Diers, B. W. (2012). Molecular mapping of soybean rust resistance in soybean accession PI 561356 and SNP haplotype analysis of the *Rpp1* region in diverse germplasm. *Theoretical and Applied Genetics* 125(6), 1339–52.

Kim, K. S., Chirumamilla, A., Hill, C. B., Hartman, G. L. and Diers, B. W. (2014). Identification and molecular mapping of two soybean aphid resistance genes in soybean PI 587732. *Theoretical and Applied Genetics* 127(5), 1251–9.

Kim, K. S., Vuong, T. D., Qiu, D., Robbins, R. T., Shannon, J. G., Li, Z. L. and Nguyen, H. T. (2016). Advancements in breeding, genetics, and genomics for resistance to three nematode species in soybean. *Theoretical and Applied Genetics* 129(12), 2295–311.

King, Z. R., Harris, D. K., Pedley, K. F., Song, Q., Wang, D., Wen, Z., Buck, J. W., Li, Z. and Boerma, H. R. (2016a). A novel *Phakopsora pachyrhizi* resistance allele (*Rpp*) contributed by PI 567068A. *Theoretical and Applied Genetics* 129(3), 517–34.

King, Z. R., Harris, D. K., Wood, E. D., Buck, J. W., Boerma, H. R. and Li, Z. (2016b). Registration of four near-isogenic soybean lines of G00-3213 for resistance to Asian soybean rust. *Journal of Plant Registrations* 10, 189–94.

Klepadlo, M., Patil, G., Vuong, T. D., Robbins, R. T., Pengyin, C. and Nguyen, H. T. (2017). High-resolution QTL mapping and haplotype analysis reveal pleiotrophic effect of soybean cyst and reniform nematode resistance gene (*in preparation*).

Kwok, P. Y. and Chen, X. (2003). Detection of single nucleotide polymorphisms. *Current Issues Molecular Biology* 5(2), 43–60.

Kwok, P. Y., Deng, Q., Zakeri, H., Taylor, S. and Nickerson, D. A. (1996). Increasing the information content of STS-based genome maps: identifying polymorphisms in mapped STSs. *Genomics* 31(1), 123–6.

La, T. C., Pathan, S. M., Vuong, T., Lee, J. D., Scaboo, A. M., Smith, J. R., Gillen, A. M., Gillman, J., Ellersieck, M. R., Nguyen, H. T. and Shannon, J. G. (2014). Effect of high-oleic acid soybean on seed oil, protein concentration, and yield. *Crop Science* 54(5), 2054–62.

Lam, H. M., Xu, X., Liu, X., Chen, W., Yang, G., Wong, F. L., Li, M. W., He, W., Qin, N., Wang, B., Li, J., Jian, M., Wang, J., Shao, G., Wang, J., Sun, S. S. and Zhang, G. (2010). Resequencing of 31 wild and cultivated soybean genomes identifies patterns of genetic diversity and selection. *Nature Genetics* 42(12), 1053–9.

Lee, G. J., Boerma, H. R., Villagarcia, M. R., Zhou, X., Carter, T. E., Li, Z. and Gibbs, M. O. (2004). A major QTL conditioning salt tolerance in S-100 soybean and descendent cultivars. *Theoretical and Applied Genetics* 109(8), 1610–19.

Lee, J. D., Yu, J. K., Hwang, Y. H., Blake, S., So, Y. S., Lee, G. J., Nguyen, H. T. and Shannon, J. G. (2008). Genetic diversity of wild soybean (*Glycine soja* Sieb. and Zucc.) accessions from South Korea and other countries. *Crop Science* 48, 606–16.

Lee, J. D., Shannon, J. G., Vuong, T. D. and Nguyen, H. T. (2009). Inheritance of salt tolerance in wild soybean (*Glycine soja* Sieb. and Zucc.) accession PI 483463. *Journal of Heredity* 100(6), 798–801.

Lenis, J. M., Gillman, J. D., Lee, J. D., Shannon, J. G. and Bilyeu, K. D. (2010). Soybean seed lipoxygenase genes: molecular characterization and development of molecular marker assays. *Theoretical and Applied Genetics* 120(6), 1139–49.

Lenis, J. M., Ellersieck, M., Blevins, D. G., Sleper, D. A., Nguyen, H. T., Dunn, D., Lee, J. D. and Shannon, J. G. (2011). Differences in ion accumulation and salt tolerance among Glycine accessions. *Journal of Agronomy and Crop Science* 197, 302–10.

Li, Z., Jakkula, L., Hussey, R. S., Tamulonis, J. P. and Boerma, H. R. (2001). SSR mapping and confirmation of the QTL from PI96354 conditioning soybean resistance to southern root-knot nematode. *Theoretical and Applied Genetics* 103(8), 1167–73.

Li, Y., Hill, C. B., Carlson, S. R., Diers, B. W. and Hartman, G. L. (2007). Soybean aphid resistance genes in the soybean cultivars Dowling and Jackson map to linkage group M. *Molecular Breeding* 19(1), 25–34.

Li, S., Smith, J. R., Ray, J. D., Frederick, R. D., Boerma, H. R. and Pedley, K. F. (2012). Identification of a new soybean rust resistance gene in PI 567102B. *Theoretical and Applied Genetics* 125, 133–42.

Li, Y. H., Zhao, S. C., Ma, J. X., Li, D., Yan, L., Li, J., Qi, X. T., Guo, X. S., Zhang, L., He, W. M., Chang, R. Z., Liang, Q. S., Guo, Y., Ye, C., Wang, X. B., Tao, Y., Guan, R. X., Wang, J. Y., Liu, Y. L., Jin,

L. G., Zhang, X. Q., Liu, Z. X., Zhang, L. J., Chen, J., Wang, K. J., Nielsen, R., Li, R. Q., Chen, P. Y., Li, W. B., Reif, J. C., Purugganan, M., Wang, J., Zhang, M. C., Wang, J. and Qiu, L. J. (2013). Molecular footprints of domestication and improvement in soybean revealed by whole genome re-sequencing. *BMC Genomics* 14, 579.

Li, Y. H., Zhou, G., Ma, J., Jiang, W., Jin, L. G., Zhang, Z., Guo, Y., Zhang, J., Sui, Y., Zheng, L., Zhang, S. S., Zuo, Q., Shi, X. H., Li, Y. F., Zhang, W. K., Hu, Y., Kong, G., Hong, H. L., Tan, B., Song, J., Liu, Z. X., Wang, Y., Ruan, H., Yeung, C. K., Liu, J., Wang, H., Zhang, L. J., Guan, R. X., Wang, K. J., Li, W. B., Chen, S. Y., Chang, R. Z., Jiang, Z., Jackson, S. A., Li, R. and Qiu, L. J. (2014). De novo assembly of soybean wild relatives for pan-genome analysis of diversity and agronomic traits. *Nature Biotechnology* 32(10), 1045–52.

Lin, Z., Hayes, B. J. and Daetwyler, H. D. (2014). Genomic selection in crops, trees and forages: A review. *Crop and Pasture Science* 65, 1177–91.

Lindblad-Toh, K., Winchester, E., Daly, M. J., Wang, D. G., Hirschhorn, J. N., Laviolette, J. P., Ardlie, K., Reich, D. E., Robinson, E., Sklar, P., Shah, N., Thomas, D., Fan, J. B., Gingeras, T., Warrington, J., Patil, N., Hudson, T. J. and Lander, E. S. (2000). Large-scale discovery and genotyping of single-nucleotide polymorphisms in the mouse. *Nature Genetics* 24(4), 381–6.

Lipka, A. E., Kandianis, C. B., Hudson, M. E., Yu, J., Drnevich, J., Bradbury, P. J. and Gore, M. A. (2015). From association to prediction: Statistical methods for the dissection and selection of complex traits in plants. *Current Opinion in Plant Biology* 24, 110–18.

Litt, M. and Luty, J. A. (1989). A hypervariable microsatellite revealed by in vitro amplification of a dinucleotide repeat within the cardiac muscle actin gene. *American Journal of Human Genetics* 44(3), 397–401.

Liu, S., Kandoth, P. K., Warren, S. D., Yeckel, G., Heinz, R., Alden, J., Yang, C., Jamai, A., El-Mellouki, T., Juvale, P. S., Hill, J., Baum, T. J., Cianzio, S., Whitham, S. A., Korkin, D., Mitchum, M. G. and Meksem, K. (2012). A soybean cyst nematode resistance gene points to a new mechanism of plant resistance to pathogens. *Nature* 492(7428), 256–60.

Liu, M., Li, S., Swaminathan, S., Sahu, B. B., Leandro, L. F., Cardinal, A. J. C., Bhattacharyya, M., Song, Q., Walker, D. R. and Cianzio, S. R. (2016). Identification of a soybean rust resistance gene in PI 567104B. *Theoretical and Applied Genetics* 129(5), 863–77.

Lorenz, A. J., Chao, S., Asoro, F. G., Heffner, E. L., Hayashi, T., Iwata, H., Smith, K. P., Sorrells, M. E. and Jannink, J. L. (2011). Genomic selection in plant breeding: Knowledge and prospects. *Advances in Agronomy* 110, 77–123.

Lu, F., Lipka, A. E., Glaubitz, J., Elshire, R., Cherney, J. H., Casler, M. D., Buckler, E. S. and Costich, D. E. (2013). Switchgrass genomic diversity, ploidy, and evolution: Novel insights from a network-based SNP discovery protocol. *Plos Genetics* 9(1), e1003215. DOI:10.1371/journal.pgen.1003215. Available from: http://journals.plos.org/plosgenetics/article?id=10.1371/journal.pgen.1003215.

Luzzi, B. M., Boerma, H. R. and Hussey, R. S. (1994). A gene for resistance to the southern root-knot nematode in soybean. *Journal of Heredity* 85, 484–6.

Ma, Y., Reif, J., Jiang, Y., Wen, Z., Wang, D., Liu, Z., Guo, Y., Wei, S., Wang, S., Yang, C., Wang, H., Yang, C., Lu, W., Xu, R., Zhou, R., Wang, R., Sun, Z., Chen, H., Zhang, W., Wu, J., Hu, G., Liu, C., Luan, X., Fu, Y., Guo, T., Han, T., Zhang, M., Sun, B., Zhang, L., Chen, W., Wu, C., Sun, S., Yuan, B., Zhou, X., Han, D., Yan, H., Li, W. and Qiu, L. (2016). Potential of marker selection to increase prediction accuracy of genomic selection in soybean (*Glycine max* L.). *Molecular Breeding* 36, 113–23.

Marino, M. A., Turni, L. A., Del Rio, S. A., Williams, P. E. and Cregan, P. B. (1995). The analysis of simple sequence repeat DNA in soybean by Capillary Gel Electrophoresis. *Theoretical and Applied Genetics* 5(1), 1–5.

Maughan, P. J., Saghi Maroof, M. A. and Buss, G. R. (1995). Microsatellite and amplified sequence length polymorphism in cultivated and wild soybean. *Genome* 38(4), 715–23.

Maughan, P. J., Smith, S., Fairbanks, D. and Jellen, E. (2011). Development, characterization, and linkage mapping of single nucleotide polymorphisms in the grain amaranths (*Amaranthus* sp.). *The Plant Genome* 4(1), 92–101.

Maughan, P. J., Smith, S. M., Rojas-Beltrán, J. A., Elzinga, D., Raney, J. A., Jellen, E. N., Bonifacio, A., Udall, J. A. and Fairbanks, D. J. (2012). Single nucleotide polymorphism identification, characterization, and linkage mapping in Quinoa. *The Plant Genome* 5(3), 114–25.

McLean, R. J. and Byth, D. E. (1980). Inheritance of resistance to rust (*Phakopsora pachyrhizi*) in soybeans. *Australian Journal of Agricultural Research* 31, 951–6.

Meksem, K., Pantazopoulos, P., Njiti, V. N., Hyten, L. D., Arelli, P. R. and Lightfoot, D. A. (2001). 'Forrest' resistance to the soybean cyst nematode is bigenic: Saturation mapping of the *Rhg1* and *Rhg4* loci. *Theoretical and Applied Genetics* 103(5), 710–17.

Mensah, C., Difonzo, C. and Wang, D. (2008). Inheritance of soybean aphid resistance in PI 567541B and PI 567598B. *Crop Science* 48(5), 1759–63.

Meuwissen, T. H. E., Hayes, B. J. and Goddard, M. E. (2001). Prediction of total genetic value using genome-wide dense marker maps. *Genetics* 157, 1819–29.

Mian, M. A. R., Hammond, R. B. and Martin, S. K. S. (2008a). New plant introductions with resistance to the soybean aphid. *Crop Science* 48(3), 1055–61.

Mian, M. A. R., Kang, S. T., Beil, S. E. and Hammond, R. B. (2008b). Genetic linkage mapping of the soybean aphid resistance gene in PI 243540. *Theoretical and Applied Genetics* 117(6), 955–62.

Miles, M. R., Frederick, R. D. and Hartman, G. L. (2006). Evaluation of soybean germplasm for resistance to *Phakopsora pachyrhizi. Plant Health Progress* 92(6), 947–52. DOI:http://dx.doi.org/10.1094/PHP-2006-0104-01-RS (Verified as of April 2017).

Monteros, M. J., Missaoui, A. M., Phillips, D. V., Walker, D. R. and Boerma, H. R. (2007). Mapping and confirmation of the 'Hyuuga' red-brown lesion resistance gene for Asian soybean rust. *Crop Science* 47(2), 829–36.

Moon, J. K., Jeong, S. C., Van, K., Saghai Maroof, M. A. and Lee, S. H. (2009). Marker-assisted identification of resistance genes to soybean mosaic virus in soybean lines. *Euphytica* 169(3), 375–85.

Morgante, M. and Olivieri, A. M. (1993). PCR-amplification microsatellites as markers in plant genetics. *Plant Journal* 3(1), 175–82.

Morgante, M., Rafalski, A., Biddle, P., Tingey, S. and Olivieri, A. M. (1994). Genetic mapping and variability of seven soybean simple sequence repeat loci. *Genome* 37(5), 763–9.

Nguyen, V. T., Vuong, T. D., Vantoai, T., Lee, J. D., Wu, X., Mian, M. a. R., Dorrance, A. E., Shannon, J. G. and Nguyen, H. T. (2012). Mapping of quantitative trait loci associated with resistance to *Phytophthora sojae* and flooding tolerance in soybean. *Crop Science* 52(6), 2481–93.

Nielsen, R., Paul, J. S., Albrechtsen, A. and Song, Y. S. (2011). Genotype and SNP calling from next-generation sequencing data. *Nature Reviews Genetics* 12(6), 443–51.

Olsen, K. M., Womace, A., Garrett, A. R., Suddith, J. I. and Purugganan, D. M. (2002). Contrasting evolutionary forces in the Arabidopsis thaliana floral development pathway. *Genetics* 160(4), 1641–50.

Papiernik, S. K., Grieve, C. M., Lesch, S. M. and Yates, S. R. (2005). Effects of salinity, imazethapyr, and chlorimuron application on soybean growth and yield. *Communications in Soil Science and Plant Analysis* 36, 951–67.

Patil, G., Do, T., Vuong, T. D., Valliyodan, B., Lee, J. D., Chaudhary, J., Shannon, J. G. and Nguyen, H. T. (2016). Genomic-assisted haplotype analysis and the development of high-throughput SNP markers for salinity tolerance in soybean. *Scientific Reports* 6, 19199.

Patil, G., Chaudhary, J., Vuong, T. D., Jenkins, B., Qiu, D., Kadam, S., Shannon, G. J. and Nguyen, H. T. (2017). Development of SNP genotyping assays for seed composition traits in soybean. *International Journal of Plant Genomics* 2017, 12 (Article ID 6572969).

Paul, C. and Hartman, G. L. (2009). Sources of soybean rust resistance challenged with single-spored isolates of *Phakopsora pachyrhizi* collected from the USA. *Crop Science* 49, 1781–5.

Pham, A. T., Lee, J. D., Shannon, J. G. and Bilyeu, K. D. (2010). Mutant alleles of FAD2-1A and FAD2-1B combine to produce soybeans with the high oleic acid seed oil trait. *BMC Plant Biology* 10, 195–208.

Pham, A. T., Lee, J. D., Shannon, J. G. and Bilyeu, K. D. (2011). A novel FAD2-1 A allele in a soybean plant introduction offers an alternate means to produce soybean seed oil with 85% oleic acid content. *Theoretical and Applied Genetics* 123(5), 793–802.

Pham, A. T., McNally, K., Abdel-Haleem, H., Roger Boerma, H. and Li, Z. (2013). Fine mapping and identification of candidate genes controlling the resistance to southern root-knot nematode in PI 96354. *Theoretical and Applied Genetics* 126(7), 1825–38.

Pham, A. T., Bilyeu, K., Chen, P. Y., Boerma, H. R. and Li, Z. L. (2014). Characterization of the *fan1* locus in soybean line A5 and development of molecular assays for high-throughput genotyping of FAD3 genes. *Molecular Breeding* 33(4), 895–907.

Phang, T. H., Shao, G. H. and Lam, H. M. (2008). Salt tolerance in soybean. *Journal of Integrative Plant Biology* 50(10), 1196–212.

Poland, J., Endelman, J., Dawson, J., Rutkoski, J., Wu, S., Manes, Y., Dreisigacker, S., Crossa, J., Sánchez-Villeda, H., Sorrells, M. and Jannink, J. L. (2012). Genomic selection in wheat breeding using genotyping-by-sequencing. *The Plant Genome* 5(3), 103–13.

Poland, J. A. and Rife, T. W. (2012). Genotyping-by-sequencing for plant breeding and genetics. *The Plant Genome* 5(3), 92–102.

Qi, X., Li, M. W., Xie, M., Liu, X., Ni, M., Shao, G., Song, C., Kay-Yuen Yim, A., Tao, Y., Wong, F. L., Isobe, S., Wong, C. F., Wong, K. S., Xu, C., Li, C., Wang, Y., Guan, R., Sun, F., Fan, G., Xiao, Z., Zhou, F., Phang, T. H., Liu, X., Tong, S. W., Chan, T. F., Yiu, S. M., Tabata, S., Wang, J., Xu, X. and Lam, H. M. (2014). Identification of a novel salt tolerance gene in wild soybean by whole-genome sequencing. *Nature Communications* 5, 4340.

Ragsdale, D. W., Landis, D. A., Brodeur, J., Heimpel, G. E. and Desneux, N. (2011). Ecology and management of the soybean aphid in North America. *Annual Review of Entomology* 56, 375–99.

Ray, J. D., Morel, W., Smith, J. R., Frederick, R. D. and Miles, M. R. (2009). Genetics and mapping of adult plant rust resistance in soybean PI 587886 and PI 587880A. *Theoretical and Applied Genetics* 119, 271–80.

Rennie, B. D. and Tanner, J. W. (1989). Fatty acid composition of oil from soybean seeds grown at extreme temperatures. *Journal of American Oil Chemistry Society* 66, 1622–4.

Rennie, B. D. and Tanner, J. W. (1991). New allele at the *fan* locus in the soybean line A5. *Crop Science* 31, 297–301.

Rennie, B. D., Zilka, J., Cramer, M. M. and Buzzell, R. I. (1988). Genetic analysis of low linolenic acid levels in the soybean line PI361088B. *Crop Science* 28, 655–7.

Reynab, N. and Sneller, C. H. (2001). Evaluation of marker-assisted introgression of yield QTL alleles into adapted soybean. *Crop Science* 41(4), 1317–21.

Robbins, R. T., Rakes, L. and Elkins, C. R. (1994). Reproduction of the reniform nematode on thirty soybean cultivars. *Journal of Nematology* 26(4 Suppl.), 659–64.

Rongwen, J., Akkaya, M. S., Bhagwat, A. A., Lavi, U. and Cregan, P. B. (1995). The use of microsatellite DNA markers for soybean genotype identification. *Theoretical and Applied Genetics* 90(1), 43–8.

Ross, A. J., Fehr, W. R., Welke, G. A. and Cianzio, S. R. (2000). Agronomic and seed traits of 1%-linolenate soybean genotypes. *Crop Science* 40, 383–6.

Rosso, M. L., Burleson, S. A., Maupi, L. M. and Rainey, K. M. (2011). Development of breeder-friendly markers for selection of MIPS1 mutations in soybean. *Molecular Breeding* 28, 127–13.

Sanger, F. and Coulson, A. R. (1975). A rapid method for determining sequences in DNA by primed synthesis with DNA polymerase. *Journal of Molecular Biology* 94(3), 441–8.

Schmutz, J., Cannon, S. B., Schlueter, J., Ma, J., Mitros, T., Nelson, W., Hyten, D. L., Song, Q., Thelen, J. J., Cheng, J., Xu, D., Hellsten, U., May, G. D., Yu, Y., Sakurai, T., Umezawa, T., Bhattacharyya, M. K., Sandhu, D., Valliyodan, B., Lindquist, E., Peto, M., Grant, D., Shu, S., Goodstein, D., Barry, K., Futrell-Griggs, M., Abernathy, B., Du, J., Tian, Z., Zhu, L., Gill, N., Joshi, T., Libault, M., Sethuraman, A., Zhang, X. C., Shinozaki, K., Nguyen, H. T., Wing, R. A., Cregan, P., Specht, J., Grimwood, J., Rokhsar, D., Stacey, G., Shoemaker, R. C. and Jackson, S. A. (2010). Genome sequence of the palaeopolyploid soybean. *Nature* 463(7278), 178–83.

Sebastian, S. A., Streit, L. G., Stephens, P. A., Thompson, J. A., Hedges, B. R., Fabrizius, M. A., Soper, J. F., Schmidt, D. H., Kallem, R. L., Hinds, M. A., Feng, L. and Hoeck, J. A. (2010). Context-specific marker-assisted selection for improved grain yield in elite soybean populations. *Crop Science* 50(4), 1196–206.

Shi, Z., Bachleda, N., Pham, A. T., Bilyeu, K., Shannon, G., Nguyen, H. and Li, Z. L. (2015a). High-throughput and functional SNP detection assays for oleic and linolenic acids in soybean. *Molecular Breeding* 35(8), 176.

Shi, Z., Liu, S., Noe, J., Arelli, P., Meksem, K. and Li, Z. (2015b). SNP identification and marker assay development for high-throughput selection of soybean cyst nematode resistance. *BMC Genomics* 16, 314.

Silva, D. C. G., Yamanaka, N., Brogin, R. L., Arias, C. A. A., Nepomuceno, A. L., Di Mauro, A. O., Pereira, S. S., Nogueira, L. M., Passianotto, A. L. L. and Abdelnoor, R. V. (2008). Molecular mapping of two loci that confer resistance to Asian rust in soybean. *Theoretical and Applied Genetics* 117(1), 57–63.

Singleton, P. W. and Bohlool, B. B. (1984). Effect of salinity on nodule formation by soybean. *Plant Physiology* 74, 72–6.

Sonah, H., O'Donoughue, L., Cober, E., Rajcan, I. and Belzile, F. (2015). Identification of loci governing eight agronomic traits using a GBS-GWAS approach and validation by QTL mapping in soya bean. *Plant Biotechnol Journal* 13(2), 211–21.

Song, Q. J., Marek, L. F., Shoemaker, R. C., Lark, K. G., Concibido, V. C., Delannay, X., Specht, J. E. and Cregan, P. B. (2004). A new integrated genetic linkage map of the soybean. *Theoretical and Applied Genetics* 109(1), 122–8.

Song, Q. J., Jia, G. F., Zhu, Y. L., Grant, D., Nelson, R. T., Hwang, E. Y., Hyten, D. L. and Cregan, P. B. (2010). Abundance of SSR motifs and development of candidate polymorphic SSR markers (BARCSOYSSR_1.0) in soybean. *Crop Science* 50(5), 1950–60.

Song, Q., Hyten, D. L., Jia, G., Quigley, C. V., Fickus, E. W., Nelson, R. L. and Cregan, P. B. (2013). Development and evaluation of SoySNP50K, a high-density genotyping array for soybean. *PLoS One* 8(1), e54985.

Song, Q., Jia, G., Quigley, C. V., Fickus, E. W., Hyten, D. L., Nelson, R. L. and Cregan, P. B. (2014). Soybean BARCSoySNP6K Beadchip: A tool for soybean genetics research. In: *Proceedings of the 22th International Plant and Animal Genome Conference, San Diego, CA*. 10–15 January 2014. Scherago International, Jersey City, NJ. Poster P306.

Sun, J., Guo, N., Lei, J., Li, L., Hu, G. and Xing, H. (2014). Association mapping for partial resistance to *Phytophthora sojae* in soybean (*Glycine max* (L.) Merr.). *Journal of Genetics* 93(2), 355–63.

Takagi, Y. and Rahman, S. M. (1996). Inheritance of high oleic acid content in the seed oil of soybean mutant M23. *Theoretical and Applied Genetics* 92, 179–82.

Tautzl, D. and Renz, M. (1984). Simple sequences are ubiquitous repetitive components of eukaryotic genomes. *Nucleic Acids Research* 12(10), 4127–38.

Tenaillon, M. I., Sawkins, M. C., Long, A. D., Gaut, R. L., Doebley, J. F. and Gaut, B. S. (2001). Patterns of DNA sequence polymorphism along chromosome 1 of maize (*Zea mays* spp. mays L.). *Proceedings of the National Academy of Sciences of the USA* 98(16), 9161–6.

Thiel, T., Michalek, W., Varshney, R. K. and Graner, A. (2003). Exploiting EST databases for the development and characterization of gene-derived SSR-markers in barley (*Hordeum vulgare* L.). *Theoretical and Applied Genetics* 106(3), 411–22.

Thomson, M. J. (2014). High-throughput SNP genotyping to accelerate crop improvement. *Plant Breeding and Biotechnology* 2, 195–212.

Torkamaneh, D., Laroche, J., Bastien, M., Abed, A. and Belzile, F. (2017). Fast-GBS: A new pipeline for the efficient and highly accurate calling of SNPs from genotyping-by-sequencing data. *BMC Bioinformatics* 18(1), 5.

Valliyodan, B., Dan, Q., Patil, G., Zeng, P., Huang, J., Dai, L., Chen, C., Li, Y., Joshi, T., Song, L., Vuong, T. D., Musket, T. A., Xu, D., Shannon, J. G., Shifeng, C., Liu, X. and Nguyen, H. T. (2016). Landscape of genomic diversity and trait discovery in soybean. *Science Reports* 6, 23598.

Van, K., Hwang, E. Y., Kim, M. Y., Park, H. J., Lee, S. H. and Cregan, P. B. (2005). Discovery of SNPs in soybean genotypes frequently used as the parents of mapping populations in the United States and Korea. *Journal of Heredity* 96(5), 529–35.

Varshney, R. K. and Tuberosa, R. (2007). Translational genomics for crop breeding, Volume I: Biotic Stresses, 1st Edn, John Wiley & Sons, Inc., Ames, IA.

Vos, P., Hogers, R., Bleeker, M., Reijans, M., Van De Lee, T., Hornes, M., Frijters, A., Pot, J., Peleman, J., Kuiper, M. and Et Al. (1995). AFLP: a new technique for DNA fingerprinting. *Nucleic Acids Research* 23(21), 4407–14.

Vuong, T. D., Sleper, D. A., Shannon, J. G. and Nguyen, H. T. (2010). Novel quantitative trait loci for broad-based resistance to soybean cyst nematode (*Heterodera glycines* Ichinohe) in soybean PI 567516C. *Theoretical and Applied Genetics* 121(7), 1253–66.

Vuong, T. D., Sleper, D. A., Shannon, J. G., Wu, X. and Nguyen, H. T. (2011). Confirmation of quantitative trait loci for resistance to multiple-HG types of soybean cyst nematode (*Heterodera glycines* Ichinohe). *Euphytica* 181(1), 101–13.

Vuong, T. D., Jiao, Y., Shannon, J. G. and Nguyen, H. T. (2013). Nematode resistance in soybean. In: Varshney, R. K. and Tuberosa, R. (Eds), *Translational Genomics for Crop Breeding, Volume I: Biotic Stresses*, (1st Edn), John Wiley & Sons, Inc., Ames, IA, pp. 95–124.

Vuong, T. D., Sonah, H., Meinhardt, C. G., Deshmukh, R., Kadam, S., Nelson, R. L., Shannon, J. G. and Nguyen, H. T. (2015). Genetic architecture of cyst nematode resistance revealed by genome-wide association study in soybean. *BMC Genomics* 16, 593.

Vuong, T. D., Walker, D. R., Nguyen, B. T., Nguyen, T. T., Dinh, H. X., Hyten, D. L., Cregan, P. B., Sleper, D. A., Lee, J. D., Shannon, J. G. and Nguyen, H. T. (2016). Molecular characterization of resistance to soybean rust (*Phakopsora pachyrhizi* Syd. & Syd.) in soybean cultivar DT 2000 (PI 635999). *PLoS One* 11(12), e0164493. DOI:10.1371/journal.pone.0164493.

Vuong, T. D., Wu, X., Pathan, M. S., Valliyodan, B. and Nguyen, H. T. (2007). Genomics approaches to soybean improvement. In: Varshney, R. K. and Tuberosa, R. (Eds), *Genomics Assisted Crop Improvement Vol. 2: Genomics Applications in Crops*, Springer, Manhattan, NY, pp. 243–79.

Walker, D. R., Boerma, H. R., Phillips, D. V., Schneider, R. W., Buckley, J. B., Shipe, E. R., Mueller, J. D., Weaver, D. B., Sikora, E. J., Moore, S. H., Hartman, G. L., Miles, M. R., Harris, D. K., Wright, D. L., Marois, J. J. and Nelson, R. L. (2011). Evaluation of USDA soybean germplasm accessions for resistance to soybean rust in the southern United States. *Crop Science* 51, 678–93.

Walker, D. R., Harris, D. K., King, Z. R., Li, Z., Phillips, D. V., Buck, J. W., Nelson, R. L. and Boerma, H. R. (2014). Soybean germplasm accession seedling reactions to soybean rust isolates from Georgia. *Crop Science* 54, 1433–47.

Wang, D. G., Fan, J. B., Siao, C. J., Berno, A., Young, P., Sapolsky, R., Ghandour, G., Perkins, N., Winchester, E., Spencer, J., Kruglyak, L., Stein, L., Hsie, L., Topaloglou, T., Hubbell, E., Robinson, E., Mittmann, M., Morris, M. S., Shen, N., Kilburn, D., Rioux, J., Nusbaum, C., Rozen, S., Hudson, T. J., Lipshutz, R., Chee, M. and Lander, E. S. (1998). Large-scale identification, mapping, and genotyping of single-nucleotide polymorphisms in the human genome. *Science* 280(5366), 1077–82.

Wang, D., Diers, B. W., Arelli, P. R. and Shoemaker, R. C. (2001). Loci underlying resistance to Race 3 of soybean cyst nematode in *Glycine soja* plant introduction 468916. *Theoretical and Applied Genetics* 103(4), 561–6.

Wang, D., Shi, J., Carlson, S. R., Cregan, P. B., Ward, R. W. and Diers, B. W. (2003). A low-cost, high-throughput polyacrylamide gel electrophoresis system for genotyping with microsatellite DNA markers. *Crop Science* 43(5), 1828–32.

Wang, J., Lin, M., Crenshaw, A., Hutchinson, A., Hicks, B., Yeager, M., Berndt, S., Huang, W. Y., Hayes, R. B., Chanock, S. J., Jones, R. C. and Ramakrishnan, R. (2009). High-throughput single nucleotide polymorphism genotyping using nanofluidic Dynamic Arrays. *BMC Genomics* 10, 561.

Wang, X. Z., Jiang, G. L., Green, M., Scott, R. A., Hyten, D. L. and Cregan, P. B. (2014). Quantitative trait locus analysis of unsaturated fatty acids in a recombinant inbred population of soybean. *Molecular Breeding* 33(2), 281–96.

Warner, K. and Fehr, W. (2008). Mid-oleic/ultra low linolenic acid soybean oil: a healthful new alternative to hydrogenated oil for frying. *Journal of the American Oil Chemists' Society* 85(10), 945.

Webb, D. M. (1996). Soybean cyst nematode resistant soybeans and methods of breeding and identifying resistant plants, Pioneer Hi-Bred International, Inc., Des Moines, IA, US Patent No. 5,491,081.

Welsh, J. and McClelland, M. (1990). Fingerprinting genomes using PCR with arbitrary primers. *Nucleic Acids Research* 18(24), 7213–8.

Wiarda, S. L., Fehr, W. R. and O'Neal, M. E. (2012). Soybean aphid (Hemiptera: Aphididae) development on soybean with *Rag1* alone, *Rag2* alone, and both genes combined. *Journal of Economic Entomology* 105(1), 252–8.

Williams, J. G., Kubelik, A. R., Livak, K. J., Rafalski, J. A. and Tingey, S. V. (1990). DNA polymorphisms amplified by arbitrary primers are useful as genetic markers. *Nucleic Acids Research* 18(22), 6531–5.

Wrather, A., Shannon, G., Balardin, R., Carregal, L., Escobar, R., Gupta, G. K., Ma, Z., Morel, W., Ploper, D. and Tenuta, A. (2010). Effect of diseases on soybean yield in the top eight producing countries in 2006. *Plant Health Progress*. Online, DOI:1094/PHP-2010–0125-01-RS. Available from: https://www.plantmanagementnetwork.org/pub/php/research/2010/loss/.

Wu, Z., Schenk-Hamlin, D., Zhan, W., Ragsdale, D. W. and Heimpel, G. E. (2004). The soybean aphid in China: A historical review. *Annals of the Entomological Society of America* 97(2), 209–18.

Wu, X., Blake, S., Sleper, D. A., Shannon, J. G., Cregan, P. and Nguyen, H. T. (2009). QTL, additive and epistatic effects for SCN resistance in PI 437654. *Theoretical and Applied Genetics* 118(6), 1093–105.

Wu, X., Ren, C., Joshi, T., Vuong, T., Xu, D. and Nguyen, H. T. (2010). SNP discovery by high-throughput sequencing in soybean. *BMC Genomics* 11, 469–79.

Xavier, A., Muir, W. M. and Rainey, K. M. (2016). Assessing predictive properties of genome-wide selection in soybeans. *G3 Genes Genomes Genetics* 6, 2611–16.

Xu, Y. (2010). Molecular breeding tools: markers and maps. In: Xu, Y. (Ed.), *Molecular Plant Breeding*, CAB International, Wallingford and Oxfordshire, UK, pp. 21–43.

Xu, X., Zeng, L., Tao, Y., Vuong, T., Wan, J., Boerma, R., Noe, J., Li, Z., Finnerty, S., Pathan, S. M., Shannon, J. G. and Nguyen, H. T. (2013). Pinpointing genes underlying the quantitative trait loci for root-knot nematode resistance in palaeopolyploid soybean by whole genome resequencing. *Proceedings of the National Academy of Sciences of the USA* 110(33), 13469–74.

Yamanaka, N., Lemos, N. G., Uno, M., Akamatsu, H., Yamaoka, Y., Abdelnoor, R. V., Braccini, A. L. and Suenaga, K. (2013). Resistance to Asian soybean rust in soybean lines with the pyramided three Rpp genes. *Crop Breeding and Applied Biotechnology* 13, 75–82.

Yamanaka, N., Morishita, M., Mori, T., Lemo, N. G., Hossain, M. M., Akamatsu, H., Kato, M. and Yamaoka, Y. (2015). Multiple Rpp -gene pyramiding confers resistance to Asian soybean rust isolates that are virulent on each of the pyramided genes. *Tropical Plant Pathology* 40, 283–90.

Yang, Z., Honda, K. H., Wang, S. and Ma, X. (2004). Re-evaluation of *Glycine soja* germplasm from north eastern China for aphid resistance. *Journal of Jilin Agricultural Science* 29(5), 3–6.

Yates, J. L., Hussey, R. S., Finnerty, S. L. and Boerma, H. R. (2009). Three soybean plant introductions possess unique resistance to peanut root-knot nematode *Crop Science* 50(1), 118–22.

Yu, N., Kim, M., King, Z. R., Harris, D. K., Buck, J. W., Li, Z. L. and Diers, B. W. (2015). Fine mapping of the Asian soybean rust resistance gene *Rpp2* from soybean PI 230970. *Theoretical and Applied Genetics* 128(3), 387–96.

Zhang, Y., Wu, K., Wyckhuys, K. A. G. and Heimpel, G. E. (2009). Effect of parasitism on flight behavior of the soybean aphid, Aphis glycines. *Biological Control* 51(3), 475–79.

Zhang, G. R., Gu, C. H. and Wang, D. C. (2010). A novel locus for soybean aphid resistance. *Theoretical and Applied Genetics* 120(6), 1183–91.

Zhang, G. R., Gu, C. H. and Wang, D. C. (2013). Mapping and validation of a gene for soybean aphid resistance in PI 567537. *Molecular Breeding* 32(1), 131–8.

Zhang, J., Song, Q., Cregan, P. B. and Jiang, G. L. (2016). Genome-wide association study, genomic prediction and marker-assisted selection for seed weight in soybean (*Glycine max*). *Theoretical and Applied Genetics* 129, 117–30.

Zhang, S., Zhang, Z., Wen, Z., Gu, C., An, Y. Q. C., Bales, C., Difonzo, C., Song, Q. and Wang, D. (2017a). Fine mapping of the soybean aphid-resistance genes *Rag6* and *Rag3c* from *Glycine soja* 85-32. *Theoretical and Applied Genetics* 130(12), 2601–15.

Zhang, S. C., Zhang, Z. N., Bales, C., Gu, C. H., Difonzo, C., Li, M., Song, Q. J., Cregan, P., Yang, Z. Y. and Wang, D. C. (2017b). Mapping novel aphid resistance QTL from wild soybean, *Glycine soja* 85-32. *Theoretical and Applied Genetics* 130(9), 1941–52.

Zhou, Z., Jiang, Y., Wang, Z., Gou, Z., Lyu, J., Li, W., Yu, Y., Shu, L., Zhao, Y., Ma, Y., Fang, C., Shen, Y., Liu, T., Li, C., Li, Q., Wu, M., Wang, M., Wu, Y., Dong, Y., Wan, W., Wang, X., Ding, Z., Gao, Y., Xiang, H., Zhu, B., Lee, S. H., Wang, W. and Tian, Z. (2015). Resequencing 302 wild and cultivated accessions identifies genes related to domestication and improvement in soybean. *Nature Biotechnology* 33(4), 408–14.

Zhu, Y. L., Song, Q. J., Hyten, D. L., Van Tassell, C. P., Matukumalli, L. K., Grimm, D. R., Hyatt, S. M., Fickus, E. W., Young, N. D. and Cregan, P. B. (2003). Single-nucleotide polymorphisms in soybean. *Genetics* 163(3), 1123–34.

Ziegle, J. S., Su, Y., Corcoran, K. P., Nie, L., Mayrand, P. E., Hoff, L. B., McBride, L. J., Kronick, M. N. and Diehl, S. R. (1992). Application of automated DNA sizing technology for genotyping microsatellite loci. *Genomics* 14(4), 1026–31.

Chapter 6

Advances in the genetic modification of soybeans

Wensheng Hou, Chinese Academy of Agricultural Sciences, China

1 Introduction

The herbicide-tolerant transgenic soybean [*Glycine max* (L.) Merr.] is one of the earliest genetically modified (GM) commercial crops. Transgenic soybean retains its position as the GM crop cultivated on the largest scale globally, with food, nutritional, industrial and pharmaceutical uses. Between 1996 and 2015, about 1030 million hectares of GM soybean were grown, and an income of more than 50 billion US$ was brought to farmers by growing GM soybean. In 2015, about 92.1 million hectares of GM soybean were grown around the world in 11 countries, accounting for 51% of total GM crop hectarage and 83% of soybean grown globally (James, 2015).

Although transgenic soybean holds a leading market share, its improved traits are quite limited. Up until 2015, 32 biotech events had been approved for cultivation and/or food/feed use since 1994. However, herbicide tolerance has consistently been the dominant trait in GM soybean. Several other traits of improved transgenic cultivars have some applications, such as insect resistance, modified oil/fatty acid and stacked traits for herbicide tolerance and insect resistance or modified oil/fatty acid. In 2015, an abiotic stress-tolerant soybean was approved for commercialization in Argentina, which will cater to the needs of farmers in marginalized and stressed conditions (James, 2015; ISAAA, 2016; Bergau, 2015).

http://dx.doi.org/10.19103/AS.2017.0034.06

Valuable genes and efficient transformation are the key factors for the improvement of GM soybean. As rice is becoming a more popular model plant, soybean studies are lagging behind somewhat in terms of transgenic lines for improving agronomic characters, possibly due to the complexity of the soybean genome and lower transformation efficiency (Xia et al., 2013). In 2010, the first soybean cultivar genomic database was developed from the entire genome sequence (Schmutz et al., 2010). A large-scale re-sequenced analysis database was subsequently reported and the patterns of genetic variation between wild and cultivated soybeans were compared (Lam et al., 2010). Recently, a large amount of soybean genome information has been resolved, and a large number of genomic, transcriptional, functional annotated sequences can now be obtained (Li et al., 2014; Zhou et al., 2015). As for many other crops, the genomic era is now underway for soybean, and these endeavours have enabled the confirmation and identification of genes and multiple loci for important agronomic traits. These valuable genes and genetic resources should facilitate future genetic modification of soybean. Meanwhile, the efficiency of soybean transformation systems has been improved. Soybean transformation was first reported in 1988, taking place as a result of *Agrobacterium tumefaciens* infection with cotyledonary node plant regeneration (Hinchee et al., 1988), or as a result of particle bombardment of meristems of immature seeds (McCabe et al., 1988). After more than 20 years, these two methods continue to be improved and developed. Although the efficient genetic transformation of soybean is not yet routine, it has produced most transgenic soybeans to date (Yamada et al., 2012; Homrich et al., 2012; Hou et al., 2014). Most excitingly of all, the CRISPR (clustered regularly interspaced short palindromic repeat)/Cas (CRISPR-associated) system, a new technology for genome editing, has been successfully used for genome engineering in many important crops in recent years (Shan et al., 2013; Jiang et al., 2013; Feng et al., 2013; Miao et al., 2013; Xie et al., 2014; Upadhyay et al., 2013; Wang et al., 2014; Liang et al., 2014; Svitashev et al., 2015). Since 2015, CRISPR/Cas9-mediated genome editing in soybean has been met with initial success (Jacobs et al., 2015; Sun et al., 2015; Cai et al., 2015; Li et al., 2015; Michno et al., 2015; Tang et al., 2016; Du et al., 2016). In 2017, the authors employed the CRISPR/Cas9 system to specifically induce targeted mutagenesis of *GmFT2a*, an integrator in the photoperiod flowering pathway in soybean. The targeted mutagenesis was stably heritable in the following T2 generation, and the homozygous *GmFT2a* mutants exhibited late flowering under both long-day and short-day conditions. This technology provides a powerful tool for accurate genetic modification and gene functional identification.

In this chapter, we describe recent advances in genetic modification of important agronomic traits and genome editing technologies for soybean.

2 Genetic modification of agronomic traits: herbicide tolerance and insect resistance

2.1 Herbicide tolerance

The most successful transgenic trait introduced into soybean is tolerance to the non-selective herbicide glyphosate (N-phosphonomethyl glycine). Glyphosate binds to and blocks the activity of 5-enolpyruvylshikimate-3-phosphate synthase (EPSPS), an enzyme of the shikimic acid pathway which produces aromatic amino acids. The Roundup Ready (RR, registered trademarks of Monsanto Technology LLC for glyphosate-tolerant crops) soybean cultivars have acquired glyphosate tolerance through the expression of the *epsps* gene from the *Agrobacterium* spp. strain CP4. The introduced glyphosate-tolerant EPSPS enzyme confers a high degree of glyphosate tolerance, and hence meets the plant's need for aromatic amino acids and other metabolites essential for plant development and growth in the presence of glyphosate (Padgette et al., 1995). The first generation of RR soybean cultivars was introduced into commercial production in 1996 and these cultivars have been planted on most soybean fields since 2004 (James, 2015). Extensive research efforts have led to the development of a second-generation glyphosate-tolerant soybean product. These cultivars were introduced into commercial production in 2007 (James, 2015). Although both transgenic events produce identical CP4 EPSPS proteins, the second-generation RR soybean has a yield advantage compared with the first in the same elite genetic background (Lundry et al., 2008). In addition, other glyphosate-tolerant soybean cultivars were commercially released through the expression of the *2mepsps* gene from *Zea mays* and the *gat4601* gene from *Bacillus licheniformis* too (Lepping et al., 2013; McNaughton et al., 2007). The double mutant EPSPS (2mEPSPS) enzyme, encoded by a modified version of the *epsps* gene from maize with a lower binding affinity for glyphosate, is highly resistant to inhibition by glyphosate. It thus allows sufficient enzyme activity for the plants to grow in the presence of herbicides that contain glyphosate (Herouet-Guicheney et al., 2009; Lepping et al., 2013). The modified glyphosate N-acetyltransferase (GAT) enzyme, encoded by the *gat4601* gene from ubiquitous soil bacteria, has excellent GAT activity and confers tolerance to glyphosate by catalysing the inactivation of glyphosate (Castle et al., 2004; McNaughton et al., 2007).

Some transgenic events tolerant to other herbicides, such as phosphinothricin (glufosinate); imidazolinone; 2,4-dichlorophenoxyacetic acid (2,4-D); sulfonylurea; isoxaflutole; dicamba and mesotrione, have also been approved for commercial release (McNaughton et al., 2007; Rouquié et al., 2010; Herman et al., 2011; Chukwudebe et al., 2012; Lepping et al., 2013; Kramer et al., 2014; James, 2015; ISAAA, 2016; Wang et al., 2016). The *bar* gene from *Streptomyces hygroscopicus* or the *pat* gene from *Streptomyces viridochromogenes* encodes for phosphinothricin acetyltransferase (PAT) which detoxifies glufosinate. The PAT enzyme acetylate L-phosphinothricin, the active isomer of the glufosinate-ammonium herbicide, results in transgenic soybean tolerant to post-emergent application of the non-selective herbicide (Zhang et al., 1999; Hérouet et al., 2005). This GM soybean variety is tolerant to the imidazolinone class of agricultural herbicides owing to the introduction of an imidazolinone-tolerant acetohydroxyacid synthase (AHAS) large subunit gene (*csr1-2*) from *Arabidopsis thaliana*. The introduction of the *csr1-2* gene allows the transformed soybean plant to grow in the presence of imidazolinone herbicides such as imazapic and imazapyr (Chukwudebe et al., 2012). 2,4-D is an effective, low-cost, broad-spectrum herbicide that controls many of the weeds that are developing resistance to glyphosate. The *add12* gene from *Delftia acidovorans* encodes for the aryloxyalkanoate dioxygenase protein, which catalyses the side chain degradation of 2,4-D herbicide. Transgenic soybean plants expressing the *add12* gene showed robust resistance to 2,4-D (Wright et al., 2010). When the *gm-hra* gene is expressed from soybean while changing 2 specific amino acids [which encodes a modified acetolactate synthase (ALS) enzyme], this confers tolerance to acetolactate synthase-inhibiting herbicides such as sulfonylurea (Mathesius et al., 2009). The enzyme p-hydroxyphenylpyruvate dioxygenase (HPPD) catalyses the formation of homogentisic acid, the aromatic precursor of plastoquinone and vitamin E. HPPD is the specific target of several herbicide families: isoxazoles, triketones and pyroxazoles. Modification of the p-hydroxyphenylpyruvate dioxygenase enzyme, encoded by *hppdPF W336* gene from *Pseudomonas fluorescens*, results in isoxaflutole tolerance in transgenic soybean (Rouquié et al., 2010). Similarly, the *avhppd-03* gene from oat (*Avena sativa*) confers tolerance to mesotrione in GM soybean (Kramer et al., 2014). Dicamba (2,6-dichloro-2-methoxybenzoic acid) is a herbicide that mimics a plant growth hormone that is effective at killing broad-leaf plants. Dicamba monooxygenase (DMO) rapidly catalyses the oxidative demethylation of dicamba to 3,6-dichlorosalicylic acid (DCSA) and formaldehyde. The dicamba-tolerant GM soybean can biosynthesize DMO encoded by the *dmo* gene from the soil bacterium *Stenotrophomonas maltophilia* strain DI-6 (Wang et al., 2016).

The introduction of herbicide resistance to soybean has resulted in a significant shift towards no-tillage practices in the America and brought about great economic and environmental benefits. The deployment of stacked traits for tolerance to different herbicides is becoming increasingly important and is most prevalent in the Americas. Herbicide resistance has also become a necessary trait in other instances of stacked genetic modification and in the selection of transgenic soybeans. Research is ongoing in this field, and there are currently many new herbicide-tolerant soybean events under development. For example, Chinese researchers recently reported that co-expression of *G2-EPSPS* and *GAT* conferred high tolerance to the herbicide glyphosate in soybean (Guo et al., 2015). Table 1 lists the main types of commercial soybean for herbicide tolerance.

2.2 Insect resistance

B. thuringiensis (*Bt*) is an entomopathogenic bacterium widely used as a biopesticide to control pest insects. Insecticidal crystal proteins (Cry proteins or δ-endotoxins) are an active component of *Bt* toxin, a biological insecticide. Cry proteins are toxic specifically to lepidopteran, dipteran and coleopteran larvae (Zhang et al., 2000). Expression of the *Bt cry* gene in soybean has proven highly effective in controlling insect pests, and resistance to lepidopteran pests was confirmed under field conditions (Stewart et al., 1996; Walker et al., 2000; Dufourmantel et al., 2005). Transgenic soybean lines expressing a synthetic *cry1Ac* gene exhibited a virtually complete efficacy against several lepidopteran pests in screenhouse and field trials (Macrae et al., 2005; Miklos et al., 2007; McPherson et al., 2009). The first insect-resistant GM soybean cultivars with the *cry1Ac* gene were introduced into commercial production in 2011 (Yu et al., 2013; James, 2015). Some transgenic events with other *cry* genes were also approved for commercial release (James, 2015). However, the discovery that insects can adapt to *Bt* Cry proteins raises concerns about long-term or high-dose use. Potential strategies for managing the development of resistance to *Bt* Cry proteins include the combination of the *Bt cry* gene and defoliating insect-resistant quantitative trait loci or other insecticidal proteins (Walker et al., 2002; Zhu et al., 2008). At present, the *Bt cry* gene is still the principal gene for insect-resistant GM soybean. Table 2 lists the main types of commercial soybean for insect resistance.

Table 1 Summary of commercial soybean for herbicide tolerance

Herbicide	Gene introduced	Gene source	Gene function	Events	References
Glyphosate	*cp4 epsps*	*Agrobacterium tumefaciens* strain CP4	Herbicide-tolerant form of 5-enolpyruvulshikimate-3-phosphate synthase (EPSPS) enzyme	GTS 40-3-2, MON87708, MON87712, MON87769, MON89788	Padgette et al. (1995) and Lundry et al. (2008)
	2mepsps	*Zea mays*	Double mutant version of 5-enolpyruvyl shikimate-3-phosphate synthase enzyme	DAS44406-6, FG72	Lepping et al. 2013
	gat4601	*Bacillus licheniformis*	Modified acetolactate synthase (ALS) enzyme	DP356043	McNaughton et al. 2007
Phosphinothricin (glufosinate)	*bar*	*Streptomyces hygroscopicus*	Phosphinothricin N-acetyltransferase (PAT) enzyme	W62, W98	Zhang et al. 1999
	pat	*Streptomyces viridochromogenes*	Phosphinothricin N-acetyltransferase (PAT) enzyme	A2704-12, A272704-21, A5547-127, A5547-35, DAS44406-6, DAS68416-4, DAS81419	Herman et al. (2011) and Lepping et al. (2013)
Imidazolinone	*csr1-2*	*Arabidopsis thaliana*	Modified acetohydroxyacid synthase large subunit (AtAHASL)	CV127	Chukwudebe et al. (2012)
2,4-D	*aad-12*	*Delftia acidovorans*	Aryloxyalkanoate dioxygenase 12 (AAD-12) protein	DAS44406-6, DAS68416-4	Lepping et al. (2013)
Sulfonylurea	*gm-hra*	*Glycine max*	Modified acetolactate synthase (ALS) enzyme	DP305423	McNaughton et al. (2007) and Mathesius et al. (2009)
Isoxaflutole	*hppdPF W336*	*Pseudomonas fluorescens* strain A32	Modified p-hydroxyphenylpyruvate dioxygenase (hppd) enzyme	FG72	Rouquié et al. (2010)
Dicamba	*dmo*	*Stenotrophomonas maltophilia* strain DI-6	Dicamba mono-oxygenase enzyme	MON87708	Wang et al. (2016)
Mesotrione	*avhppd-03*	*Avena sativa*	p-hydroxyphenylpyruvate dioxygenase	SYHTØH2	Kramer et al. (2014)

Table 2 Summary of commercial soybean for insect resistance

Insect	Gene introduced	Gene source	Gene function	Events	References
Lepidopteran	*cry1Ac*	*Bacillus thuringiensis* subsp. Kurstaki strain HD73	Cry1Ac delta-endotoxin	MON87701-DAS81419	Yu et al. (2013) and Fast et al. (2015)
	cry1F	*Bacillus thuringiensis* var. aizawai	Cry1F delta-endotoxin	DAS81419	Fast et al. (2015)
	cry1A.105	*Bacillus thuringiensis* subsp. kumamotoensis	Cry1A.105 protein which comprises the Cry1Ab, Cry1F and Cry1Ac proteins	MON87751	James (2015) and CEAR (2016)
	cry2Ab2	*Bacillus thuringiensis* subsp. kumamotoensis	Cry2Ab delta-endotoxin	MON87751	James (2015) and CEAR (2016)

3 Genetic modification of other agronomic traits

3.1 Oil composition

Soybean oil is widely used in food and in industry (such as biodiesel, lubricants, printing ink) (Yamada et al., 2012). Improvement of the oil content and its composition has long been a goal of genetic modification. Until recently, however, only some transgenic events resulting in desirable oil composition were approved for commercial release. Soybean oil is composed of palmitic, stearic, oleic, linoleic and linolenic acids (Yadav, 1996). The high level of polyunsaturated fatty acids in natural soybean oil renders the oil unstable and thus susceptible to the development of disagreeable odours and flavours. Down-regulation of the desaturation of fatty acids by ribozyme termination of RNA transcripts or RNA interference (RNAi) gene silencing decreased the polyunsaturated linoleic acid content or increased monounsaturated oleic acid. These transgenic events improve the stability of soybean oil without affecting its nutritional quality (Buhr et al., 2002; Flores et al., 2008). For example, expression of the *gm-fad2-1* gene fragment results in a greater concentration of oleic acid (18:1) by suppressing expression of the endogenous *FAD2-1* gene, which encodes an n-6 fatty acid desaturase enzyme that catalyses desaturation of 18:1 to linoleic acid (18:2). Several high-oleic-acid GM soybean cultivars have been introduced into commercial production, based on this strategy (McNaughton et al., 2008). The genes which encode fatty acid desaturase are the main targets for genetic modification of oil composition in soybean. On the other hand, ectopic expression of heterogeneous genes involved in fatty acid modification could generate other fatty acids, such as stearidonic, γ-linolenic, arachidonic, eicosapentaenoic and vernolic acids, which are minor and often undetectable fatty acids in conventional varieties (Sato et al., 2004; Kajikawa et al., 2008; Clemente and Cahoon, 2009; Yamada et al., 2012). Similarly, a transgenic event that enhances stearidonic acid (SDA) and omega-3 fatty acid (n-3) content, claimed to be beneficial to human health, has also been approved for commercial

Table 3 Summary of commercial soybean for oil composition

Oil composition	Gene introduced	Gene source	Gene function	Events	References
High-oleic-acid	*gm-fad2-1* (sense and antisense segments)	*Glycine max*	Production of endogenous delta-12 desaturase enzyme is suppressed	DP305423, 260-05	McNaughton et al. (2008)
	fatb1-A (sense and antisense segments)	*Glycine max*	Production of FATB enzymes or acyl-acyl carrier protein thioesterases is suppressed	MON87705	Wagner et al. (2011) and Hoffer et al. (2011)
	fad2-1A (sense and antisense)	*Glycine max*	Production of delta-12 desaturase enzyme is suppressed	MON87705	Wagner et al. (2011) and Hoffer et al. (2011)
Enhance stearidonic acid (SDA) and omega-3 fatty acid (ω-3) content	*Pj.D6D*	*Primula juliae*	Delta 6 desaturase protein	MON87769	James (2015) and CEAR (2016)
	Nc.Fad3	*Neurospora crassa*	Delta 15 desaturase protein	MON87769	James (2015) and CEAR (2016)

release (James, 2015; CEAR, 2016). However, although there are concerns about the value of these developments, there are few reports of increased oil content by GM methods that are not accompanied by undesirable agronomic traits (Yu et al., 2014). Table 3 lists the main types of commercial soybean for oil composition.

3.2 Amino acid composition

Soybean is an important staple source of vegetable protein in human food on account of its high quality, competitive cost, palatability and digestibility traits. Although soybean is an excellent source of protein, its nutritional quality is diminished by a low concentration of sulphur amino acids, especially methionine (Met). This limitation can result in non-specific signs of protein deficiency in humans, such as reduced resistance to disease, decreased blood proteins, and retardation of mental and physical development in young children (Galili et al., 2005). Met deficiencies in plant-derived feeds for farm animals limit animal growth, as well as affecting the quantity and quality of their yield (Tabe et al., 1998; Amir et al., 2012). Accumulation of high-Met proteins such as maize zein and bovine β-casein was induced in soybean seed under the regulation of seed-expression promoters (Maughan et al., 1999; Dinkins et al., 2001; Kim and Krishnan, 2004; Li et al., 2005), but not in sufficient quantities for nutritional improvement. For example, in the best transgenic line expressing an 11 kDa de novo protein enriched with Met, threonine, lysine and leucine, mature seed showed a limited increase of 16.2% and 65.9% in Met and cysteine, respectively, in comparison with the parent plant (Zhang et al., 2014). The accumulation of these Met-rich proteins may be limited by the absence of the proper maturation process in soybean or by the availability of sulphur-rich amino acids or of sulphur itself. Improving the pool of soluble amino acids would seem to be a reliable approach to improve the nutritional quality of soybean. Hence, in recent years, efforts have also been directed towards increasing the soluble sulphur amino acid content in soybean by manipulating its biosynthesis pathways. It has been reported that overexpression of feedback-insensitive forms of serine acetyl transferase and O-acetylserine (thiol) lyase, the last two enzymes in the Cys pathway in the cytosol, leads to increases of up to 32% in soluble Cys and up to 58–74% in protein-bound Cys in soybean seeds (Kim et al., 2012). Similarly, a Met-insensitive form of *Arabidopsis* cystathionine γ-synthase (AtD-CGS), which is the first committed enzyme of Met biosynthesis, expresses feedback-insensitive cystathionine γ-synthase in soybean seeds. This leads to a 3.8 to 7-fold increase in the level of soluble Met in developing green seeds. In addition, the total Met content increased 1.8- to 2.3-fold in the mature dry seeds (Song et al., 2013). This suggests that co-expression of Met-rich protein and of the Met biosynthesis enzyme may further improve the Met level of soybean seeds.

3.3 Improving resistance to other biotic and abiotic stresses

Soybean growth, productivity and seed quality are affected by a wide range of abiotic and biotic stresses. Biotic resistance to the soybean cyst nematode (SCN; *Heterodera glycines* Ichinohe) is a primary challenge in soybean production. Effective management of SCN relies on a combination of resistant cultivars and crop rotation. A gene from wild beet for resistance to the closely related beet cyst nematode, *Hs1*$^{pro\text{-}1}$, has enhanced SCN resistance in soybean (McLean et al., 2007). Recent studies have shown that overexpression of *GmSAMT1*, a soybean salicylic acid methyl transferase (GmSAMT1), which catalyses the conversion of salicylic acid to methyl salicylate produced a significant reduction in the development of SCN in transgenic plants (Lin et al., 2016).

Soybean mosaic virus (SMV) is the most prevalent and destructive viral pathogen in soybean-growing areas. SMV can cause serious yield losses. These have been reported to range from 35% to 50% under natural conditions in the field and from 50% to 100% during severe outbreaks (Gao et al., 2015), so virus resistance is an essential trait to be introduced. There have been some efforts to improve virus resistance in soybean by genetic modification. Overexpression of the 3-UTR region and a coat protein gene from SMV resulted in high resistance to SMV in transgenic soybean plants (Wang et al., 2001; Furutani et al., 2006). Furthermore, a pathogen-derived resistance to SMV induced by inverted repeat-SMV-*HC-Pro* genes in multiple soybean cultivars has also been reported. The helper component-protease (HC-Pro) is considered to be the suppressor apparatus for post-transcriptional gene silencing (Gao et al., 2015).

Drought is considered the major abiotic environmental stress in soybean production. Transgenic soybean plants overexpressing the *Arabidopsis* L-delta1-pyrroline-5-carboxylate reductase (*P5CR*) gene showed greater tolerance to drought and heat stresses due to an increased free proline level (De Ronde et al., 2004). Overexpression of an endogenous gene-encoding endoplasmic reticulum (ER)–resistant molecular chaperon binding protein (BiP) from soybean (soyBiPD) delayed leaf senescence during drought (Valente et al., 2009).

Salinity is another kind of abiotic stress affecting the growth of soybean. To cope with salt stress, soybean has developed several salt tolerance mechanisms. Soybean maintains ion homeostasis, adjusts in response to osmotic stress and restores osmotic balance (among other metabolic and structural adaptations) (Phang et al., 2008). Many salt tolerance–related genes in soybean have been reported (Qi et al., 2014; Guan et al., 2014), and some salt-tolerant transgenic soybean plants have been developed. For example, transgenic soybean lines overexpressing *TaNHX2* (a wheat Na^+/H^+ transporter code gene) show a higher rate of photosynthesis and higher yields under salt stress (Cao et al., 2011; Lin et al., 2015).

Alternatively, tolerance to stresses could be achieved by modulating the expression of stress-induced transcription factors (TF) and small RNAs (sRNAs) or microRNAs (miRNAs), which in turn would regulate the expression of a large number of relevant downstream genes (Agarwal et al., 2006; Ku et al., 2015; Shriram et al., 2016). For instance, overexpression of *AtDREB1A* [a member of the ethylene-responsive factors (ERF) family of TFs] in soybean may enhance drought tolerance (Polizel et al., 2011), while overexpression of *OsDREB2A* in soybean may improve tolerance to salt stress (Zhang et al., 2013).

In 2015, a drought-tolerant GM soybean was approved for commercialization in Argentina and the early food safety evaluation was completed by the FDA (US Food and Drug Administration). The Argentinian approval was the first regulatory approval for an abiotic stress tolerance trait in soybeans. The trait can provide up to a 14% yield advantage under multiple stress conditions – including drought and low-water conditions – typically found in soybean production areas (Bergau, 2015; James, 2015).

3.4 Other compounds

Vitamin E is widely used as an antioxidant in foods and oils, as a nutrient additive in poultry and cattle feeds and as a supplement in the human diet to help prevent diseases. The vitamin E family comprises tocopherols and tocotrienols (α, β, γ and δ forms). The α-tocopherol produces the highest vitamin E activity in mammals (Herbers, 2003). Enhancing the key step in the conversion of γ-tocopherol to α-tocopherol has elevated the α-tocopherol levels in the total tocopherol content of transgenic soybean seeds (Tavva et al., 2007; Chen et al., 2012).

Isoflavones are an important group of compounds that are synthesized in legumes. In addition to their role in the mediation of plant–microbe interactions, isoflavones are known as phytoestrogens and are biologically active substances associated with human health benefits such as anti-cancer effects and decreased risk of coronary heart disease. The soybean isoflavones daidzein, genistein and glycitein are structurally similar to 17β-estradiol, which explains their oestrogenic activity and identifies them as phytoestrogens (Mahmoud et al., 2014). Phenylpropanoid pathway genes were activated by expression of the maize C1 and R transcription factors in soybean seed, which decreased genistein and increased the daidzein levels, with a small overall increase in total isoflavone levels. Co-suppression of flavanone 3-hydroxylase to block the anthocyanin branch of the pathway, in conjunction with C1/R expression, resulted in higher levels of isoflavones (Yu et al., 2003). In contrast, transgenic soybeans containing three gene cassettes encoding chalcone synthase, isoflavone synthase and phenylalanine ammonia lyase produced seeds with 70% less isoflavone (Zernova et al., 2009). These results indicate that regulation of the expression of genes for phenylpropanoid biosynthesis enzymes and isoflavone-specific enzymes and interrelated transcription factor-driven genes can alter the content and composition of isoflavones.

The stacked traits of soybean, such as tolerance of different herbicides, herbicide tolerance and insect resistance, herbicide tolerance and modified fatty acids, and so on, have received more attention in recent years. Some stacked soybean cultivars have been commercially released through the transformation of multiple genes or hybridization of two events, and have been planted on a large scale since 2013 (James, 2015).

4 Genome editing technology

As a result of several genetic bottlenecks during the domestication of soybean and more recent intensive selection and breeding practices, the genetic variability of soybean germplasm has been greatly reduced (Hyten et al., 2006). In addition to the introduction of foreign genes, current efforts to expand genetic tools for breeding and gene discovery include random mutagenesis and RNAi-based approaches. However, in a highly duplicated genome such as soybean, random mutagenesis approaches often result in many lines with no phenotype due to the complementarity of many redundant genes. RNAi-based approaches such as post-transcriptional gene silencing suffer from the opposite problem, namely the difficulty of silencing individual gene copies. Instead, entire gene families are often inadvertently silenced (Kachroo et al., 2008; Curtin et al., 2011).

By contrast, targeted genome editing is an advanced approach designed to achieve efficient and precise modifications in the genomic DNA of cellular organisms. This method harnesses the DNA repair system of the cell to create modifications in the target DNA sequence. The basic approach of genome editing technology involves targeting sequence-specific nucleases (SSNs) to specific DNA sequences and inducting double-stranded breaks (DSBs) at the target site(s) within the genome. The resulting DSBs are primarily repaired by non-homologous end joining (NHEJ) or homology-directed repair (HDR). This can result in a variety of different outcomes, including site-directed mutagenesis, targeted gene replacement, targeted insertion of a foreign gene or targeted gene deletion. In other words, the DSB is usually repaired by the endogenous, error-prone NHEJ pathway, which introduces a small insertion or deletion (InDel) at the site, thus knocking out the gene. As with gene knock-in experiments, if a donor DNA fragment homologous to the flanking sequence is present the DSB is repaired through HDR using the donor DNA as a template. This results in the sequence of the donor fragment being integrated into the genome at the DSB site. The DSB repair system (NHEJ and HDR) is a ubiquitous component of all living cells. For now, the artificial nuclease with a reprogrammable recognition site is the most critical part of genome editing (Curtin et al., 2011; Kumar and Jain, 2015; Ding et al., 2016).

To date, three engineered nucleases composed of sequence-specific DNA-binding domains have been developed for genome editing in crops, including zinc finger nucleases (ZFNs), transcription activator-like effector nucleases (TALENs) and RNA-guided nucleases (RGNs) from the clustered regularly interspaced short palindromic repeat (CRISPR) and CPISPR-associated protein (Cas) system (Shan et al., 2013).

4.1 The ZFNs and TALENs systems in soybean

ZFNs and TALENs are a fusion of two enzymes, consisting of a modular DNA-binding domain and a type II restriction endonuclease (Fok1) domain. In both cases, the DNA-binding domain can be engineered to target a specific site within a DNA sequence. The nuclease domain is not sequence-specific. Both ZFNs and TALENs generate DSBs via a

FokI which non-specifically cleaves near the recognized specific sequences by binding the DNA domain (Bogdanove and Voytas, 2011; Kumar and Jain, 2015).

In ZFNs, the DNA recognition component is a fusion of modular DNA-binding proteins, which can be adapted to bind specific recognition sequences in double-stranded DNA. A single zinc finger unit consists of three or six binding modules, and each module recognizes three specific nucleotides. Two ZFNs strung together can recognize a unique 18–36 bp DNA sequence, with a 5–6 bp gap between them, making ZFNs highly target specific. ZFN dimers bind to these target sequences and induce site-specific DSBs via the catalytic action of FokI endonuclease, stimulating endogenous DNA repair pathways (Urnov et al., 2005). ZFNs have been used for genome editing in various crops, such as tobacco, maize and soybean (Kumar and Jain, 2015). Curtin et al. (2011) reported a successful case of soybean genome editing by ZFN technology. The mutation obtained exhibited efficient heritable transmission of the ZFN-induced mutation in the subsequent generation. However, no striking phenotypic alterations were observed (Curtin et al., 2011). While TALENs utilize the same FokI restriction endonuclease cleavage domain as ZFNs, the DNA recognition portion of TALENs differs. The DNA-binding components of TALENs are a combination of DNA-binding domains derived from transcription activator-like effectors (TALE) from *Xanthomonas* (Miller et al., 2011). TALENs are composed of multiple repeats of a 33–35-amino acid sequence. Each individual repeat recognizes a single nucleotide of the target DNA. The specificity of TALENs depends on a highly variable amino acid diresidue (RVD) at positions 12 and 13 of the amino acid repeat sequence (Streubel et al., 2012). Efficient genome editing has been reported in various crops, such as rice, tobacco, wheat and soybean (Haun et al., 2014; Kumar and Jain, 2015). Haun et al. (2014) reported a successful case of soybean genome editing by TALENs technology. The mutation obtained exhibited efficient heritable transmission of the TALEN-induced mutation in the subsequent generation. The fatty acid profile of the seed was dramatically changed in plants homozygous for mutations in both *FAD2-1A* and *FAD2-1B*, as well as two fatty acid desaturase two genes. The levels of oleic acid increased from 20% to 80% and linoleic acid decreased from 50% to under 4% (Haun et al., 2014). However, the construction of amino acid repeat arrays to assemble multiple repeat sequences remains a challenging task in both ZFNs and TALENs, due to complex interactions between amino acid residues and base pairs of the target sequence.

4.2 The CRISPR/Cas system in soybean

Recently, a more versatile and efficient genome editing tool has been developed based on a protein (Cas) type II adaptive immune system associated with bacterial 'clustered regularly interspaced short palindromic repeats' (CRISPR). It has been trialled in different organisms including crops (Shan et al., 2013; Jiang et al., 2013; Feng et al., 2013; Miao et al., 2013; Xie et al., 2014; Upadhyay et al., 2013; Wang et al., 2014; Liang et al., 2014; Svitashev et al., 2015). In contrast to ZFNs and TALENs, which target the genome through protein–DNA interaction, genome editing with the CRISPR/Cas is based on short RNA–DNA base pairing. The most commonly used RNA-guided nuclease (RGN) in genome editing is the Cas9 nuclease from the type II CRISPR/Cas9 of *Streptococcus pyogenes* (Jinek et al., 2012). The CRISPR/Cas system can be categorized into three types (I, II and III) based on the *Cas* genes and CRISPR sequences present. Types I and III contain multiple Cas proteins that form a complex for degrading foreign DNA/RNA. Type II directs the cleavage of targeted foreign DNA using a single Cas9 protein, which makes it the system of choice for targeted genome engineering. The main characteristic of the CRISPR/Cas9 system is the Cas9 protein, which comprises two functional domains: the RuvC-like domain and the HNH nuclease domain (Cong et al., 2013). The endonuclease Cas9 can be guided by a synthetic single-guide RNA to recognize target sequences and produce double strand breaks (DSBs) at desired target sites (Jinek et al., 2012; Cong et al., 2013). DSBs subsequently activate a series of complex DNA self-repair mechanisms in the cell and generate various site-specific genetic alterations through NHEJ or HDR. The NHEJ pathway is error-prone and typically generates insertions or deletions (indels) within the target sequence. When these indels introduce a frameshift mutation or disrupt important functional domains, the functions of the target genes will be damaged (Shan et al., 2013). The possibility of homologous recombination significantly increases when homologous DNA fragments are present during the repair process. In addition, the recombination efficiency obtained by DSBs can be improved a thousand-fold (Rouet et al., 1994). According to this principle, the HDR pathway can be used to induce precise gene targeting using homologous donor DNA as a template, such as the introduction of specific point mutations or the insertion of desired sequences in the target locus (Shan et al., 2013; Jiang et al., 2013; Feng et al., 2013). The sequence of a fragment could also be replaced using an exogenous donor template (Gratz et al., 2014; Zhao et al., 2016). The error-prone NHEJ has advantages for gene knock-out, and HDR is used for both gene replacement and gene knock-in in crops. It is simple to engineer the CRISPR/Cas9 vector because only sgRNA needs to be customized for different genomic sites and because the Cas9 protein is codon-optimized, with no need for every genome modification to be reconstructed (Shan et al., 2014). Therefore, the CRISPR/Cas9 system is a straightforward, economical and efficient genome editing technology. In 2015, the first successful cases of soybean genome editing by CRISPR/Cas9 system were reported in the hairy roots (Jacobs et al., 2015; Sun et al., 2015; Cai et al., 2015). These studies show that the CRISPR/Cas9 is a simple, efficient and highly specific genome editing tool in soybean. Li et al. (2015) then applied a Cas9–gRNA system suitable for soybean genome editing and achieved up-targeted mutagenesis through NHEJ and targeted gene integration through HDR. The integrated genes are transmitted to T1 generation and segregated according to Mendelian laws. Michno et al. (2015) developed an online tool for fast identification of CRISPR/Cas9 target loci within soybean gene models and generic DNA sequences. The Web-based tool can quickly identify a high number of potential CRISPR/Cas9 target sites, including restriction enzyme sites that can facilitate the detection of new mutations. Tang et al. (2016) validated one of the candidate genes as *Rj4*, an allele restricting nodulation gene, using CRISPR/Cas9-based gene knock-out experiments in soybean hairy roots. Du et al. (2016) have suggested that CRISPR/Cas9 is still a better choice for simultaneous editing of multiple homoeoalleles in soybean. Cai et al. (2017) employed the CRISPR/Cas9 system to specifically induce targeted mutagenesis of *GmFT2a*, an integrator in the photoperiod flowering pathway in soybean. Site-directed mutations were observed at all targeted sites by DNA sequencing analysis. T1-generation soybean plants homozygous for null alleles of the *GmFT2a* frameshift, mutated by a 1-bp insertion or short deletion, exhibited late flowering under both long-day and short-day conditions. Some 'transgene-clean' soybean plants that were homozygous for null alleles of endogenous *GmFT2a* and without any transgenic element from the T1 and T2 generations had been identified. These 'transgene-clean' mutants of *GmFT2a* may provide materials for more in-depth research into *GmFT2a* functions and the molecular mechanism of photoperiod responses in soybean. They will also contribute to soybean breeding and regional introduction. The CRISPR/Cas system is imparting new vigour to the genome editing and genetic modification in soybean.

5 Case studies

5.1 Genetically modified soybean with a higher level of Met in seeds

Soybean seeds provide an excellent source of protein for human and livestock nutrition. However, their nutritional quality is hampered by a low concentration of the essential sulphur amino acid, Met. In order to study factors that regulate Met synthesis in soybean seeds and obtain GM soybean with a higher level of Met in seeds, the authors of this chapter used the Met-insensitive form of *Arabidopsis* cystathionine γ-synthase (AtD-CGS), which is the first committed enzyme of Met biosynthesis. This gene was expressed under the control of a seed-specific promoter, legumin B4, and used to transform the soybean cultivar *Zigongdongdou* (ZD). In three transgenic lines that exhibited the highest expression level of AtD-CGS, the level of soluble Met increased significantly in developing green seeds (3.8- to 7-fold). The total Met content, which included Met incorporated into proteins, significantly increased in the mature dry seeds of these two transgenic lines by 1.8- and 2.3-fold, respectively (Song et al., 2013). We subsequently obtained the other transgenic lines in which the *AtD-CGS* gene was expressed under the control of the CaMV 35S promoter. The levels of soluble Met increased remarkably in leaves of transgenic soybeans compared to wild-type plants [6.6-fold and 7.3-fold in two transgenic ZD lines, and 3.7-fold in one transgenic *Jilinxiaoli No.1* (JX) line]. Furthermore, the total Met content was significantly increased in seeds of the transgenic ZD lines (1.5-fold to 4.8-fold increase) and in the transgenic JX lines (1.3-fold to 2.3-fold increase) than in the wild-type plants. Meanwhile, the protein content of the transgenic soybean seeds was significantly elevated compared to that of the wild type. Taken as a whole, the results obtained in this study constitute a major step towards the production of soybean seeds with higher Met levels and higher total protein levels, and, thus, with improved nutritional value.

5.2 CRISPR/Cas9-mediated genome editing in soybean

We successfully applied the type II CRISPR/Cas9 system to generate and estimate genome editing in the desired target genes in soybean. The single-guide RNA (sgRNA) and Cas9 cassettes were assembled on one vector to improve transformation efficiency, and we designed a sgRNA that targeted a transgene (*bar*) and six sgRNAs that targeted different sites of two endogenous soybean genes (*GmFEI2* and *GmSHR*). The CRISPR/Cas9 system presents relatively high mutation rates in soybean hairy roots. Targeted DNA mutations were detected in approximately 54% of the 170 transgenic hairy roots, with indel frequencies ranging from 0.6% to 95.0%. The results demonstrated that this customized CRISPR/Cas9 system shared the same efficiency for both endogenous and exogenous genes in soybean hairy roots. We also performed experiments to detect the potential of a CRISPR/Cas9 system to simultaneously edit two endogenous soybean genes using only one customized sgRNA. Overall, it was found that generating and detecting CRISPR/Cas9-mediated genome modifications in target genes of soybean hairy roots allowed us to rapidly assess the efficiency of each of the target loci. Target sites with higher efficiencies can be used for regular soybean transformation (Cai et al., 2015). Recently, some heritable *GmFT2a* (a key gene for controlling and accelerating the flowering time in soybean) mutation lines have been obtained in our lab (Cai et al., 2017). The obtained mutation showed efficient heritable transmission of the Cas9-induced mutation in the subsequent generation with typical flowering time delay characteristics.

6 Summary

6.1 Genetic modification is an effective means for increasing the levels of methionine in soybean

The results obtained in our study indicate that CGS plays a major role in Met synthesis in soybean. In the study, the total Met content increased 1.5- and 4.8-fold in the mature dry seeds of the better transgenic lines. There were no significant morphological changes in these transgenic lines. This level of increase is much higher than that in other reports of attempts to express heterologous Met-rich proteins (Dinkins et al., 2001; Li et al., 2005; Zhang et al., 2014). The level of Met in soybean seeds can be increased significantly to manipulate the Met biosynthesis pathway. The transgenic soybeans also displayed increased total protein contents in seeds, indicating that enhancing the Met content of soybeans is a practical strategy for improving both the Met and protein content of soybean. However, the increase was still not sufficient to meet the nutritional requirements of humans, livestock and poultry. On the other hand, according to our study, total Met levels were increased 1.5- to 4.8-fold, indicating that the Met elevation bottleneck was the 'source' (soluble Met) rather than the 'sink' (endogenous Met-rich proteins) in the wild-type soybeans (conventional varieties). In spite of the fact that most of the Met had been incorporated into proteins, the level of soluble Met in the transgenic seeds remained high in the transgenic lines overexpressing *AtD-CGS*. We assume that endogenous Met-rich proteins limit a new rate of protein synthesis, and thus Met cannot be incorporated further into proteins. This may suggest that the level of Met in soybean seeds can be further enhanced by co-expressing the key enzymes in the Met pathway and the Met-rich proteins.

6.2 CRISPR/Cas9-mediated genome editing provides a powerful tool for genetic modification in soybean

The CRISPR/Cas9 system has been successfully used for genome engineering in many important crops. However, this method has not been widely used in soybean. In this chapter, we have described a rapid and highly specific method of generating and detecting CRISPR/Cas9-mediated genome editing in desired target genes in soybean. To improve transformation efficiency, the sgRNA and Cas9 cassettes were assembled on one vector. The Cas9 was codon-optimized for dicotyledons. Once a desired target sequence is selected, only the DNA sequence encoding sgRNA needs replacing (Cai et al., 2015). Our results showed that the CRISPR/Cas9 system could edit both endogenous and exogenous genes in soybean hairy roots. These results are in accordance with reports from several other labs. In addition to single site or gene targets, the CRISPR/Cas9 system can be used for multiplex gene editing in soybean tissues. These results show that CRISPR/Cas9-mediated genome editing provides a powerful tool for genetic modification and genomic studies in soybean (Jacobs et al., 2015; Sun et al., 2015; Cai et al., 2015; Li et al., 2015; Michno et al., 2015; Tang et al., 2016; Du et al., 2016). Genome editing offers a timely and unique set of significant comparative advantages over conventional and transgenic crops. First, it is a precision technology due to its ability to precisely control single or multiple genes, resulting in products that do not differ from natural mutations. Secondly, viral delivery of the constructs or crossing of the transgenic plant carrying the CRISPR/Cas9 system with elite lines/cultivars can provide non-transgenic genome modifications in plants. This can address some of the ethical concerns being raised about transgenic plants. The CRISPR/Cas9 system brings together several desirable features, including simplicity, efficiency, high specificity, minimal off-target effects and amenability to multiplexing; the technique thus seems to be extremely promising in soybean crop production.

7 Future trends in research

Genome editing technology is undoubtedly one of the most powerful tools in basic research in the area of soybean genetic modification. NHEJ-mediated gene knock-outs are the most direct application of CRISPR/Cas9. The first genome-edited product to be approved and commercialized is sulfonylurea-tolerant canola (SU Canola™), which was being cultivated over 4000 hectares in the United States in 2015. Canada has also approved SU Canola™ for planting. Genome-edited canola offers farmers a new alternative for weed control in canola that will provide sound stewardship options to deal with the management of glyphosate weed resistance. Furthermore, it is an optimal fit for rotation with glyphosate-tolerant soybeans, reducing weed pressure caused by volunteer glyphosate-tolerant canola in soybean fields. Although much progress has been made in CRISPR/Cas9-based genome editing technology in the last few years, some problems remain unresolved. These include the efficiency and simplicity of HDR-mediated gene replacement and gene knock-in, side effects on nearby genes, off-target effects and the mechanism underlying the different effects of different sgRNAs on mutation efficiency. Genome editing technologies, as represented by the CRISPR/Cas9 system, can be expected to improve rapidly in order to meet these challenges. This simple, efficient, highly specific and elegant genetic scalpel will be widely applied to enhance the agricultural performance of soybean and most other crops in the near future.

Some new GM crop products could be available in the near future. The genomic era is now underway for soybean, and these scientific endeavours have enabled the confirmation and identification of genes and multiple loci for important agronomic traits. These valuable genes and genetic resource should facilitate future genetic modification of soybean. On the basis of this knowledge, the stacked traits and complex genetic traits have received more attention in recent years, such as herbicide tolerance, biotic and abiotic resistance, and protein content and composition. If a GM soybean could be developed with agronomically important traits which could not be achieved by current genetic resources, such as high yielding ability and multiple stress resistance, genetic modification approaches might be more widely accepted in soybean breeding.

8 Where to look for further information

A number of books are available covering current knowledge on soybean breeding and genetics. One example is *Genetics and Genomics of Soybean* edited by Gary Stacey (Springer Science+Business Media, 223 Spring Street LLC, New York, NY 10013, USA). On NCBI (www.ncbi.nlm.nih.gov/pubmed), the famous scientific literature websites published by the National Center for Biotechnology Information, U.S. National Library of Medicine, and most literature related to genetic modification of soybean and genome editing technology are readily available. The NCBI resource PubMed includes more than 26 million citations for biomedical literature from MEDLINE, life science journals and online books. Citations may include links to full-text content from PubMed Central and other publishers' websites. On SoyBase (www.soybase.org), it is possible to access most of the literature and databases related to genetic maps of soybean. On ISAAA (www.isaaa.org, published by the International Service for the Acquisition of Agri-biotech Applications) and CEAR (cera-gmc.org, published by the Center for Environmental Risk Assessment), it is possible to access the main information about commercial GM soybeans.

Multinational companies play a major role in the genetic modification of soybean, and it is therefore important to take account of information published by Monsanto, Dupont, Bayer CropScience, Syngenta, BASF, Dow AgroSciences and other companies. Furthermore, it is important to note new advances at some public research institutions such as the University of Georgia, USA; the University of Minnesota, USA; the University of Tennessee, USA; the Institute of Crop Sciences at the Chinese Academy of Agricultural Sciences, China; and the Institute of Genetics and Developmental Biology at the Chinese Academy of Sciences, China.

9 References

Agarwal, P. K., Agarwal, P., Reddy, M. K. and Sopory, S. K. (2006), 'Role of DREB transcription factors in abiotic and biotic stress tolerance in plants', *Plant Cell Rep.*, 25(12), 1263–74.

Amir, R., Han, T. and Ma, F. (2012), 'Bioengineering approaches to improve the nutritional values of seeds by increasing their methionine content', *Mol. Breed.*, 29(4), 915–24.

Bergau, J. (2015), 'Verdeca's HB4 stress tolerance trait completes US Food and Drug Administration early food safety evaluation', *Arcadia Biosciences, Inc.* http://www.businesswire.com/news/home/20150826005216/en/.

Bogdanove, A. J. and Voytas, D. F. (2011), 'TAL effectors: Customizable proteins for DNA targeting', *Science*, 333(6051), 1843–6.

Buhr, T., Sato, S., Ebrahim, F., Xing, A., Zhou, Y., Mathiesen, M., Schweiger, B., Kinney, A., Staswick, P. and Clemente, T. (2002), 'Ribozyme termination of RNA transcripts down-regulate seed fatty acid genes in transgenic soybean', *Plant J.*, 30(2), 155–63.

Cai, Y., Chen, L., Liu, X., Sun, S., Wu, C., Jiang, B., Han, T. and Hou, W. (2015), 'CRISPR/Cas9-mediated genome editing in soybean hairy roots', *PLoS One*, 10(8), e0136064.

Cai, Y., Chen, L., Liu, X., Guo, C., Sun, S., Wu, C., Jiang, B., Han, T. and Hou, W. (2017), 'CRISPR/Cas9-mediated targeted mutagenesis of *GmFT2a* delays flowering time in soya bean', *Plant Biotechnol. J.* - Wiley Online Library, http://onlinelibrary.wiley.com/doi/10.1111/pbi.12758/full.

Cao, D., Hou, W., Liu, W., Yao, W., Wu, C., Liu, X. and Han, T. (2011), 'Overexpression *TaNHX2* enhances salt tolerance of 'composite' and whole transgenic soybean plants', *Plant Cell Tiss. Org.*, 107(3), 541–52.

Castle, L. A., Siehl, D. L., Gorton, R., Patten, P. A., Chen, Y. H., Bertain, S., Cho, H. J., Duck, N., Wong, J., Liu, D. and Lassner, M. W. (2004), 'Discovery and directed evolution of a glyphosate tolerance gene', *Science*, 304(5674), 1151–4.

Center for environmental risk assessment (CERA). (2016), '*GM Crop Database*', ILSI Research Foundation, Washington DC. http://cera-gmc.org/index.php/GMCropDatabase.

Chen, D. F., Zhang, M., Wang, Y. Q. and Chen, X. W. (2012), 'Expression of γ-tocopherol methyltransferase gene from *Brassica napus* increased α-tocopherol content in soybean seed', *Biol. Plant.*, 56(1), 131–4.

Chukwudebe, A., Privalle, L., Reed, A., Wandelt, C., Contri, D., Dammann, M., Groeters, S., Kaspers, U., Strauss, V. and van Ravenzwaay, B. (2012), 'Health and nutritional status of Wistar rats following subchronic exposure to CV127 soybeans', *Food Chem. Toxicol.*, 50(3–4), 956–71.

Clemente, T. E. and Cahoon, E. B. (2009), 'Soybean oil: Genetic approaches for modification of functionality and total content', *Plant Physiol.*, 151(3), 1030–40.

Cong, L., Ran, F. A., Cox, D., Lin, S., Barretto, R., Habib, N., Hsu, P. D., Wu, X., Jiang, W., Marraffini, L. A. and Zhang, F. (2013), 'Multiplex genome engineering using CRISPR/Cas systems'. *Science*, 339(6121), 819–23.

Curtin, S. J., Zhang, F., Sander, J. D., Haun, W. J., Starker, C., Baltes, N. J., Reyon, D., Dahlborg, E. J., Goodwin, M. J., Coffman, A. P., Dobbs, D., Joung, J. K., Voytas, D. F. and Stupar, R. M. (2011), 'Targeted mutagenesis of duplicated genes in soybean with zinc-finger nucleases', *Plant Physiol.*, 156(2), 466–73.

De Ronde, J. A., Cress, W. A., Krüger, G. H., Strasser, R. J. and Van Staden, J. (2004), 'Photosynthetic response of transgenic soybean plants, containing an *Arabidopsis P5CR* gene, during heat and drought stress', *J. Plant Physiol.*, 161(11), 1211–24.

Ding, Y., Li, H., Chen, L. L. and Xie, K. (2016), 'Recent advances in genome editing using CRISPR/Cas9'. *Front. Plant Sci.*, 7, 703.

Dinkins, R. D., Reddy, R., Curtis, A., Meurer, C. A., Yan, B., Trick, H., Thhibaud-Nissen, F., Finer, J. J., Parrott, W. A. and Collins, G. B. (2001), 'Increased sulfur amino acids in soybean plants overexpressing the maize 15 kD zein protein', *In Vitro Cell. Dev-Pl.*, 37(6), 742–7.

Du, H., Zeng, X., Zhao, M., Cui, X., Wang, Q., Yang, H., Cheng, H. and Yu, D. (2016), 'Efficient targeted mutagenesis in soybean by TALENs and CRISPR/Cas9', *J. Biotechnol.*, 217, 90–7.

Dufourmantel, N., Tissot, G., Goutorbe, F., Garçon, F., Muhr, C., Jansens, S., Pelissier, B., Peltier, G. and Dubald, M. (2005), 'Generation and analysis of soybean plastid transformants expressing *Bacillus thuringiensis* Cry1Ab protoxin', *Plant Mol. Biol.*, 58(5), 659–68.

Fast, B. J., Schafer, A. C., Johnson, T. Y., Potts, B. L. and Herman, R. A. (2015), 'Insect-protected event DAS-81419–2 soybean (*Glycine max* L.) grown in the United States and Brazil is compositionally equivalent to nontransgenic soybean', *J. Agric. Food Chem.*, 63(7), 2063–73.

Feng, Z., Zhang, B., Ding, W., Liu, X., Yang, D. L., Wei, P., Cao, F., Zhu, S., Zhang, F., Mao, Y. and Zhu, J. K. (2013), 'Efficient genome editing in plants using a CRISPR/Cas system', *Cell Res.*, 23(10), 1229–32.

Flores, T., Karpova, O., Su, X., Zeng, P., Bilyeu, K., Sleper, D. A., Nguyen, H. T. and Zhang, Z. J. (2008), 'Silencing of GmFAD3 gene by siRNA leads to low alpha-linolenic acids (18:3) of fad3-mutant phenotype in soybean [*Glycine max* (Merr.)]', *Transgenic Res.*, 17(5), 839–50.

Furutani ,N., Hidaka, S., Kosaka, Y., Shizukawa, Y. and Kanematsu, S. (2006), 'Coat protein gene-mediated resistance to soybean mosaic virus in transgenic soybean', *Breeding Sci.*, 56(2), 119–24.

Galili, G., Amir, R., Hoefgen, R. and Hesse, H. (2005), 'Improving the levels of essential amino acids and sulfur metabolites in plants', *Biol. Chem.*, 386(9), 817–31.

Gao, L., Ding, X., Li, K., Liao, W., Zhong, Y., Ren, R., Liu, Z., Adhimoolam, K. and Zhi, H. (2015), 'Characterization of Soybean mosaic virus resistance derived from inverted repeat-SMV-*HC-Pro* genes in multiple soybean cultivars', *Theor. Appl. Genet.*, 128(8), 1489–505.

Gratz, S. J., Ukken, F. P., Rubinstein, C. D., Thiede, G., Donohue, L. K., Cummings, A. M. and O'Connor-Giles, K. M. (2014), 'Highly specific and efficient CRISPR/Cas9-catalyzed homology-directed repair in *Drosophila*', *Genetics*, 196(4), 961–71.

Guan, R., Qu, Y., Guo, Y., Yu, L., Liu, Y., Jiang, J., Chen, J., Ren, Y., Liu, G., Tian, L., Jin, L., Liu, Z., Hong, H., Chang, R., Gilliham, M. and Qiu, L. (2014), 'Salinity tolerance in soybean is modulated by natural variation in *GmSALT3*', *Plant J.*, 80(6), 937–50.

Guo, B., Guo, Y., Hong, H., Jin, L., Zhang, L., Chang, R. Z., Lu, W., Lin, M. and Qiu, L. J. (2015), 'Co-expression of G2-EPSPS and glyphosate acetyltransferase GAT genes conferring high tolerance to glyphosate in soybean', *Front. Plant Sci.*, 6, 847.

Haun, W., Coffman, A., Clasen, B. M., Demorest, Z. L., Lowy, A., Ray, E., Retterath, A., Stoddard, T., Juillerat, A., Cedrone, F., Mathis, L., Voytas, D. F. and Zhang, F. (2014), 'Improved soybean oil quality by targeted mutagenesis of the fatty acid desaturase 2 gene family', *Plant Biotechnol. J.*, 12(7), 934–40.

Herbers, K. (2003), 'Vitamin production in transgenic plants', *J. Plant Physiol.*, 160(7), 821–9.

Herman, R. A., Dunville, C. M., Juberg, D. R., Fletcher, D. W. and Cromwell, G. L. (2011), 'Performance of broiler chickens fed diets containing DAS-68416-4 soybean meal', *GM Crops*, 2(3), 169–75.

Hérouet, C., Esdaile, D. J., Mallyon, B. A., Debruyne, E., Schulz, A., Currier, T., Hendrickx, K., van der Klis, R. J. and Rouan, D. (2005), 'Safety evaluation of the phosphinothricin acetyltransferase proteins encoded by the pat and bar sequences that confer tolerance to glufosinate-ammonium herbicide in transgenic plants', *Regul Toxicol Pharmacol.*, 41(2), 134–49.

Herouet-Guicheney, C., Rouquié, D., Freyssinet, M., Currier, T., Martone, A., Zhou, J., Bates, E. E., Ferullo, J. M., Hendrickx, K. and Rouan, D. (2009), 'Safety evaluation of the double mutant 5-enol pyruvylshikimate-3-phosphate synthase (2mEPSPS) from maize that confers tolerance to glyphosate herbicide in transgenic plants', *Regul. Toxicol. Pharmacol.*, 54(2), 143–53.

Hinchee, M. A., Connor-Ward, D. V., Newell, C. A., McDonell, R. E., Sato, S. J., Gasser, C. S., Fishhoff, D. A., Re, D. B., Fraley, R. T. and Horsch, R. B. (1988), 'Production of transgenic soybean plants using *Agrobacterium*-mediated DNA transfer', *Nat. Biotechnol.*, 6(8), 915–22.

Hoffer, P., Ivashuta, S., Pontes, O., Vitins, A., Pikaard, C., Mroczka, A., Wagner, N. and Voelker, T. (2011), 'Posttranscriptional gene silencing in nuclei', *Proc. Natl. Acad. Sci. USA*, 108(1), 409–14.

Homrich, M. S., Wiebke-Strohm, B., Weber, R. L. and Bodanese-Zanettini, M. H. (2012), 'Soybean genetic transformation: A valuable tool for the functional study of genes and the production of agronomically improved plants', *Genet. Mol. Biol.*, 35(4 suppl.), 998–1010.

Hou, W. S., Lin, K. X., Chen, P., Jia, Z. W., Zhou, Y., Yu, Y. and Liu, Y. H. (2014), 'Establishment and prospect of efficient transformation systems for soybean', *Scientia Agricultura Sinica*, 47(21), 4198–210 (in Chinese with English abstract).

Hyten, D. L., Song, Q., Zhu, Y., Choi, I. Y., Nelson, R. L., Costa, J. M., Specht, J. E., Shoemaker, R. C. and Cregan, P. B. (2006), 'Impacts of genetic bottlenecks on soybean genome diversity', *Proc. Natl. Acad. Sci. USA*, 103(45), 16666–71.

ISAAA. (2016), 'GMO Approval Database', http://www.isaaa.org/gmapprovaldatabase/default.asp.

Jacobs, T. B., LaFayette, P. R., Schmitz, R. J. and Parrott, W. A. (2015), 'Targeted genome modifications in soybean with CRISPR/Cas9', *BMC Biotechnol.*, 15, 16.

James, C. (2015), *20th Anniversary (1996 to 2015) of the Global Commercialization of Biotech Crops and Biotech Crop Highlights in 2015*, Brief No.51. International Service for the Acquisition of Agri-biotech Applications, Ithaca, NY.

Jiang, W., Zhou, H., Bi, H., Fromm, M., Yang, B. and Weeks, D. P., (2013), 'Demonstration of CRISPR/Cas9/sgRNA-mediated targeted gene modification in *Arabidopsis*, tobacco, sorghum and rice', *Nucleic Acids Res.*, 41(20), e188.

Jinek, M., Chylinski, K., Fonfara, I., Hauer, M., Doudna, J. A. and Charpentier, E. (2012), 'A programmable dual-RNA-guided DNA endonuclease in adaptive bacterial immunity', *Science*, 337(6096), 816–21.

Kachroo, A., Fu, D. Q., Havens, W., Navarre, D., Kachroo, P. and Ghabrial, S. A. (2008), 'An oleic acid-mediated pathway induces constitutive defense signaling and enhanced resistance to multiple pathogens in soybean', *Mol. Plant Microbe Interact.*, 21(5), 564–75.

Kajikawa, M., Matsui, K., Ochiai, M., Tanaka, Y., Kita, Y., Ishimoto, M., Kohzu, Y., Shoji, S., Yamato, K. T., Ohyama, K., Fukuzawa, H. and Kohchi, T. (2008), 'Production of arachidonic and eicosapentaenoic acids in plants using bryophyte fatty acid Delta6-desaturase, Delta6-elongase, and Delta5-desaturase genes', *Biosci. Biotechnol. Biochem.*, 72(2), 435–544.

Kim, W. S. and Krishnan, H. B. (2004), 'Expression of an 11 kDa methionine-rich delta-zein in transgenic soybean results in the formation of two types of novel protein bodies in transitional cells situated between the vascular tissue and storage parenchyma cells', *Plant Biotechnol. J.*, 2(3), 199–210.

Kim, W. S., Chronis, D., Juergens, M., Schroeder, A. C., Hyun, S. W., Jez, J. M. and Krishnan, H. B. (2012), 'Transgenic soybean plants overexpressing O-acetylserine sulfhydrylase accumulate enhanced levels of cysteine and Bowman-Birk protease inhibitor in seeds' *Planta*, 235(1), 13–23.

Kramer, C. M., Launis, K. L., Traber, M. G. and Ward, D. P. (2014), 'Vitamin E levels in soybean (*Glycine max* (L.) Merr.) expressing a p-hydroxyphenylpyruvate gene from oat (*Avena sativa* L.)', *J. Agric. Food Chem.*, 62(15), 3453–7.

Ku, Y. S., Wong, J. W., Mui, Z., Liu, X., Hui, J. H., Chan, T. F. and Lam, H. M. (2015), 'Small RNAs in plant responses to abiotic stresses: Regulatory roles and study methods', *Int. J. Mol. Sci.*, 16(10), 24532–54.

Kumar, V. and Jain, M. (2015), 'The CRISPR-Cas system for plant genome editing: Advances and opportunities', *J. Exp. Bot.*, 66(1), 47–57.

Lam, H. M., Xu, X., Liu, X., Chen, W., Yang, G., Wong, F. L., Li, M. W., He, W., Qin, N., Wang, B., Li, J., Jian, M., Wang, J., Shao, G., Wang, J., Sun, S. S. and Zhang, G. (2010), 'Resequencing of 31 wild and cultivated soybean genomes identifies patterns of genetic diversity and selection', *Nat. Genet.*, 42(12), 1053–9.

Lepping, M. D., Herman, R. A. and Potts, B. L. (2013), 'Compositional equivalence of DAS-444Ø6-6 (AAD-12 + 2mEPSPS + PAT) herbicide-tolerant soybean and nontransgenic soybean', *J. Agric. Food Chem.*, 61(46), 11180–90.

Li, Y. H., Zhou, G., Ma, J., Jiang, W., Jin, L. G., Zhang, Z., Guo, Y., Zhang, J., Sui, Y., Zheng, L., Zhang, S. S., Zuo, Q., Shi, X. H., Li, Y. F., Zhang, W. K., Hu, Y., Kong, G., Hong, H. L., Tan, B., Song, J., Liu, Z. X., Wang, Y., Ruan, H., Yeung, C. K., Liu, J., Wang, H., Zhang, L. J., Guan, R. X., Wang, K. J., Li, W. B., Chen, S. Y., Chang, R. Z., Jiang, Z., Jackson, S. A., Li, R. and Qiu, L. J. (2014), 'De novo assembly of soybean wild relatives for pan-genome analysis of diversity and agronomic traits', *Nat. Biotechnol.*, 32(10), 1045–52.

Li, Z., Meyer, S., Essig, J. S., Liu, Y., Schapaugh, M. A., Muthukrishnan, S., Hainline, B. E. and Trick, H. N. (2005), 'High-level expression of maize γ-zein protein in transgenic soybean (*Glycine max*)', *Mol. Breed.*, 16(1), 11–20.

Li, Z., Liu, Z. B., Xing, A., Moon, B. P., Koellhoffer, J. P., Huang, L., Ward, R. T., Clifton, E, Falco, S. C. and Cigan, A. M. (2015), 'Cas9-Guide RNA directed genome editing in soybean', *Plant Physiol.*, 169(2), 960–70.

Liang, Z., Zhang, K., Chen, K. and Gao, C. (2014), 'Targeted mutagenesis in *Zea mays* using TALENs and the CRISPR/Cas system', *J. Genet. Genom.*, 41(2), 63–8.

Lin, J., Mazarei, M., Zhao, N., Hatcher, C. N., Wuddineh, W. A., Rudis, M., Tschaplinski, T. J., Pantalone, V. R., Arelli, P. R., Hewezi, T., Chen, F. and Stewart Jr., C. N. (2016), 'Transgenic soybean overexpressing GmSAMT1 exhibits resistance to multiple-HG types of soybean cyst nematode Heterodera glycines', *Plant Biotechnol. J.*, 14(11), 2100–9 doi:10.1111/pbi.12566.

Lin, K. X., Liu, X. J., Sun, S., Chen, L., Han, T. F. and Hou, W. S. (2015), 'Salt tolerance analysis of *TaNHX2* over-expression transgenic soybean', *Scientia Agricultura Sinica*, 48(20), 3998–4007 (in Chinese with English abstract).

Lundry, D. R., Ridley, W. P., Meyer, J. J., Riordan, S. G., Nemeth, M. A., Trujillo, W. A., Breeze, M. L. and Sorbet, R. (2008), 'Composition of grain, forage, and processed fractions from second-generation glyphosate-tolerant soybean, MON 89788, is equivalent to that of conventional soybean (*Glycine max* L.)', *J. Agric. Food Chem.,*. 56(12), 4611–22.

MacRae, T. C., Baur, M. E., Boethel, D. J., Fitzpatrick, B. J., Gao, A. G., Gamundi, J. C., Harrison, L. A., Kabuye, V. T., McPherson, R. M., Miklos, J. A., Paradise, M. S., Toedebusch, A. S. and Viegas, A.

(2005), 'Laboratory and field evaluations of transgenic soybean exhibiting high-dose expression of a synthetic *Bacillus thuringiensis cry1A* gene for control of Lepidoptera', *J. Econ. Entomol.*, 98(2), 577–87.

Mahmoud, A. M., Yang, W and Bosland, M. C. (2014), 'Soy isoflavones and prostate cancer: A review of molecular mechanisms', *J. Steroid Biochem. Mol.Biol.*, 140(3):116–32.

Mathesius, C. A., Barnett Jr., J. F., Cressman, R. F., Ding, J., Carpenter, C., Ladics, G. S., Schmidt, J., Layton, R. J., Zhang, J. X., Appenzeller, L. M., Carlson, G., Ballou, S. and Delaney, B. (2009), 'Safety assessment of a modified acetolactate synthase protein (GM-HRA) used as a selectable marker in genetically modified soybeans', *Regul. Toxicol. .Pharmacol.*, 55(3), 309–20.

Maughan, P. J., Philip, R., Cho, M. J., Widholm, J. M. and Vodkin, L. O. (1999), 'Biolistic transformation, expression, and inheritance of bovine β-casein in soybean (*Glycine max*)', *In Vitro Cell. Dev-Pl.*, 35(4), 344–9.

McCabe, D. E., Swain, W. F., Martinell, B. J. and Christou, P. (1988), 'Stable transformation of soybean (*Glycine max*) by particle acceleration', *Nat. Biotechnol.*, 6(8), 923–6.

McLean, M. D., Hoover, G. J., Bancroft, B, Makhmoudova, A., Clark, S. M., Welacky, T., Simmonds, D. H. and Shelp, B. J. (2007), 'Identification of the full-length *Hs1*$^{pro-1}$ coding sequence and preliminary evaluation of soybean cyst nematode resistance in soybean transformed with *Hs1*$^{pro-1}$ cDNA', *Can. J. Bot.*, 85(4), 437–41.

McNaughton, J., Roberts, M., Smith, B., Rice, D., Hinds, M., Schmidt, J., Locke, M., Brink, K., Bryant, A., Rood, T., Layton, R., Lamb, I. and Delaney, B. (2007), 'Comparison of broiler performance when fed diets containing event DP-356Ø43-5 (Optimum GAT), nontransgenic near-isoline control, or commercial reference soybean meal, hulls, and oil', *Poult. Sci.*, 86(12), 2569–81.

McNaughton, J., Roberts, M., Smith, B., Rice, D., Hinds, M., Sanders, C., Layton, R., Lamb, I. and Delaney, B. (2008), 'Comparison of broiler performance when fed diets containing event DP-3O5423–1, nontransgenic near-isoline control, or commercial reference soybean meal, hulls, and oil', *Poult. Sci.*, 87(12), 2549–61.

McPherson, R. M. and MacRae, T. C. (2009), 'Evaluation of transgenic soybean exhibiting high expression of a synthetic *Bacillus thuringiensis cry1A* transgene for suppressing lepidopteran population densities and crop injury', *J. Econ. Entomol.*, 102(4), 1640–8.

Miao, J., Guo, D., Zhang, J., Huang, Q., Qin, G., Zhang, X., Wan, J., Gu, H. and Qu, L. J. (2013), 'Targeted mutagenesis in rice using CRISPR-Cas system', *Cell Res.*, 23(10), 1233–6.

Michno, J. M., Wang, X., Liu, J., Curtin, S. J., Kono, T. J. and Stupar, R. M. (2015), 'CRISPR/Cas mutagenesis of soybean and *Medicago truncatula* using a new web-tool and a modified Cas9 enzyme', *GM Crops Food*, 6(4), 243–52.

Miklos, J. A., Alibhai, M. F., Bledig, S. A., Connor-Ward, D. C., Gao, A. G., Holmes, B. A., Kolacz, K. H., Kabuye, V. T., MacRae, T. C., Paradise, M. S., Toedebusch, A. S. and Harrison, L. A. (2007), 'Characterization of soybean exhibiting high expression of a synthetic *Bacillus thuringiensis cry1A* transgene that confers a high degree of resistance to Lepidopteran pests', *Crop Sci.*, 47(1):148–57.

Miller, J. C., Tan, S., Qiao, G., Barlow, K. A., Wang, J., Xia, D. F., Meng, X., Paschon, D. E., Leung, E., Hinkley, S. J., Dulay, G. P., Hua, K. L., Ankoudinova, I., Cost, G. J., Urnov, F. D., Zhang, H. S., Holmes, M. C., Zhang, L, Gregory, P. D. and Rebar, E. J. (2011), 'A TALE nuclease architecture for efficient genome editing', *Nat. Biotechnol.*, 29(2), 143–8.

Padgette, S. R., Kolacz, K. H., Delannay, X., Re, D. B., LaVallee, B. J., Tinius, C. N., Rhodes, W. K., Otero, Y. I., Barry, G. F., Eichholtz, D. A., Peschke, V. M., Nide, D. L., Taylor, N. B. and Kishore, G. M. (1995), 'Development, identification, and characterization of a glyphosate-tolerant soybean line', *Crop sci.*, 35(5), 1451–61.

Phang, T. H., Shao, G. and Lam, H. M. (2008), 'Salt tolerance in soybean', *J. Integr. Plant Biol.*, 50(10), 1196–212.

Polizel, A. M., Medri, M. E., Nakashima, K., Yamanaka, N., Farias, J. R., de Oliveira, M. C., Marin, S. R., Abdelnoor, R. V., Marcelino-Guimarães, F. C., Fuganti, R., Rodrigues, F. A., Stolf-Moreira,

R., Beneventi, M. A., Rolla, A. A., Neumaier, N., Yamaguchi-Shinozaki, K., Carvalho, J. F. and Nepomuceno, A. L. (2011), 'Molecular, anatomical and physiological properties of a genetically modified soybean line transformed with *rd29A:AtDREB1A* for the improvement of drought tolerance', *Genet. Mol. Res.*, 10(4), 3641–56.

Qi, X., Li, M. W., Xie, M., Liu, X., Ni, M., Shao, G., Song, C., Kay-Yuen Yim, A., Tao, Y., Wong, F. L., Isobe, S., Wong, C. F., Wong, K. S., Xu, C., Li, C., Wang, Y., Guan, R., Sun, F., Fan, G., Xiao, Z., Zhou, F., Phang, T. H., Liu, X., Tong, S. W., Chan, T. F., Yiu, S. M., Tabata, S., Wang, J., Xu, X. and Lam, H. M. (2014), 'Identification of a novel salt tolerance gene in wild soybean by whole-genome sequencing', *Nat. Commun.*, 5, 4340.

Rouet, P., Smih, F. and Jasin, M. (1994), 'Expression of a site-specific endonuclease stimulates homologous recombination in mammalian cells', *Proc. Natl. Acad. Sci. USA.*, 91(13), 6064–8.

Rouquié, D., Capt, A., Eby, W. H., Sekar, V. and Hérouet-Guicheney, C. (2010), 'Investigation of endogenous soybean food allergens by using a 2-dimensional gel electrophoresis approach', *Regul. Toxicol. Pharmacol.*, 58(3 Suppl.), S47–53.

Sato, S., Xing, A., Ye, X., Schweiger, B., Kinney, A., Graef, G. and Clemente, T. (2004), 'Production of γ-linolenic acid and stearidonic acid in seeds of marker-free transgenic soybean', *Crop Sci.*, 44(2), 646–52.

Schmutz, J., Cannon, S. B., Schlueter, J., Ma, J., Mitros, T., Nelson, W., Hyten, D. L., Song, Q., Thelen, J. J., Cheng, J., Xu, D., Hellsten, U., May, G. D., Yu, Y., Sakurai, T., Umezawa, T., Bhattacharyya, M. K., Sandhu, D., Valliyodan, B., Lindquist, E., Peto, M., Grant, D., Shu, S., Goodstein, D., Barry, K., Futrell-Griggs, M., Abernathy, B., Du, J., Tian, Z., Zhu, L., Gill, N., Joshi, T., Libault, M., Sethuraman, A., Zhang, X. C., Shinozaki, K., Nguyen, H. T., Wing, R. A., Cregan, P., Specht, J., Grimwood, J., Rokhsar, D, Stacey G., Shoemaker, R. C. and Jackson, S. A. (2010), 'Genome sequence of the palaeopolyploid soybean', *Nature*, 463(7278), 178–83.

Shan, Q., Wang, Y., Li, J., Zhang, Y., Chen, K., Liang, Z., Zhang, K., Liu, J., Xi, J. J., Qiu, J. L. and Gao, C. (2013), 'Targeted genome modification of crop plants using a CRISPR-Cas system', *Nat. Biotechnol.*, 31(8):686–8.

Shan, Q., Wang, Y., Li, J. and Gao, C. (2014), 'Genome editing in rice and wheat using the CRISPR/Cas system', *Nat. Protoc.*, 9(10), 2395–410.

Shriram, V., Kumar, V., Devarumath, R. M., Khare, T. S. and Wani, S. H. (2016), 'MicroRNAs as potential targets for abiotic stress tolerance in plants', *Front. Plant Sci.*, 7, 817.

Song, S. K., Hou, W. S., Godo, I., Wu, C. X., Yu, Y., Matityahu, I., Hacham, Y., Sun, S., Han, T. F. and Amir, R. (2013), 'Soybean seeds expressing feedback-insensitive cystathionine γ-synthase exhibit a higher content of methionine', *J. Exp. Bot.*, 64(7), 1917–26.

Stewart Jr., C. N., Adang, M. J., All, J. N., Boerma, H. R., Cardineau, G., Tucker, D. and Parrott, W. A. (1996), 'Genetic transformation, recovery, and characterization of fertile soybean transgenic for a synthetic *Bacillus thuringiensis crylAc* gene', *Plant Physiol.*, 112(1), 121–9.

Streubel, J., Blucher, C., Landgraf, A. and Boch, J. (2012), 'TAL effector RVD specificities and efficiencies', *Nat. Biotechnol.*, 30(7), 593–5.

Sun, X., Hu, Z., Chen, R., Jiang, Q., Song, G., Zhang, H. and Xi, Y. (2015), 'Targeted mutagenesis in soybean using the CRISPR-Cas9 system', *Sci. Rep.*, 5, 10342.

Svitashev, S., Young, J. K., Schwartz, C., Gao, H., Falco, S. C. and Cigan, A. M. (2015), 'Targeted mutagenesis, precise gene editing, and site-specific gene Insertion in maize using Cas9 and guide RNA', *Plant Physiol.*, 169(2), 931–45.

Tabe, L. M. and Higgins, T. J. V. (1998), 'Engineering plant protein composition for improved nutrition', *Trends Plant Sci.*, 3(7), 282–6.

Tang, F., Yang, S., Liu, J. and Zhu, H. (2016), 'Rj4, a gene controlling nodulation specificity in soybeans, encodes a thaumatin-like protein but not the one previously reported', *Plant Physiol.*, 170(1), 26–32.

Tavva, V. S., Kim, Y. H., Kagan, I. A., Dinkins, R. D., Kim, K. H. and Collins, G. B. (2007), 'Increased α-tocopherol content in soybean seed overexpressing the *Perilla frutescens γ-tocopherol methyltransferase* gene', *Plant Cell Rep.*, 26(1), 61–70.

Upadhyay, S. K., Kumar, J., Alok, A. and Tuli, R. (2013), 'RNA-guided genome editing for target gene mutations in wheat', *G3-Genes Genom. Genet.*, 3(12), 2233–8.

Urnov, F. D., Miller, J. C., Lee, Y. L., Beausejour, C. M., Rock, J. M., Augustus, S., Jamieson, A. C., Porteus, M. H., Gregory, P. D. and Holmes, M. C. (2005), 'Highly efficient endogenous human gene correction using designed zinc-finger nucleases', *Nature*. 435(7042), 646–51.

Valente, M. A., Faria, J. A., Soares-Ramos, J. R., Reis, P. A., Pinheiro, G. L., Piovesan, N. D., Morais, A. T., Menezes, C. C., Cano, M. A., Fietto, L. G., Loureiro, M. E., Aragão, F. J. and Fontes, E. P. (2009), 'The ER luminal binding protein (BiP) mediates an increase in drought tolerance in soybean and delays drought-induced leaf senescence in soybean and tobacco', *J. Exp. Bot.*, 60(2), 533–46.

Wagner, N., Mroczka, A., Roberts, P. D., Schreckengost, W. and Voelker, T. (2011), 'RNAi trigger fragment truncation attenuates soybean FAD2-1 transcript suppression and yields intermediate oil phenotypes', *Plant Biotechnol. J.*, 9(7), 723–8.

Walker, D. R., All, J. N., McPherson, R. M., Boerma, H. R. and Parrott, W. A. (2000), 'Field evaluation of soybean engineered with a synthetic *cry1Ac* transgene for resistance to corn earworm, soybean looper, velvetbean caterpillar (Lepidoptera: Noctuidae), and lesser cornstalk borer (Lepidoptera: Pyralidae)', *J. Econ. Entomol.*, 93(3), 613–22.

Walker, D., Boerma, H. R., All, J. and Parrott, W. (2002), 'Combining *cry1Ac* with QTL alleles from PI 229358 to improve soybean resistance to lepidopteran pests', *Mol. Breeding*, 9(1), 43–51.

Wang, X. Y., Eggenberger, A. L., Nutter Jr., F. W. and Hill, J. H. (2001), 'Pathogen-derived transgenic resistance to soybean mosaic virus in soybean', *Mol. Breed.*, 8(8), 119–27.

Wang, Y., Cheng, X., Shan, Q., Zhang, Y., Liu, J., Gao, C. and Qiu, J. L. (2014), 'Simultaneous editing of three homoeoalleles in hexaploid bread wheat confers heritable resistance to powdery mildew', *Nat. Biotechnol.*, 32(9), 947–51.

Wang, X., He, X., Zou, S., Xu, W., Jia, X., Zhao, B., Zhao, C., Huang, K. and Liang, Z. (2016), 'A subchronic feeding study of dicamba-tolerant soybean with the dmo gene in Sprague-Dawley rats', *Regul Toxicol Pharmacol.*, 77, 134–42.

Wright, T. R., Shan, G., Walsh, T. A., Lira, J. M., Cui, C., Song, P., Zhuang, M., Arnold, N. L., Lin, G., Yau, K., Russell, S. M., Cicchillo, R. M., Peterson, M. A., Simpson, D. M., Zhou, N., Ponsamuel, J. and Zhang, Z. (2010), 'Robust crop resistance to broadleaf and grass herbicides provided by aryloxyalkanoate dioxygenase transgenes', *Proc. Natl. Acad. Sci. USA*, 107(47), 20240–5.

Xia, Z., Zhai, H., Lü, S., Wu, H. and Zhang, Y. (2013), 'Recent achievement in gene cloning and functional genomics in soybean', *Scientific World J.*, 2013, 7, Article ID 281367.

Xie, K., Zhang, J. and Yang, Y. (2014), 'Genome-wide prediction of highly specific guide RNA spacers for CRISPR-Cas9-mediated genome editing in model plants and major crops', *Mol. Plant*, 7(5), 923–6.

Yadav, N. S. (1996), 'Genetic modification of soybean oil quality', In: Verma, D. P. S. and R. C. Shoemaker (Eds), *Soybean: Genetics, Molecular Biology and Biotechnology*. CAB International, USA, pp. 165–88.

Yamada, T., Takagi, K. and Ishimoto, M. (2012), 'Recent advances in soybean transformation and their application to molecular breeding and genomic analysis', *Breed. Sci.*, 61(5), 480–94.

Yu, O., Shi, J., Hession, A. O., Maxwell, C. A., McGonigle, B. and Odell, J. T. (2003), 'Metabolic engineering to increase isoflavone biosynthesis in soybean seed', *Phytochemistry*, 63(7), 753–63.

Yu, H., Li, Y., Li, X., Romeis, J. and Wu, K. (2013), 'Expression of Cry1Ac in transgenic *Bt* soybean lines and their efficiency in controlling lepidopteran pests', *Pest Manag. Sci.*, 69(12), 1326–233.

Yu, L., Tan, X., Jiang, B., Sun, X., Gu, S., Han, T. and Hou, W. (2014), 'A Peroxisomal long-chain Acyl-CoA synthetase from *Glycine max* involved in lipid degradation', *PLoS ONE*, 9(7), e100144.

Zernova, O. V., Lygin, A. V., Widholm, J. M. and Lozovaya, V. V. (2009), 'Modification of isoflavones in soybean seeds via expression of multiple phenolic biosynthetic genes', *Plant Physiol. Biochem.*, 47(9), 769–77.

Zhang, H. Y., Yu, Z. N. and Deng, W. X. (2000), 'Composition and ecological distribution of Cry protein and their genotypes of *Bacillus thuringiensis* isolates from warehouses in China', *J. Invertebr. Pathol.*, 76(3), 191–7.

Zhang, X. X., Tang, Y. J., Ma, Q. B., Yang, C. Y., Mu, Y. H., Suo, H. C., Luo, L. H. and Nian, H. (2013), 'OsDREB2A, a rice transcription factor, significantly affects salt tolerance in transgenic soybean', *PLoS One*, 8(12), e83011.

Zhang, Y., Schernthaner, J., Labbé, N., Hefford, M. A., Zhao, J. and Simmonds, D. H. (2014), 'Improved protein quality in transgenic soybean expressing a de novo synthetic protein, MB-16', *Transgenic Res.*, 23(3), 455–67.

Zhang, Z. Y., Xing, A. Q., Staswick, P. and Clemente, T. E. (1999), 'The use of glufosinate as a selective agent in *Agrobacterium*-mediated transformation of soybean', *Plant Cell Tiss. Org.*, 56(1), 37–46.

Zhao, Y., Zhang, C., Liu, W., Gao, W., Liu, C., Song, G., Li, W. X., Mao, L., Chen, B., Xu, Y., Li, X. and Xie, C. (2016), 'An alternative strategy for targeted gene replacement in plants using a dual-sgRNA/Cas9 design', *Sci. Rep.*, 6, 23890.

Zhou, Z., Jiang, Y., Wang, Z., Gou, Z., Lyu, J., Li, W., Yu, Y., Shu, L., Zhao, Y., Ma, Y., Fang, C., Shen, Y., Liu, T., Li, C., Li, Q., Wu, M., Wang, M., Wu, Y., Dong, Y., Wan, W., Wang, X., Ding, Z., Gao, Y., Xiang, H., Zhu, B., Lee, S. H., Wang, W. and Tian, Z. (2015), 'Resequencing 302 wild and cultivated accessions identifies genes related to domestication and improvement in soybean', *Nat. Biotechnol.*, 33(4), 408–14.

Zhu, S., Walker, D. R., Boerma, H. R., All, J. N. and Parrott, W. A. (2008), 'Effects of defoliating insect resistance QTLs and a *cry1Ac* transgene in soybean near-isogenic lines', *Theor. Appl. Genet.*, 116(4), 455–63.

Chapter 7

Advances in the drought and heat resistance of soybean

Heng Ye, Babu Valliyodan, Li Song, J. Grover Shannon, Pengyin Chen and Henry T. Nguyen, University of Missouri, USA

1 Introduction

Climate changes increase the occurrence of extreme weather patterns including irregular precipitation, higher temperature and the consequent drought and heat stresses in the global agricultural areas, which can cause significant reduction in crop production and threaten food security (Lesk et al., 2016). Yield losses of major crops under irregular weather patterns keep increasing, despite of the progressive increase in yield through breeding and management practices since the 1960s (Boyer, 2013; Rosenzweig et al., 2014; Lobell et al., 2014). It has been estimated that under the current warming scenarios crop yields would decline 30–82% by the end of the century, despite of increasing concentration of CO_2 benefiting photosynthesis (Hatfield et al., 2011).

Soybean is one of the important crops used in many products such as human food, animal feed and biofuel due to its higher protein and oil content. Soybean yield suffered from drought associated with high temperature for about 40% reduction (Specht et al., 1999). Drought and heat stress during the vegetative stage affects the growth of soybean, as it leads to stunted plant growth with considerable leaf loss in the plants. Under the drought and heat stress, soybeans in reproductive stages showed increased flower and pod abortion and in later periods of pod filling stages, prolonged stresses can result in fewer and smaller seeds with reduced seed vigour (Boyer, 1983; Chebrolu et al., 2016). It is difficult to distinguish the effects of heat stress (high temperature) from drought

http://dx.doi.org/10.19103/AS.2017.0034.07

stress in soybean. These stresses usually occur together and magnify the effects of each another. When temperature reaches above 35°C, the transpiration of soybean plants will exceed the amount of water that plants can pump up from soil, which adds the similar water-deficit stress to plants as in drought. In addition, heat stress during reproductive and pod filling stages also causes problems in pollen viability, seed abortion and seed germinability. To overcome the negative impacts of drought and heat stress in soybean, many strategies have been developed and adopted, mainly including agricultural practices and genetic improvement of soybean cultivars. Rainfall and irrigation water is being used more efficiently; however, adaptation of irrigation systems is region-limited and would substantially increase the cost of soybean production. Clearly, improving the drought and heat resistance of our soybean varieties is strategically crucial (Turner, 1996).

Drought resistance is defined as the ability of plants to survive, grow and reproduce under water-deficit conditions (Levitt, 1972; Turner, 1978). In crops, drought resistance actually refers to the ability of plants to produce an acceptable yield under the stress conditions (Turner, 2000; Fleury et al., 2010). The stress levels of drought that the crops are actually perceiving are determined by the water status of the plants in terms of water content and water potential (Turner, 1986). There are three types of mechanisms of drought resistance: escape, avoidance and tolerance (Nilsen and Orcutt, 1996; Turner, 2000; Turner et al., 2001). The escape mechanism is recognized as shorter life cycles (early maturity) to escape from the extremely hot and dry summer, but sometimes leads to a reduction in seed production due to the shorter vegetative growing periods. The avoidance mechanism refers to maintaining favourable water status under drought either by enhancing water uptake by roots or limiting water loss from transpiration. The tolerance mechanism suggests the physiological and molecular adaptation of plants perceiving dehydration stress or the trend of water loss of plant cells (Blum, 2015). The tolerance of drought is achieved via various physiological adaptations including osmotic adjustment, increasing plasticity in the cell and alterations in efficiency of photosynthesis (Taji et al., 2004). For crops, the yield under drought was supposed to be determined by the factors affecting three major components: water use, water use efficiency and harvest index (Yield = Water use × Water use efficiency × Harvest index) (Passioura, 1977, 1983). Water use is the amount of water transpired by crops and water use efficiency refers to the efficiency of crops to use water to produce dry matters. The yield of crops heavily relies on water input and the efficiency of conversion of water to biomass (Passioura, 1977, 1983). This yield framework has been extensively used in assessing yield of crops under water-limited conditions (Turner, 2000; Turner et al., 2001). To acquire more water resource and reserve the available water in plants is a key solution to improve drought resistance in crops. During drought stress, plants adopt multiple mechanisms from morphological modifications and physiological adaptations, and biochemical alternations to resist drought (Jones et al., 1981; Valliyodan and Nguyen, 2006).

Stress-resistant plants have evolved their stress adaptation mechanisms under natural selection. Exotic and wild relatives that grow under wild conditions tend to have better stress-resistant characteristics by maintaining natural variations that contribute to stress resistance (Mickelbrat et al., 2015). Some of the natural variations underlying these stress resistance traits have been lost in the current elite germplasms during domestication due to reduced selection force for these traits or linkage to some undesired agronomic traits. Incorporation of these stress resistance traits into cultivars by transferring the underlying natural variations is essential to improve the performance of cultivars under stress conditions. It can be achieved through identifying the tolerant genetic resources,

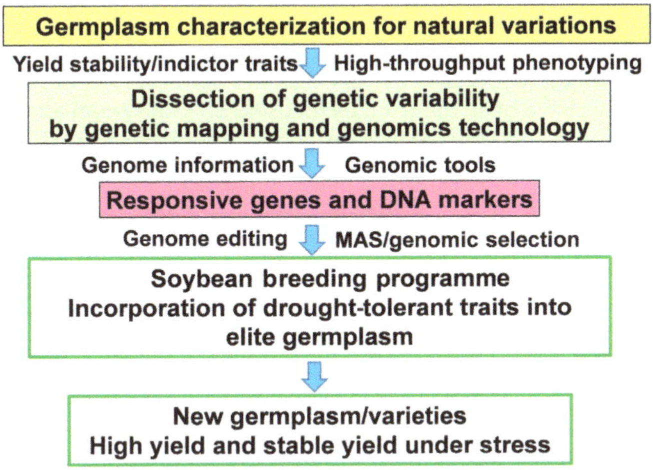

Figure 1 Strategies to improve stress tolerance in soybean (modified the figure from Fig. 6 in Foyer et al. (2016)).

discovering the underlying genetic variations and incorporating the key genetic variations into the cultivars (Fig. 1). These key natural variations can be identified through quantitative trait locus (QTL) mapping and genome-wide association studies (GWAS) (Takeda and Matsuoka, 2008; Varshney et al., 2015). With the assistance of genomic strategies, these natural variations can be identified and expeditiously incorporated into the cultivars by marker-assisted selection (MAS) and genomic selection strategies. Further characterization of genes underlying stress tolerance traits will elucidate specific biological mechanisms, which will be consequently translated to the improvement of crop (Fig. 1). Application of whole-genome sequencing/re-sequencing and high-throughput phenotyping can facilitate the above-mentioned processes.

2 Drought resistance traits: introduction, root system architectures and anatomy

2.1 Introduction to drought resistance traits

In crops, drought resistance is translated into traits enhancing yield stability rather than those increasing survivability under drought (Bartels et al., 2006; Blum, 2009; Passioura, 2010; Sinclair, 2011). These translated traits are correlated with yield under drought and have no yield penalty under non-stress conditions. The phenotyping procedures of these traits should be simple, rapid and accurate to allow high-throughput evaluation of the traits. Several of these traits have been discovered in various crops, including root systems (Bolaños and Edmeades, 1993; Uga et al., 2013; Prince et al., 2015; Aceves-García et al., 2016), carbon isotope discrimination (Turner, 1997; Condon et al., 2002; Tambussi et al., 2007; Ellsworth and Cousins, 2016), canopy temperature depression (Blum, 1988; Ludlow and Muchow, 1990; Reynolds et al., 2007, 2009; Pinto and Reynolds, 2015), photosynthesis

(Tuinstra et al., 1998; Thomas and Howarth, 2000; Shukla et al., 2004; Feller, 2016), canopy wilting (Sloane et al., 1990; Carter, 1999 and 2006; Hwang et al., 2015) and nitrogen (N)-fixation under drought (Djekoun, 1991; Sinclair and Serraj, 1995). In soybean, drought tolerance-related traits have been reported by focusing on root system architectures (RSA), slow canopy wilting and sustained N-fixation under drought (Fig. 2). These three traits were predicted to improve yield under drought >75% through the analysis of a simulation model (Sinclair et al., 2010).

2.2 RSAs and anatomy

Most of us would agree with Kramer (1969) when he made this statement about an essential characteristic of drought resistance: 'deep, wide-spreading, much-branched root system.' Root systems are usually involved in both drought avoidance and tolerance during drought stress due to the constitutive and plastic characteristics of roots. RSA refers to the shape of the roots and the physical space and the deeper and wilder root system can avoid drought stress by their ability to acquire more water resource. At the meantime, RSA is also highly plastic to environmental changes such as water deficit, which can be referred to drought tolerance and consequently lead to drought avoidance. When plants perceive water-deficit stress, roots tend to keep growing and penetrate into deeper soil layers (Hoogenboom et al., 1987; Sponghiado et al., 1989; Creelman et al., 1990; Wu et al., 1994). The ability of plants to develop deeper rooting systems under drought stress depends on the tolerance levels of the roots to water deficits. Some lines showed the ability to elongate their rooting depth significantly than some other lines under drought

Figure 2 Traits representing or contributing to drought and heat tolerance in soybean. The blue arrowheads point to root metaxylems. 'FW' stands for fast wilting and 'SW' stands for slow wilting (modified the figure from Fig. 1A in Valliyodan et al. (2016)).

stress in legumes, including soybean (Garay and Wilhelm, 1983; Sponghiado et al., 1989). Genetic diversity of RSA has evolved through geographic adaptation of plants. Benefit of a deep and heavy root system on drought tolerance has been reported in various crops, including rice (Nguyen et al., 1997; Babu et al., 2003; Kato et al., 2006; MacMillan et al., 2006; Steele et al., 2006; Bernier et al., 2009; Uga et al., 2013), maize (Tuberosa et al., 2002; Giuliani et al., 2005; Hammer et al., 2009; Landi et al., 2010; Hund et al., 2011), barley (Forster et al., 2005), wheat (Manschadi et al., 2006; Wasson et al., 2012), chickpea (Varshney et al., 2011; Chen et al., 2012) and soybean (Hudak and Patterson, 1995; Pantalone et al., 1996; Sadok and Sinclair, 2011).

Deep rooting, which is a complex trait affected by growth angle and root length (Araki et al., 2002), plays a crucial role in water uptake from deeper soils to avoid drought under water-deficit conditions. Root angle determines the direction of horizontal and vertical distribution of roots in the soil. It is recognized as an adaptive trait for drought avoidance in sorghum (Mace et al., 2012), wheat (Christopher et al., 2013) and rice (Uga et al., 2013). Root angle was identified to be correlated with rooting depth in rice (Kato et al., 2006), chickpea (Sayar et al., 2007; Kashiwagi et al., 2015) and sorghum (Singh et al., 2011). Genetic studies on root angle revealed that it is an inheritable trait and can reduce the energy inputs to penetrate in deeper horizons to access water during limited rainfall (Wasson et al., 2012; Meister et al., 2014). The first naturally occurring variation in root angle has been cloned and utilized in rice as *DEEPER ROOTING 1* (*DRO1*), which induce narrow root angle and deeper rooting system via gravity sensing (Uga et al., 2013). Under non-stress conditions, *DRO1* has no effect on yield; however, it was found to double the yield under moderate and severe drought stress (Uga et al., 2013). Currently, *DRO1* has been incorporated into elite lines to develop drought-tolerant rice varieties by MAS (Arai-Sanoh et al., 2014). In addition to deep rooting, drought stress also induces the plasticity responses of root systems by increasing the number of fibrous roots, decreasing lateral root diameter and fluctuating in root biomass (Nielsen et al., 1996; Osmont et al., 2007; Meister et al., 2014; Salazar-Henao et al., 2016). Alterations in root anatomy, such as aerenchyma formation in maize (Lynch, 2011; Burton et al., 2013), saves the energy inputs to allow improved soil penetration and exploration to compensate water deficit (Addington et al., 2006; Maseda and Fernandez, 2006).

In soybean, the improved RSA have shown to alleviate drought stress by increasing exploration for water and nutrients (Hoogenboom et al., 1987). Natural variations in RSA were reported in soybean, which was suggested to be used for improvement of drought tolerance (Carter, 1989). In the field evaluation, the drought-tolerant Japanese landrace PI 416937 have shown better fibrous rooting and exploit of a large volume of soil volume and the trait of fibrous rooting was found to be quantitatively inherited (Hudak and Patterson, 1996). Recently several soybean accessions were identified by screening of a core set (400 lines) of the USDA Germplasm Collection to have promising RSA for extensive fibrous rooting, root length or large root angle (Table 1). Genetic diversity was also observed in root anatomy of soybean and water movement through root systems was affected by change in root anatomy (Rincon et al., 2003). The examination of root anatomy of 41 soybean lines led to the discovery of variations in the number of metaxylems in roots. The number of metaxylems was found to correlate with drought tolerance in soybean, as soybean plants develop a higher number of metaxylems under drought conditions (Prince et al., 2017). The drought-tolerant lines with abundant numbers of metaxylems are listed in Table 1.

For major cereal crops, such as rice, wheat and maize, intensive QTL mapping work for RSA has been conducted (Courtois et al., 2009; Zhang, 2003; Atkinson, 2015; Tuberosa

Table 1 Genetic resources for the traits that contribute to drought and heat tolerance in soybean

Trait for drought and heat tolerance	Genetic resources of soybean[a]
Root length	PI 408105A, PI 081041, PI 090479
Root angle	PI 483463, PI 248515, PI 424088
Fibrous rooting/root area	PI 408105A, PI 416937, PI 471938, PI 603713, PI 248515
Number of metaxylems	PI 427136, 4J105-3-4, CL0J1760-6-8
Canopy wilting	PI 416937, PI 471938, Jackson, PI 567690, PI 567731, R02-1325[b], N05-7432[b], N04-9646[b], N06-7194[b], R09-1069[b], N06-7543[b], R07-5235[b], R09-1121[b], R07-7044[b], R05-5559[b]
Sustained N-fixation ability under drought	Jackson, PI 471938, PI 222547, PI 374163, PI 423886, PI 429328, PI 507039, PI 227557, PI 507414, PI 578315B, R02-1325[b], N05-7432[b], N04-9646[b], N06-7194[b], R09-1069[b], N06-7543[b], R07-5235[b], R09-1121[b], R07-7044[b], R05-5559[b]
Pollen viability and seed development	PI 587982A, DP 5915RR[b], DP 4690RR[b], DG 4950RR[b], DG 5630RR[b], Stressland[b], DP5414RR[b], DP5806RR[b], Centennial[b], DK4965RR[b], DK4868RR[b], DP3861RR[b], HBK 4920RR[b], AG4403RR[b]

[a] Lines that have promising performance for the corresponding traits.
[b] High-yielding elite lines or commercial varieties developed in the soybean breeding programmes.

et al., 2002; Giuliani et al., 2005). However, only less number of studies has been conducted in legume species, especially soybean. Initially, five QTLs associated with the fibrous rooting were mapped from an interspecific mapping population between Benning and PI 416937 containing 240 $F_{6:7}$ recombinant inbred lines (RILs) under the field conditions and the phenotypic contributions of these QTLs range from 0.07 to 0.14 (Abdel-Haleem et al., 2011). Among the five QTLs, four have donor alleles from the drought-tolerant parent PI 416937. In a bi-parental recombinant inbred population, one QTL cluster was mapped for both root length and lateral root number with the donor alleles from an older cultivar: Dunbar (Manalavan et al., 2015), which indicated that some favourable alleles for RSA could be found in some older soybean varieties and these genetic materials need consideration during breeding design (Table 1). The genetic diversity of RSA in current cultivars can be increased by reintroducing the favourable alleles from wild relatives that were lost during domestication. One study in greenhouse conditions identified four QTLs associated with surface area and root volume in a population of *Glycine max* × *Glycine soja* and the favourable alleles of the QTL were from the wild parent (PI 407162) (Prince et al., 2015). Recent report shows that one rare allele can promote lateral root growth with a potential to improve drought tolerance. GWAS of 252 phenotypically diverse soybean germplasm pinpointed this allele and this one single-nucleotide polymorphism (SNP) cause more profuse lateral rooting in soybean (unpublished results). Further screening of the USDA Germplasm Collection found that the frequency of the favourable alleles at this locus is about 0.5%, which suggests that a recent mutation caused this favourable allele which is absent in current cultivars (unpublished results). Name of the candidate gene in this locus is Glycine Lateral Root 1 (GLR1) and it encodes a leucine-rich protein. Functional characterization of GLR1 shows increased lateral roots in a soybean hairy root transgenic system. As discussed above, several QTL for RSA in soybean have been identified; however, the relationship between these QTL and drought tolerance have not yet been evaluated. Therefore, the evaluation of these QTL/genes in drought tolerance under real drought scenarios is necessary.

3 Other traits for drought resistance and heat tolerance

3.1 Water use efficiency and canopy wilting under drought

In addition to enhancing ability to access more water resource to achieve drought resistance, improving water use efficiency is another promising strategy to overcome drought stress in the field. The essential factor to improve water use efficiency is to conserve water in plants and reduce the unnecessary transpiration lost (Turner, 2000; Turner et al., 2001). In soybean, researchers and breeders made efforts to look for traits associated with water use efficiency, which can allow screening for drought tolerance on a large scale. In 1990, a phenotype of slow canopy wilting under drought and heat stresses was observed in PI 416937, the Japanese drought-tolerant landrace in the maturity group (MG) VI (Sloane et al., 1990). The canopy of PI 416937 showed significant delayed leaf wilting than other drought-sensitive lines indicating that the PI 416937 was not subjected to drought stress compared with other lines. Further characterization revealed that PI 416937 has larger root surface and rooting area than drought-sensitive checks (Hudak and Patterson, 1995; Pantalone et al., 1996). However, field evaluation found a contradictory phenomenon that PI 416937 actually used a significantly less amount of water than the drought-sensitive checks (Hudak and Patterson, 1996). These results suggested that PI 416937 must employ water conservation strategy to improve water use efficiency to resist drought stress. Further research on water conservation aspects of PI 416937 identified that this line can limit its transpiration rate under vapour pressure deficit (VPD) above 2.0 KPa compared with other drought-sensitive genotypes (Tanaka et al., 2010; Ries et al., 2012; Fletcher et al., 2007). Later on, another line (PI 471938) at MG VI from Nepal with the slow canopy wilting phenotype was identified (Carter, 1999, 2006). Interestingly, this line exhibited a different physiological mechanism compared to PI 416937. It cannot limit its own transpiration rate under high VPD, instead, it can maintain a higher leaf-turgor level than other drought-sensitive checks (Sadok et al., 2012). These two lines offered breeding resources for improving drought tolerance in late MGs of soybean. Recently, two additional slow canopy wilting landraces (PI 567690 and PI 567731) at MG III were identified after the evaluation of a core set of 250 soybean germplasm lines for canopy wilting and drought tolerance (Pathan et al., 2014). These two lines shared similar physiological mechanisms of limiting transpiration rate under high VPD as PI 416937 (unpublished results) and offered breeding materials for early MGs in soybean.

QTL mapping identified genomic regions associated with slow canopy wilting traits and associated DNA markers were developed for MAS in breeding. Initial QTL mapping work was conducted using a RIL population derived from Benning × PI 416937. Seven QTL were mapped in that population and the one on chromosome 12 (*qSW-Gm12*) was identified as the major across all five independent environments with a phenotypic contribution up to 27% (Abdel-Haleem et al., 2012). The following QTL mapping in five populations with 15 environments (site-years) were identified: a total of 13 QTL were mapped for canopy wilting with varied phenotypic contributions (R^2: 0.04 to 0.29) and the majority of these QTL (11/13) have donor alleles for slow wilting from either PI 416937 and Jackson (soybean cultivar) (Hwang et al., 2015). Among the 13 QTL, eight were further confirmed and their confidence intervals were refined through meta-QTL analysis (Hwang et al., 2016). Most of the mapped QTL were found to be unstable across independent environments and populations. For instance, the major QTL on chromosome 12 was not detected in any populations or site-years reported by Hwang et al. (2015, 2016). Therefore, QTL confirmation should be performed to validate each individual QTL. The complexity of the canopy wilting trait also indicates that stacking all confirmed QTL by MAS or genomic selection is necessary to recover the drought tolerance performance shown in the original drought-tolerant exotic PIs.

3.2 Sustained N-fixation under drought

N-fixation is highly sensitive to drought stress (Djekoun, 1991; Adams et al., 2016). Grain legumes (soybean and cowpea) transport ureides from nodules to shoots, which make them more sensitive to drought stress than the other legumes (peanut and chickpea) transporting amides (Sinclair and Serraj, 1995). In the grain legume systems, ureides accumulate in leaves and nodules during drought stress and impose a negative feedback to inhibit nitrogenase activity for N-fixation (Serraj et al., 1999, 2001; Vadez et al., 2000; Ladrera et al., 2007). Shoot ureide concentration was identified to correlate with drought tolerance and used as another indicator of drought tolerance (de Silva et al., 1996; Serraj and Sinclair, 1996a,b; Purcell et al., 1998). The drought-tolerant soybean variety Jackson maintained nitrogen-fixation activity under water-deficit conditions and have low leaf ureide concentration (Sall and Sinclair, 1991; Serraj and Sinclair, 1996a,b). The two drought-tolerant and slow canopy witling lines, PI 416937 and PI 471938, were also found to have low leaf ureide concentration and be able to sustain nitrogen-fixation under water-deficit conditions (Patterson and Hudak, 1996; Sadok et al., 2012). These findings confirmed that the ureides and nitrogen content in the shoots are a reliable indicator of the nitrogen-fixation activity in soybean and can represent the tolerance level of the plants (Sinclair and Serraj, 1995; Serraj et al., 1999, 2001). The commonly used methods of measuring ureides and nitrogen content were well adopted by using colorimetric (Young and Conway, 1942) and ^{15}N-isotope (McClure et al., 1980) methods. These methods can allow screening drought tolerance performance in the field at a large scale, as they are relatively much easier to be performed than yield evaluation under drought. In 1990s, Sinclair et al. (2000) screened 3081 soybean PIs for sustained nitrogen-fixation, in which eight PIs were identified to have best performance in sustained nitrogen-fixation under drought and have promising drought tolerance related to yield stability (Table 1). All these soybean genotypes have been used in breeding programmes to develop drought-tolerant elite lines (Devi et al., 2014) and the new drought-tolerant lines can protect yield under drought more than 40% (Dr. Chen, University of Missouri, pers. comm.).

QTL mapping work has been conducted for shoot ureide and nitrogen concentration in both bi-parental populations and GWAS in diverse lines. In a cross KS4893 × Jackson, two QTL associated with shoot ureide concentration on Chro. 9 and 19 and two QTL associated with shoot nitrogen concentration on Chro. 13 and 17 were identified under water stress condition with phenotypic contributions ranging from 0.11 to 0.31 (Hwang et al., 2013). The two QTL on Chro. 13 and 19 with larger effects have donor alleles from the drought-tolerant parent (Jackson) and the other two on Chro. 9 and 17 have donor alleles from KS4895. These four QTL and the associated DNA markers can be used for MAS in the breeding programmes for drought tolerance in soybean. In the same study, more QTL for shoot ureide and nitrogen concentration were identified under irrigated conditions, but no QTL was found to be overlapped under both stress and non-stress conditions (Hwang et al., 2013). Although QTL associated with N-fixation indicator traits under both stress and non-stress conditions were mapped in this study, the reduction in N-fixation under drought compared to optimum conditions was not analysed, which would reflect the actual effects of drought on N-fixation. Recently, a GWAS for a set of 374 diverse lines at MG IV mapped 54 and 19 putative loci associated with shoot ureides (Ray et al., 2015) and nitrogen concentrations (Dhanapal et al., 2015), respectively. The large number of loci through GWAS indicated the complexity of N-fixation under drought and suggested that genomic selection should be better suited to improve this trait.

3.3 Pollination and seed development under extreme heat

Reproduction of soybean is sensitive to high temperatures (>35°C) (Salem et al., 2007). Heat stress during reproductive stages of pollination and seed development significantly decrease soybean yield (Prasad et al., 1999; Redden et al., 2014; Kebede et al., 2012; Siebers et al., 2015). Increased flower and pod abortion and reduced seed germinability was observed in soybean plants subjective to extreme heat (Boyer, 1983; Salem et al., 2007; Chebrolu et al., 2016). Fattened and collapsed pollens were observed in soybean under heat stress, which resulted in lower pollen viability and fertilization rates (Salem et al., 2007). Soybean plants exposed to high temperature (38/28°C) showed 22.7% reduction in pollen germination and consequently had about 35% reduction in pot setting and the anatomical changes in pollens under heat stress were observed (Djanaguiraman et al., 2013). Evaluation of 44 soybean genotypes from MGs III to IV for heat tolerance helped to categorize into heat-tolerant, -intermediate and -sensitive groups based on pollen viability. Among the 44 genotypes, thirteen were identified as most heat tolerant and can be used in breeding programmes for heat tolerance at the reproductive stages (Table 1).

Seed development is much more vulnerable than vegetative tissues to heat stress. Heat stress at the pod filling stages of soybean results in seed with less vigour, poor germination and increased incidence of pathogen infection (Hatfield et al., 2011). Genetic diversity and significant differences in germinability under heat stress between the heat-tolerant and -sensitive lines were reported (Chebrolu et al., 2016). Germination of seeds from the heat-sensitive genotype reduced by 50% under 36/24°C and completely inhibited under 42/26°C compared to normal conditions (28/22°C) (Chebrolu et al., 2016). In contrast, seed germinability from the heat-tolerant genotypes was unaffected under 36/24°C, and was reduced by 75% under 42/26°C treatment compared with normal conditions (Chebrolu et al., 2016). These identified heat-tolerant lines are good targets for gene discovery for heat tolerance and soybean breeding programmes.

4 Genomic resources and strategies

Modern tools and approaches are needed to facilitate crop improvement under the context of adverse effects of drought and heat stress on crop production. Recently, high-resolution genome information has been started to be adopted by breeders for germplasm characterization, genetic dissection of agronomic traits and prediction for breeding value. Along with the development of next-generation sequencing techniques, sequencing and re-sequencing of plant genomes have expanded greatly. The re-sequencing of selected populations of several plant species including *Arabidopsis*

(The 1001 genomes consortium, 2016), rice (The 3000 rice genomes project, 2014), cucumber (Qi et al., 2013) and sorghum (Mace et al., 2013) have been performed based on the benefit of new sequencing techniques. At the same, more reference genomes for crops have been built (Goff et al., 2002; Schnable et al., 2009; Paterson et al., 2009; Huang et al., 2009; Xu et al., 2011; Chalhoub et al., 2014). High-resolution genomic information and large number of SNP markers were generated from these resources for each species.

A complete reference genome for soybean was available in 2010 (Schmutz et al., 2010) and several projects were initiated to develop soybean genomic resources that can be utilized in the genetic analysis and trait discovery. Initially, 31 diverse soybean lines were re-sequenced (5× genome coverage), which revealed higher linkage disequilibrium compared to other crops and identified higher genetic diversity in wild soybean than the cultivated accessions (Lam et al., 2010). Later on, a larger population of 302 diverse soybean lines was re-sequenced (11× genome coverage) and enhanced the genome variation matrix (Zhou et al., 2015), which can greatly assist in trait-marker association analysis and characterization of genetic diversity. Recently, a core set of 106 soybean germplasm lines were sequenced at deeper genome coverage (17× genome coverage) to support the trait discovery and soybean crop improvement programmes (Valliyodan et al., 2016). This re-sequencing information became a foundation for deeper and comprehensive analysis of haplotype blocks underlying major genes, such as haplotype analysis of salinity tolerance gene in soybean (Patil et al., 2016). Genomic resources generated from the recent re-sequencing of diverse germplasm sets provided powerful tools for characterizing soybean genetic diversity and building strong foundation for trait/gene discovery to accelerate future breeding for elite cultivars.

An initial version of SNP detection assays was developed to identify a large number of SNPs and a better resolution genetic map, after sequencing a number of small DNA fragments (Hyten et al., 2010). Later on, the re-sequencing of reduced representation libraries from six cultivated and two wild soybean accessions identified more SNPs and helped to develop an Illumina Infinium BeadChip with 50 000 SNPs for soybean (Song et al., 2013). The 50 000 SNPs information is publicly available and widely used to construct high-resolution genetic linkage maps, GWAS and haplotype analysis. Benefiting from the new cost-effective sequencing platforms, high-throughput genotyping by sequencing methodologies were developed, which includes both restriction digest-based and skim whole-genome sequencing (low genome coverage)-based genotyping methods. In soybean, the skim whole-genome sequencing-based genotyping method significantly improved the resolution of linkage maps and facilitated in pinpointing candidate genes underlying target traits, for example, root-knot nematode resistance (Xu et al., 2013).

Following the release of soybean reference genome in 2010, reference genomes for other legume crops, including pigeon pea (Varshney et al., 2012; Singh et al., 2012), chickpea (Varshney et al., 2013; Jain et al., 2013; Parween et al., 2015), common bean (Schmutz et al., 2014) and peanut (Bertioli et al., 2016) were built. Comparative legume genome analysis utilizing these genomic resources helped to identify changes in genome structures and genome synteny. Cloning of genes or genomic regions for drought and heat tolerance in one legume species extended into other legume crops by comparing genome structures and synteny.

5 Advances in breeding for drought and heat resistance

Stress escape mechanism has been utilized to resist drought and heat stresses in soybean production. Growing areas of soybeans of MGs III–IV keep increasing in Southern US, where normally MGs VI–VII are grown. Adoption of the earlier MG soybeans can shift the reproductive and pod filling stages into early and late June to avoid the drought and heat stress in August. Breeding progress for drought- and heat-resistant varieties is still limited due to the complexity of the resistance mechanisms. In the United States, soybean breeders have been putting efforts to incorporate the major traits discussed above into high-yielding elite lines to develop stress-resistant varieties. Currently, the slow canopy wilting phenotype has been used as an indicator to screen drought resistance in soybean (Charlson et al., 2009) and several drought-tolerant elite lines for Southern US MGs (V–VIII) have been developed (Devi et al., 2014). The newly developed drought-tolerant elite lines listed in Table 1 were estimated to protect yield up to 50% under drought stress (Dr. Shannon, University of Missouri, pers. comm.). The identified genotypes with sustained N-fixation under drought have also been used in breeding programmes to develop drought-tolerant elite lines (Devi et al., 2014) and the new drought-tolerant lines can protect yield under drought more than 40% (Dr. Chen, University of Missouri, pers. comm.). Currently, soybean breeders are also working on to combine all these drought and heat resistance traits to maximize the soybean performance under stress conditions.

6 Conclusions and future perspectives

The sustainability of crop production and food security are being threatened by the increasing unpredictability and severity of drought and heat stresses due to global climate changes. Genetic variations for stress tolerance in natural populations reflect the evolution of plants in the diverse ecological conditions and are being identified in crop species. Introduction of these adapted natural genetic variations into breeding programmes can enrich current genetic diversity of stress tolerance and improve yield under stresses. Screening germplasm for stress tolerance traits has provided donor resources with potential to improve yield under stress conditions. Identification of the genomic regions underlying these natural genetic variations and transferring the favourable alleles into elite germplasm with the assistance of genomic technologies will speed up the genetic improvement of stress tolerance in soybean. QTL associated with drought tolerance traits such as RSA, canopy wilting and N-fixation ability under drought were identified in soybean and associated DNA markers have been developed for MAS. QTL or genes that enable stable yield under the stress conditions can be lost during domestication processes due to its linkage to negative loci or pleiotropic effects on yield drag. The subsequent fine-mapping are still desired to determine whether the yield drag effects are due to pleiotropy or linkage to other negative loci. Screening a large population of segregants can break the tight linkage with the assistance of MAS. In addition, cloning and functional characterization of genes underlying the natural genetic variations can lead to understanding of molecular and physiological pathways of stress adaptation, which will direct further improvement in stress tolerance in soybean.

High-throughput and cost-effective phenotyping and genotyping technologies are foundations of genomics-assisted breeding programmes. High-throughput phenotyping platforms that monitor RSA and canopy status are essential to advance our knowledge on drought and heat tolerance. High-throughput phenotyping of RSA has focused on field-based approaches in crops such as maize (Trachsel et al., 2011; York et al., 2015), rice (Wissuwa et al., 2016) and wheat (Richard et al., 2015), with a recent shift in legume species, including soybean, common bean and cowpea (Manalavan et al., 2015, Prince et al., 2015; Burridge et al., 2016). The adoption of high-throughput image analysis using digital imaging of root traits further improved the efficiency and accuracy of the phenotyping of RSA (Bucksch et al., 2014; Das et al., 2015). Canopy spectral reflectance data can be measured with remote sensing, imaging and digital processing in a high-throughput and cost-effective manner. Some stress-related traits, such as canopy temperature, chlorophyll content, photosynthesis rate and transpiration rate, can be represented by using canopy spectral reflectance data. Certain wavelengths in the canopy reflectance data, such as the near-infrared wavelengths (730–1305 nm), was found to be highly correlated with agronomic traits (Christenson et al., 2015), which offers a new high-throughput phenotyping method to evaluate those agronomic traits including stress tolerance.

Benefiting from the aggressive development of next-generation sequencing technologies, construction of high-resolution genomic/genetic maps and development of DNA markers for phenotypic traits are much more affordable. Precise allelic mining for the stress tolerance traits can be achieved utilizing the accumulating genomic information. The targeted genome editing by CRISPR/Cas9 system (Jacobs et al., 2015; Sun et al., 2015) can enable further modification of these alleles to maximize the performance of soybean plants under stress conditions. In addition to soybean genome information, the release of reference genomes and continuous building-up re-sequencing information of various legume crops can strengthen the comparative legume genomics for discovery and characterization of key genes and gene families involved in stress tolerance. Integration of these genomic resources and technologies into the next-generation breeding strategies will accelerate the genetic improvement of yield under stress conditions for crops, including soybean.

7 Acknowledgements

We acknowledge funding support from the United Soybean Board towards the soybean genomics and breeding research programme at the University of Missouri.

8 References

Abdel-Haleem, H., Lee, G. J., Boerma, R. H. 2011. Identification of QTL for increased fibrous roots in soybean. *Theoretical and Applied Genetics* 122, 935–946.

Abdel-Haleem, H., Carter Jr., T. E., Purcell, L. C., King, C. A., Ries, L. L., Chen, P., Schapaugh, Jr. W., Sinclair, T. R., Boerma, H. R. 2012. Mapping of quantitative trait loci for canopy-wilting trait in soybean (Glycine max L. Merr). *Theoretical and Applied Genetics* 125, 837–46.

Aceves-García, P., Álvarez-Buylla, E. R., Garay-Arroyo, A., García-Ponce, B., Muñoz, R., Sánchez Mde, L. 2016. Root architecture diversity and meristem dynamics in different populations of arabidopsis thaliana. *Frontiers in Plant Science* 7, 858.

Adams, M. A., Turnbull, T. L., Sprent, J. I. and Buchmann N. 2016. Legumes are different: leaf nitrogen, photosynthesis, and water use efficiency. *Proceedings of the National Academy of Sciences of the United States of America* 113, 4098–103.

Addington, R. N., Donovan, L. A., Mitchell, R. J., Vose, J. M., Pecot, S. D., Jack, S. B., Hacke, U. G., Sperry, J. S. and Oren, R. 2006. Adjustments in hydraulic architecture of Pinus palustris maintain similar stomatal conductance in xeric and mesic habitats. *Plant, Cell & Environment* 29, 535–45.

Arai-Sanoh, Y., Takai, T., Yoshinaga, S., Nakano, H., Kojima, M., Sakakibara, H., Kondo, M. and Uga, Y. 2014. Deep rooting conferred by DEEPER ROOTING 1 enhances rice yield in paddy fields. *Science Reports* 4, 5563.

Araki, H., Morita, S., Tatsumi, J. and Iijima, M. 2002. Physiomorphological analysis on axile root growth in upland rice. *Plant Production Science* 5, 286–93.

Atkinson, J. A., Wingen, L. U., Griffiths, M. and et al. 2015. Phenotyping pipeline reveals major seedling root growth QTL in hexaploid wheat. *Journal of Experimental Botany* 66, 2283–92.

Babu, R. C., Nguyen, B. D., Chamarerk, V. and et al. 2003. Genetic analysis of drought resistance in rice by molecular markers: association between secondary traits and field performance. *Crop Science* 43, 1457–69.

Bartels, D., Ditzer, A. and Furini, A. 2006. What can we learn from resurrection plants? In J.-M. Ribaut (Ed.), *Drought Adaptation in Cereals*. The Haworth Press, Inc, Binghamton, NY, pp. 599–622.

Bernier, J., Kumar, A., Spaner, D., Verulkar, S., Mandal, N. P., Sinha. P. K., Peeraju. P., Dongre, P. R., Mahto, R. N. and Atlin, G. N. 2009. Characterization of the effect of rice drought tolerance qtl12.1 over a range of environments in the Philippines and eastern India. *Euphytica* 166, 207–17.

Bertioli, D. J., Cannon, S. B., Froenicke, L. and et al. 2016. The genome sequences of Arachis duranensis and Arachis ipaensis, the diploid ancestors of cultivated peanut. *Nature Genetics* 48, 438–46.

Blum, A. 1988. *Plant Breeding for Stress Environments. Plant Breeding for Stress Environments*. CRC Press, Boca Raton, Florida.

Blum, A. 2009. Effective use of water (EUW) and not water-use efficiency (WUE) is the target of crop yield improvement under drought stress. *Field Crops Research* 112, 119–23.

Blum, A. 2016. Stress, strain, signaling, and adaptation – not just a matter of definition. *Journal of Experimental Botany*, 67, 563–6.

Bolaños, J., Edmeades, G. O. 1993. Eight cycles of selection for drought tolerance in lowland tropical maize. II. Responses in reproductive behavior. *Field Crops Research* 31, 253–68.

Boyer, J. S. 1983. Environmental stress and crop yields. In C. D. Raper and P. J. Kramer (Eds), *Crop Reaction to Water and Temperature Stresses in Humid, Temperate Climates*. Westview Press, Boulder, USA, pp. 3–7.

Boyer, J. S., Byrneb, P., Cassmand, K. G. and et al. 2013. The U.S. drought of 2012 in perspective: a call to action. *Global Food Security* 2, 139–43.

Bucksch, A., Burridge, J., York, L. M., Das, A., Nord, E., Weitz, J. S. and Lynch, J. P. 2014. Image-based high-throughput field phenotyping of crop roots. *Plant Physiology* 166, 470–86.

Burridge, J., Jochuaa, C. N., Buckschc, A. and Lynch, J. P. 2016. Legume shovelomics: high-throughput phenotyping of common bean (Phaseolus vulgaris L.) and cowpea (Vigna unguiculata subsp, unguiculata) root architecture in the field. *Field Crops Research* 192, 21–32.

Burton, A. L., Lynch, J. P. and Brown, K. M. 2013. Spatial distribution and phenotypic variation in root cortical aerenchyma of maize (Zea mays L.). *Plant Soil* 367, 263–74.

Carter Jr., T. E. 1989. Breeding for drought tolerance in soybean: where do we stand? In A. J. Pascale (Ed.), *Proceeding World Soybean Conference IV*. Buenos Aires, Argentina, pp. 1001–8.

Carter Jr., T. E., De Souza, P. I. and Purcell, L. C. 1999. Recent advances in breeding for drought and aluminum resistance in soybean. In H. Kauffman (Ed.), *Proc.World Soybean Conf. VI Chicago*, IL, 4–7 August 1999. Superior Print., Champaign, IL, pp. 106–25.

Carter Jr., T. E., Orf, J., Purcell, L., Specht, J., Boerma, H. R., Chen, P., Sinclair, T., Rufty, T. 2006. Tough times, tough plants – New soybean genes defend against drought and other stresses. In

Proceedings Soybean Seed Res. Conf., 33rd, Chicago, IL. 5–8 December 2006. CD-ROM. Am. Seed Trade Assoc., Alexandria, VA.

Chalhoub, B., Denoeud, F., Liu, S. and et al. 2014. Plant genetics. Early allopolyploid evolution in the post-Neolithic Brassica napus oilseed genome. *Science* 345, 950–3.

Charlson, D. V., Bhatnagar, S., King, C. A., Ray, J. D., Sneller, C. H., Carter Jr., T. E. and Purcell, L. C. 2009. Polygenic inheritance of canopy wilting in soybean [*Glycine max*, (L.) Merr.]. *Theoretical and Applied Genetics* 119, 587–94.

Chebrolu, K. K., Fritschi, F. B., Ye, S., Krishnan, H. B., Smith, J. R. and Gillman, J. D. 2016. Impact of heat stress during seed development on soybean seed metabolome. *Metabolomics* 12, 28.

Chen, L., Ren, F., Zhou, L., Wang, Q. Q., Zhong, H., Li, X. B. 2012. The Brassica napus calcineurin B-like 1-CBL-interacting protein kinase 6 (CBL1/CIPK6) component is involved in the plant response to abiotic stress and ABA signaling. *Journal of Experimental Botany* 63, 6211–22.

Christenson, B. S., Schapaugh, W. T., An, N., Price, K. P. and Fritz, A. K. 2015 Characterizing changes in soybean spectral response curves with breeding advancements. *Crop Science* 54, 1585–97.

Christopher, J., Christopher, M., Jennings, R., Jones, S., Fletcher, S., Borrell, A., Manschadi, A. M., Jordan, D., Mace, E. and Hammer, G. 2013. QTL for root angle and number in a population developed from bread wheats (Triticum aestivum) with contrasting adaptation to water-limited environments. *Theoretical and Applied Genetics* 126, 1563–74.

Condon, A. G., Richards, R. A., Rebetzke, G. J. and Farquhar, G. D. 2002. Improving intrinsic water-use efficiency and crop yield. *Crop Science* 42, 122–31.

Courtois, B., Ahmadi, N., Khowaja, F., Price, A. H., Rami, J.-F, Frouin, J., Hamelin, C. and Ruiz, M. 2009. Rice root genetic architecture: meta-analysis from a drought QTL database. *Rice* 2, 115–28.

Creelman, R. A., Mason, H. S., Bensen, R. J., Boyer, J. S. and Mullet, J. E. 1990. Water deficit and abscisic acid cause differential inhibition of shoot *versus* root growth in soybean seedlings. *Plant Physiology* 92, 205–14.

Das, A., Schneider, H., Burridge, J., Ascanio, A. K., Wojciechowski, T., Topp, C. N., Lynch, J. P., Weitz, J. S. and Bucksch, A. 2015. Digital imaging of root traits (DIRT): a high-throughput computing and collaboration platform for field-based root phenomics. *Plant Methods*. doi: 10.1186/s13007-015-0093-3.

de Silva, M., Purcell, L. C. and King, C. A. 1996. Soybean petiole ureide response to water deficits and decreased transpiration. *Crop Science* 36, 611–16.

Devi, M. J., Sinclair, T. R., Chen, P. and Carter, T. E. 2014. Evaluation of elite southern maturity soybean breeding lines for drought tolerant traits. *Agronomy Journal* 106, 1947–54.

Dhanapal, A. P., Ray, J. D., Singh, S. K., Hoyos-Villegas, V., Smith, J. R., Purcell, L. C., Andy King, C., Cregan, P. B., Song, Q. and Fritschi, FB. 2015. Genome-wide association study (GWAS) of carbon isotope ratio (δ13C) in diverse soybean [*Glycine max* (L.) Merr.] genotypes. *Theoretical and Applied Genetics* 128, 73–91.

Djanaguiraman, M., Prasad, P. V.V., Boyle, D. L., Schapaugh, W. T. 2013. Soybean pollen anatomy, viability and pod set under high temperature stress. *Journal of Agronomy and Crop Science* 199, 171–7.

Djekoun, A., Planchon, D. 1991. Water status effect on dinitrogen fixation and photosynthesis in soybean. *Agronomy Journal* 83, 316–22.

Ellsworth, P. Z., Cousins, A. B. 2016. Carbon isotopes and water use efficiency in C4 plants. *Current Opinion in Plant Biology* 31, 155–61.

Feller, U. 2016. Drought stress and carbon assimilation in a warming climate: reversible and irreversible impacts. *Journal of Plant Physiology*. doi: 10.1016/j.jplph.2016.04.002.

Fletcher, A. L., Sinclair, T. R. and Allen, L. H. 2007. Transpiration responses to vapour pressure deficit in well-watered 'slow-wilting' and commercial soybean. *Environmental and Experimental Botany* 61, 145–51.

Fleury, D., Jefferies, S., Kuchel, H. and Langridge, P. 2010. Genetic and genomic tools to improve drought tolerance in wheat. *Journal of Experimental Botany* 61, 3211–22.

Forster, B. P., Thomas, W. T. B. and Chloupek, O. 2005. Genetic controls of barley root systems and their associations with plant performance. *Aspects of Applied Biology* 73, 199–204.

Foyer, C. H., Lam, H. M., Nguyen, H. T. and et al. 2016. Neglecting legumes has compromised human health and sustainable food production. *Nature Plants* 2, 16112.
Garay, A. F. and Wilhelm, W. W. 1983. Root system characteristics of two soybean isolines undergoing water stress conditions. *Agronomy Journal* 75, 973–7.
Giuliani, S., Sanguineti, M. C., Tuberosa, R., Bellotti, M., Salvi, S. and Landi, P. 2005. Root-ABA1, a major constitutive QTL, affects maize root architecture and leaf ABA concentration at different water regimes. *Journal of Experimental Botany* 56, 3061–70.
Goff, S. A., Ricke, D., Lan, T. H. and et al. 2002. A draft sequence of the rice genome (Oryza sativa L. ssp. japonica). *Science* 296, 92–100.
Guan, R., Qu, Y., Guo, Y. and et al. 2014. Salinity tolerance in soybean is modulated by natural variation in GmSALT3. *Plant Journal* 80, 937–50.
Hammer, G., Dong, Z., Mclean, G., Doherty, A., Messina, C., Schussler, J., Zinselmeier, C., Paszkiewicz, S. and Cooper, M. 2009. Can changes in canopy and/or root system architecture explain historical maize yield trends in the U.S. Corn Belt? *Crop Science* 49, 299–312.
Hatfield, J. L., Boote, K. J., Kimball, B. A., Ziska, L. H., Izaurralde, R. C., Ort, D., Thomson, A. M. and Wolfe, D. 2011. Climate impacts on agriculture: implications for crop production. *Agronomy Journal* 103, 2.
Hoogenboom, G., Huck, M. G. and Peterson, C. M. 1987. Root growth rate of soybean as affected by drought stress. *Agronomy Journal* 79, 607–14.
Huang, S., Li, R., Zhang, Z. and et al. 2009. The genome of the cucumber, *Cucumis sativus* L. *Nature Genetics* 41, 1275–81.
Hudak, C. M. and Patterson, R. P. 1995. Vegetative growth analysis of a drought resistant soybean plant introduction. *Crop Science* 35, 464–71.
Hudak, C. M. and Patterson, R. P. 1996. Root distribution and soil moisture depletion pattern of a drought-resistant soybean plant introduction. *Agronomy Journal* 88, 478–85.
Hund, A., Reimer, R. and Messmer, R. 2011. A consensus map of QTLs controlling the root length of maize. *Plant and Soil* 344, 143–58.
Hwang, S., King, C. A., Davies, M. K. and et al. 2013. QTL Analysis of shoot ureide and nitrogen concentrations in soybean (Glycine max L.[Merr.]). *Crop Science* 53, 2421–33.
Hwang, S., King, C. A., Ray, J. D. and et al. 2015. Confirmation of delayed canopy wilting QTLs from multiple soybean mapping populations. *Theoretical and Applied Genetics* 128, 2047–65.
Hwang, S., King, C. A., Chen, P. and et al. 2016. Meta-analysis to refine map position and reduce confidence intervals for delayed-canopy-wilting QTLs in soybean. *Molecular Breeding* 36, 91.
Hyten, D. L., Cannon, S. B., Song, Q., Weeks, N., Fickus, E. W., Shoemaker, R. C., Specht, J. E., Farmer, A. D., May, G. D. and Cregan, P. B. 2010. High-throughput SNP discovery through deep resequencing of a reduced representation library to anchor and orient scaffolds in the soybean whole genome sequence. *BMC Genomics* 11, 38.
Hyten, D. L., Song, Q., Zhu, Y., Choi, I. Y., Nelson, R. L., Costa, J. M., Specht, J. E., Shoemaker, R. C. and Cregan, P. B. 2006. Impacts of genetic bottlenecks on soybean genome diversity. *Proceedings of the National Academy of Sciences of the United States of America* 103, 16666–71.
Jacobs, T., LaFayette, P. R., Schmitz, R. J., Parrott, W. A. 2015. Targeted genome modifications in soybean with CRISPR/Cas9. *BMC Biotechnology* 15, 16. doi: 10.1186/s12896-015-0131-2.
Jain, M., Misra, G., Patel, R. K. and et al. 2013. A draft genome sequence of the pulse crop chickpea (Cicer arietinum L.). *Plant Journal* 74, 715–29.
Jones, M. M., Turner, N. C., Osmond, C. B. 1981. Mechanisms of drought resistance. In L. G. Paleg and D. Aspinall (Eds), *The Physiology and Biochemistry of Drought Resistance in Plants*. New York, Academic Press.
Kashiwagi, J., Krishnamurthy, L., Purushothaman, R., Upadhyaya, H. D., Gaur, P. M., Gowda, C. L. L., Ito, O. and Varshney, R. K. 2015. Scope for improvement of yield under drought through the root traits in chickpea (*Cicer arietinum L.*). *Field Crops Research* 170, 47–54.
Kato, Y., Abe, J., Kamoshita, A. and Yamagishi, J. 2006. Genotypic variation in root growth angle in rice (*Oryza sativa L.*) and its association with deep root development in upland fields with different water regimes. *Plant and Soil* 287, 117–29.

Kebede, H., Fisher, D. K. and Young, L. D. 2012. Determination of moisture deficit and heat stress tolerance in corn using physiological measurements and a low-cost microcontroller based monitoring system. *Journal of Agronomy Crop Science* 198, 118–29.
Kramer, P. J. 1969. *Plant and Soil Water Relationships: A Modern Synthesis*. McGrawHill, New York, NY, p. 482.
Ladrera, R., Marino, D., Larrainzar, E., Gonzalez, E. M. and Arrese-Igor, C. 2007 Reduced carbon availability to bacteroids and elevated ureides in nodules, but not in shoots, are involved in the nitrogen fixation response to early drought in soybean. *Plant Physiology* 145, 539–46.
Lam, H. M., Xu, X. and Liu, X. et al. 2010. Resequencing of 31 wild and cultivated soybean genomes identifies patterns of genetic diversity and selection. *Nature Genetics* 42, 1053–9.
Landi, P., Giuliani, S., Salvi, S., Ferri, M., Tuberosa, R. and Sanguineti, M. C. 2010. Characterization of root-yield-1.06, a major constitutive QTL for root and agronomic traits in maize across water regimes. *Journal of Experimental Botany* 61, 3553–62.
Lesk, C., Rowhani, P. and Ramankutty, N. 2016. Influence of extreme weather disasters on global crop production. *Nature* 529, 84–7.
Levitt, J. 1972. *Responses of Plants to Environmental Stresses*. Academic Press, New York.
Lobell, D. B., Roberts, M. J., Schlenker, W., Noah, B., Little, B. B., Rejesus, R. M. and Hammer, G. L. 2014. Greater sensitivity to drought accompanies maize yield increase in the U.S. Midwest. *Science* 344, 516–19.
Ludlow, M. M. and Muchow, R. C. 1990. A critical evaluation of traits for improving crop yields in water-limited environments. *Advances in Agronomy* 43, 107–53.
Lynch, J. P. 2011. Root phenes for enhanced soil exploration and phosphorus acquisition: tools for future crops. *Plant Physiology* 156, 1041–9.
Mace, E. S., Singh, V., Van Oosterom, E. J., Hammer, G. L., Hunt, C. H. and Jordan, D. R. 2012. QTL for nodal root angle in sorghum (Sorghum bicolor L. Moench) co-locate with QTL for traits associated with drought adaptation. *Theoretical and Applied Genetics* 124, 97–109.
Mace, E. S., Tai, S., Gilding, E. K. and et al. 2013. Whole-genome sequencing reveals untapped genetic potential in Africa's indigenous cereal crop sorghum. *Nature Communication*. doi: 10.1038/ncomms3320.
MacMillan, K., Emrich, K., Piepho, H. P., Mullins, C. E. and Price, A. H. 2006. Assessing the importance of genotype x environmental interaction for root traits in rice using a mapping population II: conventional QTL analysis. *Theoretical and Applied Genetics* 113, 953–64.
Manavalan, L. P., Prince, S. J., Musket, T. A. and et al. 2015. Identification of Novel QTL Governing Root Architectural Traits in an Interspecific Soybean Population. *Plos One*. doi.org/10.1371/journal.pone.0120490.
Manschadi, A. M., Christopher, J., Devoil, P. and Hammer, G. L. 2006. The role of root architectural traits in adaptation of wheat to water-limited environments. *Functional Plant Biology* 33, 823–37.
Maseda, P. H. and Fernandez, R. J. 2006. Stay wet or else: three ways in which plants can adjust hydraulically to their environment. *Journal of Experimental Botany* 57, 3963–77.
McClure, P. R., Israel, D. W. and Volk, R. J. 1980. Evaluation of the relative ureide content of xylem sap as an indicator of N2 fixation in soybeans: greenhouse studies. *Plant Physiology* 66, 720–5.
Meister, R., Rajani, M. S., Ruzicka, D. and Schachtman, D. P. 2014. Challenges of modifying root traits in crops for agriculture. *Trends in Plant Science* 19, 779–88.
Mickelbart, M. V., Hasegawa, P. M. and Bailey-Serres, J. 2015. Genetic mechanisms of abiotic stress tolerance that translate to crop yield stability. *Nature Reviews Genetics* 16, 237–51.
Nguyen, H., Babu, R. C. and Blum, A. 1997. Breeding for drought resistance in rice Physiology and molecular genetics considerations. *Crop Science* 37, 1426–34.
Nielsen, K. L., Lynch, J. P. and Weiss, H. N. 1997. Fractal geometry of bean root systems: correlations between spatial and fractal dimension. *American Journal of Botany* 84, 26–33.
Nilsen, E. T. and Orcutt, D. M. 1996. *Physiology of Plants Under Stress – Abiotic Factors*. John Wiley & Sons, New York.

Oosterhuis, D. M., Scott, H. D., Hampton, R. E. and Wullschleter, S. D. 1990. Physiological response of two soybean [*Glycine max*, L. Merr] cultivars to short-term flooding. *Environmental and Experimental Botany* 30, 85–93.
Osmont, K. S., Sibout, R. and Hardtke, C. S. 2007. Hidden branches: developments in root system architecture. *Annual Review of Plant Biology* 58, 93–113.
Pantalone, V. R., Rebetzke, G. J., Burton, J. W. and Carter, T. E. 1996. Phenotypic Evaluation of Root Traits in Soybean and Applicability to Plant Breeding. *Crop Science* 36, 456–9.
Parween, S., Nawaz, K., Roy, R. and et al. 2015. An advanced draft genome assembly of a desi type chickpea (Cicer arietinum L.). *Scientific Reports*. doi: 10.1038/srep12806.
Passioura, J. B. 2010. Scaling up: the essence of effective agricultural research. *Functional Plant Biology* 37, 585–91.
Passioura, J. B. 1983. Root and drought resistance. *Agricultural Water Management* 7, 265–80.
Passioura, J. P. 1983. Grain yield, harvest index and water use of wheat. *Journal of the Australian Institute of Agricultural Science* 43, 117–20.
Paterson, A. H., Bowers, J. E., Bruggmann, R. and et al. 2009. The Sorghum bicolor genome and the diversification of grasses. *Nature* 457, 551–6.
Pathan, S. M., Lee, J. D., Sleper, D. A. and et al. 2014. Two soybean plant introductions display slow leaf wilting and reduced yield loss under drought. *Journal of Agronomy and Crop Science* 200, 231–6.
Patil, G., Do, T., Vuong, T. D., Valliyodan, B., Lee, J. D., Chaudhary, J., Shannon, G. J. and Nguyen, H. T. 2016. Genomic-assisted haplotype analysis and the development of high-throughput SNP markers for salinity tolerance in soybean. *Scientific Reports*. doi: 10.1038/srep19199.
Patterson, P. R. and Hudak, C. M. 1996. Drought-avoidant soybean germplasm maintains nitrogen fixation capacity under water stress. *Plant and Soil* 186, 39–43.
Pinto, R. S. and Reynolds, M. P. 2015. Common genetic basis for canopy temperature depression under heat and drought stress associated with optimized root distribution in bread wheat. *Theoretical and Applied Genetics* 128, 575–85.
Prasad, P. V. V., Craufurd, P. Q. and Summerfield, R. J. 1999. Fruit number in relation to pollen production and viability in groundnut exposed to short episodes of heat stress. *Annals of Botony* 84, 381–6.
Prince, S. J., Song, L., Qiu, D. and et al. 2015. Genetic variants in root architecture-related genes in wild soybeans, a potential resource to improve cultivated soybean. *BMC Genomics*. doi: 10.1186/s12864-015-1334-6.
Prince, S. J., Murphy, M., Mutava, R. N., Durnell, L. A., Valliyodan, B., Shannon, J. G. and Nguyen, H. T. 2017. Root xylem plasticity to improve water use and yield in water-stressed soybean. *Journal of Experimental Botony*. doi: 10.1093/jxb/erw472.
Purcell, L. C., Serraj, R., de Silva, M., Sinclair, T. R. and Bona, S. 1998. Ureide concentration of field-grown soybean in response to drought and the relationship to nitrogen fixation. *Journal of Plant Nutrition* 21, 949–66.
Qi, J., Liu, X., Shen, D. and et al. 2013. A genomic variation map provides insights into the genetic basis of cucumber domestication and diversity. *Nature Genetics* 45, 1510–15.
Ray, J. D., Dhanapal, A. P., Singh, S. K. and et al. 2015. Genome-wide association study of ureide concentration in diverse maturity group IV soybean [*Glycine max* (L.) Merr.] accessions. *G3: Genes|Genomes|Genetics* 5, 2391–403.
Redden, R. J., Hatfield, J. L., Vara Prasad, P. V., Ebert, A. W., Yadav, S. S. and O'Leary, G. J. 2014. Temperature, climate change, and global food security. In Franklin, K. and Wigge, P. (Eds), *Temperature and Plant Development*. Wiley, New York, pp. 181–202.
Reynolds, M., Dreccer, F. and Trethowan, R. 2007. Drought-adaptive traits derived from wheat wild relatives and landraces. *Journal of Experimental Botany* 58, 177–86.
Reynolds, M. P., Manes, Y., Izanloo, A. and Langridge, P. 2009. Phenotyping for physiological breeding and gene discovery in wheat. *Annals of Applied Biology* 155, 309–320.
Richard, C. A., Hickey, L. T., Fletcher, S., Jennings, R., Chenu, K. and Christopher, J. T. 2015. High-throughput phenotyping of seminal root traits in wheat. *Plant Methods*. doi: 10.1186/s13007-015-0055-9.

Ries, L. L., Purcell, L. C., Carter Jr., T. E., Edwards, J. T. and King, C. A. 2012. Physiological traits contributing to differential canopy wilting in soybean under drought. *Crop Science* 52, 272–81.

Rincon, C. A., Raper, C. D., Patterson Jr., R. P. 2003. Genotypic differences in root anatomy affecting water movement through roots of soybean. *International Journal of Plant Science* 164, 543-51.

Rosenzweig, C., Elliott, J., Deryng, D. and et al. 2014. Assessing agricultural risks of climate change in the 21st century in a global gridded crop model intercomparison. *Proceedings of the National Academy of Sciences of the United States of America* 111, 3268–73.

Sadok, W. and Sinclair, T. R. 2011. Crops yield increase under water – limited conditions: review of recent physiological advances for soybean genetic improvement. *Advances in Agronomy* 113, 313–37

Sadok, W., Matthew, G. E., Raza, A. W. S and Sinclair, T. R. 2012. Basis of slow wilting phenotype in soybean PI 471938. *Crop Science* 52, 1261–9.

Salazar-Henao, J. E., Vélez-Bermúdez, I. C. and Schmidt, W. 2016. The regulation and plasticity of root hair patterning and morphogenesis. *Development* 143, 1848–58.

Salem, M. A., Kakani, V. G., Kot, S. and Reddy, K. R. 2007. Pollen-based screening of soybean genotypes for high temperature. *Crop Science* 47, 219–31.

Sall, K. and Sinclair, T. R. 1991. Soybean genotypic differences in sensitivity of symbiotic nitrogen fixation to soil dehydration. *Plant and Soil* 133, 31–7.

Sayar, R., Khemira, H. and Kharrat, M. 2007. Inheritance of deeper root length and grain yield in half-diallel durum wheat (Triticum durum) crosses. *Annals of Applied Biology* 151, 213–20.

Schmutz, J., McClean, P. E., Mamidi, S. and et al. 2014. A reference genome for common bean and genome-wide analysis of dual domestications. *Nature Genetics* 46, 707–13.

Schmutz, J., Cannon, S. B., Schlueter, J. and et al. 2010. Genome sequence of the palaeopolyploid soybean. *Nature* 463, 178–83.

Schnable, P. S., Ware, D., Fulton, R. S. and et al. 2009. The B73 maize genome: complexity, diversity, and dynamics. *Science* 326, 1112–15.

Serraj, R. and Sinclair, T. R. 1996a. Processes contributing to N2-fixation insensitivity to drought in the soybean cultivar Jackson. *Crop Science* 36, 961–8.

Serraj, R. and Sinclair, T. R. 1996b. Inhibition of nitrogenase activity and nodule oxygen permeability by water deficit. *Journal of Experimental Botany* 47, 1067–73.

Serraj, R., Vadez, V., Denison, R. F. and Sinclair, T. R. 1999. Involvement of ureides in nitrogen fixation inhibition in soybean. *Plant Physiology* 119, 289–96.

Serraj, R., Vadez, V. and Sinclair, T. R. 2001. Feedback regulation of symbiotic N2 fixation under drought stress. *Agronomie* 21, 621–6.

Shukla, A. K., Ladha, J. K., Singh, V., Dwivedi, B., Balasubramanian, V., Gupta, R. K., Sharma, S., Singh, Y., Pathak, H. and Pandey, P. 2004. Calibrating the leaf color chart for nitrogen management in different genotypes of rice and wheat in a systems perspective. *Agronomy Journal* 96, 1606–21.

Siebers, M. H., Yendrek, C. R., Drag, D. and et al. 2015. Heat waves imposed during early pod development in soybean (Glycine max) cause significant yield loss despite a rapid recovery from oxidative stress. *Global Change Biology* 21, 3114–25.

Sinclair, T. R. and Serraj, R. 1995. Dinitrogen fixation sensitivity to drought among grain legume species. *Nature* 378, 344.

Sinclair, T. R., Purcell, L. C., Vadez, V., Serraj, R., King, C. A. and Nelson, R. 2000. Identification of soybean genotypes with N2 fixation tolerance to water deficits. *Crop Science* 40, 1803–9.

Sinclair, T. R., Messina, C. D., Beatty, A. and Samples, M. 2010. Assessment across the United States of the benefits of altered soybean drought traits. *Agronomy Journal* 102, 475–82.

Sinclair, T. R. 2011. Challenges in breeding for yield increase for drought. *Trends in Plant Science* 16, 289–93.

Singh, N. K., Gupta, D. K., Jayaswal, P. K. and et al. 2012. The first draft of the pigeonpea genome sequence. *Journal of Plant Biochemistry and Biotechnology* 21, 98–112.

Singh, V., van Oosterom, E. J., Jordan, D. R., Hunt, C. H. and Hammer, G. L. 2011. Genetic variability and control of nodal root angle in sorghum. *Crop Science* 51, 2011–20.

Sloane, R. J., Patterson, R. P. and Carter, T. E. 1990. Field drought tolerance of a soybeanplant introduction. *Crop Science* 30, 118–23.

Song, Q., Hyten, D. L., Jia, G., Quigley, C. V., Fickus, E. W., Nelson, R. L. and Cregan, P. B. 2013. Development and evaluation of SoySNP50K, a high- density genotyping array for soybean. *PLoS One* 8, e0054985.

Specht, J. E., Hume, D. J. and Kumudini, S. V. 1999. Soybean yield potential- A genetic and physiological perspective. *Crop Science* 39, 1560–70.

Sponchiado, B. N., White, J. W., Castillo, J. A. and Jones, P. G. 1989. Root growth of four common bean cultivars in relation to drought tolerance in environments with contrasting soil types. *Experimental Agriculture* 25, 249–57.

Steele, K. A., Price, A. H., Shashidhar, H. E. and Witcombe, J. R. 2006. Marker-assisted selection to introgress rice QTLs controlling root traits into an Indian upland rice variety. *Theoretical and Applied Genetics* 112, 208–21.

Sun, X., Hu, Z., Chen, R., Jiang, Q., Song, G., Zhang, H. and Xi, Y. 2015. Targeted mutagenesis in soybean using the CRISPR-Cas9 system. *Scientific Reports* 5, 10342.

Taji, T., Seki, M., Satou, M., Sakurai, T., Kobayashi, M., Ishiyama, K., Narusaka, Y., Narusaka, M., Zhu, J. K. and Shinozaki, K. 2004. Comparative genomics in salt tolerance between Arabidopsis and Arabidopsis-related halophyte salt cress using Arabidopsis microarray. *Plant Physiology* 135, 1697–709.

Takeda, S. and Matsuoka, M. 2008. Genetic approaches to crop improvement: responding to environmental and population changes. *Nature Reviews Genetics* 9, 444–57.

Tambussi, E. A., Bort, J., Guiamet, J. J., Nogués, S. and Araus, J. L. 2007. The photosynthetic role of ears in C3 cereals: metabolism, water use efficiency and contribution to grain yield. *Critical Reviews in Plant Sciences* 26, 1–16.

Tanaka, Y., Fujii, K. and Shiraiwa, T. 2010. Variability of leaf morphologyand stomatal conductance in soybean [*Glycine max* (L.) Merr.] cultivars. *Crop Science* 50, 2525–32.

The 1001 Genome Consortium. 2016. 1135 genomes reveal the global pattern of polymorphism in *Arabidopsis thaliana*. *Cell* 166, 1–11.

The 3000 Rice Genome Project. 2014. The 3000 rice genomes project. 3K RGP *GigaScience* 3, 7.

The Arabidopsis Genome Initiative. 2000. Analysis of the genome sequence of the flowering plant *Arabidopsis thaliana*. *Nature* 408, 796–815.

Thomas, H. and Howarth, C. J. 2000. Five ways to stay green. *Journal of Experimental Botany* 5, 329–37.

Trachsel, S., Kaeppler, S. M., Brown, K. M. and Lynch, J. P. 2011. Shovelomics: high throughput phenotyping of maize (Zea mays L.) root architecture in the field. *Plant and Soil* 341, 75–87.

Tuberosa, R., Salvi, S., Sanguineti, M. C., Landi, P., Maccaferri, M. and Conti, S. 2002. Mapping QTLs regulating morpho-physiological traits and yield: case studies, shortcomings and perspectives in drought-stressed maize. *Annals of Botany* 89, 941–63.

Tuinstra, M. R., Ejeta, G. and Goldsbrough, P. B. 1998. Evaluation of nearisogenic sorghum lines contrasting for QTL markers associated with drought tolerance. *Crop Science* 38, 835–42.

Turner, N. C., Begg, J. E. and Tonnet, M. L. 1978. Osmotic adjustment of sorghum and sunflower crops in response to water deficits and its influence on the water potential at which stomata close. *Australian Journal of Plant Physiology* 5, 597–608.

Turner, N. C. 1986. Crop water deficits: a decade of progress. *Advances in Agronomy* 39, 1–51.

Turner, N. C. 1996. Further progress in crop water relations. *Advances in Agronomy* 58, 293–338.

Turner, N. C. 2000. Drought resistance: a comparison of two frameworks. In Saxena, N. P., Johansen, C., Chauhan, Y. S. and Rao, R. C. N. (Eds), *Management of Agriculture and Drought: Agronomic and Genetic Options*. Oxford and IBH, New Delhi.

Turner, N. C., Wright, G. C. and Siddique, K. H. M. 2001. Adaptation of grain legumes (pulses) to water-limited environments. *Advances in Agronomy* 71, 193–231.

Uga, Y., Sugimoto, K., Ogawa, S. and et al. 2013. Control of root system architecture by DEEPER ROOTING 1 increases rice yield under drought conditions. *Nature Genetics* 45, 1097–102.

Vadez V. 2000. Ureide degradation pathways in intact soybean leaves. *Journal of Experimental Botany* 51, 1459–65. https://doi.org/10.1093/jexbot/51.349.1459

Valliyodan, B., Qiu, Q., Patil, G. and et al. 2016. Landscape of genomic diversity and trait discovery in soybean. *Scientific Reports*. doi: 10.1038/srep23598.

Valliyodan, B. and Nguyen, H. T. 2006. Understanding regulatory networks and engineering for enhanced drought tolerance in plants. *Current Opinions in Plant Biology* 9, 189–95.

Varshney, R. K., Chen, W., Li. Y. and et al. 2012. Draft genome sequence of pigeonpea (Cajanus cajan), an orphan legume crop of resource-poor farmers. *Nature Biotechnology* 30, 83–9.

Varshney, R. K., Pazhamala, L., Kashiwagi, J., Gaur, P. M., Krishnamurthy, L. and Hoisington, D. 2011. Genomics and physiological approaches for root trait breeding to improve drought tolerance in chickpea (*Cicer arietinum L.*). In A. Costa de Oliveira and R. K. Varshney (Eds), *Root Genomics*.Springer, Dordrecht, The Netherlands, pp. 233–50.

Varshney, R. K., Singh, V. K., Hickey, J. M. and et al. 2015. Analytical and decision support tools for genomics-assisted breeding. *Trends Plant Science* 15, S1360–85.

Varshney, R. K., Song, C., Saxena, R. K. and et al. 2013. Draft genome sequence of chickpea (Cicer arietinum) provides a resource for trait improvement. *Nature Biotechnology* 31, 240–6.

Wasson, A. P., Richards, R. A., Chatrath, R., Misra, S. C., Sai Prasad, S. V., Rebetzke, G. J., Kirkegaard, J. A., Christopher, J. and Watt, M. 2012. Traits and selection strategies to improve root systems and water uptake in water-limited wheat crops. *Journal of Experimental Botany* 63, 3485–98.

Wissuwa, M., Kretzschmar, T., Rose, T. J. 2016. From promise to application: root traits for enhanced nutrient capture in rice breeding. *Journal of Experimental Botany* 67, 3605–15.

Wu, Y., Spollen, W. G., Sharp, R. E. and et al. 1994. Root growth maintenance at low water potentials. *Plant Physiology* 106, 607–15.

Xu, X., Pan, S., Cheng, S. and et al. 2011. Genome sequence and analysis of the tuber crop potato. *Nature* 475, 189–95.

Xu, X., Zeng, L., Tao, Y. and et al. 2013. Pinpointing genes underlying the quantitative trait loci for root-knot nematode resistance in palaeopolyploid soybean by whole genome resequencing. *Proceedings of the National Academy of Sciences of the United States of America* 110, 13469–74.

York, L. M. and Lynch, J. P. 2015. Intensive field phenotyping of maize (Zea mays L.) root crowns identifies phenes and phene integration associated with plant growth and nitrogen acquisition. *Journal of Experimental Botany* 66, 5493–505.

Young, E. G. and Conway, C. F. 1942. On the estimation of allantoin by the RiminiSchryver Schryver reaction. *The Journal of Biological Chemistry* 142, 839–53.

Zhang, Y.-J., Lynch, J. P. and Brown, K. M. 2003 Ethylene and phosphorous availability have interacting yet distinct effects on root hair development. *Journal of Experimental Botany* 54, 2351–61.

Zhou, Z. K., Jiang, Y., Wang, Z. and et al. 2015. Resequencing 302 wild and cultivated accessions identifies genes related to domestication and improvement in soybean. *Nature Biotechnology* 33, 408–14.

Chapter 8

Towards improving the salt tolerance of soybean

Ailin Liu, Wai-Lun Cheung, Wai-Shing Yung, Carol Lee, Fuk-Ling Wong, Kit-Wah Siu and Hon-Ming Lam, The Chinese University of Hong Kong, China

1 Introduction

Soybean is one of the major crops worldwide and its production has been impacted by multiple environmental factors, including salinity (Phang et al. 2008). Salinity hampers the growth and yield of soybean by decreasing water efficiency inside the plant cell and it also causes sodium (Na^+) and chloride (Cl^-) toxicity (Miransari 2015). The world production of soybean is about 260 million tons, utilizing over 100 million hectares of farmland (Anon. 2012b). In 2015, the total area devoted to soybean production in United States was 33.4 million hectares, and its yield reached 3.2 metric ton/hectare (SoyStats: http://soystats.com). Twenty percent of irrigated land, which produces 40% of the world's food, has been affected by salinity (Pimentel et al. 2004). The yield of soybean decreases when soil salinity exceeds 5 dS/m (Ashraf 1994). Numerous studies have shown that salt stress leads to a reduction of more than 50% in soybean production (Essa 2002; Hosseini et al. 2002). The USDA estimation of the potential economic impact of saline soil on soybean production shows that the annual economic impact on soybeans yield loss was up to US$7.88 million in upper James River, Virginia (Anon. 2012a). Species within the genus *Glycine* show a spectrum of salt tolerance capacity, with the tolerance threshold to sodium chloride (NaCl) ranging from 3.0 to 17.5 g/L in the perennial *Glycine* species and ranging from 5.2 to 8.0 g/L in *Glycine max* cultivars (Lu et al. 2009). As a major food crop, it is crucial to maintain and even improve soybean production, and to achieve this goal within a sustainable agricultural system.

http://dx.doi.org/10.19103/AS.2017.0034.08

Damage due to salinity on soybean is observed at different developmental stages. A salinity level at above 0.1% significantly reduced the germination rate while lower salt conditions (0.05% and 0.1%) delayed the germination of soybean seeds (Abel and MacKenzie 1964). Seedlings of cultivated soybean are also very sensitive to salt stress. At 330 mM NaCl, even though the germination rate of *Glycine max* L. cv Williams can reach 81%, the seedling growth rate was almost zero (Hosseini et al. 2002). Besides, high salinity impacts negatively on the agronomic traits of soybean, including reduction in the number of internodes, pods, branches and leaves, as well as reduction in height, weight and biomass (Abel and MacKenzie 1964; Chang et al. 1994). Salinity also hampers growth, osmotic adjustment and pressure, sodium ion exclusion and water extraction in the root (An et al. 2001, 2002). Moreover, salt stress strongly hinders nodule initiation by inhibiting the deformation of root hair, which is the natural response upon receiving the Nod factor from rhizobia (Duzan et al. 2004). It reduces the overall number, weight and function of nodules, thus lowering the efficiency of nitrogen fixation (Singleton and Bohlool 1984). In addition, salt stress lowers the protein content in seeds (Chang et al. 1994).

The most common salt stresses in nature are caused by sodium salts, particularly NaCl. Normally, salt stress is first encountered by the root system. It impairs plant growth by inducing osmotic stress as a result of reduced water availability and increased ion toxicity induced by solute imbalance. To survive, plants have evolved mechanisms to combat salt stress, mainly by avoiding such stress by different morphological and physiological adjustments, and by enhancing the internal adaptive molecular mechanisms (see Fig. 1) (Xu et al. 2016; Ren et al. 2016; Li et al. 2006a, 2006b; Chen et al. 2015; Weisany et al. 2011; El-sabagh et al. 2015b; Arshi et al. 2012; Wei et al. 2016).

Structural adaptations could be effective means to combat salt stress. Cell wall remodelling is a common phenomenon in plants under salt stress. In cultivated soybean, salt treatment induced stem cuticle thickening as a result of a higher density of cutin and trichome (Dolatabadian et al. 2011). Salt-exposed cultivated soybean showed increased contents of total phenolics, lignin, the lignin-modifying enzyme phenylalanine ammonia lyase (PAL) and cell wall-bound peroxidases (Neves et al. 2010). The accumulation of peroxidases together with lignin can harden the cell wall mechanical strength by cross-linking glycoproteins and phenolic compounds (Le Gall et al. 2015). Cell wall structure

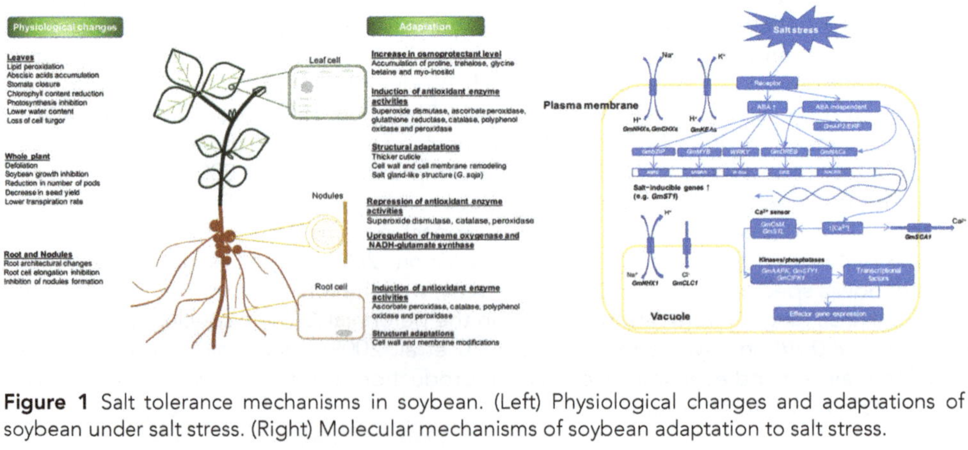

Figure 1 Salt tolerance mechanisms in soybean. (Left) Physiological changes and adaptations of soybean under salt stress. (Right) Molecular mechanisms of soybean adaptation to salt stress.

remodelling was also achieved by increasing the production of a proline-rich cell wall protein SbPRP (Soybean proline-rich protein) (He et al. 2002).

Cell membrane composition could also get damaged under severe salt stress as a result of the leakage of electrolytes and organic compounds (Shao et al. 1993). A salt-tolerant soybean cultivar could stabilize the cell membrane structure by increasing the phospholipid-to-galactolipid ratio (Yu et al. 2004). Overexpressing the plasma membrane intrinsic protein, GmPIP1;6, in the soybean cultivar Williams 82 conferred salt tolerance, with high Na^+ exclusion in the shoot and increased seed yield (Zhou et al. 2014). Moreover, some proteins in the plasma membrane can serve as receptors in the salt stress signal transduction pathway. *GmSCA1*, a gene for a plasma membrane-localized Ca^{2+}-ATPase, was found to be induced rapidly in soybean exposed to salt stress (Chung et al. 2000).

Unlike cultivated soybean, wild soybean (*G. soja*) could develop salt gland-like structures in the leaf and stem for salt secretion (Lu et al. 1998). The glandular hairs discovered on the leaves may also function in salt secretion. Studies showed that the shape of glandular hair had the characteristics of typical salt glands, such as big vacuoles, dense cytoplasm, large amounts of mitochondria and chloroplasts, and thicker cell wall. X-ray microanalysis detected that Na^+-related contents were different among glandular cells, epidermal cells and mesophyll cells in leaves, which suggested the ability of salt secretion by *G. soja* (Lu et al. 1999; San and Zhou 2003).

Soybean also undergoes osmotic adjustments under salt stress via physiological adaptations. For instance, it accumulates osmoprotectants, that is, proline, glycine betaines, trigonelline and pinitol, to act as compatible solutes to reduce the cellular osmotic potential without disturbing normal metabolic reactions (Hasegawa et al. 2000). Besides, late embryogenesis abundant (LEA) proteins might serve as antioxidants, membrane and protein stabilizers, and as potential space fillers, which fill in the cell as plasticizers or mechanical buffers to restraint protein aggregation (Chakrabortee et al. 2007; Goyal et al. 2005), and prevent cellular collapse under salt stress (Tunnacliffe and Wise 2007).

A number of soybean homologs of ion transporters play significant roles in the maintenance of ion homeostasis. They include tonoplast-localized plant Na^+-H^+ exchangers, *GmNHX1*, which sequester Na^+ from the cytoplasm to the vacuole to reduce ion toxicity (Li et al. 2006a). *GmCHX1*, a cation/proton transporter, was identified in a major quantitative trait locus (QTL) of salt tolerance; a *GmCHX1*-overexpressing soybean showed high salt tolerance (Qi et al. 2014). The plasma membrane-localized cations/H^+ antiporter, *GmCAX1*, functions as an antiporter of Na^+, K^+ and Li^+ by transporting these cations out of the cell through the plasma membrane or into the vacuole to enhance plant salt tolerance (Luo et al. 2005). The tonoplast-localized Cl^- channel, *GmCLC1*, sequesters Cl^- into the vacuole; the *GmCLC1*-transgenic soybean had improved tolerance to salt stress (Wei et al. 2016; Li et al. 2006a). It was found that the reduction in Na^+, K^+, Li^+ and Cl^- in the cytoplasm alleviated the detrimental effects of salt stress (Lauchli and Wieneke 1979; Wieneke and Lauchli 1980; Rathert et al. 1981). It has been reported that *G. max* and *G. soja* have different salt stress tolerance because of their differential preferences in utilizing Na^+ and Cl^- ions. In *G. max*, the salt-sensitive cultivars accumulate high amounts of sodium and chloride in their leaves, while salt-tolerant cultivars could maintain lower Na^+ and Cl^- contents in leaves (Li et al. 2006b; Essa 2002). Luo et al. (2005) found that, compared to two *G. soja* varieties, *G. max* cultivars are more sensitive to Cl^- toxicity. On the contrary, *G. max* varieties show higher tolerance to Na^+ toxicity than *G. soja*. Another study observed that the accumulation of Na^+ in the roots of *G. soja* is higher than in those of *G. max*, while there is less Na^+ accumulation in *G. soja* leaves than in *G. max*

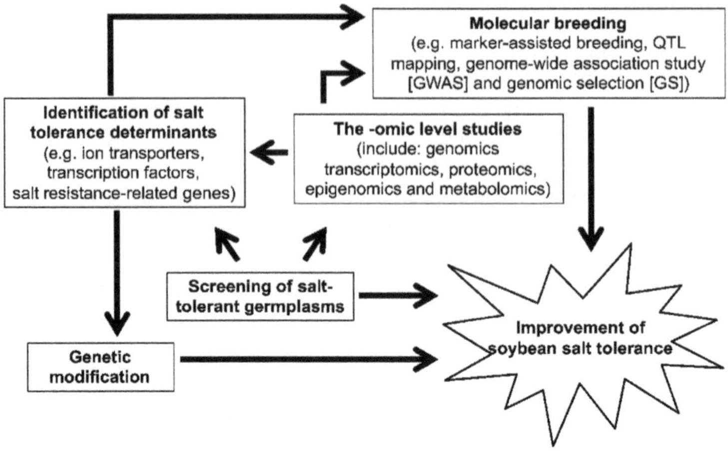

Figure 2 Strategies for improving salt tolerance in soybean.

leaves, indicating that wild soybean can sequester the Na^+ in the roots, and prevent the transportation of Na^+ to the leaves to protect them from salt damage (Xue et al. 2014). Evidence also showed that *G. soja* BB52 grown in Yellow River Delta, China, could maintain lower Na^+ and Cl^- contents in both leaves and roots than cultivar ZH13 after salt treatment (Chen et al. 2013).

The restoration of oxidative balance is another possible mechanism of salt tolerance in soybean. Salt stress could impair the scavenging of reactive oxygen species (ROS), leading to undesirably high levels of ROS (Weisany et al. 2012). By elevating the amounts and activities of various antioxidants, such as superoxide dismutase (SOD) (Phang et al. 2008), the production rate of ROS could be reduced, thus restoring the oxidative balance and minimizing cellular damage caused by salt-induced oxidative stress.

In recent decades, research and advanced genome-wide '-omics-based' approaches such as genomics, proteomics, transcriptomics and epigenomics have facilitated a deeper understanding of the complex molecular mechanism of salt tolerance. This chapter introduces the current strategies being employed to analyse salt stress tolerance in soybean and develop salt-tolerant soybean, for example, by screening salt-tolerant soybean germplasms to explore the genetic diversity and discover new sources of genetic variations (Fig. 2). The identification of salt-tolerant determinants in the soybean germplasms opens the door to improving salt stress tolerance by genetic engineering and molecular-assisted breeding. The -omic-level approaches have generated a huge amount of data in the genome for the understanding of molecular mechanisms of salt tolerance.

2 Soybean germplasm resources for salt tolerance

2.1 Screening for salt-tolerant germplasms

Soybean germplasms are important sources of genetic information. Genetic diversity among soybean germplasms provides a big genetic resource for bringing back the beneficial traits that were lost during the domestication process by crop breeding. There are more than 90 000 soybean germplasms collected by various seed banks worldwide. China has the largest collection of soybean germplasms in the world, with about 28 000 accessions of *G. max* (data from http://www.most.gov.cn), followed by the United States with about 21 000 accessions of *G. max* (data from GENESYS-PGR).

To screen germplasms with salt-tolerant traits, different approaches have been developed. Traditionally, seeds were germinated and irrigated with diluted seawater during their development. Effects of the seawater, including germination rate and necrosis on soybean leaves, were then evaluated at different developmental stages (Shao et al. 1994). A similar screening method is still being used today but with a simpler and more cost-effective setup. Seedlings are grown hydroponically and challenged with different concentrations of NaCl, and visible foliar symptoms such as chlorosis are examined several days after planting (Valencia et al. 2008). The measure of germination percentage is a rapid and simple method which is widely used to screen soybean genotypes, but the salt tolerance capacity at the vegetative, flowering and maturity stages should also be measured using different approaches as part of the screening process (Arzani 2008). As salt-tolerant seeds are essential to the production of soybean, the germination-rated trait was widely studied under salt stress (Kan et al. 2015). For young seedlings, salt tolerance can be measured by the salt damage index (Qi et al. 2014). Salt tolerance rating according to the degree of leaf chlorosis and the leaf chlorophyll content have also been applied to discover the salt tolerance QTL in soybean seedlings (Guan et al. 2014a; Hamwieh et al. 2011). The number of branches per plant, number of pods per plant, 100 seed weight and seed yield per plant can all be used to determine the impacts of salt stress at the reproductive stage (El-Sabagh et al. 2015a).

Classical screening methods are based on the assessment of yield in response to salt stress. However, due to the fact that variable degrees of salt tolerance are found in different germplasms, quantifiable methods have recently been applied to screening. Relative shoot dry weight (RSDW) is one of the semi-quantitative methods that measures the dry weight of the salt-treated soybean shoots relative to the untreated ones, and classifies soybean into four different categories: tolerant (>80%), moderately tolerant (60–80%), moderately susceptible (40–60%) and susceptible (<40%) (Mannan et al. 2012; Khan et al. 2013).

2.2 Functional genes for salt tolerance

In other crops such as rice (Moncada et al. 2001), barley (Yun et al. 2005) and tomato (Fulton et al. 2000), beneficial alleles have been identified in wild germplasms and utilized in cultivar improvement. Similarly, new soybean cultivars could be generated through utilizing the genetic resources stored in diverse germplasms. Since both Na^+ and Cl^- bring toxicity and cause ionic stress to soybean, scientists have been focusing on discovering their respective transporters and other novel genes related to exporting these ions in the past decades. One classical example is a dominant *Ncl* allele from cultivar Lee which follows the Mendelian inheritance (Abel 1969). It confers salt tolerance to soybean via its function in chloride exclusion. Recently, this regulatory locus has also been found to have specificities towards sodium and potassium ions under salt stress in soybean, thus conferring salt tolerance by decreasing the accumulation of these ions in shoots (Do et al. 2016). The same dominant alleles related to salt tolerance, *GmSALT3* from cultivar Tiefeng 8 and *GmCHX1* from wild soybean W05, are both located on chromosome 3 (Guan et al.

2014b; Qi et al. 2014). *GmSALT3* (or *GmCHX1*) was reported to be related to cation/H^+ exchanger proteins that help expel excess intracellular sodium ions, as suggested by the significant reduction in sodium ion accumulation after salt treatment as compared to those carrying a recessive copy of this allele in the corresponding locus (Liu et al. 2016; Guan et al. 2014b). Another single dominant allele, *Ncl2* (non-allelic to *Ncl*), that governs the salt tolerance of *G. soja* via an unknown mechanism has also been reported (Lee et al. 2009).

3 Integrating -omic approaches for investigating salt tolerance in soybean

3.1 The soybean genome

The application of next-generation sequencing technology on the soybean genome has provided genomic resources for isolating gene families related to salt stress responses. The genome of the cultivated soybean, *G. max* Wm82.a2.v1, which was first sequenced by a whole-genome shotgun approach in 2010 (Schmutz et al. 2010), now serves as the reference genome. Later, 17 wild and 14 cultivated soybean genomes were re-sequenced using the Illumina Genome Analyzer II platform (Lam et al. 2010). Several more sequencing projects on a considerable number of cultivated and wild soybean accessions were performed and the soybean genome was subsequently updated (Table 1).

3.2 Transcriptomic analyses on salt stress

In addition to the soybean genome, transcriptomic analysis is frequently used to identify candidate genes related to salt stress responses. Currently, over 1.4 million expressed sequence tags (ESTs) in soybean (*G. max*) have been submitted to the GenBank dbEST (https://www.ncbi.nlm.nih.gov/nucest/). Through examining the consensus sequences in publicly available EST databases or specific datasets generated using libraries constructed from salt-stressed soybean plants, transcription factor family members related to abiotic stress responses in other model organisms such as *Arabidopsis* can be identified in soybean (Chen et al. 2007; Granger et al. 2002; Meng et al. 2007; Tian et al. 2004). Differential expressions of selected members between soybean under stressed and non-stressed conditions can be verified using RT-PCR. Functional analyses can then be performed by generating transgenic plants. This efficient approach was applied in the characterization of the roles played by AP2/ERF (*GmERF3*), bZIP (*GmbZIP44*, *GmbZIP62* and *GmbZIP78*), WRKY (*GmWRKY54*), NAC (*GmNAC11* and *GmNAC20*), MYB (*GmMYB76*, *GmMYB92* and *GmMYB177*) and DREB (*GmDREB2*) transcription factors in salt stress responses (Li et al. 2005; Liao et al. 2008a, 2008b, 2008c; Zhang et al. 2009; Hao et al. 2011; Zhou et al. 2008).

Besides, cDNA-amplified fragment length polymorphism (AFLP) was also used to identify and sequence differentially expressed genes (Breyne and Zabeau 2001). These genes were then grouped according to their predicted protein functions for further studies. For example, a study on *G. max* cv. Enrei under salt stress uncovered a differentially expressed gene, *GmTDF-5*, which may encode a bZIP transcription factor (Aoki et al. 2005). The same dataset was also instrumental in a study on the nucleus-localized *GmOSBP*, which was hypothesized to be involved in a signalling mechanism (Li et al. 2008). Another

Table 1 List of soybean varieties and tissues reported in -omic studies related to salt stress

Omic study	Soybean varieties	Tissues	References
Transcriptome	Lee	Shoots and roots of seedlings	Umezawa et al. 2002
Transcriptome	Kefeng1	Leaves of seedlings	He et al. 2003; Tian et al. 2004
Transcriptome	Enrei	Seedlings	Aoki et al. 2005
Transcriptome	RI lines of Jindou No. 23 x Huibuzhi	Leaves of seedlings	Wei et al. 2009
Transcriptome	HJ-1	Leaves and roots of seedlings	Fan et al. 2013
Transcriptome	Williams 82	Roots of seedlings	Belamkar et al. 2014
Proteome	Enrei	Leaves, hypocotyls and roots of seedlings	Aghaei et al. 2008; Sobhanian et al. 2010
Proteome	N2899	Germinated seeds	Xu et al. 2011
Proteome	Pusa-37	Leaves of seedlings	Hakeem et al. 2012
Proteome	Lee68	Germinated seeds; roots and leaves of seedlings	Xu et al. 2011, Ma et al. 2012, 2014
Proteome	Jackson	Roots and leaves of seedlings	Ma et al. 2012, 2014
Proteome	Yunhe	Cotyledon and embryos	Yin et al. 2015
Proteome	PR91M10	Embryonic axes of germinating seeds	Fercha et al. 2016
Metabolome	C08, W05 and their RI lines	Leaves of seedlings and seeds	Luet al. 2013; Wuet al. 2008

cDNA-AFLP analysis led to the characterization of *GmPHD2*, which enhanced salt tolerance in transgenic *Arabidopsis* if overexpressed (Wei et al. 2009a). Moreover, cDNA-AFLP was used to identify induced gene expressions specific to ionic stress or osmotic stress due to salt treatment (Umezawa et al. 2002). With the availability of next-generation sequencing technology, high-throughput RNA-seq was also used for expression profiling and candidate identification (Fan et al. 2013). For example, the expression profiles of the homeodomain leucine zipper (HD-Zip) gene family members under salt stress at several time points were reported using this technique (Belamkar et al. 2014). The same dataset was also used for analysing the expressions of *EF1a* and *WRKY* genes (da Cruz Saraiva et al. 2016; Song et al. 2016). Most of the available sequencing data related to salt stress is shown in Table 1.

3.3 Proteomic analyses of salt stress

Transcriptomic analyses can only provide insights into salt response mechanisms at the gene expression level. A different technique is required to understand what happens at the protein level. Proteomic analyses (which involve performing a two-dimensional gel electrophoresis of protein extracts, followed by the excision of differentially expressed proteins from the gel, as detected with the use of software and the identification of these proteins by mass spectrometry) can identify differential protein productions in soybean under salt stress. Various soybean varieties have been used for this purpose (Table 1). Proteomic analyses of soybean hypocotyl and root under salt stress revealed a concentration-dependent accumulation of free proline under NaCl stress (Aghaei et al. 2009). Glutathione-*S*-transferase (GST), which is responsible for the protection against ROS, was upregulated in the seedling roots, leaves, germinated seeds and embryonic axes of germinating seeds of several varieties under salt stress (Aghaei et al. 2009; Fercha et al. 2016; Gill and Tuteja 2010; Sobhanian et al. 2010; Xu et al. 2011). Under salt stress, the levels of metabolism-related proteins were reduced in soybean roots, hypocotyls, leaves and glyceraldehyde-3-phosphate dehydrogenase was downregulated, indicating that the metabolism of glucose through glycolysis is important to overcome the salinity stress (Sobhanian et al. 2010). Recently, Fercha et al. (2016) identified 141 differentially expressed proteins using high-sensitivity nano HPLC-MS/MS, and most of the enriched salt-responsive proteins could be divided into three functional categories: disease/defence response, protein destination and storage, and primary metabolism. A review of soybean proteomic analysis under abiotic stress provided a comprehensive picture of protein regulation networks primarily involved in stress tolerance mechanism (Hossain et al. 2013). In addition to several proteins found to be important for salt tolerance, a considerable number of previously uncharacterized proteins were also identified. Validation and further analyses of their roles may provide new information about salt stress response mechanisms.

3.4 Epigenomic studies on salt stress

Epigenetic modification is a heritable form of gene regulation, which changes gene expressions through DNA methylation/demethylation and chromatin/histone modification without altering the coding sequence of the gene (Fazzari and Greally 2004). It contributes to the regulation of gene expression and plant development under salt stress (Song et al. 2012). In eukaryotes, DNA methylation occurs on the C-5 position of cytosine and affects the transcription rate through changing the affinity of DNA towards transcription factors or altering the packing density of the chromatin. Histone modifications, including methylation, acetylation, ubiquitylation, phosphorylation, SUMOylation, ADP-ribosylation and citrullination (Chan et al. 2012), affect histone-DNA interaction, accessibility of DNA and recruitment of regulatory proteins, thus altering gene expression.

It is known that environmental stresses induce DNA methylation changes in plants (Zhu 2008). Since as early as 1994, scientists have been studying the soybean genomic structure and DNA methylation patterns of duplicated sequences (Zhu et al. 1994). The methylation polymorphisms between 20 wild soybean accessions and 27 soybean cultivars were compared. It was found that the soybean cultivars contain higher levels of both methylation-insensitive and methylation-sensitive polymorphisms than wild soybeans (Zhong et al. 2009). In 2014, wild and cultivated soybeans in the northern Huang Huai region of China were selected for comparison and an epigenetic divergence among the wild populations was observed. It implies that natural selection might have acted on epigenetic variations in a similar manner to that on genes (Yan et al. 2014). Bisulphite sequencing down to single base pairs allow us to study the genome-wide DNA methylation patterns. Recently, whole-genome methylation patterns have been made available for soybean. Besides, understanding the salt-induced epigenetic process gives us clues about the regulation mechanisms to combat salt stress in plants. A study on DNA methylation and histone modifications of 49 salt-responsive transcription factor genes using bisulphite sequencing suggested that the methylation status of some of these genes were changed under salt stress. Also, the upregulation of two transcription factor genes under salt stress was found to be correlated with reduced DNA methylation level and increased levels of histone H3K4 trimethylation or H3K9 acetylation (Song et al. 2012). In addition, previous studies found that the salt-inducible transcriptional factor *GmPHD5* in soybean can recognize H3K4me2/3 and interact with an acetyltransferase *GmGNAT1* which can preferentially acetylate H3K14 (Wu et al. 2011).

3.5 Metabolomic analyses of salt tolerance

Metabolomic analyses of data obtained from mass spectrometry using salt-tolerant and -sensitive varieties have been performed to identify the major metabolites contributing to, or caused by, the variations in salt tolerance (Table 1). Lu et al. (2013) reported the content levels of 98 metabolites in the leaves of wild soybean (W05) and cultivated soybean (C08), and concluded that higher levels of glutathione and abscisic acid confer salt tolerance in W05. Also, genistin and group B saponins were identified as contributors to salt tolerance (Wu et al. 2008).

4 Molecular breeding of salt-tolerant soybean

To develop salt-tolerant soybean against both hyperosmotic stress and ion toxicity, marker-assisted breeding and genetic engineering have been recognized as efficient ways to accelerate the development of new, more productive and better adapted cultivars (Deshmukh et al. 2014; Gupta and Huang 2014; Sudaric et al. 2008; Xu and Crouch 2008). Tolerance to salt stress depends on many factors (developmental stages, types, intensity and combination of stress, etc.), and the results are a combination of acclimation responses (phenotypic plasticity) and adaptation mechanisms under each specific condition.

Since the 2000s, a series of studies consistently revealed a major QTL for salinity tolerance located on the linkage group N (chromosome 3) around the SSR markers, Sat_091, Satt255, Satt237, Sat_304 and Satt339, in different genetic backgrounds (Xu 2012; Hamwieh and Xu 2008; Hamwieh et al. 2011; Lee et al. 2004). It accounts for 31.5–68.7% of the phenotypic variance (Xu 2012). Recently, using *de novo* assembly, re-sequencing and map-based cloning, the dominant gene, *Glyma03g32900*, a cation/H^+ exchanger protein primarily responsible for salt tolerance was identified in this QTL. It was named *GmCHX1* from a wild soybean, *G. soja* Siebold and Zucc. (Qi et al. 2014), and *GmSALT3* from a cultivated soybean, *G. max* (L.) Merr. (Guan et al. 2014b). One of the major bottlenecks of the marker-assisted breeding is the lack of appropriate markers linked to the major QTLs. As technology advances, scientists suggested a combined approach of QTL mapping/genome-wide association study (GWAS) and genomic selection (GS) that can make maximum gains in breeding values (Deshmukh et al. 2014). This combined approach has been successfully demonstrated (Patil et al. 2016). Using high-quality (15x) whole-genome re-sequencing on 106 diverse soybean lines, they identified single-nucleotide polymorphisms (SNPs) associated with major structural variants and allelic variations in the promoter and genic regions of *GmCHX1* and developed a panel of breeder-friendly

Kompetitive allele-specific polymerase chain reaction (KASPar) assays. About 104 soybean lines and an interspecific biparental population (F8) from PI483463 x Hutcheson were used for the validation of these identified SNP markers, and a strong correlation was observed between the genotype and salt treatment phenotype. Over 91% of the salt-tolerant/-sensitive genotypes and more than 98% of the structural variants can be identified by these SNP markers (Ibid.). The availability of efficient, high-throughput and cost-effective marker assays for salinity tolerance in soybean can greatly contribute to the development of salt-tolerant varieties.

Nowadays, the development of genetically modified (GM) salt-tolerant soybean is still limited to the laboratory, although the technique has been intensively applied to enhance soybean's tolerance to herbicides/insects, and the GM herbicide/insect-tolerant soybeans make up about 93% of the planted soybeans in the United States (Fernandez-Cornejo et al. 2014); there are still no commercially available GM salt-tolerant soybeans on the market. An enormous number of salt stress response genes in soybean have been isolated and characterized in the past decade, among which several have also been proven to be able to enhance salt tolerance via transgenic studies (Table 2). However, since salt tolerance in soybean is not regulated by just one or two genes, it increases the difficulty in developing GM salt-tolerant soybeans. Taking advantage of the achievements in genomic, transcriptomic, proteomic and metabolomic technologies, our understanding of the salt tolerance mechanisms in soybean has increased significantly. Combining the newly discovered key regulators of salt stress responses and new biotechnology tools would allow us to generate GM salt-tolerant soybeans with good agronomic traits. To bring GM salt-tolerant soybeans from the laboratory to the field, proper translational research should be performed to evaluate improved varieties in field conditions (Nelissen et al. 2014).

5 Future trends and conclusion

In view of the increasing need for soybean as a cash crop worldwide, we see a need to increase the yield of cultivated soybean and expand the area of land available for its cultivation. However, soybean cultivation and productivity is threatened by severe climate conditions and salinity stress (IPCC 2012). In recent decades, there have been more and more studies on the molecular mechanisms of salt tolerance. Advanced research approaches in morphological, cellular, biochemical and molecular markers have been utilized to exploit the genetic diversity of wild and cultivated soybeans and to breed salt-tolerant soybean cultivars (Zhou et al. 2015). Some strategies to breed salt-tolerant soybeans are shown in Fig. 2, but there are still lots of challenges that hinder the success of breeding salt-tolerant crops including soybean.

The soybean germplasm is a natural resource to broaden the current soybean genetic base (Chung and Singh 2008). Introducing and generating novel germplasm sources with hybridization programmes will provide continuous development of cultivars adaptive to biotic and abiotic stresses, as well as increase genetic diversities. There are thousands of soybean breeding lines and hundreds of elite cultivars being produced yearly via soybean hybridization programmes all over the world (Sammour 2011). However, the number of germplasms containing good-quality salt stress tolerance traits is still less than sufficient.

Table 2 Salt-related genes characterized by plant or plant cell transformation

Putative function	Soybean variety	Gene name	Abiotic and biotic stress	Recipient plants/cells	References
Transcription factor					
bZIP transcription factor	*G. max*	*GmbZIP1*	Salt, drought and freezing tolerance	*Arabidopsis*, tobacco, wheat	(Gao et al. 2011)
bZIP transcription factor	*G. max*	*GmbZIP44*	Salt tolerance, freezing tolerance	*Arabidopsis*	(Liao et al. 2008c)
bZIP transcription factor	*G. max*	*GmbZIP62*	Salt tolerance, freezing tolerance	*Arabidopsis*	(Liao et al. 2008c)
bZIP transcription factor	*G. max*	*GmbZIP78*	Salt tolerance, freezing tolerance	*Arabidopsis*	(Liao et al. 2008c)
bZIP transcription factor	*G. max*	*GmbZIP110*	Salt tolerance	Soybean (hairy root), *Arabidopsis*	(Xu et al. 2016)
bZIP transcription factor	*G. max*	*GmbZIP132*	Salt tolerance	*Arabidopsis*	(Liao et al. 2008a)
Dehydration-responsive element-binding protein	*G. max*	*GmDREB3*	Salt, drought and freezing tolerance	*Arabidopsis*	(Chen et al. 2009)
Ethylene-responsive factor	*G. max*	*GmERF3*	Salt, pathogen and drought tolerance	Tobacco	(Zhang et al. 2009)
Ethylene-responsive factor	*G. max*	*GmERF4*	Salt and drought tolerance	Tobacco	(Zhang et al. 2010)
Ethylene-responsive factor	*G. max*	*GmERF7*	Salt tolerance	Tobacco	(Zhai et al. 2013)
Ethylene-responsive factor	*G. max*	*GmERF089*	Salt and drought tolerance	Tobacco	(Zhang et al. 2008)
MYB transcription factor	*G. max*	*GmMYB12B2*	Salt and UV stress tolerance	*Arabidopsis*	(Li et al. 2016)
MYB transcription factor	*G. max*	*GmMYB76*	Salt and freezing tolerance	*Arabidopsis*	(Liao et al. 2008b)
MYB transcription factor	*G. max*	*GmMYB92*	Salt and freezing tolerance	*Arabidopsis*	(Liao et al. 2008b)
MYB transcription factor	*G. max*	*GmMYB177*	Salt and freezing tolerance	*Arabidopsis*	(Liao et al. 2008b)
NAC transcription factor	*G. max*	*GmNAC11*	Salt tolerance	*Arabidopsis*	(Hao et al. 2011)

(*Continued*)

Table 2 (*Continued*)

Putative function	Soybean variety	Gene name	Abiotic and biotic stress	Recipient plants/cells	References
NAC transcription factor	*G. max*	*GmNAC20*	Salt and freezing tolerance	*Arabidopsis*	(Wang et al. 2015; Hao et al. 2011)
Plant homeodomain finger protein	*G. max*	*GmPHD2*	Salt tolerance	*Arabidopsis*	(Wei et al. 2009b)
Trihelix transcription factor	*G. max*	*GmGT-2A*	Salt, drought and freezing tolerance	*Arabidopsis*	(Xie et al. 2009)
Trihelix transcription factor	*G. max*	*GmGT-2B*	Salt, drought and freezing tolerance	*Arabidopsis*	(Xie et al. 2009)
WRKY transcription factor	*G. max*	*GmWRKY27*	Salt and drought tolerance	Soybean (hairy root)	(Wang et al. 2015)
WRKY transcription factor	*G. max*	*GmWRKY54*	Salt and drought tolerance	*Arabidopsis*	(Zhou et al. 2008)
Ion transporters					
Na^+/H^+ antiporter	*G. max*	*GmsSOS1*	Salt tolerance	*Arabidopsis*	(Nie et al. 2015)
Na^+/H^+ antiporter	*G. max*	*GmNHX1*	Salt tolerance	Tobacco BY-2 cells	(Li et al. 2006a)
Ca^{2+} ATPase	*G. soja*	*GsACA1*	Salt and carbonate alkaline stress tolerance	Alfalfa	(Sun et al. 2016)
Cation/H^+ exchanger	*G. soja*	*GmCHX1*	Salt tolerance	Soybean (hairy root), tobacco BY-2 cells	(Qi et al. 2014; Guan et al. 2014b; Do et al. 2016a)
Chloride/proton antiporter	*G. max*	*GmCLC1*	Salt tolerance	Soybean (hairy root), tobacco BY-2 cells	(Wei et al. 2016)
High-affinity K^+ transporter	*G. max*	*GmHKT1*	Salt tolerance	Tobacco	(Chen et al. 2011)
Cell signalling					
G-type lectin S-receptor-like serine/threonine protein kinase	*G. soja*	*GsSRK*	Salt tolerance	*Arabidopsis*	(Sun et al. 2013a)

(*Continued*)

Table 2 (*Continued*)

Putative function	Soybean variety	Gene name	Abiotic and biotic stress	Recipient plants/cells	References
Receptor-like serine/threonine protein kinases	*G. soja*	*GsRLCK*	Salt and drought tolerance	*Arabidopsis*	(Sun et al. 2013b)
Serine/threonine protein kinase	*G. max*	*GmWNK1*	Salt and osmotic stresses tolerance	*Arabidopsis*	(Wang et al. 2011)
Halotolerance protein	*G. max*	*GmHAL3a*	Salt tolerance	*Arabidopsis*	(Guo et al. 2016)
Halotolerance protein	*G. max*	*GmHAL3b*	Salt tolerance	*Arabidopsis*	(Guo et al. 2016)
Calcineurin B-like protein	*G. max*	*GmCBL1*	Salt and drought tolerance	*Arabidopsis*	(Li et al. 2012)
Calcium-dependent calmodulin-binding receptor-like kinase	*G. soja*	*GsCBRLK*	Salt and ABA tolerance	*Arabidopsis*	(Yang et al. 2010)
Calmodulin	*G. max*	*GmCaM4*	Salt and pathogen tolerance	Soybean	(Rao et al. 2014)
Jasmonate ZIM-domain protein	*G. soja*	*GsJAZ2*	Salt and alkali stress tolerance	*Arabidopsis*	(Zhu et al. 2012)
Metabolic enzyme					
Myo-inositol-*O*-methyl transferase	*G. max*	*GmIMT1*	Salt and ABA tolerance	*Arabidopsis*	(Wang et al. 2014)
2-hydroxyisoflavone synthase	*G. max*	*GmIFS1*	Salt tolerance	Soybean (hairy root), tobacco	(Jia et al. 2017)
Aldehyde dehydrogenase-related	*G. max*	*GmTP55*	Salt, drought and oxidative stress tolerance	*Arabidopsis*, tobacco	(Rodrigues et al. 2006)
Cytochrome P450 monooxygenases	*G. max*	*GmCYP82A3*	Salt, pathogen and osmotic stress tolerance	*Nicotiana benthamiana*	(Yan et al. 2016)
Flavone synthase II/ cytochrome P450 monooxygenases	*G. max*	*GmFNSII-1*	Salt tolerance	Soybean (hairy root)	(Yan et al. 2013)
Flavone synthase II/ cytochrome P450 monooxygenases	*G. max*	*GmFNSII-2*	Salt tolerance	Soybean (hairy root)	(Yan et al. 2013)
Germin gene/ oxalate oxidase	*G. max*	*GmGER 9*	Salt tolerance	Tobacco	(Lu et al. 2010)
Glutathione S-transferase	*G. max*	*GmGSTL1*	Salt tolerance	*Arabidopsis*, tobacco BY-2 cells	(Chan and Lam 2014)

(*Continued*)

Table 2 (*Continued*)

Putative function	Soybean variety	Gene name	Abiotic and biotic stress	Recipient plants/cells	References
Myo-inositol-1-phosphate synthase	*G. soja*	*GsMIPS2*	Salt tolerance	*Arabidopsis*	(Nisa et al. 2016)
Protein degradation					
DA1-like/ubiquitin-binding protein	*G. soja*	*GsoDA1-1*	Salt tolerance	*Arabidopsis*	(Zhao et al. 2015)
DA1-like/ubiquitin-binding protein	*G. soja*	*GsoDA1-2*	Salt tolerance	*Arabidopsis*	(Zhao et al. 2015)
DA1-like/ubiquitin-binding protein	*G. soja*	*GsoDA1-3*	Salt tolerance	*Arabidopsis*	(Zhao et al. 2015)
DA1-like/ubiquitin-binding protein	*G. soja*	*GsoDA1-7*	Salt tolerance	*Arabidopsis*	(Zhao et al. 2015)
DA1-like/ubiquitin-binding protein	*G. soja*	*GsoDA1-8*	Salt tolerance	*Arabidopsis*	(Zhao et al. 2015)
Ubiquitin-conjugating enzyme	*G. max*	*GmUBC2*	Salt and drought tolerance	*Arabidopsis*	(Zhou et al. 2010)
Other functions					
Nictaba-like lectins	*G. max*	*GmNLL1*	Salt and pathogen tolerance	*Arabidopsis*	(Van Holle et al. 2016)
Nictaba-like lectins	*G. max*	*GmNLL2*	Salt and pathogen tolerance	*Arabidopsis*	(Van Holle et al. 2016)
Chilling inducible, small peptide	*G. max*	*GmCHI*	Salt, drought and freezing tolerance	*Arabidopsis*	(Cheng et al. 2009)
Plasma membrane intrinsic protein	*G. max*	*GmPIP1;6*	Salt tolerance	Soybean	(Zhou et al. 2014)

Traditional methods to screen soybean in response to salt tolerance are time-consuming and low in efficiency (Valencia et al. 2008).

In addition, soybean germplasms contain a wide range of phenotypic variations which are controlled and altered by genes and environmental changes. While the phenotypic data cannot reveal the underlying genetic variations directly (Sammour 2011), the discovery of molecular markers has made it possible to detect the genetic and environmental components of variations more accurately. Molecular markers for screening salt-tolerant lines have been applied to several crop plants (Witcombe et al. 2008). DNA markers such as RAPD, AFLP, RFLP, SSR and SNP markers are more stable and reliable for identifying soybean genotypes with better salt tolerance (Pathan et al. 2007). The challenge now is how to use these markers efficiently to identify the useful soybean variations with high salt tolerance. As there are thousands of genetic source collections, it is difficult to narrow down the germplasms. To overcome this, we need to optimize the protocols for phenotype mining, and develop faster and cheaper sequencing platforms to detect allelic variations.

The -omic technologies, including genome assembly, transcriptomics, proteomics and metabolomics, enable a more comprehensive approach in studying plant genetic patterns under salt stress. Also, by combining these approaches with bioinformatics, researchers can maximize the use of available information from lots of soybean databases and constructed networks. The reference genome of *G. max* (cultivated soybean) provides important information and is a powerful tool for soybean research and breeding programmes. Owing to the complexity and highly repetitive nature of the soybean genome, some genomic regions are still not covered in this reference genome and gaps are still present between aligned scaffolds (Chan et al. 2012). *G. soja*, containing higher genomic diversity, will allow us to detect new useful genes or alleles. The draft genome of wild soybean has been published, but it also has many gaps. Therefore, a more complete *G. soja* genome by sequencing longer contigs at increasing depths is still needed. Besides the genome assembly, the transcriptome data under salt stress generated by next-generation sequencing could uncover the molecular basis of salt stress on a wider scale, provide general expression patterns of all transcripts and thus allow the mining of putative genes related with salt stress. In recent years, epigenetic regulations appear to be more and more important for plants coping with abiotic stresses. Studies have shown that epigenetic changes in DNA methylation and histone modifications can be induced by different abiotic stresses (Song et al. 2012; Liu et al. 2015), and the changes can be inherited by the next generation (Migicovsky et al. 2014; Iglesias and Cerdán 2016; Chinnusamy and Zhu 2009). In soybean, the genome-wide DNA methylation patterns of soybean cultivar Heinong 44 and the correlations with gene expressions and sRNAs in normal growing conditions have been clearly elucidated (Song et al. 2013). Differentially methylated regions (DMR) from parent lines are heritable to the recombinant inbred lines in three contexts: CG, CHG and CHH (Schmitz et al. 2013). Nevertheless, the DNA methylation and histone modification regulatory mechanisms under salt stress still remain largely unknown.

To generate soybean varieties more adapted to salt stress, more research and studies in plant and cell physiology, genetics and molecular breeding are required. After elucidating the molecular mechanisms and genomic information, a large number of genes have been identified and some of them are being used to develop salt-tolerant transgenic plants. The technology of *Agrobacterium*-mediated soybean transformation has been hampered by low transformation efficiency, time consumption and transgenic line instability. It still depends on the ability to combine efficient transformation with regeneration techniques (Homrich et al. 2012). Functional analyses of salt stress-related putative genes mined from genomic and QTL analyses were mostly studied only in the model plant, *Arabidopsis*, and thus lacked the final confirmation based on a homologous expression system. In 2007, *Agrobacterium rhizogenes*-mediated root transformation was developed, and almost 100% of the infected soybean plants formed hairy roots within 1 month (Kereszt et al. 2007). The root transformation of soybean partially solves the homologous expression issue but is only applicable in roots. At the same time, soybean mutant resources produced by antisense or RNAi-mediated transcriptional or post-transcriptional gene silencing (Nunes et al. 2006; Ibrahim et al. 2011; Subramanian et al. 2005) are also important for molecular breeding. Another alternative solution is to perform functional analyses by virus-induced gene silencing technology (Nagamatsu et al. 2007; Zhang and Ghabrial 2006). The CRISPR/Cas9 system, a powerful and precise tool for genome editing, has been established in soybean successfully by root transformation (Bortesi and Fischer 2015). Last but not least, apart from putting effort into producing higher salt-tolerance transgenic soybean lines, applying the transgenic lines to field testing needs to receive more attention, including a greater emphasis on developing a widely applicable set of parameters for measuring salt stress.

6 Where to look for further information

Munoz, N., Liu, A., Kan, L., Li, M. W., Lam, H. M., Potential uses of wild germplasms of grain legumes for crop improvement, *International journal of molecular sciences* 2017, 18(2), 328.

Foyer, C.H., Lam, H.M., Nguyen, H.T., Siddique, K.H., Varshney, R.K., Colmer, T.D., Cowling, W., Bramley, H., Mori, T.A., Hodgson, J.M., et al., Neglecting legumes has compromised human health and sustainable food production, *Nature plants* 2016, *2*, 16112.

Li, M.W., Xin, D., Gao, Y., Li, K.P., Fan, K., Munoz, N.B., Yung, W.S., Lam, H.M., Using genomic information to improve soybean adaptability to climate change, *Journal of Experimental Botany* 2016, erw348.

Kumawat, G., Gupta, S., Ratnaparkhe, M. B., Maranna, S., Satpute, G. K., QTLomics in soybean: A way forward for translational genomics and breeding, *Frontiers in Plant Science* 2016, 7, 1852.

Smýkal, P., Coyne, C. J., Ambrose, M. J., Maxted, N., Schaefer, H., Blair, M. W., Berger, J., Greene, S. L., Nelson, M. N., Besharat, N., Legume crops phylogeny and genetic diversity for science and breeding, *Critical Review in Plant Science* 2015, 34, 43–104.

Deshmukh, R., umira Sonah, H., Patil, G., Chen, W., Prince, S., Mutava, R., Vuong, T., Valliyodan, B., Nguyen, H. T., Integrating omic approaches for abiotic stress tolerance in soybean, *Frontiers in Plant Science* 2014, 5, 244.

Phang, T. H., Shao, G., Lam, H. M., Salt tolerance in soybean, *Journal of Integrative Plant Biololy*, 2008, 50(10), 1196–212.

7 Acknowledgements

This work was supported by grants from Hong Kong Research Grants Council (Collaborative Research Fund: CUHK3/CRF/11G; Area of Excellence: AoE/M-403/16), CUHK VC Discretionary Fund VCF2014004 and the Lo Kwee-Seong Biomedical Research Fund to H.-M. Lam. We thank Mr. Kejing Fan and Mr. Yishu Gao for organizing the references and Ms. Jee-Yan Chu for copyediting the manuscript.

8 References

Abel, G. H. and MacKenzie, A. J. (1964), 'Salt tolerance of soybean varieties (*Glycine max* L. Merrill) during germination and later growth', *Crop Sci.*, 4(2), 157–61.

Abel, G. H. (1969), 'Inheritance of the capacity for chloride inclusion and chloride exclusion by soybeans', *Crop Sci.*, 9(6), 697–8.

Aghaei, K., Ehsanpour, A., Shah, A. and Komatsu, S. (2009) 'Proteome analysis of soybean hypocotyl and root under salt stress', *Amino acids*, 36(1), 91–8.

An, P., Inanaga, S., Cohen, Y., Kafkafi, U. and Sugimoto, Y. (2002), 'Salt tolerance in two soybean cultivars', *J. Plant Nutr.*, 25(3), 407–23.
An, P., Inanaga, S., Kafkafi, U., Lux, A. and Sugimoto, Y. (2001), 'Different effect of humidity on growth and salt tolerance of two soybean cultivars', *Biologia plantarum*, 44(3), 405–10.
Anon. (2012a), 'Economic impact of saline soils in upper James River', United States Department of Agriculture, https://www.nrcs.usda.gov/Internet/FSE_DOCUMENTS/nrcs141p2_035584.pdf (accessed 17 March 2017)
Anon. (2012b), 'Monthly soybean supply & demand round-up', Food and Agricultural Organization of the United Nations, http://www.fao.org/fileadmin/templates/est/COMM_MARKETS_MONITORING/Oilcrops/Documents/MSSDR/MSSDR_Aug__12_rev.pdf (accessed 17 March 2017)
Aoki, A., Kanegami, A., Mihara, M., Kojima, T., Shiraiwa, M. and Takahara, H. (2005), 'Molecular cloning and characterization of a novel soybean gene encoding a leucine-zipper-like protein induced to salt stress', *Gene*, 356, 135–45.
Arshi, A., Ahmad, A., Aref, I. M. and Iqbal, M. (2012), 'Comparative studies on antioxidant enzyme action and ion accumulation in soybean cultivars under salinity stress', *J. Environ. Biol.*, 33(1), 9–20.
Arzani, A. (2008), 'Improving salinity tolerance in crop plants: a biotechnological view', *In Vitro Cell Dev. Biol. Plant*, 44(5), 373–83.
Ashraf, M. (1994), 'Breeding for salinity tolerance in plants', *Crit. Rev. Plant Sci.*, 13(1), 17–42.
Belamkar, V., Weeks, N. T., Bharti, A. K., Farmer, A. D., Graham, M. A. and Cannon, S. B. (2014), 'Comprehensive characterization and RNA-Seq profiling of the HD-Zip transcription factor family in soybean (*Glycine max*) during dehydration and salt stress', *BMC Genomics*, 15, 950, doi:10.1186/1471-2164-15-950.
Bortesi, L. and Fischer, R. (2015), 'The CRISPR/Cas9 system for plant genome editing and beyond', *Biotechnol. Adv.*, 33(1), 41–52.
Breyne, P. and Zabeau, M. (2001), 'Genome-wide expression analysis of plant cell cycle modulated genes', *Curr. Opin. Plant Biol.*, 4(2), 136–42.
Chakrabortee, S., Boschetti, C., Walton, L. J., Sarkar, S., Rubinsztein, D. C. and Tunnacliffe, A. (2007), 'Hydrophilic protein associated with desiccation tolerance exhibits broad protein stabilization function', *Proc. Natl. Acad. Sci. U. S. A.*, 104(46), 18073–8.
Chan, C. and Lam, H. M. (2014), 'A putative lambda class glutathione S-transferase enhances plant survival under salinity stress', *Plant Cell Physiol.*, 55(3), 570–9.
Chan, C., Qi, X., Li, M. W., Wong, F. L. and Lam, H. M. (2012), 'Recent developments of genomic research in soybean', *J. Genet. Genomics*, 39 (7), 317–24.
Chang, R. Z., Chen, Y. W., Shao, G. H. and Wan, C. W. (1994), 'Effect of salt stress on agronomic characters and chemical quality of seeds in soybean', *Soybean Sci.*, 13, 101–5.
Chen, H. T., Xin, C., Wu, B. Y., Yuan, X. X., Zhang, H. M., Cui, X. Y. and Liu, X. Q. (2015), 'Whole-genome identification and expression analysis of K^+ efflux antiporter (KEA) and Na^+/H^+ antiporter (NHX) families under abiotic stress in soybean', *J. Integr. Agr.*, 14(6), 1171–83.
Chen, H., He, H. and Yu, D. (2011), 'Overexpression of a novel soybean gene modulating Na^+ and K^+ transport enhances salt tolerance in transgenic tobacco plants', *Physiol. Plant*, 141(1), 11–18.
Chen, M., Wang, Q. Y., Cheng, X. G., Xu, Z. S., Li, L. C., Ye, X. G., Xia, L. Q. and Ma, Y. Z. (2007), '*GmDREB2*, a soybean DRE-binding transcription factor, conferred drought and high-salt tolerance in transgenic plants', *Biochem. Biophys. Res. Commun.*, 353(2), 299–305.
Chen, M., Xu, Z., Xia, L., Li, L., Cheng, X., Dong, J. and Ma, Y. (2009), 'Cold-induced modulation and functional analyses of the DRE-binding transcription factor gene, *GmDREB3*, in soybean (*Glycine max* L.)', *J. Exp. Bot.*, 60(1), 121–35.
Chen, P., Yan, K., Shao, H. and Zhao, S. (2013), 'Physiological mechanisms for high salt tolerance in wild soybean (*Glycine soja*) from yellow river delta, China: photosynthesis, osmotic regulation, ion flux and antioxidant capacity', *PloS One*, 8 (12), e83227.
Cheng, L., Huan, S., Sheng, Y., Hua, X., Shu, Q., Song, S. and Jing, X. (2009), 'GMCHI, cloned from soybean [*Glycine max* (L.) Meer.], enhances survival in transgenic Arabidopsis under abiotic stress', *Plant Cell Rep.*, 28(1), 145–53.

Chinnusamy, V. and Zhu, J. K. (2009), 'Epigenetic regulation of stress responses in plants', *Curr. Opin. Plant Biol.*, 12 (2), 133–9.

Chung, G. and Singh, R. J. (2008), 'Broadening the genetic base of soybean: a multidisciplinary approach', *Crit. Rev. Plant Sci.*, 27(5), 295–341.

Chung, W. S., Lee, S. H., Kim, J. C., Do Heo, W., Kim, M. C., Park, C. Y., Park, H. C., Lim, C. O., Kim, W. B. and Harper, J. F. (2000), 'Identification of a calmodulin-regulated soybean Ca^{2+}-ATPase (SCA1) that is located in the plasma membrane', *Plant Cell*, 12(8), 1393–407.

da Cruz Saraiva, K. D., Oliveira, A. E. R., dos Santos, C. P., Lima, K. T. L., de Sousa, J. M., de Melo, D. F., and Costa, J. H. (2016), 'Phylogenetic analysis and differential expression of EF1α genes in soybean during development, stress and phytohormone treatments', *Mol. Gen. Genet.*, 291(4), 1505–1522.

Deshmukh, R., Umira Sonah, H., Patil, G., Chen, W., Prince, S., Mutava, R., Vuong, T., Valliyodan, B. and Nguyen, H. T. (2014), 'Integrating omic approaches for abiotic stress tolerance in soybean', *Front. Plant. Sci.*, 5, 244, doi:10.3389/fpls.2014.00244.

Do, T. D., Chen, H., Hien, V. T. T., Hamwieh, A., Yamada, T., Sato, T., Yan, Y., Cong, H., Shono, M. and Suenaga, K. (2016), '*Ncl* synchronously regulates Na^+, K^+, and Cl^- in soybean and greatly increases the grain yield in saline field conditions', *Sci. Rep.*, 6, 19147.

Dolatabadian, A., Sanavy, S. A. M. M. and Ghanati, F. (2011), 'Effect of salinity on growth, xylem structure and anatomical characteristics of soybean', *Not. Sci. Biol.*, 3(1), 41–5.

Duzan, H. M., Zhou, X., Souleimanov, A. and Smith, D. L. (2004), 'Perception of *Bradyrhizobium japonicum* Nod factor by soybean *Glycine max* (L.) Merr. root hairs under abiotic stress conditions', *J. Exp. Bot.*, 55(408), 2641–6.

El-sabagh, A., Islam, M. S., Ueda, A., Saneoka, H. and Barutçular, C. (2015a), 'Increasing reproductive stage tolerance to salinity stress in soybean', *J. Agron. Crop Sci.*, 48(5):738–45.

El-sabagh, A., Sorour, S., Ueda, A. and Saneoka, H. (2015b), 'Evaluation of salinity stress effects on seed yield and quality of three soybean cultivars', *Azarian J. Agric.*, 2(5), 138–41.

Essa, T. (2002), 'Effect of salinity stress on growth and nutrient composition of three soybean (*Glycine max* L. Merrill) cultivars', *J. Agron. Crop Sci.*, 188(2), 86–93.

Fan, X. D., Wang, J. Q., Yang, N., Dong, Y. Y., Liu, L., Wang, F. W., Wang, N., Chen, H., Liu, W. C. and Sun, Y. P. (2013), 'Gene expression profiling of soybean leaves and roots under salt, saline-alkali and drought stress by high-throughput Illumina sequencing', *Gene*, 512(2), 392–402.

Fazzari, M. J. and Greally, J. M. (2004), 'Epigenomics: beyond CpG islands', *Nature Rev. Genet.*, 5(6), 446–55.

Fercha, A., Capriotti, A. L., Caruso, G., Cavaliere, C., Stampachiacchiere, S., Zenezini Chiozzi, R. and Laganà, A. (2016), 'Shotgun proteomic analysis of soybean embryonic axes during germination under salt stress', *Proteomics*, 16(10), 1537–46.

Fernandez-Cornejo, J., Wechsler, S., Livingston, M. and Mitchell, L. (2014), *Genetically engineered crops in the United States*, United States Department of Agriculture.

Fulton, T., Grandillo, S., Beck-Bunn, T., Fridman, E., Frampton, A., Lopez, J., Petiard, V., Uhlig, J., Zamir, D. and Tanksley, S. (2000), 'Advanced backcross QTL analysis of a *Lycopersicon esculentum* × *Lycopersicon parviflorum* cross', *Theor. Appl. Genet.*, 100(7), 1025–42.

Gao, S. Q., Chen, M., Xu, Z. S., Zhao, C. P., Li, L., Xu, H. J. and Ma, Y. Z. (2011), 'The soybean *GmbZIP1* transcription factor enhances multiple abiotic stress tolerances in transgenic plants', *Plant Mol. Biol.*, 75(6), 537–53.

Gill, S. S. and Tuteja, N. (2010), 'Reactive oxygen species and antioxidant machinery in abiotic stress tolerance in crop plants', *Plant Physiol. Biochem.*, 48(12), 909–30.

Goyal, K., Walton, L. J. and Tunnacliffe, A. (2005), 'LEA proteins prevent protein aggregation due to water stress', *Biochem. J.*, 388(1), 151–7.

Granger, C., Coryell, V., Khanna, A., Keim, P., Vodkin, L. and Shoemaker, R. C. (2002), 'Identification, structure, and differential expression of members of a BURP domain containing protein family in soybean', *Genome*, 45(4), 693–701.

Guan, R., Chen, J., Jiang, J., Liu, G., Liu, Y., Tian, L., Yu, L., Chang, R. and Qiu, L. (2014a), 'Mapping and validation of a dominant salt tolerance gene in the cultivated soybean (*Glycine max*) variety Tiefeng 8', *Crop J.*, 2(6), 358–65.

Guan, R., Qu, Y., Guo, Y., Yu, L., Liu, Y., Jiang, J., Chen, J., Ren, Y., Liu, G., Tian, L., Jin, L., Liu, Z., Hong, H., Chang, R., Gilliham, M. and Qiu, L. (2014b), 'Salinity tolerance in soybean is modulated by natural variation in *GmSALT3*', *Plant J.*, 80(6), 937–50.

Guo, N., Wang, M. X., Xue, C. C., Xue, D., Xu, J. Y., Wang, H. T. and Zhao, J. M. (2016), 'Over-expression of *GmHAL3* modulates salt stresses tolerance in transgenic Arabidopsis', *J. Plant Biol.*, 59(5), 444–55.

Gupta, B. and Huang, B. (2014), 'Mechanism of salinity tolerance in plants: physiological, biochemical, and molecular characterization', *Int. J. Genom.*, 2014, 701596, doi:10.1155/2014/701596.

Hakeem, K. R., Khan, F., Chandna, R., Siddiqui, T. O. and Iqbal, M. (2012), 'Genotypic variability among soybean genotypes under NaCl stress and proteome analysis of salt-tolerant genotype', *Appl. Biochem. Biotechnol.*, 168(8), 2309–29.

Hamwieh, A. and Xu, D. (2008), 'Conserved salt tolerance quantitative trait locus (QTL) in wild and cultivated soybeans', *Breed. Sci.*, 58(4), 355–9.

Hamwieh, A., Tuyen, D., Cong, H., Benitez, E., Takahashi, R. and Xu, D. (2011), 'Identification and validation of a major QTL for salt tolerance in soybean', *Euphytica*, 179(3), 451–9.

Hao, Y. J., Wei, W., Song, Q. X., Chen, H. W., Zhang, Y. Q., Wang, F., Zou, H. F., Lei, G., Tian, A. G. and Zhang, W. K. (2011), 'Soybean NAC transcription factors promote abiotic stress tolerance and lateral root formation in transgenic plants', *Plant J.*, 68(2), 302–13.

Hasegawa, P. M., Bressan, R. A., Zhu, J. K. and Bohnert, H. J. (2000), 'Plant cellular and molecular responses to high salinity', *Annu. Rev. Plant Physiol. Plant Mol. Biol.*, 51, 463–99.

Hassanzadehdelouei, M., Vazin, F. and Nadaf, J. (2013), 'Effect of salt stress in different stages of growth on qualitative and quantitative characteristics of cumin (*Cuminum cyminum* L.)', *Cercet. Agron. Mold.*, 46(1), 89–97.

He, C. Y., Zhang, J. S. and Chen, S. Y. (2002), 'A soybean gene encoding a proline-rich protein is regulated by salicylic acid, an endogenous circadian rhythm and by various stresses', *Theor. Appl. Genet.*, 104(6), 1125–31.

Homrich, M. S., Wiebke-Strohm, B., Weber, R. L. M. and Bodanese-Zanettini, M. H. (2012), 'Soybean genetic transformation, A valuable tool for the functional study of genes and the production of agronomically improved plants', *Genet. Mol. Biol.*, 35(4), 998–1010.

Hossain, Z., Khatoon, A. and Komatsu, S. (2013), 'Soybean proteomics for unraveling abiotic stress response mechanism', *J. Proteome Res.*, 12(11), 4670–84.

Hosseini, M. K., Powell, A. A. and Bingham, I. J. (2002), 'Comparison of the seed germination and early seedling growth of soybean in saline conditions', *Seed Sci. Res.*, 12(3), 165–72.

Ibrahim, H. M., Alkharouf, N. W., Meyer, S. L., Aly, M. A., Gamal El-Din Ael, K., Hussein, E. H. and Matthews, B. F. (2011), 'Post-transcriptional gene silencing of root-knot nematode in transformed soybean roots', *Exp. Parasitol.*, 127(1), 90–9.

Iglesias, F. M. and Cerdán, P. D. (2016), 'Maintaining epigenetic inheritance during DNA replication in plants', *Front. Plant Sci.*, 7, 38, doi:10.3389/fpls.2016.00038.

Jia, T., An, J., Liu, Z., Yu, B. and Chen, J. (2017), 'Salt stress induced soybean *GmIFS1* expression and isoflavone accumulation and salt tolerance in transgenic soybean cotyledon hairy roots and tobacco', *Plant Cell Tiss. Organ Cult.*, 128(2), 469–77.

Kan, G., Zhang, W., Yang, W., Ma, D., Zhang, D., Hao, D., Hu, Z. and Yu, D. (2015), 'Association mapping of soybean seed germination under salt stress', *Mol. Genet. Genom.*, 290(6), 2147–62.

Kereszt, A., Li, D., Indrasumunar, A., Nguyen, C. D., Nontachaiyapoom, S., Kinkema, M. and Gresshoff, P. M. (2007), 'Agrobacterium rhizogenes-mediated transformation of soybean to study root biology', *Nat. Protoc.*, 2(4), 948–52.

Khan, F., Hakeem, K. R., Siddiqi, T. O. and Ahmad, A. (2013), 'RAPD markers associated with salt tolerance in soybean genotypes under salt stress', *Appl. Biochem. Biotech.*, 170(2), 257–72.

Lam, H. M., Xu, X., Liu, X., Chen, W., Yang, G., Wong, F. L., Li, M. W., He, W., Qin, N., Wang, B. Li, J., Jian, M., Wang, J., Shao, G., Wang, J., Sun, S. S. and Zhang, G. (2010), 'Resequencing of

31 wild and cultivated soybean genomes identifies patterns of genetic diversity and selection', *Nat. Genet.*, 42(12), 1053–9.

Lauchli, A. and Wieneke, J. (1979), 'Studies on growth and distribution of Na^+, K^+ and Cl^- in soybean varieties differing in salt tolerance', *Z Pflanz Bodenkunde*, 142(1), 3–13.

Le Gall, H., Philippe, F., Domon, J.-M., Gillet, F., Pelloux, J. and Rayon, C. (2015), 'Cell wall metabolism in response to abiotic stress', *Plants*, 4(1), 112–66.

Lee, G., Boerma, H., Villagarcia, M., Zhou, X., Carter Jr., T., Li, Z. and Gibbs, M. (2004), 'A major QTL conditioning salt tolerance in S-100 soybean and descendent cultivars', *Theor. Appl. Genet.*, 109(8), 1610–19.

Lee, J. D., Shannon, J. G., Vuong, T. D. and Nguyen, H. T. (2009), 'Inheritance of salt tolerance in wild soybean (*Glycine soja* Sieb. and Zucc.) accession PI483463', *J. Hered.*, 100(6), 798–801.

Li, D. Y., Inoue, H., Takahashi, M., Kojima, T., Shiraiwa, M. and Takahara, H. (2008), 'Molecular homologue from soybean', *Gene*, 407(1), 12–20.

Li, W. Y. F., Wong, F. L., Tsai, S. N., Phang, T. H., Shao, G. H. and Lam, H. M. (2006a), 'Tonoplast-located *GmCLC1* and *GmNHX1* from soybean enhance NaCl tolerance in transgenic bright yellow (BY)-2 cells', *Plant Cell Environ.*, 29(6), 1122–37.

Li, X. P., Tian, A. G., Luo, G. Z., Gong, Z. Z., Zhang, J. S. and Chen, S. Y. (2005), 'Soybean DRE-binding transcription factors that are responsive to abiotic stresses', *Theor. Appl. Genet.*, 110(8), 1355–62.

Li, X. W., Wang, Y., Yan, F., Li, J. W., Zhao, Y., Zhao, X., Zhai, Y. and Wang, Q. Y. (2016), 'Overexpression of soybean R2R3-MYB transcription factor, *GmMYB12B2*, and tolerance to UV radiation and salt stress in transgenic Arabidopsis', *Genet. Mol. Res.*, 15(2), gmr6573, doi:10.4238/gmr.15026573.

Li, X., An, P., Inanaga, S., Eneji, A. E. and Tanabe, K. (2006b), 'Salinity and defoliation effects on soybean growth', *J. Plant Nutr.*, 29(8), 1499–508.

Li, Z. Y., Xu, Z. S., He, G. Y., Yang, G. X., Chen, M., Li, L. C. and Ma, Y. Z. (2012), 'Overexpression of soybean *GmCBL1* enhances abiotic stress tolerance and promotes hypocotyl elongation in Arabidopsis', *Biochem. Biophys. Res. Commun.*, 427(4), 731–6.

Liao, Y., Zhang, J. S., Chen, S. Y. and Zhang, W. K. (2008a), 'Role of soybean *GmbZIP132* under abscisic acid and salt stresses', *J. Integr. Plant. Biol.*, 50(2), 221–30.

Liao, Y., Zou, H. F., Wang, H. W., Zhang, W. K., Ma, B., Zhang, J. S. and Chen, S. Y. (2008b), 'Soybean *GmMYB76*, *GmMYB92*, and *GmMYB177* genes confer stress tolerance in transgenic Arabidopsis plants', *Cell Res.*, 18(10), 1047–60.

Liao, Y., Zou, H. F., Wei, W., Hao, Y. J., Tian, A. G., Huang, J., Liu, Y. F., Zhang, J. S. and Chen, S. Y. (2008c), 'Soybean *GmbZIP44, GmbZIP62* and *GmbZIP78* genes function as negative regulator of ABA signaling and confer salt and freezing tolerance in transgenic Arabidopsis', *Planta*, 228(2), 225–40.

Liu, X., Luo, M., Yang, S. and Wu, K. (2015) Role of epigenetic modifications in plant responses to environmental stresses. In *Nuclear Functions in Plant Transcription, Signaling and Development*, Pontes, O. and Jin, H. (Eds), pp 81–92, Springer, New York city, United States.

Liu, Y., Yu, L., Qu, Y., Chen, J., Liu, X., Hong, H., Liu, Z., Chang, R., Gilliham, M., Qiu, L. and Guan, R. (2016), '*GmSALT3*, which confers improved soybean salt tolerance in the field, increases leaf Cl- exclusion prior to Na^+ exclusion but does not improve early vigor under salinity', *Front. Plant Sci.*, 7, 1485, doi:10.3389/fpls.2016.01485.

Lu, J., Liu, Y., Hu, B. and Zhuang, B. (1998), 'The discovery of salt gland-like structure in *Glycine soja*', *Chin. Sci. Bull.*, 43, 2074–8.

Lu, J., Liu, Y., Hu, B. and Zhuang, B. (1999), 'Salt glands in *Glycine soja* L. China', *Chin. Sci. Bull.*, 44(10), 923–6.

Lu, K. X., Cao, B. H., Feng, X. P., He, Y. and Jiang, D. A. (2009), 'Photosynthetic response of salt-tolerant and sensitive soybean varieties', *Photosynthetica*, 47(3), 381–7.

Lu, M., Han, Y. P., Gao, J. G., Wang, X. J. and Li, W. B. (2010), 'Identification and analysis of the germin-like gene family in soybean', *BMC genomics*, 11(1), 620.

Lu, Y., Lam, H., Pi, E., Zhan, Q., Tsai, S., Wang, C. and Ngai, S. (2013), 'Comparative metabolomics in *Glycine max* and *Glycine soja* under salt stress to reveal the phenotypes of their offspring', *J. Agric. Food Chem.*, 61(36), 8711–21.

Luo, G. Z., Wang, H. W., Huang, J., Tian, A. G., Wang, Y. J., Zhang, J. S. and Chen, S. Y. (2005), 'A putative plasma membrane cation/proton antiporter from soybean confers salt tolerance in Arabidopsis', *Plant Mol. Biol.*, 59(5), 809–20.

Ma, H., Song, L., Huang, Z., Yang, Y., Wang, S., Wang, Z. and Xiao, L. (2014), 'Comparative proteomic analysis reveals molecular mechanism of seedling roots of different salt tolerant soybean genotypes in responses to salinity stress', *EuPA Open Proteom.*, 4, 40–57.

Ma, H., Song, L., Shu, Y., Wang, S., Niu, J., Wang, Z. and Ma, H. (2012), 'Comparative proteomic analysis of seedling leaves of different salt tolerant soybean genotypes', *J. Proteomics*, 75(5), 1529–46.

Mannan, M. A., Karim, M. A., Haque, M. M., Khaliq, Q. A., Higuchi, H. and Nawata, E. (2012), 'Response of soybean to salinity I. Genotypic variations in salt tolerance at the vegetative stage', *Trop. Agr. Dev.*, 56(4), 117–22.

Margueron, R. and Reinberg, D. (2010), 'Chromatin structure and the inheritance of epigenetic information', *Nat. Rev. Genet.*, 11(4), 285–96.

Meng, Q., Zhang, C., Gai, J. and Yu, D. (2007), 'Molecular cloning, sequence characterization and tissue-specific expression of six NAC-like genes in soybean (*Glycine max* (L.) Merr.)', *J. Plant Physiol.*, 164(8), 1002–12.

Migicovsky, Z., Yao, Y. and Kovalchuk, I. (2014), 'Transgenerational phenotypic and epigenetic changes in response to heat stress in *Arabidopsis thaliana*', *Plant Signal Behav.*, 9(2), e27971.

Miransari, M. (2015), *Abiotic and Biotic Stresses in Soybean Production, Soybean Production, vol 1*, Academic Press, Cambridge, Massachusetts, United States.

Moncada, P., Martinez, C., Borrero, J., Châtel, M., Gauch Jr., H., Guimaraes, E., Tohme, J. and McCouch, S. R. (2001), 'Quantitative trait loci for yield and yield components in an *Oryza sativa* × *Oryza rufipogon* BC2F2 population evaluated in an upland environment', *Theor. Appl. Genet.*, 102(1), 41–52.

Nagamatsu, A., Masuta, C., Senda, M., Matsuura, H., Kasai, A., Hong, J. S., Kitamura, K., Abe, J. and Kanazawa, A. (2007), 'Functional analysis of soybean genes involved in flavonoid biosynthesis by virus-induced gene silencing', *Plant Biotechnol. J.*, 5(6), 778–90.

Nelissen, H., Moloney, M. and Inzé, D. (2014), 'Translational research, from pot to plot', *Plant Biotechnol. J.*, 12(3), 277–85.

Neves, G., Marchiosi, R., Ferrarese, M., Siqueira-Soares, R. and Ferrarese-Filho, O. (2010), 'Root growth inhibition and lignification induced by salt stress in soybean', *J. Agron Crop Sci.*, 196(6), 467–73.

Nie, W. X., Xu, L. and Yu, B. J. (2015), 'A putative soybean *GmsSOS1* confers enhanced salt tolerance to transgenic Arabidopsis sos1–1 mutant', *Protoplasma*, 252(1), 127–34.

Nisa, Z. U., Chen, C., Yu, Y., Chen, C., Mallano, A. I., Xiang-bo, D. and Yan-ming, Z. (2016), 'Constitutive overexpression of myo-inositol-1-phosphate synthase gene (*GsMIPS2*) from *Glycine soja* confers enhanced salt tolerance at various growth stages in Arabidopsis', *J. Northeast Agric. Univ.*, 23(2), 28–44.

Nunes, A. C., Vianna, G. R., Cuneo, F., Amaya-Farfán, J., de Capdeville, G., Rech, E. L. and Aragão, F. J. (2006), 'RNAi-mediated silencing of the myo-inositol-1-phosphate synthase gene (*GmMIPS1*) in transgenic soybean inhibited seed development and reduced phytate content', *Planta*, 224(1), 125–32.

Pathan, M. S., Lee, J.-D., Shannon, J. G. and Nguyen, H. T. (2007), Recent advances in breeding for drought and salt stress tolerance in soybean. In *Advances in Molecular Breeding Toward Drought and Salt Tolerant Crops*, Jenks, M. A., Hasegawa, P. M. and Jain, S. M. (Eds), pp 739–73. Springer, New York city, United States.

Patil, G., Do, T., Vuong, T. D., Valliyodan, B., Lee, J.-D., Chaudhary, J., Shannon, J. G. and Nguyen, H. T. (2016), 'Genomic-assisted haplotype analysis and the development of high-throughput SNP markers for salinity tolerance in soybean', *Sci. Rep.*, 6, 19199.

Phang, T. H., Shao, G. and Lam, H. M. (2008), 'Salt tolerance in soybean', *J. Integr. Plant Biol.*, 50(10), 1196–212.

Pimentel, D., Berger, B., Filiberto, D., Newton, M., Wolfe, B., Karabinakis, E., Clark, S., Poon, E., Abbett, E. and Nandagopal, S. (2004), 'Water resources: agricultural and environmental issues', *BioScience*, 54(10), 909–18.

Qi, X., Li, M. W., Xie, M., Liu, X., Ni, M., Shao, G., Song, C., Yim, A. K. Y., Tao, Y., Wong, F. L., Isobe, S., Wong, C. F., Wong, K. S., Xu, C., Li, C., Wang, Y., Guan, R., Sun, F., Fan, G., Xiao, Z., Zhou, F., Phang, T. H., Liu, X., Tong, S. W., Chan, T. F., Yiu, S. M., Tabata, S., Wang, J., Xu, X. and Lam, H. M. (2014), 'Identification of a novel salt tolerance gene in wild soybean by whole-genome sequencing', *Nat. Commun.*, 5, 4340.

Rao, S. S., El-Habbak, M. H., Havens, W. M., Singh, A., Zheng, D., Vaughn, L. and Ghabrial, S. A. (2014), 'Overexpression of *GmCaM4* in soybean enhances resistance to pathogens and tolerance to salt stress', *Mol. Plant Pathol.*, 15(2), 145–60.

Rathert, G., Doering, H. W. and Witt, J. (1981), 'Influence of extreme K-Na ratios and high substrate salinity on plant-metabolism of crops differing in salt tolerance', *J. Plant Nutr.*, 4(2), 131–41.

Ren, S., Lyle, C., Jiang, G. L. and Penumala, A. (2016), 'Soybean salt tolerance 1 (*GmST1*) reduces ROS production, enhances ABA sensitivity, and abiotic stress tolerance in *Arabidopsis thaliana*', *Front. Plant. Sci.*, 7, 445, doi:10.3389/fpls.2016.00445.

Rodrigues, S. M., Andrade, M. O., Gomes, A. P. S., DaMatta, F. M., Baracat-Pereira, M. C. and Fontes, E. P. (2006), 'Arabidopsis and tobacco plants ectopically expressing the soybean antiquitin-like ALDH7 gene display enhanced tolerance to drought, salinity, and oxidative stress', *J. Exp. Bot.*, 57(9), 1909–18.

Sammour, R. H. (2011), Genetic diversity and allele mining in soybean germplasm. In: *Soybean - Genetics and Novel Techniques for Yield Enhancement*, Krezhova, D. (Ed.), InTech, Croatia –European.

San, Z. and Zhou, K. F. (2003), 'Discussion on the problem of salt gland of *Glycine soja*', *Acta Bot. Sin.*, 45(5), 574–80.

Schmitz, R. J., He, Y., Valdés-López, O., Khan, S. M., Joshi, T., Urich, M. A., Nery, J. R., Diers, B., Xu, D. and Stacey, G. (2013), 'Epigenome-wide inheritance of cytosine methylation variants in a recombinant inbred population', *Genome Res.*, 23(10), 1663–74.

Schmutz, J., Cannon, S. B., Schlueter, J., Ma, J., Mitros, T., Nelson, W., Hyten, D. L., Song, Q., Thelen, J. J. and Cheng, J. (2010), 'Genome sequence of the palaeopolyploid soybean', *Nature*, 463(7278), 178–83.

Shao, G., Wan, C. and Li, S. (1994), 'Preliminary study on the physiology of soybean tolerance to salt stress at germinating stage', *Crops*, 6, 25–7.

Shao, H., Wan, C., Chang, R. and Chen, Y. (1993), 'Preliminary study on the damage of plasma membrane caused by salt stress', *Crops*, 1, 39–40.

Singleton, P. W. and Bohlool, B. B. (1984), 'Effect of salinity on nodule formation by soybean', *Plant Physiol.*, 74(1), 72–6.

Sobhanian, H., Razavizadeh, R., Nanjo, Y., Ehsanpour, A. A., Jazii, F. R., Motamed, N. and Komatsu, S. (2010), 'Proteome analysis of soybean leaves, hypocotyls and roots under salt stress', *Proteome Sci.*, 8(1),19.

Song, H., Wang, P., Lin, J.-Y., Zhao, C., Bi, Y. and Wang, X. (2016), 'Genome-wide identification and characterization of WRKY gene family in peanut', *Front. Plant. Sci.*, 7, 534. doi:10.3389/fpls.2016.00534.

Song, Q. X., Lu, X., Li, Q. T., Chen, H., Hu, X. Y., Ma, B., Zhang, W. K., Chen, S. Y. and Zhang, J. S. (2013), 'Genome-wide analysis of DNA methylation in soybean', *Mol. Plant*, 6(6), 1961–74.

Song, Y., Ji, D., Li, S., Wang, P., Li, Q. and Xiang, F. (2012), 'The dynamic changes of DNA methylation and histone modifications of salt responsive transcription factor genes in soybean', *PLoS One*, 7(7), e41274.

Subramanian, S., Graham, M. Y., Yu, O. and Graham, T. L. (2005), 'RNA interference of soybean isoflavone synthase genes leads to silencing in tissues distal to the transformation site and to enhanced susceptibility to *Phytophthora sojae*', *Plant Physiol.*, 137(4), 1345–53.

Sudaric, A., Vrataric, M., Rajcan, I., Duvnjak, T. and Volenik, M. (2008), 'Application of molecular markers in parental selection in soybean', *Acta Agron. Hung.*, 56(4), 393–8.

Sun, M., Jia, B., Cui, N., Wen, Y., Duanmu, H., Yu, Q. and Zhu, Y. (2016), 'Functional characterization of a *Glycine soja* Ca2+ ATPase in salt–alkaline stress responses', *Plant Mol. Biol.*, 90(4–5), 419–34.

Sun, X. L., Yu, Q. Y., Tang, L. L., Ji, W., Bai, X., Cai, H. and Zhu, Y. M. (2013a), '*GsSRK*, a G-type lectin S-receptor-like serine/threonine protein kinase, is a positive regulator of plant tolerance to salt stress', *J. Plant Physiol.*, 170(5), 505–15.

Sun, X., Sun, M., Luo, X., Ding, X., Ji, W., Cai, H. and Zhu, Y. (2013b), 'A *Glycine soja* ABA-responsive receptor-like cytoplasmic kinase, *GsRLCK*, positively controls plant tolerance to salt and drought stresses', *Planta*, 237(6), 1527–45.

Tian, A. G., Wang, J., Cui, P., Han, Y. J., Xu, H., Cong, L. J., Huang, X. G., Wang, X. L., Jiao, Y. Z. and Wang, B. J. (2004), 'Characterization of soybean genomic features by analysis of its expressed sequence tags', *Theor. Appl. Genet.*, 108(5), 903–13.

Tunnacliffe, A. and Wise, M. J. (2007), 'The continuing conundrum of the LEA proteins', *Naturwissenschaften*, 94(10), 791–812.

Umezawa, T., Mizuno, K. and Fujimura, T. (2002), 'Discrimination of genes expressed in response to the ionic or osmotic effect of salt stress in soybean with cDNA-AFLP', *Plant Cell Environ.*, 25(12), 1617–25.

Valencia, R., Chen, P., Ishibashi, T. and Conatser, M. (2008), 'A rapid and effective method for screening salt tolerance in soybean', *Crop Sci.*, 48(5), 1773–9.

Van Holle, S., Smagghe, G. and Van Damme, E. J. (2016), 'Overexpression of nictaba-like lectin genes from *Glycine max* confers tolerance toward *Pseudomonas syringae* infection, aphid infestation and salt stress in transgenic Arabidopsis plants', *Front. Plant Sci.*, 7, 1590, doi:10.3389/fpls.2016.01590.

Wang, F., Chen, H. W., Li, Q. T., Wei, W., Li, W., Zhang, W. K. and Man, W. Q. (2015), '*GmWRKY27* interacts with *GmMYB174* to reduce expression of *GmNAC29* for stress tolerance in soybean plants', *Plant J.*, 83(2), 224–36.

Wang, H. T., Guo, N., Zhao, J. M., Karthikeyan, A., Xue, D., Xue, C. C. and Xing, H. (2014), 'Cloning and expression analysis of a stress-induced *GmIMT1* gene in soybean (*Glycine max*)', *Genet. Mol. Res.*, 13(1), 806–18.

Wang, Y., Suo, H., Zhuang, C., Ma, H. and Yan, X. (2011), 'Overexpression of the soybean *GmWNK1* altered the sensitivity to salt and osmotic stress in Arabidopsis', *J. Plant Physiol.*, 168(18), 2260–7.

Wei, P., Wang, L., Liu, A., Yu, B. and Lam, H. M. (2016), '*GmCLC1* confers enhanced salt tolerance through regulating chloride accumulation in soybean', *Front. Plant Sci.*, 7, 1082, doi:10.3389/fpls.2016.01082.

Wei, W., Huang, J., Hao, Y. J., Zou, H. F., Wang, H. W., Zhao, J. Y., Liu, X. Y., Zhang, W. K., Ma, B. and Zhang, J. S. (2009), 'Soybean *GmPHD*-type transcription regulators improve stress tolerance in transgenic Arabidopsis plants', *PLoS One*, 4(9), e7209.

Weisany, W., Sohrabi, Y., Heidari, G., Siosemardeh, A. and Ghassemi-Golezani, K. (2012), 'Changes in antioxidant enzymes activity and plant performance by salinity stress and zinc application in soybean ('*Glycine max*'L.)', *Plant Omi. J.*, 5(2), 60–7.

Weisany, W., Sohrabi, Y., Heidari, G., Siosemardeh, A. and Ghassemi-Golezani, K. (2011), 'Physiological responses of soybean ('*Glycine max*'L.) to zinc application under salinity stress', *Aust. J. Crop Sci.*, 5(11), 1441–7.

Wieneke, J. and Lauchli, A. (1980), 'Effects of salt stress on distribution of Na^+ and some other cations in 2 soybean varieties differing in salt tolerance', *Z Pflanz Bodenkunde*, 143(1), 55–67.

Witcombe, J. R., Hollington, P. A., Howarth, C. J., Reader, S. and Steele, K. A. (2008), 'Breeding for abiotic stresses for sustainable agriculture', *Philos. T R Soc. B*, 363(1492), 703–16.

Wu, T., Pi, E. X., Tsai, S. N., Lam, H. M., Sun, S. M., Kwan, Y. W. and Ngai, S. M. (2011), '*GmPHD5* acts as an important regulator for crosstalk between histone H3K4 di-methylation and H3K14 acetylation in response to salinity stress in soybean', *BMC Plant Biol.*, 11(1), 178.

Wu, W., Zhang, Q., Zhu, Y. M., Lam, H. M., Cai, Z. W. and Guo, D. J. (2008), 'Comparative metabolic profiling reveals secondary metabolites correlated with soybean salt tolerance', *J. Agr. Food Chem.*, 56(23), 11132–8.

Xie, Z. M., Zou, H. F., Lei, G., Wei, W., Zhou, Q. Y., Niu, C. F. and Zhang, J. S. (2009), 'Soybean trihelix transcription factors *GmGT-2A* and *GmGT-2B* improve plant tolerance to abiotic stresses in transgenic Arabidopsis', *PLoS One*, 4(9), e6898.

Xu, X. Y., Fan, R., Zheng, R., Li, C. M. and Yu, D. Y. (2011), 'Proteomic analysis of seed germination under salt stress in soybeans', *J. Zhejiang Univ. Sci. B*, 12(7), 507–17.

Xu, Y. B. and Crouch, J. H. (2008), 'Marker-assisted selection in plant breeding, From publications to practice', *Crop Sci.*, 48(2), 391–407.

Xu, Z., Ali, Z., Xu, L., He, X., Huang, Y., Yi, J., Shao, H., Ma, H. and Zhang, D. (2016), 'The nuclear protein *GmbZIP110* has transcription activation activity and plays important roles in the response to salinity stress in soybean', *Sci. Rep.*, 6, 20366.

Xu, D. (2012), 'Genetic studies on saline and sodic tolerances in soybean', *Breed. Sci.*, 61(5), 559–65.

Xue, Z., Zhao, S., Gao, H. and Sun, S. (2014), 'The salt resistance of wild soybean (*Glycine soja* Sieb. et Zucc. ZYD 03262) under NaCl stress is mainly determined by Na^+ distribution in the plant', *Acta Physiol. Plant*, 36(1), 61–70.

Yan, J., Wang, B., Jiang, Y., Cheng, L. and Wu, T. (2013), '*GmFNSII*-controlled soybean flavone metabolism responds to abiotic stresses and regulates plant salt tolerance', *Plant Cell Physiol.*, 55(1), 74–86.

Yan, Q., Cui, X., Lin, S., Gan, S., Xing, H. and Dou, D. (2016), '*GmCYP82A3*, a soybean cytochrome P450 family gene involved in the jasmonic acid and ethylene signaling pathway, enhances plant resistance to biotic and abiotic stresses', *PloS One*, 11(9), e0162253.

Yan, X., Lui, X., Li, J., Zhao, H., Li, Q., Wang, Y., Yuan, C., Zhang, L. and Dong, Y. (2014), 'Genetic and epigenetic diversity of wild and cultivated soybean in local populations in Northern Huang Huai region of China', *Plant Omics*, 7(6), 415.

Yang, L., Ji, W., Zhu, Y., Gao, P., Li, Y., Cai, H. and Guo, D. (2010), '*GsCBRLK*, a calcium/calmodulin-binding receptor-like kinase, is a positive regulator of plant tolerance to salt and ABA stress', *J. Exp. Bot.*, 61(9), 2519–33.

Yin, Y., Yang, R., Han, Y. and Gu, Z. (2015), 'Comparative proteomic and physiological analyses reveal the protective effect of exogenous calcium on the germinating soybean response to salt stress', *J. Proteomics*, 113, 110–26.

Yu, B., Lam, H., Shao, G. and Liu, Y. (2004), 'Effects of salinity on activities of H^+-ATPase, H^+-PPase and membrane lipid composition in plasma membrane and tonoplast vesicles isolated from soybean (*Glycine max* L.) seedlings', *J. Environ. Sci. (China)*, 17(2), 259–62.

Yun, S., Gyenis, L., Hayes, P., Matus, I., Smith, K., Steffenson, B. and Muehlbauer, G. (2005), 'Quantitative trait loci for multiple disease resistance in wild barley', *Crop Sci.*, 45(6), 2563–72.

Zhai, Y., Wang, Y., Li, Y., Lei, T., Yan, F., Su, L., Li, X., Zhao, Y., Sun, X., Li, J. and Wang, Q. (2013), 'Isolation and molecular characterization of *GmERF7*, a soybean ethylene-response factor that increases salt stress tolerance in tobacco', *Gene*, 513(1), 174–83.

Zhang, C. and Ghabrial, S. A. (2006), 'Development of bean pod mottle virus-based vectors for stable protein expression and sequence-specific virus-induced gene silencing in soybean', *Virology*, 344(2), 401–11.

Zhang, G., Chen, M., Chen, X., Xu, Z., Guan, S., Li, L. C., Li, A., Guo, J., Mao, L. and Ma, Y. (2008), 'Phylogeny, gene structures, and expression patterns of the ERF gene family in soybean (*Glycine max* L.)', *J. Exp. Bot.*, 59(15), 4095–107.

Zhang, G., Chen, M., Chen, X., Xu, Z., Li, L., Guo, J. and Ma, Y. (2010), 'Isolation and characterization of a novel EAR-motif-containing gene *GmERF4* from soybean (*Glycine max* L.)', *Mol. Biol. Rep.*, 37(2), 809–18.

Zhang, G., Chen, M., Li, L., Xu, Z., Chen, X., Guo, J. and Ma, Y. (2009), 'Overexpression of the soybean *GmERF3* gene, an AP2/ERF type transcription factor for increased tolerances to salt, drought, and diseases in transgenic tobacco', *J. Exp. Bot.*, 60(13), 3781–96.

Zhao, M., Gu, Y., He, L., Chen, Q. and He, C. (2015), 'Sequence and expression variations suggest an adaptive role for the DA1-like gene family in the evolution of soybeans', *BMC Plant Biol.*, 15(1), 120.

Zhong, X. F., Wang, Y. M., Liu, X. D., Gong, L., Ma, Y., Qi, B., Dong, Y. S. and Liu, B. (2009), 'DNA methylation polymorphism in annual wild soybean (*Glycine soja* Sieb. et Zucc.) and cultivated soybean (*G. max* L. Merr.)', *Can. J. Plant Sci.*, 89(5), 851–63.
Zhou, G. A., Chang, R. Z. and Qiu, L. J. (2010), 'Overexpression of soybean ubiquitin-conjugating enzyme gene *GmUBC2* confers enhanced drought and salt tolerance through modulating abiotic stress-responsive gene expression in Arabidopsis', *Plant Mol. Biol.*, 72(4–5), 357–67.
Zhou, L., Wang, C., Liu, R., Han, Q., Vandeleur, R. K., Du, J., Tyerman, S. and Shou, H. (2014), 'Constitutive overexpression of soybean plasma membrane intrinsic protein *GmPIP1*; 6 confers salt tolerance', *BMC Plant Biol.*, 14(1), 181.
Zhou, Q. Y., Tian, A. G., Zou, H. F., Xie, Z. M., Lei, G., Huang, J., Wang, C. M., Wang, H. W., Zhang, J. S. and Chen, S. Y. (2008), 'Soybean WRKY-type transcription factor genes, *GmWRKY13*, *GmWRKY21*, and *GmWRKY54*, confer differential tolerance to abiotic stresses in transgenic Arabidopsis plants', *Plant Biotechnol. J.*, 6(5), 486–503.
Zhou, Z., Jiang, Y., Wang, Z., Gou, Z., Lyu, J., Li, W., Yu, Y., Shu, L., Zhao, Y., Ma, Y., Fang, C., Shen, Y., Liu, T., Li, C., Li, Q., Wu, M., Wang, M., Wu, Y., Dong, Y., Wan, W., Wang, X., Ding, Z., Gao, Y., Xiang, H., Zhu, B., Lee, S. H., Wang, W. and Tian, Z. (2015), 'Resequencing 302 wild and cultivated accessions identifies genes related to domestication and improvement in soybean', *Nat. Biotechnol.*, 33(4), 408–14.
Zhu, D., Cai, H., Luo, X., Bai, X., Deyholos, M. K., Chen, Q. and Zhu, Y. (2012), 'Over-expression of a novel JAZ family gene from *Glycine soja*, increases salt and alkali stress tolerance', *Biochem. Biophys. Res. Commun.*, 426(2), 273–9.
Zhu, T., Schupp, J. M., Oliphant A. and Keim, P. (1994), 'Hypomethylated sequences: characterization of the duplicate soybean genome', *Mol. Gen. Genet.*, 244(6), 638–45.
Zhu, J. K. (2008), 'Epigenome sequencing comes of age', *Cell*, 133(3), 395–7.

Chapter 9

Advances in flood-tolerant varieties of soybean

C. Wu, L. Mozzoni and W. Hummer, University of Arkansas, USA; P. Chen, G. Shannon, H. Ye and H. T. Nguyen, University of Missouri, USA; G. Kaur and J. Orlowski, Mississippi State University, USA; T. Carter, USDA-ARS, USA; and B. Buckley, Louisiana State University, USA

1 Introduction

Flooding is a major abiotic stress on crop and forage production worldwide, affecting approximately 10% of global agricultural acres (Patel et al., 2014) and causing anything from negligible harm to complete crop losses. The term 'flooding' can be used to describe the inundation by water of all or part of a plant (Bailey-Serres et al., 2012), and is categorized into waterlogging, partial submergence and complete submergence, based on the water level affecting the crop. Waterlogging is the complete saturation of the soil pores with water, exposing the root system of the plant to anaerobic conditions. Partial submergence occurs when the soil pores are saturated and a layer of water forms at the surface of the soil, inundating both the root zone and a part of the above-ground portion of the plant. Complete submergence occurs when the entire plant, including roots and shoots, is covered by water. In current agricultural production systems, partial submergence is the most common scenario worldwide (Ahmed et al., 2013; Striker, 2012).

Soybean [*Glycine max* (L.) Merr.] is one of the most important row crops, with high levels of protein, oil and carbohydrates in seed (Thomas et al., 2003; Bellaloui et al., 2013; Waqas et al., 2014), making it an excellent grain for the food and feed industry. However, soybean production can be severely impacted under adverse environmental conditions, including flooding stress. In the United States, flooding ranks second to drought as the most significant abiotic stress. Flooding can cause major economic damage, as was the case in 2011 when an outbreak in the Midwest region of the United States caused over

http://dx.doi.org/10.19103/AS.2017.0034.09

$1.6 billion in losses of corn and soybean crops (Bailey-Serres et al., 2012; Voesenek and Bailey-Serres, 2013). In the Mississippi Delta region (Missouri, Arkansas, Mississippi and Louisiana), annual soybean seed yields are typically reduced by up to 25% due to natural flooding or to flood-irrigation techniques (VanToai et al., 2010). Moreover, National Aeronautics and Space Administration (NASA) weather simulation models predict a 30% increase in heavy precipitation events by the year 2030, suggesting that flooding will pose an even more serious challenge to food production worldwide (Rosenzweig et al., 2002).

Most soybean varieties are vulnerable to flooding stress (Hou and Thseng, 1991; VanToai et al., 2010). Although flooding is at times inevitable, there are two main mitigation strategies to reduce the risk of damage caused by flooding. The first strategy involves agronomic practices that can help maintain aeration at the root zone, and the second one involves breeding strategies to select varieties that are high-yielding and less sensitive to flooding. The main objective of this chapter is to summarize recent developments in soybean agronomic-, genetic- and breeding research, with the aim of mitigating flooding stress in soybean production.

2 Flooding stress and damage

Flooding can be caused by heavy, prolonged periods of rainfall or by overflow of rivers, streams or creeks. Excessive irrigation, or rainfall after an irrigation event, can also cause flooding. The physical properties of soil, such as texture (for instance, clay-rich soils) or subsurface compacted soil (hardpan), can contribute to flooding conditions, as they affect the percolation and lateral movement of water. Zero-grade fields and poorly drained soils are likely to become flooded with excessive rain or irrigation. Under flooded conditions, soil pores become filled with water, reducing the diffusion of oxygen in the soil and resulting in hypoxic (oxygen-deficient) and later anoxic (oxygen-absent) conditions. The diffusion of oxygen is reduced by 10^4 times in water as compared to its diffusion in air-filled soil pores. Oxygen is an electron acceptor in the microbial decomposition of organic carbon compounds in soil, but a reduction in the availability and diffusion of oxygen due to flooding can cause soil microbes to shift to other electron acceptors such as nitrate (NO^{-3}), manganese (Mn^{+4}), iron (Fe^{+3}) compounds, SO_2 and CO_2 (Patrick and Jugsujinda, 1992). In flooded soils, therefore, more phytotoxic and reduced forms of mineral ions are accumulated.

The anaerobic condition of soil can be assessed by measuring soil redox potential, which varies according to the type and amount of organic matter, the nature and content of electron acceptors, soil temperature, soil pH and aeration (Ponnamperuma, 1972), demonstrating values ranging from +900 to −300 mV. Aerated soils have a redox potential typically greater than +350 mV, and flooded soils less than +350 mV (Husson, 2013). This variation in redox potential due to flooding occurs rather rapidly, as demonstrated by Balakhnina et al. (2010), reporting a change in soil redox potential from +543 to +70 mV within a few hours of a flooding event, and a corresponding restoration of redox potential within a few days after drainage. In addition, flooding damage increases with rising air and soil temperature. Trought and Drew (1982) reported increased plant damage as soil temperature rose in flooded soils, caused by an increased oxygen requirement for respiration, and a decreasing soil redox potential. In addition, Maekawa et al. (2011) reported that soil redox potential and temperature affect the recovery of nitrogen-fixating nodules when soybeans are waterlogged or flooded.

Under flooded conditions, soil pH approaches neutrality, thus increasing the availability of nutrients such as ammonium, P, K, Ca, Mg, Fe, Mn and Si (Sahrawat, 2008). However, increased Fe and Mn availability in soil solution can result in phytotoxicity (Scott et al., 1990), and prolonged flooding also reduces sulphate to hydrogen sulphide, which is phytotoxic (Ponnamperuma, 1972). In addition, flooded soil conditions can cause N losses via leaching, runoff, erosion and denitrification. These losses may result in N deficiency in crops post-flooding, negatively affecting their recovery from flooding damage (Bacanamwo and Purcell, 1999a).

Soybean response to flooding varies by soil types. Scott et al. (1989) found that soybean yield was reduced by 124 and 157 kg ha^{-1} with each day of flooding at V4 and R2 growth stages, respectively, in a fine-textured clay soil. In contrast, yield reduction rates were 53 and 101 kg ha^{-1} day^{-1} of flooding at V4 and R2 growth stages, respectively, in a coarse-textured silt loam soil (Scott et al., 1989). It has been reported that flooding in clay soils causes a greater reduction in leaf N (Scott et al., 1990) and N and K (Rhine et al., 2010) as compared to flooding in a silt loam soil. It can be hypothesized, then, that these variations in yield and elemental composition might be due to the different abilities of these soils to dry after flooding.

Previous studies have demonstrated that flooding can cause many chemical and physical changes in soil, and that soil properties can influence the degree of damage on soybeans upon flooding. This suggests that there is potential to explore and exploit environmental variation and genotype-by-environment (G × E) interactions when conducting breeding for soybean flooding tolerance. The interactions have not yet been extensively explored, and further research on environmental classification and G × E interactions is necessary to advance understanding of flooding tolerance genetics and breeding.

2.1 Flooding damage to soybean

Flooding deprives plants of oxygen, carbon dioxide and light for photosynthesis (Jackson et al., 2009). Respiration in the plant roots is the main metabolic process affected by oxygen deficiency under saturated soil conditions. Lack of oxygen for respiration results in reduced root growth and also favours the formation of phytotoxins, which further restrict root growth and can lead to root decay (Armstrong, 1979; Blom and Voesenek, 1996; Vartapetian and Jackson, 1997; Voesenek et al., 2004; Bailey-Serres and Voesenek, 2008; Wegner, 2010). Flooding also reduces water and nutrient uptake by plant roots (Kramer, 1951) and affects nitrogen fixation (Maekawa et al., 2011) as well as overall N elemental accumulation (Board, 2008). Moreover, Bacanamwo and Purcell (1999b) observed a 26% decrease in leaf greenness after 21 days of flooding, and Wu et al. (2017a) reported a 35.6% and 30.8% reduction in foliar chlorophyll content from flooding at V5 and R1 growth stages, respectively (Fig. 1).

Flooding severely damages soybean seeds and affects germination (Hou and Thseng, 1991). Wuebker et al. (2001) observed soybean seed damage after just one hour of flooding, and Wu et al. (2017b) reported a steep linear decrease over the first 48 h of flooding, where seed germination rate dropped from 80% to 10%. This period of early crop establishment appears to be the most sensitive to flooding stress.

Flooding also reduces the shoot and root dry weight, canopy height, leaf area, nodule number, nodule dry weight and relative ureide-N content (Youn et al., 2008) of established soybean plants. Soybean canopy height decreased by 29–36% when flooded for 14 days at V4 growth stage, as compared to the non-flooded control, but flooding at R2 growth stage reduced canopy height by 38% compared to flooding at V4 growth stage (Scott

Figure 1 Variation of soybean leaf colour under flooding stress.

et al., 1989). Scott et al. (1989) also reported that flooding at R2 stage reduced soybean growth rate by 25–35%, as compared to flooding at V4 stage. These results suggest that soybean flooded at vegetative stage has more time to recover through the remainder of the vegetative period, as compared to flooding during reproductive stages very close to the critical seed-filling period.

Multiple studies have reported reductions in soybean yield due to flooding. Oosterhuis et al. (1990) reported that soybean seed yield was reduced by 17–43% and 50–56% when flooding was imposed at V4 and R2 growth stages, respectively. A flood-susceptible cultivar, 'Centennial', was found to have a 93% yield loss when flooded at R3 stage for seven days (Linkemer et al., 1998). Sullivan et al. (2001) revealed that three days of flooding at the V2–V3 growth stage reduced seed yield by 20%, while six days of flooding decreased yield by 93%. Beutler et al. (2014) observed soybean grain yield reduce by 17% and 29% at R2 and R5 growth stages after 16 days of flooding, while increasing flooding duration to 32 days caused yield reduction of 41% at R2 stage and 36% at R5 stage. Shannon et al. (2005) reported a 40% yield reduction in a flood-tolerant soybean group, compared to an 80% reduction in a flood-sensitive group. Similarly, Rhine et al. (2010) reported a 39% yield loss in flood-tolerant cultivars compared to a 77% yield loss in sensitive soybean cultivars during severe flooding conditions. In addition, Wu et al. (2017a) reported a 50% yield reduction in a flood-tolerant soybean group versus a 70% yield reduction in a flood-sensitive group after a three-day flooding. Studying yield components, Linkemer et al. (1998) found that yield losses due to flooding at V2 stage resulted from plants having fewer branches, yield reduction at R1 stage was due to reduced pod numbers, and harvest losses at R5 were due to smaller seed size.

3 Responses of soybean to flooding

3.1 Morphological responses

Studies show that plants develop a suite of anatomical and morphological mechanisms to adapt to a submerged environment (Armstrong, 1979; Kozlowski, 1984; Vartapetian and Jackson, 1997; Seago et al., 2005; Striker et al., 2005; Colmer and Voesenek, 2009). Methods of plant adaptation and evolution under flooding stress include the generation of aerenchyma in tissues, adventitious root formation, fast petiole extension, stem hypertrophy, hyponastic growth, stomata closing, reduction of transpiration and inhibition of photosynthesis (Justin and Armstrong, 1987; Laan et al., 1990; Naidoo and Mundree, 1993; Armstrong et al., 1994; Grimoldi et al., 1999; Cox et al., 2003; Striker et al., 2005).

Bacanamwo and Purcell (1999b) indicated that adaptation and acclimation of soybean to flooding stress involved preferential allocation of photosynthates, development of porous adventitious roots and formation of aerenchyma. Flood-induced aerenchyma and the development of adventitious roots in soybean facilitate the transportation of atmospheric O_2 to the submerged roots, and are critical for soybean flooding tolerance (Valliyodan et al. 2014; Kim et al., 2015). Shimamura et al. (2003) also reported that soybeans can develop secondary aerenchyma, a white-coloured, spongy tissue found in the stem, hypocotyl, tap root and adventitious roots under flooded conditions. After two weeks of flooding, plants with well-developed aerenchyma had a hypocotyl porosity greater than 30%, versus 10% porosity in plants with no aerenchyma. The number and size of aerenchyma were higher, and the development of aerenchyma cells was different in a flood-tolerant line (Valliyodan et al. 2014; Kim et al., 2015).

Development of adventitious roots as a primary morphological response to flooding in soybean was reported by Henshaw et al. (2007). A four-week flooding event resulted in adventitious roots, displaying greater dry weights, lengths, diameters, volumes and surface areas at harvest compared with those from two-week flooding treatments (Henshaw et al., 2007). Kim et al. (2015) also reported that adventitious roots were better developed in the flood-tolerant soybean line than in the flood-susceptible line. Miura et al. (2012) reported that the root length density of wild soybean increased during a waterlogging treatment at the primary leaf stage, resulting in a greater ability to recover from waterlogging damage than cultivated soybean.

3.2 Physiological responses

The physiological and biochemical responses of plants to flooding are triggered by several molecules and compounds, including oxygen, nitric oxide, reactive oxygen species, haemoglobin and plant hormones (Dat et al., 2004). Oxygen is the terminal acceptor of the mitochondrial electron transport chain in eukaryotes, and a complete lack of oxygen severely affects gene expression, metabolism and plant growth and development (Bailey-Serres and Chang, 2005). The absence of oxygen blocks mitochondria from producing adenosine triphosphate (ATP), resulting in the plant going through anaerobic glycolysis and fermentation – much less efficient ways of producing energy compared to mitochondrial respiration. Examples of plant-response mechanisms that manage post-hypoxia/anoxia oxidative stress include an increased availability of soluble sugars, greater activity of glycolytic pathways and fermentation enzymes, and the involvement of antioxidant defences (Sairam et al., 2008). Plant hormones such as ethylene, abscisic acid, cytokinin, gibberellic acid, auxin and salicylic acid are involved in signalling cascades of flooding stress, in which the ethylene signalling pathway plays a crucial regulating role (Dat et al., 2004; Kreuzwieser and Rennenberg, 2014). In soybean, the rate of ethylene production increases in line with an increased duration of flooding (Kadempir et al., 2014). Yin et al. (2014a) reported that ethylene signalling regulated soybean tolerance in the initial stages of flooding through protein phosphorylation in the root tips. Kim et al. (2015) revealed that soybean's production of ethylene under flooding was significantly higher in a flood-tolerant line than in flood-sensitive and non-flooded plants. In addition, soybean endogenous hormones, such as abscisic acid, gibberellic acid and salicylic acid, also showed significant variations between flood-tolerant lines and flood-susceptible lines (Kim et al., 2015).

Nitrogen accumulation and N_2 fixation are considerably reduced under flooded conditions (Buttery, 1987; Youn et al., 2008), as observed by a significant decrease in the level of leghemoglobin on nodules under flooding stress (Sanchez et al., 2011), and the negligible levels of shoot ureide after a 21-day flooding on non-N fertilized soybeans (Thomas and Sodek, 2005). Soybean nitrogen fixation decreased within seven days of a flooding event, and both the onset and rate of decrease of nitrogen fixation were more pronounced than those of biomass accumulation for the same genotypes (Bacanamwo and Purcell, 1999a).

3.3 Molecular responses

Current molecular biology and biotechnology methods have improved abilities of understanding of soybean responses to flooding at the molecular level. Komatsu et al. (2009) reported that soybean plants increased mRNA and protein levels of glycolysis-related enzymes and activated glycolysis and fermentation cascades under flooding stress. Hashiguchi et al. (2009) showed that expression levels of soybean genes associated with alcohol fermentation, ethylene biosynthesis, pathogen defence and cell loosening were significantly up-regulated after 12 h of flooding. Flooding also inhibited the tricarboxylic acid cycle by down-regulating the expression of relevant genes (Komatsu et al., 2011). Tougou et al. (2012) confirmed that both alcohol dehydrogenase (Adh) mRNA and protein increased temporarily in soybean root tips under flooding stress. They also indicated that the *GmAdh2* gene induced variations in glycolysis and alcohol fermentation, which improved tolerance to flooding stress (Tougou et al., 2012). Komatsu et al. (2013) revealed that flooding stress changed the expression of 921 genes, the abundance of 75 proteins and the accumulation of 92 metabolites in soybean seedlings. In addition, Nanjo et al. (2011a) and Tamang et al. (2014) showed that more than 5000 soybean genes were differentially expressed under flooding stress, with an observed up-regulation of the genes related to photosynthesis, glycolysis, fermentation, cell wall loosening, Ser-Gly-Cys group amino acid synthesis, regulation of transcription, ubiquitin-mediated protein degradation, disease resistance, protein kinase signalling genes and cell death. Concomitantly, the group of genes that were observed to be down-regulated included genes related to cell wall synthesis, secondary metabolism, protein synthesis and folding, metabolite transport, cell organization, chromatin structure synthesis and degradation of aspartate family amino acid (Nanjo et al., 2011a,b; Tamang et al., 2014). Yin et al. (2014a) identified 114 phosphoproteins in the root tip of soybean, 34 of which were significantly regulated after 3 h of flooding stress, and where the mRNA expression levels of sucrose-phosphate-synthase-1F and eukaryotic-translation-initiation-factor-4G, which has homology to programmed-cell-death-4 protein, were down-regulated. Moreover, Yin et al. (2014b) and Yin and Komatsu (2015) indicated that soybean calreticulin expression was up-regulated, and expressions of citrate lyase and heat-shock-protein-70 were down-regulated in the initial phase of flooding. In addition, calcium-related signal transduction played a regulating role in soybean's early responses to flooding stress. Oh and Komatsu (2015) reported that S-adenosylmethionine-synthetase is involved in the regulation of soybean response to flooding stress.

More and more molecular mechanisms of soybean responses to flooding are continually being uncovered. These and future studies on epigenetic modifications are critical to gain a new level of understanding that could help conduct targeted genome modifications to enhance soybean response to flooding beyond those naturally occurring in the soybean genetic pool.

4 Agronomic management strategies

Agronomic management is an important strategy that could increase soybean survival rates and reduce yield loss under flooding stress. Agronomic tools are designed to increase aeration of the root zone, by either increasing drainage or raising the planting beds, or alternatively focusing on nutrient management to help speedy recovery of the crop after the flooding stress has receded. These approaches to agronomic management have shown beneficial effects and could be used in soybean cultivation, particularly in areas with flat farm land with heavy soil and abundant precipitation, as part of a risk-aversion strategy.

4.1 Use of drainage systems

Proper drainage is necessary for controlling waterlogging or flooding. Draining excess water from the soil improves root growth, crop emergence, plant stands, and consequently, crop yields (Blevins et al., 1996; Nelson et al., 2011, 2012). Drainage systems may include open field drains for surface drainage or tile drainage systems for subsurface water drainage. Use of a tile drainage system in poorly drained claypan soils in Missouri allowed soybean planting to occur 17 days earlier, and increased yields by 9–22% compared to the non-drained control (Nelson et al., 2011). However, the response to drainage varied according to soybean cultivar, and requires further research.

4.2 Raised bed cultivation

Permanent raised bed and furrow systems can improve drainage by maintaining a favourable soil moisture content that can increase crop yields (Velmurugan et al., 2016). Raised beds enhance soil's physical properties, including higher macroporosity and infiltration, lower bulk density and higher aggregate stability (Roth et al., 2005; Hassan et al., 2005). Limiting traffic in the furrows prevents soil compaction in the beds, which further allows better soil conditions for infiltration, root growth and surface/subsurface drainage. Bakker et al. (2005) reported that raised beds increased grain yields by 18% for different crops (oats, canola, barley, wheat, field peas, lupins) on poorly drained soils, and also improved soil structure as indicated by lower bulk densities and higher infiltration rates on raised beds. Hassan et al. (2005) reported that raised beds increased net benefit by 54% for corn, along with a 30% increase in yields and 24% lower weed infestation.

4.3 Nutrient management

Flooding also results in nutrient losses, deficiencies or toxicity, and a concomitant reduction in crop growth and yields. Nitrogen is usually lost through leaching, denitrification, erosion or runoff under flooded soil conditions. Plants show N deficiency because of soil losses and due to reduced nutrient uptake by the roots under flooded conditions. Multiple studies have shown the beneficial effects of N additions for recovery from flooding stress in different crops including corn (Ritter and Beer, 1969; Forde, 2002); wheat (Wu et al., 2013, 2014); cotton (Zhigou and Derrick, 2012); canola (Habibzadeh et al., 2013); forage grasses such as bahiagrass, limpograss and maidencane (Sigua et al., 2012); and soybean (Thomas and Zodek, 2005).

5 Breeding soybean for flooding tolerance

The breeding goal for flooding tolerance is to attain high yield potential in the absence of flooding while maintaining yield performance under stress. The foundational elements for any breeding programme are to have genetic variability for the desired trait and a reliable method for measuring the trait in the progenies.

5.1 Genetics of flooding tolerance in soybean

Flooding stress is a strong driver for adaptation and evolution (Jackson and Colmer, 2005). Genetic diversity to adapt to flooding tolerance may have evolved via natural or artificial selection as a result of geographic and environmental adaptation, as seen in cereal crops including rice (Xu and Mackill, 1996; Xu et al., 2006), wheat (Huang et al., 1994; Setter et al., 1999), barley (Takeda and Fukuyama, 1986; Setter et al., 1999; Pang et al., 2004; Zhou, 2011) and maize (Tang et al., 2005; Qiu et al., 2007).

Soybean was domesticated from its annual wild relative (*Glycine soja*) in China about 5000 years ago (Carter et al., 2004) and is generally considered sensitive to flooding (Oosterhuis et al., 1990; Russell et al., 1990; VanToai et al., 1993, 1994, 2010); however, some soybean landraces from Asia and southeastern China have adapted to flooding stress due to long-term rotation with rice paddy fields, and exhibit more flooding tolerance than soybean germplasm from other areas (Ralph, 1983). However, VanToai et al. (1994) reported that genetic variability for flooding tolerance also exists among American soybean varieties.

5.2 Genetic variation and QTL mapping

The existence of genetic variability for soybean flooding tolerance was reported more than three decades ago, and subsequent studies have expanded the list of donor sources available for crop improvement (Table 1) (Stanley et al., 1980; Oosterhuis et al., 1990; Russel et al., 1990; VanToai et al., 1994). Efforts have been made to systematically characterize the genetic variation in the soybean germplasm collection; for instance, Shannon et al. (2005) screened a set of 350 soybean germplasm lines for flooding tolerance at early reproductive stages and observed a wide yield gap between susceptible and tolerant lines, reporting that, as a whole, exotic soybean plant introductions (PIs) have much higher flooding tolerance than US cultivars.

In addition, genetic mapping of soybean populations has been conducted to identify genomic regions associated with flooding tolerance (Table 2). Githiri et al. (2006) mapped seven quantitative trait loci (QTLs; *ft1* to *ft7*) for yield under flooding stress with donor alleles from Misuzudaizu, with a major QTL near Satt100; however, the major QTL was tightly linked with flowering time, and the other QTLs were not repeatable across years. The higher yield under flooding stress for lines having the proposed major QTL may be affected by the longer recovery period before the reproductive stages due to delayed flowering. Wang et al. (2008) reported three QTLs (*sub-1*, *sub-3* and *sub-5*) associated with soybean flooding tolerance located in the regions Satt531, Satt648 and Satt038 on chromosomes 1, 5 and 18, respectively, explaining 4.4–7.6% of the total phenotypic variation. In addition, Sayama et al. (2009) developed 96 F_6 RILs from a cross of 'Peking', a flood-tolerant line with a black seed coat, and 'Tamahomare', an elite yellow-seeded variety commonly used in Japan, and identified four stable

Table 1 Soybean flood-tolerant germplasm reported by previous studies

Germplasm	Maturity group	Origin	Reference
Edison	MG III	USA	VanToai et al. (1994)
GR 8836	MG III	USA	VanToai et al. (1994)
CX 415	MG IV	USA	VanToai et al. (1994)
Archer	MG II	USA	VanToai et al. (2001)
Misuzudaizu	–	Japan	Githiri et al. (2006)
Kefeng No. 1	–	China	Wang et al. (2008)
Peking	–	China	Sayama et al. (2009)
Nam Vang	74–83 days	Vietnam	VanToai et al. (2010)
VND2	76–83 days	China	VanToai et al. (2010)
ATF15-1	79–88 days	Australia	VanToai et al. (2010)
PI 408105A	MG IV	South Korea	Shannon et al. (2005)
PI 574476A	MG IV	Vietnam	Shannon et al. (2005)
91210-350	MG V	USA	Henshaw et al. (2007)
91210-316	MG V	USA	Henshaw et al. (2007)

QTLs – *Sft1*, *Sft2*, *Sft3* and *Sft4* – located on chromosomes 12, 8, 4 and 2, respectively (Table 2).

VanToai et al. (2001) mapped one QTL on chromosome 18 associated with the simple sequence repeat marker Sat_064 in two recombinant inbred populations, having Archer soybean variety as the flood-tolerance donor. Furthermore, this QTL was confirmed by Cornelious et al. (2005) in near isogenic lines (NILs) derived from heterogeneous inbred families at the F_6 generation. The NILs with the tolerant allele produced an average of 60.9% flooding yield index compared with only 32.6% flooding yield index of the NILs with the sensitive allele. Five additional flooding tolerance QTLs were mapped to be associated with markers Satt485 on chromosome 3; Satt385 on chromosome 5; and Satt160, Satt252 and Satt269 on chromosome 13 through bulk-segregation analysis and partial linkage mapping (Cornelious et al., 2005). The tolerant alleles of these two QTLs were also inherited from Archer (Table 2). These research efforts also suggest that flooding tolerance might be associated with resistance to *Phytophthora sojae* in soybean. *Phytophthora sojae*, being an oomycete, needs free water for spore dispersion. Flooding waters are ideal conditions for infection to occur. Unsurprisingly, some of the significant markers associated with flooding damage scores were found to be collocated with *Rps* genes or with QTLs for partial resistance to *Phytophthora sojae* (Van Toai et al., 2001; Cornelious et al., 2005). Moreover, two new QTLs (FTS-11 and FTS-13) were mapped on chromosomes 11 and 13 in the cross of S99-2281 (elite) × PI 408105A (exotic) associated with both flooding damage score and flooding yield index (Nguyen et al., 2012). The QTL on chromosome 13 was found to be a major QTL, with an R^2 of 18% at multiple locations and years, and be associated with partial resistance to *Phytophthora sojae* (Nguyen et al., 2012). These results suggest that incorporating flooding tolerance traits into elite backgrounds would also enhance resistance to root fungal diseases such as *Phytophthora sojae*.

Table 2 QTLs associated with flooding tolerance in soybean

QTL[a]	Donor parent	Chr.[b]	Nearest marker[c]	R^{2d}	Reference
Chr.3	Archer	3	Satt485	–	Cornelious et al. (2005)
Chr.5	Archer	5	Satt385	0.10	Cornelious et al. (2005)
Chr.13	Archer	13	Satt160/252/269	0.16	Cornelious et al. (2005)
Chr.18	Archer	18	Sat_064	–	VanToai et al. (2001) and Cornelious et al. (2005)
ft1	Misuzudaizu	6	Satt100	0.49	Githiri et al. (2006)
ft2	Misuzudaizu	2	Satt282	0.07	Githiri et al. (2006)
ft3	Misuzudaizu	19	A489	0.10	Githiri et al. (2006)
ft4	Misuzudaizu	11	A520	0.11	Githiri et al. (2006)
ft5	Misuzudaizu	14	A685b	0.10	Githiri et al. (2006)
ft6	Misuzudaizu	7	A715	0.07	Githiri et al. (2006)
ft7	Misuzudaizu	10	Satt477	0.07	Githiri et al. (2006)
sub-1	Kefeng No. 1	1	Satt531	0.08	Wang et al. (2008)
sub-3	Kefeng No. 1	5	Satt648	0.06	Wang et al. (2008)
sub-5	Kefeng No. 1	18	Satt038	0.06	Wang et al. (2008)
Sft 1	Peking	12	Sat_175	–	Sayama et al. (2009)
Sft 2	Peking	8	Satt187	–	Sayama et al. (2009)
Sft 3	Peking	4	Satt338	–	Sayama et al. (2009)
Sft 4	Peking	2	Sat_279	–	Sayama et al. (2009)
FTS-11	PI 408105A	11	BARC-016279-02316	0.18	Nguyen et al. (2012)
FTS-13	PI 408105A	13	BARC-024569-04982	0.14–0.18	Nguyen et al. (2012)

[a] Name of the QTL generated in the corresponding reference or named afterwards in this review.
[b] Chromosome.
[c] Nearest DNA marker of the QTL.
[d] Phenotypic variation that can be explained by the QTL in the original mapping populations.

5.3 Screening methods for flooding tolerance

Many studies have evaluated soybean flooding tolerance in the field using yield-trial type methodologies (Oosterhuis et al., 1990; Russel et al. 1990; VanToai et al., 1994, 2010), typically involving multiple planted rows per plot and large plot dimensions. The defects of these screening methodologies are their high cost and the low reproducibility of yield of a single-location trial. VanToai et al. (2010) studied the correlation of yield and flooding score between field yield trials and greenhouse experiments; while they did not see a correlation of yield between field and greenhouse, they reported a good correlation of flooding score ranking between both methodologies.

Shannon et al. (2005) implemented a hill-plot screening (Fig. 2), where a small number of seeds (5–12) were planted in a very short row (2–5 cm). While the planting distance between two hill plots could be adjusted to the user's needs, Shannon et al. (2005) used

Figure 2 Soybean flood-tolerant germplasm selection through hill-plot screening method.

0.6 m between hill plots in a row and used a 10- to 15-cm sheet of water for flooding; they later drained the plot area and two weeks later scored the plots visually using a scale of one (no apparent damage) to five (all plants dead). The benefits of implementing hill-plot screening are the significant reduction in area needed for the screening, the facilitation of flooding management, the increase of soil uniformity for the trial and the concomitant cost reductions from smaller trial areas. A disadvantage is that it may require specialized machinery for planting, otherwise the plots must be planted by hand.

Another simple and effective screening method was developed by the University of Arkansas (Wu et al., 2017a), where a 3-m row plot (Fig. 3) is planted with 100 seeds and flooded with a sheet of water of 10–16 cm for 6–12 days, and plant foliar damage and plant survival rates are scored using a one to nine scale (one being <10% plants in a plot, nine being >85% plants in a plot with flooding injury) for symptoms of foliar damage or death, respectively. The duration of

Figure 3 Soybean flood-tolerant germplasm selection through row plot screening method.

the flooding event is variable to the physiological stage of the crop, due to a higher sensitivity of soybeans to flooding stress at R1 compared to V5 growth stage.

5.4 Active breeding efforts for flooding tolerance

Research has shown that introducing the flooding tolerance trait into elite soybean lines can prevent substantial yield reduction during flooding stress (Scott et al., 1989; Shannon et al., 2005; Rhine et al., 2010; VanToai et al., 2010), and therefore, breeding flood-tolerant soybean cultivars could be a very successful approach to managing waterlogging risk. Recently, the University of Missouri soybean breeding programme has developed several flood-tolerant lines through marker-assisted selection (MAS), utilizing the flood-tolerant accession PI 408105A. The lines have comparable yield potential to commercial checks under non-stress conditions, outperform the commercial checks by up to 1000 kg ha^{-1} under severe flooding conditions and are in the process of advanced-stage testing in preparation for commercial release (Shannon, pers. comm., 2017).

In addition, many wild soybeans, such as PI 407171, PI 562561, PI 479751, PI 424082, PI 407229, PI 366124, PI 424116 and PI 407038, exhibiting flooding tolerance (Fig. 4) are being introduced into high-yield cultivars via back-crossing or forward-crossing methods (Shannon and Nguyen, pers. comm., 2017). Simultaneously, the University of Arkansas soybean breeding programme has actively screened lines and bred for flooding tolerance for several years, with more than 2000 soybean genotypes evaluated in field screening over seven years. Although most genotypes were sensitive to flooding, 52 soybean genotypes showed flooding tolerance, 11 of which had consistent ratings over four to six years of continuous screenings (Wu et al. 2017c), thus becoming the flooding-tolerant sources for the University of Arkansas breeding programme. As of 2017, a total of 35 advanced breeding lines were developed and entered in yield trials, tested under normal production practices and also rated for flooding stress damage. Further selection, testing and breeding cycles will be needed for the successful commercial release of a flood-tolerant cultivar (Mozzoni, pers. comm., 2017).

Figure 4 Wild soybeans (*Glycine soja*) exhibited more flooding tolerance than soybean (*Glycine max*).

6 Summary

The development of flood-tolerant soybean cultivars through plant breeding is a key strategy for managing flooding stress in soybean production. In nature, genetic variations exist, but are limited in response to flooding stress, and new methods of flooding tolerance are required for breeding efforts, alongside reliable, reproducible and cost-effective screening methods. Most screening for flooding tolerance to date has been conducted upon reproductive or late vegetative stages. However, further refinement of emergency screening methodologies is necessary because most flooding damage occurs due to severe weather during the planting season in the spring. A simple, rapid and reliable greenhouse screen protocol may also be helpful. In addition, implementing imaging and remote sensing methodologies for high-throughput phenotyping will help reach the scale needed to screen lines at earlier breeding stages, and will be a perfect tool for the development of large training data sets for genomic selection efforts.

Although major QTL were detected and significant progress has been made using an MAS approach, the number of QTLs, percentage of variation explained and distribution throughout the genome (Table 2) point to a highly quantitative trait. These types of traits are difficult to breed using MAS alone, but are prime candidates for genomic selection efforts. Training sets need to be developed and models validated so that breeding programmes can make faster genetic gains on such an expensive-to-measure trait. However, to do so, programmes need to develop a significant volume of materials screened for the trait in a fast and effective manner, requiring a critical investment in genotyping progenies. The cost of marker technologies is starting to fall, and the availability of service providers is expanding, thus making genomic selection for flooding tolerance a plausible reality in the not-so-distant future.

In addition, greater understanding of the molecular mechanisms of soybean response to flooding through transdisciplinary approaches involving genomics, proteomics and metabolomics is required. This knowledge would allow progress to go beyond the 'black box' approach of genomic selection to a targeted approach, such as genetic transformation or gene editing. The newer CRISPR gene editing technology is an approach that could be used to modify key elements in the pathways involved in flooding tolerance to develop new commercial products.

Further development of agronomic tools is also important for flooding management. Additional studies on soil science will help refine our understanding of the interactions of soil's physical and chemical properties with soybean cultivars, expanding the opportunities to exploit cultivar responses to soil properties. Similarly, the implementation of planting technologies that allow for variable variety selection could pose an interesting agronomical solution – placing 'racehorse' elite varieties in areas with lower risk of flooding, and a flood-tolerant variety in higher-risk areas. Finally, since there is an overlap of at least some *Phytophthora* tolerant/resistant QTLs with flooding tolerant QTLs, agronomic studies on the interaction of seed fungicide treatments with genotypes expressing various degrees of flooding tolerance/susceptibility would identify another management option for the reduction of flooding stress risk.

In summary, flooding is a very complex abiotic stress, and no silver bullet will solve the problem with a simple and economical approach. The best strategy for managing flooding may be to rely on an integrated approach, including flood-tolerant cultivar development via conventional breeding efforts, a supplemental trait resulting from either transgenic or gene editing technologies, and advanced agronomic management solutions that can minimize flooding risk and maximize farm revenue.

7 References

Ahmed, F., Rafii, M., Ismail, M. R., Juraimi, A. S., Rahim, H., Asfaliza, R. and Latif, M. A. (2013). 'Waterlogging tolerance of crops: Breeding, mechanism of tolerance, molecular approaches, and future prospects'. *BioMed Research International* 2013, 1–10.

Armstrong, W. (1979). 'Aeration in higher plants'. *Advances in Botanical Research* 7, 225–332.

Armstrong, W., Strange, M., Cringle, S. and Beckett, P. (1994). 'Microelectrode and modelling study of oxygen distribution in roots'. *Annals of Botany* 74, 287–99.

Bacanamwo, M. and Purcell, L. C. (1999a). 'Soybean dry matter and N accumulation responses to flooding stress, N sources and hypoxia'. *Journal of Experimental Botany* 50, 689–96.

Bacanamwo, M. and Purcell, L. C. (1999b). 'Soybean root morphological and anatomical traits associated with acclimation to flooding'. *Crop Science* 39, 143–9.

Bailey-Serres, J. and Chang, R. (2005). 'Sensing and signalling in response to oxygen deprivation in plants and other organisms'. *Annals of Botany* 96, 507–18.

Bailey-Serres, J. and Voesenek, L. (2008). 'Flooding stress: Acclimations and genetic diversity'. *Annual Review of Plant Biology* 59, 313–39.

Bailey-Serres, J., Lee, S. C. and Brinton, E. (2012). 'Waterproofing crops: Effective flooding survival strategies'. *Plant Physiology* 160, 1698–709.

Bakker, D., Hamilton, G., Houlbrooke, D. and Spann, C. (2005). 'The effect of raised beds on soil structure, waterlogging, and productivity on duplex soils in Western Australia'. *Soil Research* 43, 575–85.

Balakhnina, T. I., Bennicelli, R. P., Stepniewska, Z., Stepniewski, W. and Fomina, I. R. (2010). 'Oxidative damage and antioxidant defense system in leaves of *Vicia faba major* L. cv. Bartom during soil flooding and subsequent drainage'. *Plant Soil* 327, 293–301.

Bellaloui, N., Hu, Y., Mengistu, A., Kassem, M. A. and Abel, C. A. (2013). 'Effects of foliar boron application on seed composition, cell wall boron, and seed δ15N and δ13C isotopes in water-stressed soybean plants'. *Frontiers in Plant Science* 4, 270.

Beutler, A. N., Giacomeli, R., Alberto, C. M., Silva, V. N., da Silva Neto G. F., Machado, G. A. and Santos, A. T. L. (2014). 'Soil hydric excess and soybean yield and development in Brazil'. *Australian Journal of Crop Science* 8, 1461.

Blevins, D., Wilkison, D., Kelly, B. and Silva, S. (1996). 'Movement of nitrate fertilizer to glacial till and runoff from a claypan soil'. *Journal of Environmental Quality* 25, 584–93.

Blom, C. and Voesenek, L. (1996). 'Flooding: The survival strategies of plants'. *Trends in Ecology & Evolution* 11, 290–5.

Board, J. (2008). 'Waterlogging effects on plant nutrient concentrations in soybean'. *Journal of Plant Nutrition* 31, 828–38.

Buttery, B. (1987). 'Some effects of waterlogging and supply of combined nitrogen on soybean growth'. *Canadian Journal of Plant Science* 67, 69–77.

Carter, T., Nelson, R., Sneller, C. and Cui, Z. (2004). 'Soybeans: Improvement production and uses'. *Agronomy* 16, 303–416.

Colmer, T. and Voesenek, L. (2009). 'Flooding tolerance: Suites of plant traits in variable environments'. *Functional Plant Biology* 36, 665–81.

Cornelious, B., Chen, P., Chen, Y., Leon, N. D., Shannon, J. and Wang, D. (2005). 'Identification of QTLs underlying water-logging tolerance in soybean'. *Molecular Breeding* 16, 103–12.

Cox, M. C., Millenaar, F. F., Van Berkel, Y. E. D. J., Peeters, A. J. and Voesenek, L. A. (2003). 'Plant movement. Submergence-induced petiole elongation in Rumex palustris depends on hyponastic growth'. *Plant Physiology* 132, 282–91.

Dat, J. F., Capelli, N., Folzer, H., Bourgeade, P. and Badot, P.-M. (2004). 'Sensing and signalling during plant flooding'. *Plant Physiology and Biochemistry* 42, 273–82.

Forde, B. G. (2002). 'Local and long-range signaling pathways regulating plant responses to nitrate'. *Annual Review of Plant Biology* 53, 203–24.

Githiri, S., Watanabe, S., Harada, K. and Takahashi, R. (2006). 'QTL analysis of flooding tolerance in soybean at an early vegetative growth stage'. *Plant Breeding* 125, 613–18.

Grimoldi, A., Insausti, P., Roitman, G. and Soriano, A. (1999). 'Responses to flooding intensity in Leontodon taraxacoides'. *New Phytologist* 141, 119–28.

Habibzadeh, F., Sorooshzadeh, A., Pirdashti, H. and Modarres-Sanavy, S. A. M. (2013). 'Alleviation of waterlogging damage by foliar application of nitrogen compounds and tricyclazole in canola'. *Australian Journal of Crop Science* 7, 401.

Hashiguchi, A., Sakata, K. and Komatsu, S. (2009). 'Proteome analysis of early-stage soybean seedlings under flooding stress'. *Journal of Proteome Research* 8, 2058–69.

Hassan, I., Hussain, Z. and Akbar, G. (2005). 'Effect of permanent raised beds on water productivity for irrigated maize–wheat cropping system'. In: Roth, C. H., Fischer, R. A. and Meisner, C. A. (Eds), *Evaluation and Performance of Permanent Raised Bed Cropping Systems in Asia, Australia and Mexico*. Australian Centre for International Agricultural Research (ACIAR) Proceedings No. 121, Griffith, NSW, Australia, pp. 59–65.

Henshaw, T., Gilbert, R., Scholberg, J. and Sinclair, T. (2007). 'Soya Bean (*Glycine max* L. Merr.) genotype response to early-season flooding: I. Root and nodule development'. *Journal of Agronomy and Crop Science* 193, 177–88.

Hou, F. and Thseng, F. (1991). 'Studies on the flooding tolerance of soybean seed: Varietal differences'. *Euphytica* 57, 169–73.

Huang, B., Johnson, J. W., Nesmith, S. and Bridges, D. C. (1994). 'Growth, physiological and anatomical responses of two wheat genotypes to waterlogging and nutrient supply'. *Journal of Experimental Botany* 45, 193–202.

Husson, O. (2013). 'Redox potential (Eh) and pH as drivers of soil/plant/microorganism systems: A transdisciplinary overview pointing to integrative opportunities for agronomy'. *Plant and Soil* 362, 389–417.

Jackson, M. and Colmer, T. (2005). 'Response and adaptation by plants to flooding stress'. *Annals of Botany* 96, 501–5.

Jackson, M. B., Ishizawa, K. and Ito, O. (2009). 'Evolution and mechanisms of plant tolerance to flooding stress'. *Annals of Botany* 103, 137–42.

Justin, S. and Armstrong, W. (1987). 'The anatomical characteristics of roots and plant response to soil flooding'. *New Phytologist* 106, 465–95.

Kadempir, M., Galeshi, S., Soltani, A. and Ghaderifar, F. (2014). 'The Effect of flooding and nutrition levels on reproductive growth stages of aerenchyma formation and ethylene production in soybean (*Glycine max* L)'. *International Journal of Advanced Biological and Biomedical Research* 2, 487–95.

Kim, Y.-H., Hwang, S.-J., Waqas, M., Khan, A. L., Lee, J.-H., Lee, J.-D., Nguyen, H. T. and Lee, I.-J. (2015). 'Comparative analysis of endogenous hormones level in two soybean (*Glycine max* L.) lines differing in waterlogging tolerance'. *Frontiers in Plant Science* 6, 714.

Komatsu, S., Wada, T., Abaléa, Y., Nouri, M.-Z., Nanjo, Y., Nakayama, N., Shimamura, S., Yamamoto, R., Nakamura, T. and Furukawa, K. (2009). 'Analysis of plasma membrane proteome in soybean and application to flooding stress response'. *Journal of Proteome Research* 8, 4487–99.

Komatsu, S., Thibaut, D., Hiraga, S., Kato, M., Chiba, M., Hashiguchi, A., Tougou, M., Shimamura, S. and Yasue, H. (2011). 'Characterization of a novel flooding stress-responsive alcohol dehydrogenase expressed in soybean roots'. *Plant Molecular Biology* 77, 309–22.

Komatsu S., Han, C., Nanjo, Y., Altaf-Un-Nahar, M., Wang, K., He, D. and Yang, P. (2013). 'Label-free quantitative proteomic analysis of abscisic acid effect in early-stage soybean under flooding'. *Journal of Proteome Research* 12, 4769–84.

Kozlowski, T. (1984). 'Extent, cause, and impact of flooding'. In: Kozlowski, T. (Ed.), *Flooding and Plant Growth*. Academic Press, New York, NY, pp. 1–7.

Kramer, P. J. (1951). 'Causes of injury to plants resulting from flooding of the soil'. *Plant Physiology* 26, 722.

Kreuzwieser, J. and Rennenberg, H. (2014). 'Molecular and physiological responses of trees to waterlogging stress'. *Plant, Cell & Environment* 37, 2245–59.

Laan, P., Tosserams, M., Blom, C. and Veen, B. (1990). 'Internal oxygen transport in Rumex species and its significance for respiration under hypoxic conditions'. *Plant and Soil* 122, 39–46.

Linkemer, G., Board, J. E. and Musgrave, M. E. (1998). 'Waterlogging effects on growth and yield components in late-planted soybean'. *Crop Science* 38, 1576–84.

Maekawa, T., Shimamura, S. and Shimada, S. (2011). 'Effects of short-term waterlogging on soybean nodule nitrogen fixation at different soil reductions and temperatures'. *Plant Production Science* 14, 349–58.

Miura, K., Ogawa, A., Matsushima, K. and Morita, H. (2012). 'Root and shoot growth under flooded soil in wild groundnut (*Glycine soja*) as a genetic resource of waterlogging tolerance for soybean (*Glycine max*)'. *Pakistan Journal of Weed Science Research* 18, 427–3.

Naidoo, G. and Mundree, S. (1993). 'Relationship between morphological and physiological responses to waterlogging and salinity in *Sporobolus virginicus* (L.) Kunth'. *Oecologia* 93, 360–6.

Nanjo, Y., Maruyama, K., Yasue, H., Yamaguchi-Shinozaki, K., Shinozaki, K. and Komatsu, S. (2011a). 'Transcriptional responses to flooding stress in roots including hypocotyl of soybean seedlings'. *Plant Molecular Biology* 77, 129–44.

Nanjo, Y., Skultety, L., Uváčková, L. U., Klubicová, K. N., Hajduch, M. and Komatsu, S. (2011b). 'Mass spectrometry-based analysis of proteomic changes in the root tips of flooded soybean seedlings'. *Journal of Proteome Research* 11, 372–85.

Nelson, K., Smoot, R. and Meinhardt, C. (2011). 'Soybean response to drainage and subirrigation on a claypan soil in Northeast Missouri'. *Agronomy Journal* 103, 1216–22.

Nelson, K. A., Meinhardt, C. G. and Smoot, R. L. (2012). 'Soybean cultivar response to subsurface drainage and subirrigation in Northeast Missouri'. *Crop Management* 11. https://doi.org/10.1094/cm-2012-0320-03-rs.

Nguyen, V. T., Vuong, T. D., VanToai, T., Lee, J. D., Wu, X., Mian, M. A. R., Dorrance, A. E., Shannon, J. G. and Nguyen, H. T. (2012). 'Mapping of quantitative trait loci associated with resistance to and flooding tolerance in soybean'. *Crop Science* 52, 2481.

Oh, M. W. and Komatsu, S. (2015). 'Characterization of proteins in soybean roots under flooding and drought stresses'. *Journal of Proteomics* 114, 161–81.

Oosterhuis, D., Scott, H., Hampton, R. and Wullschleger, S. (1990). 'Physiological responses of two soybean [*Glycine max* (L.) Merr] cultivars to short-term flooding'. *Environmental and Experimental Botany* 30, 85–92.

Pang, J., Zhou, M., Mendham, N. and Shabala, S. (2004). 'Growth and physiological responses of six barley genotypes to waterlogging and subsequent recovery'. *Crop and Pasture Science* 55, 895–906.

Patel, P. K., Singh, A. K., Tripathi, N., Yadav, D. and Hemantaranjan, A. (2014). 'Flooding: Abiotic constraint limiting vegetable productivity'. *Advances in Plants and Agriculture Research* 1(3), 00016. http://dx.doi.org/10.15406/apar.2014.01.00016.

Patrick, W. and Jugsujinda, A. (1992). 'Sequential reduction and oxidation of inorganic nitrogen, manganese, and iron in flooded soil'. *Soil Science Society of America Journal* 56, 1071–3.

Ponnamperuma, F. N. (1972). 'The chemistry of submerged soils'. *Advances in Agronomy* 24, 29–96.

Qiu, F., Zheng, Y., Zhang, Z. and Xu, S. (2007). 'Mapping of QTL associated with waterlogging tolerance during the seedling stage in maize'. *Annals of Botany* 99, 1067–81.

Ralph, W. (1983). 'Soybeans respond to controlled waterlogging'. *Rural Research* 120, 4–8.

Rhine, M. D., Stevens, G., Shannon, G., Wrather, A. and Sleper, D. (2010). 'Yield and nutritional responses to waterlogging of soybean cultivars'. *Irrigation Science* 28, 135–42.

Ritter, W. and Beer, C. (1969). 'Yield reduction by controlled flooding of corn'. *Transactions of the ASAE* 12, 46–7.

Rosenzweig, C. E., Tubiello, F., Goldberg, R., Mills, E. and Bloomfield, J. (2002). 'Increased crop damage in the U.S. from excess precipitation under climate change'. *Global Environmental Change* 12, 197–202.

Roth, C. H., Fischer, R. A., Piggin, C. and Meyer, W. (2005). 'Evaluation and performance of permanent raised bed systems in Asia, Mexico and Australia: A synopsis'. In: *Proceedings of a Workshop on Evaluation and Performance of Permanent Raised Bed Systems in Asia, Mexico and Australia.* Australian Centre for International Agricultural Research (ACIAR) Proceedings No. 121, Griffith, NSW, Australia, pp. 200–8.

Russell, D. A., Wong, D. M.-L. and Sachs, M. M. (1990). 'The anaerobic response of soybean'. *Plant Physiology* 92, 401–7.
Sahrawat, K. (2008). 'Soil fertility advantages of submerged rice cropping systems: A review'. *Journal of Sustainable Agriculture* 31, 5–23.
Sairam, R., Kumutha, D., Ezhilmathi, K., Deshmukh, P. and Srivastava, G. (2008). 'Physiology and biochemistry of waterlogging tolerance in plants'. *Biologia Plantarum* 52, 401–12.
Sanchez, C., Tortosa, G., Granados, A., Delgado, A., Bedmar, E. J. and Delgado, M. J. (2011). 'Involvement of *Bradyrhizobium japonicum* denitrification in symbiotic nitrogen fixation by soybean plants subjected to flooding'. *Soil Biology and Biochemistry* 43, 212–17.
Sayama, T., Nakazaki, T., Ishikawa, G., Yagasaki, K., Yamada, N., Hirota, N., Hirata, K., Yoshikawa, T., Saito, H. and Teraishi, M. (2009). 'QTL analysis of seed-flooding tolerance in soybean (*Glycine max* [L.] Merr.)'. *Plant Science* 176, 514–21.
Scott, H., DeAngulo, J., Daniels, M. and Wood, L. (1989). 'Flood duration effects on soybean growth and yield'. *Agronomy Journal* 81, 631–6.
Scott, H., DeAngulo, J., Wood, L. and Pitts, D. (1990). 'Influence of temporary flooding at three growth stages on soybeans grown on a clayey soil'. *Journal of Plant Nutrition* 13, 1045–71.
Seago Jr., J. L., Marsh, L. C., Stevens, K. J., Soukup, A., Votrubova, O. and Enstone, D. E. (2005). 'A re-examination of the root cortex in wetland flowering plants with respect to aerenchyma'. *Annals of Botany* 96, 565–79.
Setter, T., Burgess, P., Waters, I. and Kuo, J. (1999). 'Genetic diversity of barley and wheat for waterlogging tolerance in Western Australia'. *Proceedings of the 9th Australian Barley Technical Symposium*. Australian Barley Technical Symposium Inc, 12–16 September 1999, Melbourne, Australia.
Shannon, J., Stevens, W., Wiebold, W., McGraw, R., Sleper, D. and Nguyen, H. (2005). 'Breeding soybeans for improved tolerance to flooding'. *Proceedings of the 30th Soybean Research Conference*. American Seed Trade Association, 7 December 2005, Chicago, IL.
Shimamura, S., Mochizuki, T., Nada, Y. and Fukuyama, M. (2003). 'Formation and function of secondary aerenchyma in hypocotyl, roots and nodules of soybean (*Glycine max*) under flooded conditions'. *Plant and Soil* 251, 351–9.
Sigua, G., Williams, M., Grabowski, J., Chase, C. and Kongchum, M. (2012). 'Effect of flooding duration and nitrogen fertilization on yield and protein content of three forage species'. *Agronomy Journal* 104, 791–8.
Stanley, C., Kaspar, T. and Taylor, H. (1980). 'Soybean top and root response to temporary water tables imposed at three different stages of growth'. *Agronomy Journal* 72, 341–6.
Striker, G. G. (2012). 'Flooding stress on plants: Anatomical, morphological and physiological responses'. In: Mworia, J. K. (Ed.), *Botany*. InTech Open Access Publishers, Rijeka, Croatia, pp. 3–28. doi:10.5772/2245.
Striker, G. G., Insausti, P., Grimoldi, A. A., Ploschuk, E. L. and Vasellati, V. (2005). 'Physiological and anatomical basis of differential tolerance to soil flooding of *Lotus corniculatus* L. and *Lotus glaber* Mill'. *Plant and Soil* 276, 301–11.
Sullivan, M., VanToai, T., Fausey, N., Beuerlein, J., Parkinson, R. and Soboyejo, A. (2001). 'Evaluating on-farm flooding impacts on soybean'. *Crop Science* 41, 93–100.
Takeda, K. and Fukuyama, T. (1986). 'Variation and geographical distribution of varieties for flooding tolerance in barley seeds'. *Barley Genetics Newsletter (USA)* 16, 28–9.
Tamang, B. G., Magliozzi, J. O., Maroof, M. S. and Fukao, T. (2014). 'Physiological and transcriptomic characterization of submergence and reoxygenation responses in soybean seedlings'. *Plant, Cell & Environment* 37, 2350–65.
Tang, W., Zhang, Z., Zou, X. and Zheng, Y. (2005). 'Functional genomics of maize submergence tolerance and cloning of the related gene Sicyp51'. *Science in China Series C: Life Sciences* 48, 337–45.
Thomas, A. L. and Sodek, L. (2005). 'Development of the nodulated soybean plant after flooding of the root system with different sources of nitrogen'. *Brazilian Journal of Plant Physiology* 17, 291–7.

Thomas, J., Boote, K., Allen, L., Gallo-Meagher, M. and Davis, J. (2003). 'Elevated temperature and carbon dioxide effects on soybean seed composition and transcript abundance'. *Crop Science* 43, 1548–57.

Tougou, M., Hashiguchi, A., Yukawa, K., Nanjo, Y., Hiraga, S., Nakamura, T., Nishizawa, K. and Komatsu, S. (2012). 'Responses to flooding stress in soybean seedlings with the alcohol dehydrogenase transgene'. *Plant Biotechnology* 29, 301–5.

Trought, M. and Drew, M. (1982). 'Effects of waterlogging on young wheat plants (*Triticum aestivum* L.) and on soil solutes at different soil temperatures'. *Plant and Soil* 69, 311–26.

Valliyodan, B., VanToai T. T., Alves J. D., de Fátima P., Goulart, P., Lee, J. D., Fritschi, F. B., Rahman, M. A., Islam, R., Shannon, J. G. and Nguyen, H. T. (2014). 'Expression of root-related transcription factors associated with flooding tolerance of soybean (*Glycine max*)'. *International Journal of Molecular Sciences* 15, 17622–43.

VanToai, T., St Martin, S., Zhang, J. and Yu, S. (1993). 'RAPD markers of flooding tolerant Chinese soybean germplasm'. *Soybean Genetics Newsletter (USA)* 20, 153–9.

VanToai, T., Beuerlein, A., Schmitthenner, S. and St Martin, S. (1994). 'Genetic variability for flooding tolerance in soybeans'. *Crop Science* 34, 1112–15.

VanToai, T. T., St Martin, S. K., Chase, K., Boru, G., Schnipke, V., Schmitthenner, A. F. and Lark, K. G. (2001). 'Identification of a QTL associated with tolerance of soybean to soil waterlogging'. *Crop Science* 41, 1247–52.

VanToai, T. T., Hoa, T. T. C., Hue, N. T. N., Nguyen, H. T., Shannon, J. G. and Rahman, M. A. (2010). 'Flooding tolerance of soybean [*Glycine max* (L.) Merr.] germplasm from Southeast Asia under field and screen-house environments'. *The Open Agriculture Journal* 4, 38–46.

Vartapetian, B. B. and Jackson, M. B. (1997). 'Plant adaptations to anaerobic stress'. *Annals of Botany* 79, 3–20.

Velmurugan, A., Swarnam, T., Ambast, S. and Kumar, N. (2016). 'Managing waterlogging and soil salinity with a permanent raised bed and furrow system in coastal lowlands of humid tropics'. *Agricultural Water Management* 168, 56–67.

Voesenek, L. and Bailey-Serres, J. (2013). 'Flooding tolerance: O_2 sensing and survival strategies'. *Current Opinion in Plant Biology* 16, 647–53.

Voesenek, L., Rijnders, J., Peeters, A., Van de Steeg, H. and De Kroon, H. (2004). 'Plant hormones regulate fast shoot elongation under water: From genes to communities'. *Ecology* 85, 16–27.

Wang, F., Zhao, T. J., Yu, D. Y., Chen, S. Y. and Gai, J. Y. (2008). 'Inheritance and QTL analysis of submergence tolerance at seedling stage in soybean'. *Acta Agronomica Sinica* 34, 748–53.

Waqas, M., Khan, A. L., Kang, S.-M., Kim, Y.-H. and Lee, I.-J. (2014). 'Phytohormone-producing fungal endophytes and hardwood-derived biochar interact to ameliorate heavy metal stress in soybeans'. *Biology and Fertility of Soils* 50, 1155–67.

Wegner, L. H. (2010). 'Oxygen transport in waterlogged plants'. In: Mancuso, S. and Shabala, S. (Eds), *Waterlogging Signalling and Tolerance in Plants*. Springer, Berlin, Germany, pp. 3–22.

Wu, J. D., Li, J. C., Wang, C. Y., Wei, F. Z., Zhang, Y. and Wu, W. M. (2013). 'Effects of spraying foliar nitrogen on activities of key regulatory enzymes involved in protein formation in winter wheat suffered post-anthesis high temperature and waterlogging'. *Journal of Food, Agriculture & Environment* 11, 668–73.

Wu, J. D., Li, J. C., Wei, F. Z., Wang, C. Y., Zhang, Y. and Sun, G. (2014). 'Effects of nitrogen spraying on the post-anthesis stage of winter wheat under waterlogging stress'. *Acta Physiologiae Plantarum* 36, 207–16.

Wu, C., Zeng, A., Chen, P., Florez-Palacios, L., Hummer, W., Mokua, J., Klepadlo, M., Yan, L., Ma, Q. and Cheng, Y. (2017a). 'An effective field screening method for flood tolerance in soybean'. *Plant Breeding* 5, 710–19.

Wu, C., Chen, P., Hummer, W., Zeng, A. and Klepadio, M. (2017b). 'Effect of flood stress on soybean seed germination in the field'. *American Journal of Plant Science* 8, 53–68.

Wu, C., Chen, P., Zeng, A., Hummer, W., Mokua, J., Shannon, J. G. and Nguyen, H. T. (2017c). 'Evaluation and development of flood-tolerant genotypes and cultivars in soybean'. *Plant Breeding* 6, 913–23.

Wuebker, E. F., Mullen, R. E. and Koehler, K. (2001). 'Flooding and temperature effects on soybean germination'. *Crop Science* 41, 1857–61.
Xu, K. and Mackill, D. J. (1996). 'A major locus for submergence tolerance mapped on rice chromosome 9'. *Molecular Breeding* 2, 219–24.
Xu, K., Xu, X., Fukao, T., Canlas, P., Maghirang-Rodriguez, R., Heuer, S., Ismail, A. M., Bailey-Serres, J., Ronald, P. C. and Mackill, D. J. (2006). 'Sub1A is an ethylene-response-factor-like gene that confers submergence tolerance to rice'. *Nature* 442, 705–8.
Yin, X. and Komatsu, S. (2015). 'Quantitative proteomics of nuclear phosphoproteins in the root tip of soybean during the initial stages of flooding stress'. *Journal of Proteomics* 119, 183–95.
Yin, X., Sakata, K. and Komatsu, S. (2014a). 'Phosphoproteomics reveals the effect of ethylene in soybean root under flooding stress'. *Journal of Proteome Research* 13, 5618–34.
Yin, X., Sakata, K., Nanjo, Y. and Komatsu, S. (2014b). 'Analysis of initial changes in the proteins of soybean root tip under flooding stress using gel-free and gel-based proteomic techniques'. *Journal of Proteomics* 106, 1–16.
Youn, J. T., Van, K., Lee, J. E., Kim, W. H., Yun, H. T., Kwon, Y. U., Ryu, Y. H. and Lee, S. H. (2008). 'Waterlogging effects on nitrogen accumulation and N2 fixation of supernodulating soybean mutants'. *Journal of Crop Science and Biotechnology* 11, 111–18.
Zhigou, Z. and Derrick, M. O. (2012). 'Physiological mechanism of nitrogen mediating cotton (*Gossypium hirsutum* L.) seedlings growth under water-stress conditions'. *American Journal of Plant Sciences*3(6), Article ID: 20019, 10p.
Zhou, M. (2011). 'Accurate phenotyping reveals better QTL for waterlogging tolerance in barley'. *Plant Breeding* 130, 203–8.

Part 2

Cultivation techniques

Chapter 10

Modelling the effects of temperature and photoperiod on soybean reproductive development

H. Yang, University of Nebraska, USA

1 Introduction

Soybean (*Glycine max* L.) is a short-day plant species whose floral induction requires the daylength, or photoperiod, to be shorter than a critical value (Johnson et al., 1960; Bernard, 1972). The rate of soybean development, especially the time to first flowering, has long been recognized to be sensitive to temperature and photoperiod (Garner and Allard, 1930; Steinberg and Garner, 1936; Johnson et al., 1960; Hesketh et al., 1973; Major et al., 1975a; Summerfield and Wilcox, 1978; Cregan and Hartwig, 1984; Wilkerson et al., 1989; Summerfield et al., 1993, 1998; Cober et al., 2001; Salmerón and Purcell, 2016). In addition, the response of the soybean developmental rate to photoperiod during post-flowering stages has also been observed (Johnson et al., 1960; Major et al., 1975a; Summerfield et al., 1993; Asumadu et al., 1998). The impact of the photoperiod on soybean yield is also significant, with delayed planting often resulting in yield loss due to shortened vegetative growth and grain filling (Board et al., 1990; Board and Settimi, 1986). In general, higher temperature favours the development rate in both vegetative and reproductive stages, while a longer photoperiod has a negative effect on reproductive stages.

Most soybean cultivars have a so-called quantitative photoperiod response, that is the cultivar will eventually flower even when the photoperiod is longer than the critical length, while some cultivars have a qualitative or obligate response to photoperiod and will not flower if the photoperiod is longer than the critical value (Hadley et al., 1984; Upadhyay et al., 1994).

http://dx.doi.org/10.19103/AS.2017.0034.10

Soybean cultivars are classified into maturity groups (MGs) using 13 Roman numerals, and fractional MGs further designate the maturity difference within a group (Heatherly and Roger, 2004). The soybean maturity rating is based on adaptation to the conventional planting practices of a geographic area, although cultivars of two or three MGs can often be grown successfully at one location. Six loci, that is E1–E6, have been reported to control soybean time to flowering and maturity (Cober et al., 2001, 2014). Low MGs are generally more suited to high latitudes and a cooler environment, while the opposite is true for high MGs. Soybean cultivars are also classified into indeterminate and determinate types based on stem growth (Heatherly and Roger, 2004). Cultivars that are classified in MGs IV and lower are generally of an indeterminate type, while those classified in MGs of V and above are of a determinate type. Semi-determinate types of soybean cultivars also exist as a result of genetic variation (Thompson et al., 1997). For a determinate type of soybean, the growth of the main stem length ceases after flowering, although branching continues throughout R1–R5 (Board and Settimi, 1986). In contrast, an indeterminate stem type allows continuous stem elongation and node production after flowering, producing a more tapered canopy shape (Thompson et al., 1997; Heatherly and Roger, 2004).

The most widely used soybean phenology staging system was developed by Fehr and Caviness (1977). In this system, major vegetative stages include emergence when cotyledons appear above the soil surface (V0), V1 when the unifoliate leaf is fully open, and V2 to Vn when the first and last trifoliate leaves on the main stem are fully open (i.e. edges are not touching). The major reproductive stages include R1 (beginning flower) marked by the first open flower on the main stem, R3 (beginning pod) when a pod of at least 5 mm in length appears at one of the four uppermost nodes on the main stem, R5 (beginning seed) when seed filling begins, R6 when seed filling becomes linear and R7 (physiological maturity) when one matured pod of brown or tan colour appears at any node on the main stem. While staging the crop in the field, a stage is declared when 50% of the plants have reached that particular stage. For modelling purposes, the end of floral induction, or R0, is used extensively. Soybean phenological stages are under tight genetic control. Under given cultivar/location/planting date combinations, values are consistent across years with low variation. Consequently, phenological data are frequently averaged across years in field studies (Board and Settimi, 1986). Other studies have demonstrated the consistency of phenological stages across years for similar cultivar/location/planting date combinations (Board and Harville, 1996).

1.1 Background to soybean-specific modelling

The development of crop-physiology-based simulation models for soybean or generic crops started in the 1980s. Several generations of soybean models have since been developed and used in research, education and production. SOYGRO (Wilkerson et al., 1983; Jones et al., 1989; Piper et al., 1996), which evolved later into CROPGRO-Soybean in the DSSAT package (Boote et al., 1998, 2003; Sau et al., 1999; Ruiz-Nogueira et al., 2001; Wang et al., 2003; Pedersen et al., 2004; Salmeron et al., 2016), can be regarded as one of the first-generation soybean-specific models and probably still has the most influence on the development of other soybean models. Other first-generation soybean models include the World Food Study, or WOFOST model (Diepen et al., 1989; Eitzinger et al., 2004; Boogaard et al., 2011, 2013), a crop-generic model that covers soybean, the Sinclair soybean models (Sinclair, 1986; Muchow and Sinclair, 1986; Sinclair et al.,

1987, 1991, 2003; Soltani and Sinclair, 2012), the Erosion-Productivity Impact Calculator, or EPIC model (Williams et al., 1984; Liu et al., 2007), a soil- and water-erosion-oriented model that includes a generic crop growth module, and the Soyphen model (Hodges and French, 1985) for soybean phenology only. The second-generation soybean models, most of which are crop generic with the inclusion of soybean, include CropSyst (Stockle et al., 1994, 2003), the Agricultural Production Systems sIMulator, or APSIM model (Wang et al., 2002; Keating et al., 2003; Mohanty et al., 2012; Archontoulis et al., 2014; Chen et al., 2016), and the Simulateur mulTIdisciplinaire pour les Cultures Standard, or STICS model (Brisson et al., 1998, 2003; Beaudoin et al., 2008; Jego et al., 2010). More recently developed, or third-generation soybean models, include SOYDEV (Setiyono et al., 2007, 2010a,b; Torrion et al., 2011), MONICA (Nendel et al., 2011, 2014) and AquaCrop (Raes et al., 2009; Steduto et al., 2009; Foster et al., 2017) models.

Robust simulation of crop phenology is critical in order to simulate crop development, growth and final yield, because phenology marks not only important changes in crop morphology but also the shift in internal physiological functions (Soltani and Sinclair, 2012; Jones et al., 1991). In crop models, changes in phenology mark the transition of mathematical functions and/or parameter values used to describe many biological processes. For instance, when most cereal crops come to flowering, vegetative growth will cease and seeds will set, followed by the exclusive allocation of photosynthate to seeds (Jones and Kiniry, 1986; Yang et al., 2004, 2017; Egli, 2010). For soybeans, however, flowering is not a distinct one-time event. Instead, it lasts a few weeks from R1 until the start of effective seed filling at R5 and reproductive growth overlays with vegetative growth. The reliable prediction of crop phenology is also important for crop management in the field, as phenology is used by many producers to schedule field operations (e.g. fertilizer applications, chemical spraying, etc.). However, the accurate simulation of soybean phenology has been a scientific challenge (Jones et al., 1991; Setiyono et al., 2007; Archontoulis et al., 2014; Cera et al., 2017). Although the qualitative understanding of the roles of temperature and photoperiod on soybean development is relatively clear, quantitative predictions of phenology, especially for reproductive stages under diverse management practices (e.g. planting date) and environmental conditions (location, temperature regime), are often far less satisfactory, especially in environments or management practices (e.g. date of planting) that are different to those used for model development or without model calibration (Archontoulis et al., 2014; Cera et al., 2017; Salmeron and Purcell, 2016). Using data of MGs from 00 to IX, Hodges and French (1985) reported the root mean square error (RMSE) of 7–11 days for the model-predicted R1 stage and 16–17 days for the R7 stage. Cera et al. (2017) tested the SOYDEV model as implemented in the SoySim model (Setiyono et al., 2010a,b) without calibration in southern Brazil for MGs of IV and over, and found errors in model predictions in the range of 10–16 days for R3 and R7 stages and up to 25 days for the R5 stage. Even after optimization of SOYGRO (version 5.42) and CROPGRO (version 3.0) models using observed data that covered MGs from 00 to VIII and latitudes from 18.03 to 45.00, Piper et al. (1996) reported an average RMSE of four days for the predicted date of R1, and 5–7 days for the R7 stage.

Many models require the calibration of parameters, using experimental observations under the same or similar conditions as those to be predicted (Jones et al., 1989; Piper et al., 1996; Boote et al., 2003; Keating et al., 2003; Boogaard et al., 2013; Archontoulis et al., 2014). Part of the reason is that cultivars change over time, and as a result, some of the model parameters need to be adjusted to reflect the changing crop genetics (Jones et al., 1989; Boote et al., 2003). The other reason is some model functions or parameter values do not always work well in new environments, as suggested by Hadley et al. (1984), Cera et al. (2017) and Battisti et al. (2017). As soybean development, especially the start of flowering and the start of seed filling, is sensitive to both temperature and photoperiod, the mathematical functions and choice of cardinal values of the functions are critical to the success of predicting phenology and yield. The model calibration of major phenological stages can indeed lead to more accurate simulation (Jego et al., 2010; Ruiz-Nogueira et al., 2001). However, for applications that predict outcomes under changing conditions or different environments (e.g. consequences of climate change or different planting dates), calibration is not always an option, making model results less reliable or even misleading.

Most models, including mathematical functions only or their implementation in computer software, that have been reported in publications performed well within the boundary conditions in which they were developed or calibrated (Wang et al., 2003; Torrion et al., 2011). The real question may be how robust they are when applied outside their original boundary conditions and how to make them more robust in less familiar territories. It is not uncommon for different models to produce significantly different results when simulating the same cases, as reported in van den Berg et al. (2002) and Bassu et al. (2014). Due to uncertainties of the current models, an ensemble of models has been proposed as a remedy when making serious predictions for unfamiliar situations, for example consequences of climate change (Palmer et al., 2004; Rotter et al., 2011; Rosenzweig et al., 2013a,b).

The objective of this chapter is to review the major approaches involved in the modelling of soybean reproductive stages, including their features, comparisons and the challenges and remaining scientific questions, with a focus on the functions of temperature, photoperiod and their interactions. The review only includes models that have been used more extensively in publications, and their formulation description was accessible to the author. Notations of mathematical symbols have been unified to keep them consistent throughout the chapter and, as a result, they may differ from the original publications.

2 Observed responses of soybean reproductive stages to temperature and photoperiod

Responses of soybean reproductive development to temperature and photoperiod were observed and reported as early as the 1930s (Garner and Allard, 1930; Steinberg and Garner, 1936). Garner and Allard (1930) found that first flowering (R1) was retarded by low temperature for soybean cultivars of vastly different maturities. The same study also reported that the duration from germination to R1 became the same when cultivars of different maturities were exposed to the same artificially shortened photoperiod, of 9.5–12 h, and the optimal photoperiod (P_{opt}), below which the development of flowering is at a maximum when the temperature is optimal, was around 12 h for some late maturity cultivars but shorter for early maturity cultivars. Steinberg and Garner (1936) reported that the photoperiod beyond which flowering would be greatly delayed or even indefinite, termed the critical photoperiod (P_{crt}), ranged from 17 to 13 h for early and late maturity cultivars, respectively. Similar findings and conclusions were reported in many later studies (Johnson et al., 1960; Major et al., 1975b; Wilkerson et al., 1989;

Summerfield et al., 1993, 1998; Cober et al., 2001). Note that the term P_{crt} has also been used to mean P_{opt} in some publications, for example Jones et al. (1989, 1991) and Cober et al. (2001).

2.1 Observed responses to temperature

The qualitative nature of the response of soybean reproductive development to temperature has been relatively consistent. At a photoperiod shorter than P_{opt}, the soybean development rate towards R1 responds positively with temperature, that is the rate increases with increasing temperature (Garner and Allard, 1930; Hadley et al., 1984; Seddigh et al., 1989), and the response is linear (Hadley et al., 1984; Upadhyay et al., 1994; Sinclair et al., 1991; Bastidas et al., 2008). Such responses are also valid for almost the entire growth cycle, including both vegetative and reproductive stages (Jones et al., 1991; Soltani and Sinclair, 2012). In addition, earlier maturity cultivars are more responsive to temperature than the later ones (Major et al., 1975b; McBlain et al., 1987). Although some studies, for example Hadley et al. (1983) and Seddigh et al. (1989), suggest that night temperature is more important in controlling soybean development, most studies have used the daily mean temperature (T_{mean}) to represent the temperature effect (Garner and Allard, 1930; Brown and Chapman, 1960; Hesketh et al., 1973; Hadley et al., 1984; Bastidas et al., 2008).

Quantitative attributes of the response of development rate to temperature include the base temperature (T_{base}) below which development ceases, start and end optimal temperatures (T_{opt1}, T_{opt2}, respectively) within which the development rate is optimal (or maximum), and maximum temperature (T_{max}) beyond which development pauses (Boote et al., 1998, 2003; Jones et al., 1989, 1991; Keating, et al., 2003; Archontoulis et al., 2014). Direct experimental measurements of those attributes are rare, and most reported values are mathematically derived based on experimental data. Upadhyay et al. (1994) estimated a T_{base} of 6–12°C and the range of T_{opt1} and T_{opt2} at around 30°C for diverse soybean cultivars, while Hadley et al. (1983) reported a T_{base} of 4.6–10°C and recommended 8°C as a generic value. Steinberg and Garner (1936) reported a T_{opt2} to be 28°C, while Hesketh et al. (1973) suggested 12°C and 30°C for T_{opt1} and T_{opt2}, respectively.

2.2 Observed responses to photoperiod

The response of soybean reproductive development to photoperiod is more complex. At an optimal temperature, most soybean cultivars exhibit a maximum development rate toward R1 when the photoperiod is shorter than P_{opt}, and development slows down when the photoperiod becomes longer than P_{opt} (Steinberg and Garner, 1936; Johnson et al., 1960; Criswell and Hume, 1972; Lawn and Byth, 1973; Major et al., 1975a,b; Major, 1980; Wilkerson et al., 1989; Summerfield and Roberts, 1984, 1993, 1998). Moreover, early maturity cultivars have a weaker response than later ones, and very early maturity cultivars (e.g. MG 00) can even be neutral to the photoperiod (Criswell and Hume, 1972; Summerfield and Minchin, 1976; Nissly et al., 1981; Cregan and Hartwig, 1984; Upadhyay et al., 1994). In addition, the response is reported to be linear for earlier maturity cultivars but curvilinear or even combination of linear and quadratic for the later ones (Major, 1980; Cregan and Hartwig, 1984). Although most soybean cultivars have a quantitative response to the photoperiod and will eventually flower even when the photoperiod becomes longer than P_{crt}, some cultivars have an obligate (qualitative) response to the photoperiod and will not flower if the photoperiod is longer than P_{crt} (Hadley et al., 1983; Upadhyay et al., 1994).

Another aspect to be considered when studying the soybean response to photoperiod is the phases during which the effect exists. There are conflicting reports about when the soybean plant first becomes photoperiod sensitive. Some studies, for example Major and Johnson (1977) and Shanmugasundaran and Tsou (1978), reported a period of insensitivity to photoperiod of up to nine days after emergence, whereas Wilkerson et al. (1989) found among six MGs that the plant was sensitive to the photoperiod as early as the unifoliate leaf stage, and Wang et al. (2002) reported that the response began once seedlings were exposed to light. In contrast, reports are more consistent on the strong response of soybeans to the photoperiod during the seedling stage, at least V1 stage, when the first trifoliate leafs are fully expanded (Major and Johnson, 1977; Hodges and French, 1985; Ellis et al., 1992; Collinson et al., 1993). This is also when soybean floral induction takes place, which lasts several days (Seddigh et al., 1989; Wilkerson et al., 1989; Heatherly and Elmore, 2004).

In comparison, research on photoperiod response after floral induction, especially after R1, has received less attention than that on response in phases prior to R1. Research by Johnson et al. (1960), Major et al. (1975a), Board and Settimi (1986) and Summerfield et al. (1993) showed that soybeans remained photoperiod responsive during R1 and R5. While Asumadu et al. (1998) suggested that the response of soybeans to the photoperiod ends before the appearance of the last flower, Kantolic and Slafer (2001, 2005) reported the longer duration of R3–R6 due to a longer photoperiod. Thomas and Raper (1976) and Cure et al. (1982) suggested that the photoperiod also affects the duration of seed filling (i.e. R5–R7).

Quantitative attributes of the response of the soybean development rate to photoperiod include P_{opt} and P_{crt}. Early maturity cultivars, namely those that are adapted to high latitudes and normally experience a longer photoperiod after emergence, tend to have greater P_{opt}, while the opposite is true for late maturity cultivars that are adapted to low latitudes and normally experience a relatively shorter photoperiod after emergence (Major et al., 1975b; McBlain et al., 1987; Upadhyay et al., 1994). There are few direct reports of P_{opt} and P_{crt} values from experiments and most are estimated mathematically using experimental data. Steinberg and Garner (1936) found P_{opt} to range from 8 to 13 h over a wide range of maturities and P_{crt} to range from 13 to 17 h for late and early maturity cultivars, respectively. Cregan and Hartwig (1984) found P_{opt} to be 15 h for earlier maturity cultivars (<MG III) and <11 h for late maturity cultivars (>MG III). In addition, Hadley et al. (1983) and Upadhyay et al. (1994) suggested that P_{opt} and P_{crt} may also be affected by temperature.

Soybean responses to temperature and photoperiod are believed to be independent of each other due to their separate control at the gene level (Buzzell, 1971; Buzzell and Voldeng, 1980; Major, 1980; Upadhyay et al., 1994). As a result, their effects are additive. When one factor is at an optimal level, the reproductive development rate depends solely on the other factor. However, there is considerable variation in photoperiod response due to genetic variation among cultivars of different maturities as well as similar maturities (Major et al., 1975b; McBlain et al., 1987; Kuroda et al., 1998). In addition, short maturity cultivars (adapted to high latitudes) have a stronger response to temperature but weaker response to photoperiod than long maturity cultivars (adapted to low latitudes) (Major et al., 1975; McBlain et al., 1987).

3 Modelling approaches for soybean reproductive stages

3.1 Major et al. (1975a) model

Major et al. (1975a), some of the pioneers in the mathematical modelling of soybean reproductive development in response to temperature and photoperiod, proposed a multiplicative effect of daily photoperiod (P, hrs) and daily mean temperature (T, °C) on the daily soybean reproductive development rate (R, ≥0) for any phases from emergence to R7:

$$R = f(P) * f(T) \tag{1}$$

where $f(P)$ and $f(T)$ are P and T functions, respectively. Note that R is the inverse of the duration in days of the phase and thus has the unit of d^{-1}. They further proposed curvilinear functions for $f(P)$ and $f(T)$:

$$f(P) = a_1(P - P_{crt}) + a_2(P - P_{crt})^2 \tag{2}$$

$$f(T) = b_1(T - T_{base}) + b_2(T - T_{base})^2 \tag{3}$$

where *a* and *b* sets of coefficients are regression coefficients, respectively, and $R = 0$ when either $P = P_{crt}$ or $T = b_{base}$.

The duration (D) from stage S_1 to stage S_2 is obtained by the integration of R over the period of S_1 to S_2:

$$D = \int_{S_1}^{S_2} R * dt \tag{4}$$

For daily computation, $dt = 1$. The unit and value of D can be days with a numerical value larger than 1 if R has a unit of day, or dimensionless with a value of 1 if R is dimensionless with a value of a fraction. As the coefficients in Eqs. (1) to (4) are specific for a given phase, their values must be derived from the fitting of observed data.

The feature of this method is the consideration of the photoperiod effect for all stages of soybean development. Its weakness is the requirement of curve fitting to obtain the six coefficients in Eqs. (2) and (3). Based on the results of ten cultivars of MGs I to V planted on different days and locations, the estimated values of P_{crt} and T_{base} can differ significantly for the same cultivars but at different stages. This may indicate either soybean biological variation at different stages or an artefact due to the use of the same form of the equation for different stages. Another sign of a possible artefact is the change of sign of the coefficients of the same cultivars but at different stages. In addition, the multiplicative effect of photoperiod and temperature shown in Eq. (1) does not allow scenarios where soybeans continue to develop and flower even under very long day conditions, as suggested by Hadley et al. (1984) and Roberts and Summerfield (1987).

3.2 Hadley et al. (1984) model

Hadley et al. (1984) proposed a linear function to describe the effect of photoperiod and temperature on the development rate (R) towards stage R1 for soybean cultivars that are photoperiod sensitive:

$$R = a + b * P + c * T \quad (5)$$

where *a*, *b* and *c* are coefficients. Using 55 sets of data of different cultivars and growth conditions, Hadley et al. (1984) reported the very good fit of Eq. (5) with r^2 ranging from 0.90 to 0.99 in controlled environment chambers.

Equation (5) differs from Eq. (1) in that: (1) it considers *P* and *T* as an additive role instead of a multiplicative role as in Eq. (1); (2) it maintains a minimum development rate under the conditions of very long daylength along with very low temperature when the roles of *P* and *T* are nil and (3) it is proposed to only work over the period from planting to R1. Equation (5) is similar to Eqs. (2) and (3) in that the coefficients of *a*, *b* and *c* must be obtained from regression or optimization through iteration using observed data, thus limiting its capability to predict soybean development under different conditions without calibration.

3.3 Jones and Liang (1978) model

Jones and Liang (1978) proposed the following model to describe soybean stages from planting to primary leaf, primary leaf to floral initiation and floral initiation to R1:

$$X_i = (X_{i-1} + N_i) * K * (1 - A + \Delta N) \quad (6)$$

where X_i represents the balance between a promotion and inhibition system on day *i*, in which X_i increases in the dark and decreases in the daylight, N_i is the night length, *K* is a temperature-dependent coefficient (≤ 1) and *A* is an auxiliary variable (see below). Flowering is initiated when X_i becomes greater than *R*, a predetermined, temperature-dependent variable. Coefficients *A*, *K* and *R* are estimated using the following equations:

$$A = a_1 + b_1 * T - c_1 * T^2 \quad (7)$$

$$K = a_2 + b_2 * T - c_2 * T^2 \quad (8)$$

$$R = a_3 + b_3 * T - c_3 * T^2 \quad (9)$$

where *a*, *b* and *c* are regression coefficients.

This model treats floral initiation and flowering (R1) as two separate events, but does not consider stages after R1. In addition, the mathematical functions are more complex than those in Major et al. (1975a) and Hadley et al. (1984), and thus are more difficult to calibrate.

3.4 Sinclair–Horie model

Sinclair et al. (1991) adopted the logistic model reported in Horie et al. (1986) for rice (*Oryza sativa* L.) to describe the responses of soybean flowering (R1) to temperature and photoperiod. It consists of three equations:

$$R = R_{max} * f(P) * f(T) \quad (10)$$

in which

$$f(T) = \frac{1}{\{1 + \exp[-A(T - T_h)]\}} \quad (11)$$

$$f(P) = 1 - \exp[B(P - P_{crt})] \text{ if } P \leq P_{opt}, \text{and } f(P) = 0 \text{ if } P > P_{opt} \quad (12)$$

where R_{max} is the maximum development rate (i.e. the reciprocal of days from sowing to flowering under optimal photoperiod and temperature conditions), A is a coefficient that presents the slope of the response, T_h is the temperature at which $f(T) = 0.5$ and B is a coefficient representing photoperiod responsiveness. The graphic representation of Eqs. (11) and (12) is shown in Fig. 1.

Using data from a 3-year field trial with 13 cultivars of MGs from III to IX and 10 planting dates, Sinclair et al. (1991) compared the fitting of the data to Eq. (5) [Hadley et al. (1984) model] and Eqs. (11) and (12), and found the latter to be more robust (the mean r^2 is 0.69 for the former and 0.81 for the latter). The same study also reported that the value of coefficient A in Eq. (11) increased from 0.10 for MG III to 0.34 for MG IX, the value of coefficient B in Eq. (12) decreased from 0.58 for MG III to 0.40 for MG IX, and P_{crt} decreased from 16.9 h for MG III to 14.3 h for MG IX. In contrast, the trend of R_{max} and T_h among the cultivars was less clear. The logistic model differs from the common linear models in that there are no definite cardinal values that define the turning points of temperature and photoperiod responses. Instead, it uses the two coefficients A and B to define the respective responsiveness, which have no clear biophysical definitions.

3.5 SOYGRO model

SOYGRO simulates all major aspects of soybean development and growth, including phenology, canopy expansion and dry matter accumulation; however, only reproductive stage development is discussed below. The early version of SOYGRO (version 4.2) as reported in Wilkerson et al. (1983) does not simulate phenology. Instead, major phenological dates are user inputs. Phenology simulation appeared in later versions in Wilkerson et al. (1985) and Jones et al. (1989).

Soybean reproductive staging in SOYGRO includes a juvenile phase (from emergence to V1 for most cultivars) during which the plant is photoperiod insensitive, followed by photoperiod-sensitive phases, including R0, R1, R3, R5 and R7. The model uses the concept of photothermal days (*PTD*), the combined effect of photoperiod and temperature (T), to measure the soybean development rate (R, ≤ 1) on a daily basis, similar to Eqs. (1) and (4):

$$R = f(N) * f(T) \quad (13)$$

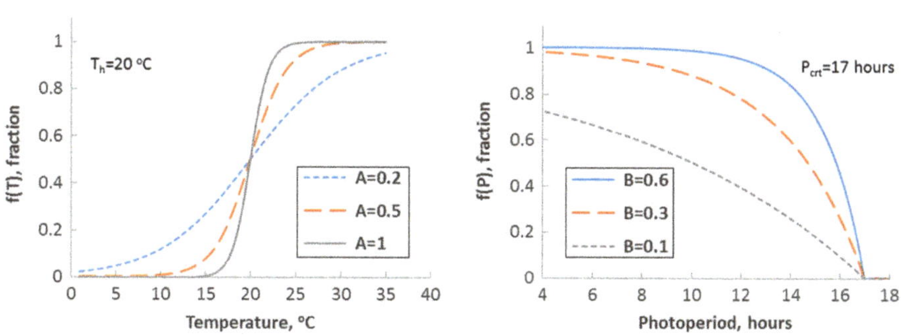

Figure 1 Temperature (left panel) and photoperiod (right panel) response of duration from soybean sowing to flowering (R1) using different values for A and B in Eqs. (11) and (12), respectively.

$$PTD = \int_{S_1}^{S_2} R * dt \tag{14}$$

where N is the night length and T refers to night temperature. The temperature function $f(T)$ follows a piecewise function as illustrated in Fig. 2. The value of $f(T)$ is 0 with T around 8°C, then increases almost linearly to the maximum of 1 at around 20°C and maintains this rate until about 28°C before dropping to zero at about 42°C. Note that the start and end temperatures of optimal development for photoperiod-insensitive stages in Fig. 2 are different from those of photoperiod-sensitive stages.

The photoperiod response of reproductive development is illustrated in Fig. 3. The duration of a reproductive phase will be minimum (i.e. the maximum relative development rate of 1) at an optimal night length (VARNO in Fig. 3, 12 h in this case) or longer, and the relative duration will then increase linearly to a threshold (VARTH in Fig. 3, that is the relative development rate becomes 1/VARTH) when night length decreases to the critical level (5 h in this case) or shorter. Note that the value of $f(N)$ in Eq. (13) equals $1/f_N$ in Fig. 3. Both $f(N)$ and $f(T)$ in Eq. (13) have a value between 0 and 1, with 1 for the maximum positive effect and 0 for no effect. For R to be 1 thus requires optimal conditions for both temperature and photoperiod. A reproductive stage arrives when PTD reaches a predetermined threshold.

VARNO and VARTH define the photoperiod sensitivity in SOYGRO, and their values vary most with cultivar maturities that are among the six coefficients, as shown in Fig. 4. In addition, SOYGRO uses four other genetic coefficients to describe the photoperiod effect on reproductive development under optimal temperature, including: (1) VARTHR-4, days of floral induction (i.e. from unifoliate leaf stage or V1 to R0); (2) VARTHR-6, days from R1 to R3; (3) VARTHR-8, days from R1 to R5 and (4) VARTHR-10, days from R1 to R7.

Although the SOYGRO model comes with generic values of the six genetic coefficients for most MGs, calibrations were required in many studies (Egli and Bruening, 1992; Colson et al., 1995; Boote et al., 1996; Piper et al., 1996, 1998). Calibration of the coefficients requires observed dates of sowing, V1, R1, R3, R5 and R7, and involves iteration procedures to minimize simulation errors. SOYGRO also assumes the same temperature response

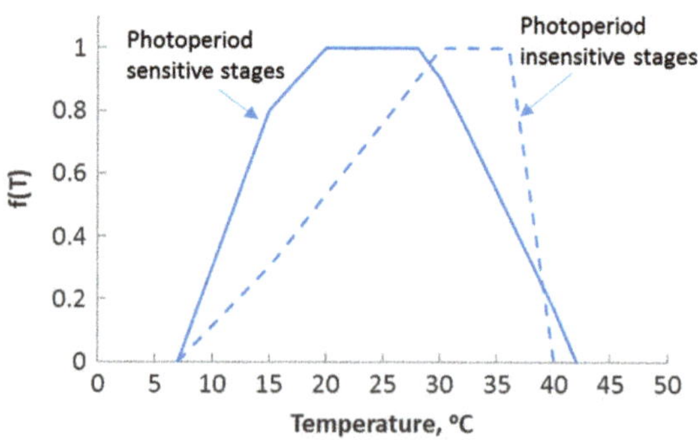

Figure 2 Temperature response for photoperiod-sensitive (solid line) and photoperiod-insensitive (dashed line) stages of soybean development used in SOYGRO. Note that night time mean temperature is used for photoperiod-sensitive stages, whereas daily mean temperature is used for photoperiod-insensitive stages. Source: Adapted from Jones et al. (1989).

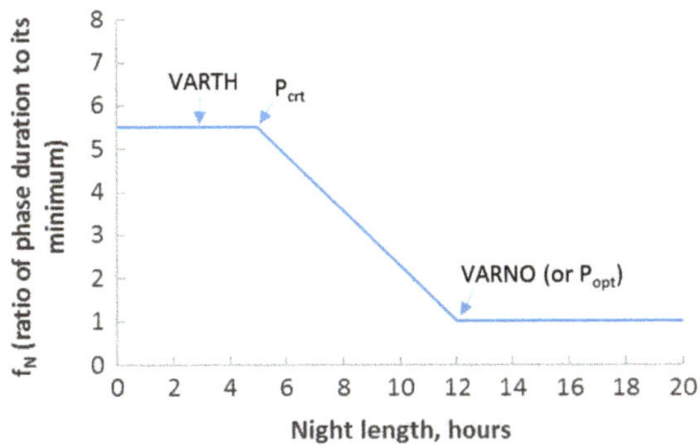

Figure 3 General relationship of relative duration of a soybean reproductive phase with night length used in SOYGRO. The values of VARTH, P_{crt} and VARNO (equivalent to P_{opt}) may differ among soybean cultivars. Source: Adapted from Jones et al. (1989). $1/f_N$ equals f(N) in Eq. (10).

function as in Fig. 2 for all stages and all cultivars, but direct experimental proof is still lacking. Nevertheless, calibration of the six major genetic coefficients provides the flexibility to adapt the model to different situations although with the constraints of limited power of projections for new conditions. Another interesting and potential controversial point of SOYGRO's formulation of the photoperiod response is that all post-R1 reproductive stages, including R3, R5 and R7, are linked directly back to R1, instead of as a successive relay one after another. This strategy may avoid the potential issue of errors from previous stages passing to the next ones. However, it does not follow the mechanism of sequential processes and could, at least potentially, result in the situation of a later stage occurring before its predecessor, as Cera et al. (2017) reported for the SOYDEV model following a similar strategy. Moreover, SOYGRO also assumes the same photoperiod sensitivity, as defined by VARNO and VARTH in Fig. 3, for all phases after R1, including R1–R3, R1–R5 and R1–R7. This indeed reduces the number of parameters in the model and may make parameter calibration easier, but may not capture all soybean biological behaviour with

Figure 4 General relationship of VARNO (i.e. optimal nightlength for soybean floral induction, left panel) and VARTH (i.e. the ratio of duration in days from end of juvenile stage to floral initiation under long photoperiod to that under short photoperiod, right panel) with soybean maturity group used in the SOYGRO model (version 5.42). Source: Adapted from Jones et al. (1989).

regard to photoperiod response. One sign of the deficiency of the SOYGRO formulation of temperature and photoperiod responses is that the reproductive development rate is too slow for very late soybean planting relative to its regular planting time, as reported by SOYGRO developers (Jones et al., 1991).

3.6 CROPGRO model

The CROPGRO model, as described in Boote et al. (1998), is a consolidation and unification of three models for legume crops, including SOYGRO (for soybeans), PNUTGRO (for peanuts) and BEANGRO (for drybeans). As such, it inherited some of the functions from each of the three models but kept crop-specific functions and parameters in separate external files. Each phase of crop development has its own unique developmental accumulator (or threshold value) that starts from a unique prior endpoint stage. Each stage is differentially sensitive to temperature and photoperiod (as well as water stress and nitrogen deficiency). The duration of a developmental phase is measured by 'physiological days', that is the number of calendar days under optimal conditions of temperature and photoperiod (and water and nitrogen).

The major difference between CROPGRO for soybeans and SOYGRO in the simulation of reproductive development is that CROPGRO uses two sets of cardinal temperatures for reproductive stages before R5 and after R5 (Fig. 5), whereas SOYGRO uses only one set. This strategy made CROPGRO more flexible and robust in simulating soybean reproductive stages under diverse conditions, for example late planting or high latitudes. Of particular interest is the mathematically derived very low T_{base} (−48°C) for reproductive stages after R5, based on the linear extrapolation of the response line. As the plant will not survive such a low temperature, such a T_{base} value only serves a mathematical purpose rather than a biophysical explanation. Alternatively, an extra section could be added to the line so that a more realistic T_{base} could be used. Also note that the cardinal temperatures for both reproductive stages are different from those for vegetative stages

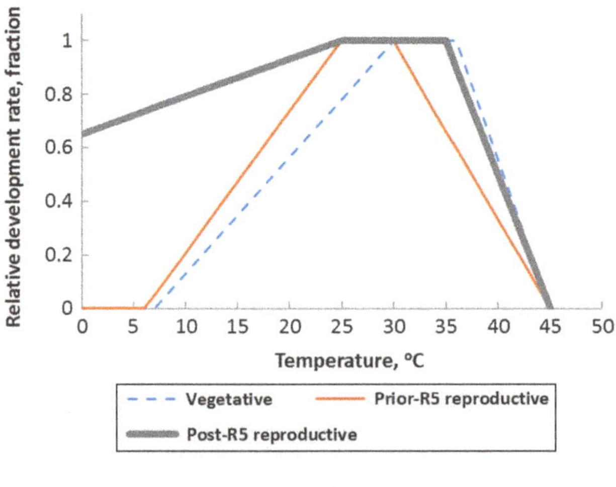

Figure 5 Temperature response of relative development rate of reproductive stages before R5 and after R5, and vegetative stages in CROPGRO. Source: Adapted from Boote et al. (1998).

in Fig. 5. Compared with SOYGRO, CROPGRO is more robust in simulating soybean phenology under a wide range of conditions, as reported by Piper et al. (1996), but at the expense of more parameters to calibrate. In addition, the high number of parameters can make model application more difficult when few data are available for calibration (Robertson et al., 2002).

3.7 APSIM model

APSIM is a generic crop model for farming system simulation, and as such it simulates a wider range of crops, including cereal crops and legumes (including soybeans) (Robertson et al., 2002; Wang et al., 2002; Keating et al., 2003). It uses a unified modelling framework, and all plant species follow the same physiological principles for growth and development. The main differences for different crops are the threshold values and shapes of their response functions. The legume module in APSIM, as described in Robertson et al. (2002), uses thermal time to drive phenological development where thermal time is calculated using three cardinal temperatures: T_{base}, T_{opt} and T_{max}. A stage arrives when the cumulative thermal time reaches a predetermined threshold value. For development phases that are sensitive to photoperiod, however, their thermal time thresholds are modified by the photoperiod effect during the phases. The cardinal values of photoperiod sensitivity include P_{opt} and P_{crt}.

As a later model, relative to SOYGRO and CROPGRO, APSIM for legumes differs from the former in four aspects related to phenology development. First, APSIM has only one development phase that is sensitive to photoperiod – from the end of the juvenile phase to floral initiation, whereas SOYGRO and CROPGRO treat all post-juvenile phases as photoperiod sensitive. Second, APSIM tracks reproductive stages sequentially one after another (Archontoulis et al., 2014), instead of tracking all post-R1 stages from R1 as in SOYGRO. Third, APSIM allows users to specify up to five 'x/y pairs' of the photoperiod response of a phase development rate and the model linearly interpolates between two adjacent points (Archontoulis et al., 2014), instead of using a fixed shape of response function. This makes it more flexible in dealing with diverse cultivars and conditions. Fourth, APSIM uses the triangle temperature response (Fig. 6) to describe the temperature response in crop development without the plateau, as in Figs. 2 and 5.

3.8 Stewart–Cober model

Stewart et al. (2003) reported a model that was an extension of the earlier work by Cober et al. (2001). This model describes the separate effects as well as interaction of temperature (*T*) and photoperiod (*P*) on the soybean development rate (*R*) towards R1. The functions are:

$$R = b\left[(T - T_{base}) - C_F\left(P - P_{opt}\right)\right] \tag{15}$$

in which

$$C_F = C - d\left(T_D - T\right) \tag{16}$$

where b and d are genetic coefficients, C is a photoperiod coefficient and T_D is a constant and set at 28°C. Note that T must be $\geq T_{base}$ and $\leq T_{max}$, and P must be $\geq P_{opt}$ and $\leq P_{crt}$. When

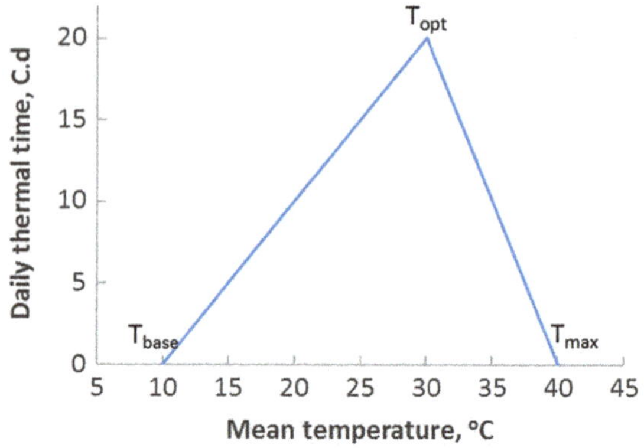

Figure 6 Temperature response of crop development in APSIM. Source: Adapted from Archontoulis et al. (2014).

using the concept of PTD, R can be replaced by PTD and *b* be removed from Eq. (15). Values of all coefficients and cardinal parameters in Eqs. (15) and (16) should be obtained from experiments or fitting observed phenological dates to the equations. Stewart et al. (2003) further decomposed *C* as the summation of effects from seven individual alleles (C_j):

$$C = \sum_{j=1}^{7} C_j \tag{17}$$

where *j* represents the six loci (E1, E2, E3, E4, E5 and E7), plus the effect of unknown loci represented by a seventh term (E?). Each component (C_j) has a positive value when dominant alleles are present and a value of zero when recessive alleles are present, except for the seventh term, which was a simple positive number.

Essentially, Eq. (16) introduces the T×P interaction and links T and P through the genetic coefficients d and genetic composition term C. Based on field data under natural light of 29 indeterminate near-isogenic lines, Stewart et al. (2003) reported r^2 of 0.92 between observed and calculated dates of R1 when not decomposing *C*, and r^2 of 0.89 when decomposing *C* to individual allele effects using Eq. (17). The decomposition of *C* resulted in individual values of C_j for the seven alleles, which indicate the different importance of the individual alleles in controlling the photoperiod effect. They further reported a linear positive correlation between the value of *C* and the number of loci with dominant alleles. So far, this model has only been tested for the phase from emergence to R1, and it is still unknown how well it may work for post-R1 phases.

3.9 WOFOST, CropSyst and STICS models

The WOFOST model, as described for version 7.1.2 in Boogaard et al. (2011), uses probably the most simplified method to account for the temperature and photoperiod responses in crop development, as shown in Fig. 7. Its temperature response did not include the suppressive effect at temperatures higher than the optimal as seen in SOYGRO, CROPGRO and APSIM models. This may not be a significant issue considering the rareness of having daily mean temperatures higher than T_{max}, for example 30°C. WOFOST treats photoperiod response in the same way as SOYGRO, but simpler than CROPGRO and less

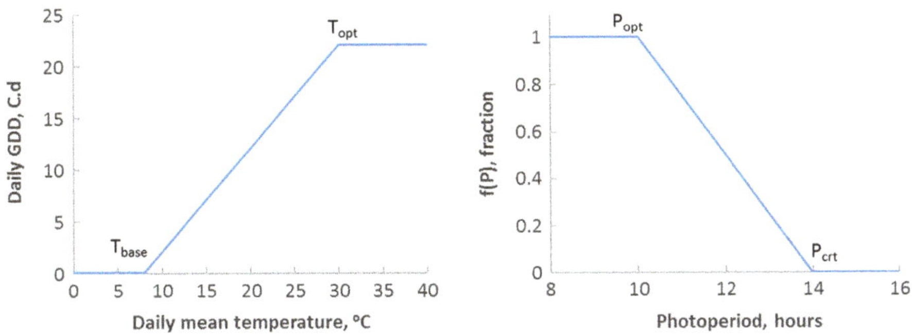

Figure 7 General response of temperature (left panel) and photoperiod (right panel, for short-day crops) in crop development in the WOFOST model. The cardinal values in the graphs vary among crop types and cultivars. Source: Adapted from Boogaard et al. (2011).

flexible than APSIM. In addition, WOFOST only considers the effect of photoperiod until the flowering stage.

The CropSyst model, as described in Stockle et al. (2003), is similar to WOFOST in accounting for the temperature effect using daily growing degree day (GDD) accumulation based on the concept of T_{base} and T_{opt}. The photoperiod effect on crop development, however, is not specifically formulated but requires the user to consider adjustment of the total GDD required for a phase. In addition, water stress in CropSyst is considered to be an accelerator for GDD accumulation due partly to the elevated canopy temperature under crop water stress.

The STICS model, as described in Brisson et al. (2003), is similar to WOFOST in its accounting for temperature and photoperiod effects. Different from WOFOST, STICS uses crop temperature, a corrected air temperature, for GDD calculation. This makes it similar to CropSyst in accounting for the effect of water stress on crop development rate.

3.10 SOYDEV model

SOYDEV (Setiyono et al., 2007) is a soybean phenology model that was implemented later in the SoySim model (Setiyono et al., 2010a,b). Many of its parameters were derived for MGs from 0.8 to 4.2 and the model was calibrated under conditions of the western Corn Belt of the USA. The model attempted to reduce the number of genetic parameters that are normally required by models like CROPGRO to facilitate model applications. Similar to CROPGRO, it treats most stages from the end of the juvenile phase as responsive to both temperature and photoperiod, but with the exception of the phase from cotyledon leaf (V0) to the end of floral induction (R0), which is insensitive to temperature. Like SOYGRO, SoySim tracks the duration of major reproductive phases, including R5 and R7, back to R1, instead of as successive phases. Its simulation of the daily reproductive development rate follows Eq. (10), but the effects of temperature f(T) and photoperiod f(P) are different. The f(T) is based on a non-linear beta function as used in Wang and Engel (1998):

$$f(T) = \frac{2(T - T_{base})^{\alpha} * (T_{opt} - T_{base})^{\alpha} - (T - T_{base})^{2\alpha}}{(T_{opt} - T_{base})^{2\alpha}} \tag{18}$$

in which

$$\alpha = \frac{\ln(2)}{\ln\left(T_{max} - \frac{T_{base}}{T_{opt}} - T_{base}\right)} \tag{19}$$

and $f(T) = 0$ if $T_{min} < T < T_{max}$
where T is the daily mean temperature and α determines the shape of the beta function. The comparison of Eq. (18) with the functions used in CROPGRO (Fig. 5) and WOFOST (Fig. 7, left panel) is shown in Fig. 8. The three methods are very close to each other except when T is greater than T_{opt}. SoySim's $f(P)$ function is a modified beta function of Yin et al. (1995):

$$f(P) = \left[\left(\frac{P - P_{opt}}{3} + 1\right)\left(\frac{P_{crt} - P}{P_{crt} - P_{opt}}\right)^{(P_{crt} - P_{opt})/3}\right]^{\alpha} \tag{20}$$

in which

$$\alpha = \frac{\ln(2)}{\ln\left[\left(\frac{P_{crt} - P_{opt}}{3}\right) + 1\right]} \tag{21}$$

and $f(P) = 0$ if $P > P_{crt}$ and $f(P) = 1$ if $P < P_{opt}$. Another feature of SoySim is that it uses generalized functions to estimate several genetic coefficients for a given MG to facilitate model applications to a wider range of MGs without calibration. These coefficients are P_{opt}

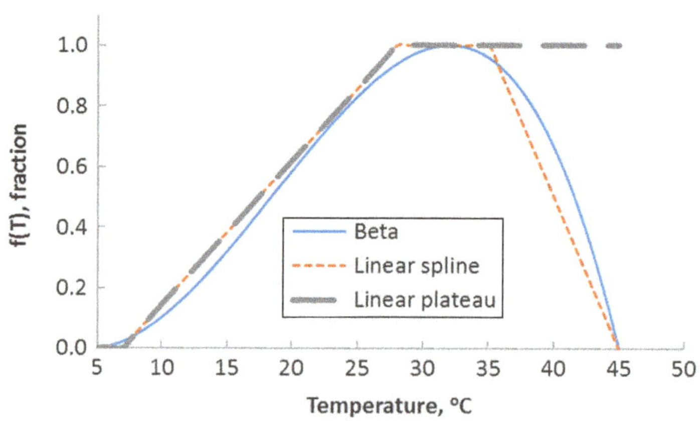

Figure 8 Comparison of three types of f(T) functions: beta (in SoySim), linear plateau (in WOFOST) and linear spline (in CROPGRO). The cardinal points are T_{base} = 5, T_{opt} = 32 and T_{max} = 45°C for the beta function; T_{base} = 7 and T_{opt} = 28°C for the linear plateau function; and T_{base} = 7, T_{opt1} = 28, T_{opt2} = 35 and T_{max} = 45°C for the linear spline function.

and P_{crt} (both in hours) for the phase of R0–R1, R_{max} (d^{-1}) of R0–R1, R1–R5 and R1–R7. The equations are:

$$P_{opt} = 12.759 - 0.388MG - 0.058MG^2 \quad \text{for R0 to R1} \tag{22}$$

$$P_{crt} = 27.275 - 0.493MG - 0.066MG^2 \quad \text{for R0 to R1} \tag{23}$$

$$R_{max} = \frac{0.2551 + 0.0965 * MG}{1 + 2.1107 * MG} \quad \text{for R0 to R1} \tag{24}$$

$$R_{max} = 0.075 - 0.0085 * MG \quad \text{for R1 to R5} \tag{25}$$

$$R_{max} = \frac{0.0563 + 0.0228 * MG}{1 + 1.9683 * MG} \quad \text{for R1 to R7} \tag{26}$$

Equations (22) and (23) were derived from data of Cregan and Hartwig (1984), which were for experiments conducted in photoperiod houses under several constant photoperiods using 12 genotypes with MGs from 00 to VIII. Equations (24)–(26) were based on field data of Setiyono et al. (2007) for MGs of 3.8 and lower. The graphic representation of Eq. (20) is shown in Fig. 9 for MGs I, II, III and IV. For general applications, SOYDEV, as implemented in SoySim (Setiyono et al., 2010a,b), requires only input settings of cultivar MG, stem termination type (indeterminate or semi-determinate) and sowing date, while other input settings (e.g. plant population, early growth conditions) are not related to phenology but yield. Other model parameters related to phenology that can be modified include: R_{max} for phases from emergence to V1, node appearance, V0–R0, R0–R1, R1–R5, R1–R7; stage indicators (i.e. the fraction between two ends of a phase) for R2 and R3 between R1 and R5, R3 between R1 and R5, and R6 between R1 and R7; cardinal temperatures (T_{min}, T_{opt} and T_{max}) for phases of emergence to V1, node appearance, R0–R1, R1–R5 and R1–R7; and cardinal photoperiod values (P_{opt}, P_{crt}) for phases of V0–R0, R1–R5 and R1–R7.

Equation (20) differs from those in Figs. 1, 3 and 7 in that it behaves more like a linear response initially after P becomes longer than P_{opt} but gradually changes to more

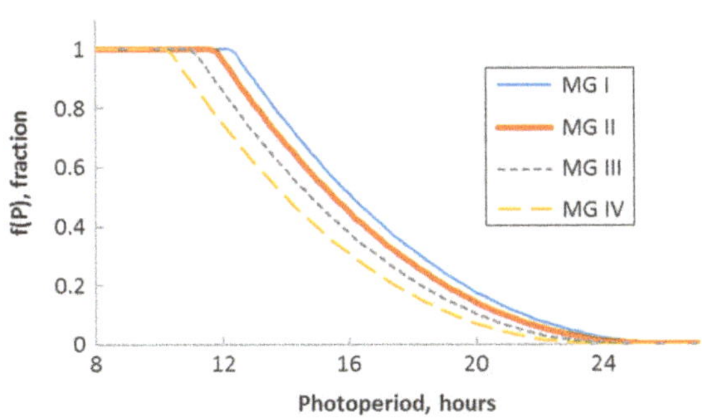

Figure 9 Photoperiod response of soybean development in the SoySim model. The four lines represent MGs from I and IV whose respective values of P_{opt} and P_{crt} were generated by Eqs. (19) and (20). Source: Adapted from Setiyono et al. (2007).

curvilinear when P approaches P_{opt}. As a side effect of the formulation, the value of P_{crt} exceeds 24 h for MGs IV and shorter, which may be explained as maintaining a minimum development rate at a daylength of 24 h. Based on the data of multiple year × cultivar × sowing day combinations with MGs 0.8–4.2 in Lincoln, Nebraska, USA, Setiyono reported RMSE of <2 days for a long-term trial and <4 days in a cultivar × sowing date trial for all major development stages after cultivar-specific calibration. When using generic coefficient values, however, RMSE increased to six and nine days in two years. Torrion et al. (2011) reported excellent results of model simulation using data from both farmers' fields and research plots. However, when the model was tested in southern Brazil for cultivars of MG ranging from 4.8 to 8.2 including both indeterminate and semi-determinate cultivars and under different sowing dates, the simulation errors were much larger. Moreover, some stages were predicted to occur even earlier than its predecessor in some cases (e.g. R6 occurred before R5). The causes for the larger errors for longer MGs, different planting dates and outside the environment where the model was originally developed and calibrated, could be the treatment of independent phenological development from R1 for R5, R6 and R7 or the use of a simplified relationship of R_{max} and P_{opt} and P_{crt} with MG.

4 Discussion of models

The diverse approaches employed in modelling the effects of temperature and photoperiod on soybean reproductive development raise questions regarding their strengths and weaknesses, when to best use them and what are the remaining scientific questions. As modelling is a simplified representation of a real system or process, there will never be a perfect match between model results and observations. On the other hand, users expect satisfactory and meaningful results from models. Depending on the purpose of model development and application, the best approach can differ. For instance, the simple approach used in the WOFOST model as shown in Fig. 7 can be used if an application does not require high accuracy or sensitivity to photoperiod response, whereas models like CROPGRO and APSIM may be a better option in more complex situations but with sufficient data for calibration. At present, the major scientific challenge is predicting soybean phenology in relatively untested situations, for example significant change of crop sequences in environments where a crop can be planted within a wide-ranging time window (e.g. in the tropics and subtropics), and scenarios of climate change. Those cases bring in complex combinations of change in temperature regime, shift in photoperiod at different stages and the possibility of new cultivars.

There are several scientific issues that are still unclear and controversial in the models reviewed earlier. The first one is whether the photoperiod response exists after R1. Some models like SOYGRO, CROPGRO and SOYDEV treat all phases after R1 as photoperiod sensitive, while models like WOFOST and APSIM consider photoperiod sensitivity only for phases prior to R1. Experimental conclusions are also inconsistent on this issue (Johnson et al., 1960; Major et al., 1975a; Thomas and Raper, 1976; Cure et al., 1982; Summerfield et al., 1993; Setiyono et al., 2007). This could partly be due to the narrow range of photoperiod variation in experiments, making it difficult to detect the effect, or other factors (e.g. temperature) masking the photoperiod effect.

The second question is the nature of temperature and photoperiod response: linear vs non-linear or curvilinear functions. Linear functions have fewer parameters and are easier to use, whereas non-linear or curvilinear functions may be more flexible in dealing with diverse situations. Moreover, a curvilinear relationship can often be simplified to a linear function within a narrow range of the independent variable. To a certain extent, the choice depends on the range of conditions of applications. As a result, a linear function would work adequately over a range of conditions, but may deviate too much when used beyond the range in which it was established. On the other hand, curvilinear functions may better capture the underlying processes, making them more robust scientifically but at the cost of more data requirement for development, parameterization and testing (Major, 1980; Cregan and Hartwig, 1984; Boote, et al., 1998).

The third question is the additive vs multiplicative effect of temperature and photoperiod; experimental evidence is not conclusive. For instance, research at the gene level suggests a separate and independent effect of temperature and photoperiod (Buzzell, 1971; Buzzell and Voldeng, 1980; Major, 1980; Upadhyay et al., 1994), while other studies, for example Cober et al. (2001) and Stewart et al. (2003), suggest the existence of interactions between the two factors, which can also be interpreted as dependence of the photoperiod effect on temperature. The difference in the conclusion may also be due to the use of different mathematical functions to explain the observed variations. One advantage of the additive function is that it allows a minimum development rate at temperatures below T_{base} and photoperiod beyond P_{crt}, as observed in experiments for some cultivars, for example in Major et al. (1975a).

The fourth question regards the values of the cardinal parameters, including T_{base}, T_{opt}, T_{max}, P_{opt} and P_{crt}, as they are supposed to be physically measurable. However, it is not always feasible to do so in conventional experimental setups and they are frequently estimated mathematically using data points from other ranges of conditions. This would not be an issue if the function were still valid at the cardinal points. Unfortunately, many functions still lack direct experimental proof that this is the case. For instance, T_{base} is difficult to observe under field conditions and few experiments have been conducted to address this question. A side effect of deriving cardinal values mathematically is that their values depend on the underlying functions and may lose biophysical meaning. For instance, the T_{base} for the phase of R5–R7 in Fig. 5 has a value of −48°C, and P_{crt} in Fig. 9 has values exceeding 24 h. In those cases, the cardinal values serve more as a mathematical convenience for modelling rather than directly revealing biophysical attributes. The dependence of parameter values on the functions makes direct comparisons across models of those values less meaningful. This issue also adds to the difficulties of experimental verification and confirmation. Another question related to cardinal parameters is, what if they have the same values at different stages of crop development. Models using the same values for different stages often experience poor predictions in situations that are significantly different from their usual applications (e.g. SOYGRO in Jones et al., 1989), whereas models that use a different set of values can experience significantly improved performance (e.g. CROPGRO in Boote et al., 1998).

The fifth question concerns model calibration. Many models require parameter calibration when used in conditions that are new to a model. One of the issues is how to ensure that the calibration has targeted the correct parameters. Although some models provide automated procedures for calibration, the results are not always as intended because of the possibility of internal error cancellation or compensation, as reported by Salmeron and Purcell (2016) and Battisti et al. (2017). In addition, calibration may not always be feasible in all applications because of a lack of observation data, and this is particularly true for projections under unfamiliar conditions.

It is always a compromise and trade-off in the choice of functions for model development. Simple functions are easy to use but may lack robustness and wider applicability. In contrast, more complex functions could potentially be more powerful in predicting outcomes under a wider range of scenarios and explaining underlying processes, but their requirement for input setup may be beyond the access of many applications. Sometimes modelling can be more of an art than science. For applications that demand less mechanistic processes, empiric modelling can be valuable and more practical.

As plant behaviour is determined by genetics and gene expression in the environment, a more fundamental approach of modelling is at the gene level, as attempted by Stewart et al. (2003) and Yin and Struik (2008). This requires quantification of plant genetic attributes that can be measured independent of environmental conditions. As expression of some genes depends on environmental conditions, an effective quantification scheme is also needed. This type of research requires the consolidation of biotechnology, bioinformatics and conventional field experimentation.

5 Conclusion

In conclusion, there are diverse mathematical approaches to describe the effects of temperature and photoperiod on soybean reproductive development. Their robustness depends on their intended applications. Most models require parameter calibration when applied outside the range of conditions of their original development. To improve the current modelling of temperature and photoperiod effect on soybean development, quantitative research of integrated modelling and experimentation at the gene level may hold the final key.

6 References

Archontoulis, S. V., Miguez, F. E. and Moore, K. J. 2014. A methodology and an optimization tool to calibrate phenology of short-day species included in the APSIM PLANT model: Application to soybean. *Environmental Modelling & Software*, 62, 465–77.

Asumadu, H., Summerfield, R. J., Ellis, R. H. and Qi, A. 1998. Variation in the durations of the photoperiod-sensitive and photoperiod-insensitive phases of post-first flowering development in maturity isolines of soyabean *Glycine max* (L.) Merrill 'Clark'. *Annals of Botany*, 82, 773–8.

Bassu, S., Brisson, N., Durand, J. L., Boote, K., Lizaso, J., Jones, J. W., Rosenzweig, C., Ruane, A. C., Adam, M., Baron, C. and Basso, B. 2014. How do various maize crop models vary in their responses to climate change factors? *Global Change Biology*, 20, 2301–20.

Bastidas, A. M., Setiyono, T. D., Dobermann, A., Cassman, K. G., Elmore, R. W., Graef, G. L. and Specht, J. E. 2008. Soybean sowing date: The vegetative, reproductive, and agronomic impacts. *Crop Science*, 48, 727–40.

Battisti, R., Sentelhas, P. C. and Boote, K. J. 2017. Inter-comparison of performance of soybean crop simulation models and their ensemble in southern Brazil. *Field Crops Research*, 200, 28–37.

Beaudoin, N., Launay, M., Sauboua, E., Ponsardin, G. and Mary, B. 2008. Evaluation of the soil crop model STICS over 8 years against the 'on farm' database of BruyFres catchment. *European Journal of Agronomy*, 29, 46–57.

Bernard, R. L. 1972. Two genes affecting stem termination in soybeans. *Crop Science*, 12, 235–9.

Board, J. E. and Harville, B. G. 1996. Growth dynamics during the vegetative period affects yield of narrow-row, late-planted soybean. *Agronomy Journal*, 88, 567–72.

Board, J. E. and Settimi, J. R. 1986. Photoperiod effect before and after flowering on branch development in determinate Soybean. *Agronomy Journal*, 78, 995–1002.
Board, J. E., Harville, B. G. and Saxton, A. M. 1990. Branch dry weight in relation to yield increases in narrow-row soybean. *Agronomy Journal*, 82, 540–4.
Boogaard, H. L., de Wit, A. J. W., te Roller, J. A. and van Diepen, C. A. 2011. User's guide for the WOFOST Control Center 1.8 and WOFOST 7.1.3 crop growth simulation model. Alterra, Wageningen University, Wageningen, the Netherlands.
Boogaard, H., Wolf, J., Supit, I., Niemeyer, S. and van Ittersum, M. 2013. A regional implementation of WOFOST for calculating yield gaps of autumn-sown wheat across the European Union. *Field Crops Research*, 143, 130–42.
Boote, K. J., Jones, J. W. and Pickering, N. B. 1996. Potential uses and limitations of crop models. *Agronomy Journal*, 88, 704–16.
Boote, K. J., Jones, J. W., Hoogenboom, G. and Pickering, N. B. 1998. The CROPGRO model for grain legumes. In: Tsuji, G. Y., Hoogenboom, G. and Thornton, P. K. (Eds), *Understanding Options for Agricultural Production*. Springer Netherlands, Dordrecht, the Netherlands, pp. 99–128.
Boote, K. J., Jones, J. W., Batchelor, W. D., Nafziger, E. D. and Myers, O. 2003. Genetic coefficients in the CROPGRO-Soybean model: Links to field performance and genomics. *Agronomy Journal*, 95, 32–51.
Brisson, N., Mary, B., Ripoche, D., Jeuffroy, M. H., Ruget, F., Nicoullaud, B., Gate, P., Devienne-Barret, F., Antonioletti, R., Durr, C., Richard, G., Beaudoin, N., Recous, S., Tayot, X., Plenet, D., Cellier, P, Machet, J. M., Meynard, J. M. and Delecolle R. 1998. STICS: A generic model for the simulation of crops and their water and nitrogen balances. I. Theory and parameterization applied to wheat and corn. *Agronomy Journal*, 18, 311–46.
Brisson, N., Gary, C., Justes, E., Roche, R., Mary, B., Ripoche, D., Zimmer, D., Sierra, J., Bertuzzi, P., Burger, P., Bussière, F., Cabidoche, Y. M., Cellier, P., Debaeke, P., Hénault, C., Maraux, F., Seguin, B. and Sinoquet, H. 2003. An overview of the crop model STICS. *European Journal of Agronomy*, 18, 309–32.
Brown, D. M. and Chapman, L. J. 1960. Soybean ecology. II. Development-temperature-moisture relationships from field studies. *Agronomy Journal*, 52, 496–9.
Buzzell, R. L. 1971. Inheritance of a soybean flowering response to fluorescent-daylength conditions. *Canadian Journal of Genetic Cytology*, 13, 703–7.
Buzzell R. I. and Voldeng H. D. 1980. Inheritance of insensitivity to long daylength. *Soybean Genetics Newsletter*, 7, 26–9.
Cera, J. C., Streck, N. A., Yang, H. S., Zanon, A. J., de Paula, G. M. and Lago, I. 2017. Extending the evaluation of the SoySim model to soybean cultivars with high maturation groups. *Field Crops Research*, 201, 162–74.
Chen, C., Lawes, R., Fletcher, A., Oliver, Y., Robertson, M., Bell, M. and Wang, E. 2016. How well can APSIM simulate nitrogen uptake and nitrogen fixation of legume crops? *Field Crops Research*, 187, 35–48.
Cober, E. R., Stewart, D. W. and Voldeng, H. D. 2001. Photoperiod and temperature responses in early-maturing, near-isogenic soybean lines. *Crop Science*, 41, 721–7.
Cober, E. R., Curtis, D. F., Stewart, D. W. and Morrison, M. J. 2014. Quantifying the effects of photoperiod, temperature and daily irradiance on flowering time of soybean isolines. *Plants*, 3, 476–97.
Collinson, S. T., Summerfield, R. J., Ellis, R. H. and Roberts, E. H. 1993. Durations of the photoperiod-sensitive and photoperiod-insensitive phases of development to flowering in 4 cultivars of soybean Glycine-max (L) Merrill. *Annals of Botany*, 71, 389–94.
Colson, J., Bouniols, A. and Jones, J. W. 1995. Soybean reproductive development: Adapting a model for European cultivars. *Agronomy Journal*, 87, 1129–39.
Cregan, P. B. and Hartwig, E. E. 1984. Characterization of flowering response to photoperiod in diverse soybean genotypes. *Crop Science*, 24, 659–62.
Criswell, J. G. and Hume, D. J. 1972. Variation in sensitivity to photoperiod among early maturing soybean strains. *Crop Science*, 12, 657–60.

Cure, J. D., Patterson, R. P., Raper Jr., C. D. and Jackson, W. A. 1982. Assimilate distribution in soybeans as affected by photoperiod during seed development. *Crop Science*, 22, 1245–50.
Diepen, C. A. van, Wolf, J., Keulen, H. van and Rappoldt, C. 1989. WOFOST: A simulation model of crop production. *Soil Use and Management*, 5(1), 16–24.
Egli, D. B. 2010. SOYPOD: A model of fruit set in soybean. *Agronomy Journal*, 102, 39–47.
Egli, D. B. and Bruening, W. 1992. Planting date and soybean yield: Evaluation of environmental effects with a crop simulation model: SOYGRO. *Agricultural and Forest Meteorology*, 62, 19–29.
Eitzinger, J., Trnka, M., Hösch, J., Zalud, Z. and Dubrovský, M. 2004. Comparison of CERES, WOFOST and SWAP models in simulating soil water content during growing season under different soil conditions. *Ecological Modelling*, 171, 223–46.
Ellis, R. H., Collinson, S. T., Hudson, D. and Patefield, W. M., 1992. The analysis of reciprocal transfer experiments to estimate the durations of the photoperiod-sensitive and photoperiod-insensitive phases of plant development – an example in soya bean. *Annals of Botany*, 70, 87–92.
Fehr, W. R. and Caviness, C. E. 1977. Stages of soybean development. Cooperative Extension Service, Agriculture and Home Economics Experiment Station Iowa State University, Ames, IA.
Foster, T., Brozović, N., Butler, A. P., Neale, C. M. U., Raes, D., Steduto, P., Fereres, E. and Hsiao, T. C. 2017. AquaCrop-OS: An open source version of FAO's crop water productivity model. *Agricultural Water Management*, 181, 18–22.
Garner, W. W. and Allard, H. A. 1930. Photoperiodic response of soybeans in relation to temperature and other environmental factors. *Journal of Agricultural Research*, 41, 719–35.
Hadley, P., Roberts, E. H., Summerfield, R. J. and Minchin, F. R. 1983. A quantitative model of reproductive development in cowpea [*Vigna unguiculata* (L) Walp.] in relation to photoperiod and temperature, and implications for screening germplasm. *Annals of Botany* 51, 531–43.
Hadley, P., Roberts, E. H., Summerfield, R. J. and Minchin, F. R. 1984. Effects of temperature and photoperiod on flowering in soya bean [*Glycine max* (L.) Merrill]: A quantitative model. *Annals of Botany*, 53, 669–81.
Heatherly, L. G. and Elmore, R. W. H. 2004. Managing inputs for peak production. In: R. Boerma and J. E. Specht (Eds), *Soybeans: Improvement, Production, and Uses*. American Society of Agronomy, Madison, WI, pp. 451–536.
Hesketh, J. D., Myhre, D. L. and Willey, C. R. 1973. Temperature control of time intervals between vegetative and reproductive events in soybeans. *Crop Science*, 13, 250–4.
Hodges, T. and French, V. 1985. Soyphen: Soybean growth stages modeled from temperature, daylength, and water availability. *Agronomy Journal*, 77, 500–5.
Horie, T., Nakagawa, H. and Kira, T. 1986. Modeling and prediction of the developmental processes of rice: I. Dynamic model for predicting the flowering date. *Japanese Journal of Crop Science*, 55, 214–15.
Jego, G., Pattey, E., Bourgeois, G., Morrison, M. J., Drury, C. F., Tremblay, N. and Tremblay, G. 2010. Calibration and performance evaluation of soybean and spring wheat cultivars using the STICS crop model in Eastern Canada. *Field Crops Research*, 117, 183–96.
Johnson, H. W., Borthwick, H. A. and Leffel, R. C. 1960. Effects of photoperiod and time of planting on rates of development of the soybean in various stages of the life cycle. *Botanical Gazette*, 122, 77–95.
Jones, J. W., Boote, K. J., Hoogenboom, G., Jagtap, S. S. and Wilkerson, G. G. 1989. SOYGRO V5.42: Soybean crop growth simulation model user guide. Florida Agricultural Experiment Station journal no. 8304, University of Florida, Gainesville, FL, 75p.
Jones, J. W., Boote, K. J., Jagtap, S. S. and Mishoe, J. W. Hanks, R. J. and Ritchie, J. T. 1991. Soybean development. Modeling plant and soil systems. Agronomy Monograph no. 31, ASA-CSSA-SSSA, Madison, WI.
Jones, C. A. and Kiniry, J. R. 1986. *CERES-Maize: A Simulation Model of Maize Growth and Development*. Texas A&M University Press, College Station, TX.
Jones, P. G. and Lang, D. R. 1978. Simulation of the phenology of soybeans. *Agricultural Systems*, 3, 295–311.

Kantolic, A. G. and Slafer, G. A. 2001. Photoperiod sensitivity after flowering and seed number determination in indeterminate soybean cultivars. *Field Crops Research*, 72, 109–18.

Kantolic, A. G. and Slafer, G. A. 2005. Reproductive development and yield components in indeterminate soybean as affected by post-flowering photoperiod. *Field Crops Research*, 93, 212–22.

Keating, B. A., Carberry, P. S., Hammer, G. L., Probert, M. E., Robertson, M. J., Holzworth, D., Huth, N. I., Hargreaves, N. G., Meinke, H., Hochman, Z., McLean, G., Verburg, K., Snow, V., Dimes, J. P., Silburn, M., Wang, E., Brown, S., Bristow, K. L., Asseng, S., Chapman, S., McCown, R. L., Freebairn, D. M. and Smith, C. J. 2003. An overview of APSIM, a model designed for farming systems simulation. *European Journal of Agronomy*, 18, 267–88.

Kuroda, T., Saitoh, K., Mahmood, T. and Yanagawa, K. 1998. Differences in flowering habit between determinate and indeterminate types of soybean. *Plant Production Science*, 1(1), 18–24.

Lawn, R. J. and Byth, D. E. 1973. Response of soyabeans to planting data on south-western Queensland. 1. Influence of photoperiod and temperature on phasic development patterns. *Australian Journal of Agricultural Research*, 24, 67–80.

Liu, J., Williams, J. R., Zehnder, AJ. B. and Yang, H. 2007. GEPIC – modelling wheat yield and crop water productivity with high resolution on a global scale. *Agricultural Systems*, 94, 478–93.

Major, D. J. 1980. Photoperiod response characteristics controlling flowering of nine crop species. *Canadian Journal of Plant Science*, 60, 777–84.

Major, D. J. and Johnson D. R. 1977. A Review of the Effect of daylength and temperature on development of Soybeans (Glycine Max (L.) Merrill). Lethbridge Research Station Mimeo Report 2. Agricultural Canada Research Station, Lethbridge, Alberta, Canada.

Major, D. J., Johnson, D. R., Tanner, J. W. and Anderson, I. C. 1975a. Effects of daylength and temperature on soybean development. *Crop Science*, 15, 174–9.

Major, D. J., Johnson, D. R. and Luedders, V. D. 1975b. Evaluation of eleven thermal unit methods for predicting soybean development. *Crop Science*, 15, 172–4.

McBlain, B. A., Hesketh, J. D. and Bernard, R. L. 1987. Genetic effects on reproductive phenology in soybean isolines differing in maturity genes. *Canadian Journal of Plant Science*, 67, 105–15.

Mohanty, M., Probert, M. E., Reddy, K. Sammi, Dalal, R. C., Mishra, A. K., Subba Rao, A., Singh, M. and Menzies, N. W. 2012. Simulating soybean-wheat cropping system: APSIM model parameterization and validation. *Agriculture, Ecosystems & Environment*, 152, 68–78.

Muchow, R. C. and Sinclair, T. R. 1986. Water and nitrogen limitations in soybean grain production II. Field and model analyses. *Field Crops Research*, 15, 143–56.

Nendel, C., Berg, M., Kersebaum, K. C., Mirschel, W., Specka, X., Wegehenkel, M., Wenkel, K. O. and Wieland, R. 2011. The MONICA model: Testing predictability for crop growth, soil moisture and nitrogen dynamics. *Ecological Modelling*, 222, 1614–25.

Nendel, C., Kersebaum, K. C., Mirschel, W. and Wenkel, K. O. 2014. Testing farm management options as climate change adaptation strategies using the MONICA model. *European Journal of Agronomy*, 52(Part A), 47–56.

Nissly, L. R., Bernard, R. L. and Hittle, C. N. 1981. Variations in photoperiod sensitivity for time of flowering and maturity among soybean strains of Maturity Group III. *Crop Science*, 21, 833–6.

Palmer, T. N., Doblas-Reyes, F. J., Hagedorn, R., Alessandri, A., Gualdi, S., Andersen, U., Feddersen, H., Cantelaube, P., Terres, J. M., Davey, M. and Graham, R. 2004. Development of a European multimodel ensemble system for seasonal-to-interannual prediction (DEMETER). *Bulletin of the American Meteorological Society*, 85, 853–72.

Pedersen, P., Boote, K. J., Jones, J. W. and Lauer, J. G. 2004. Modifying the CROPGRO-Soybean model to improve predictions for the upper Midwest. *Agronomy Journal*, 96, 556–64.

Piper, E. L., Boote, K. J., Jones, J. W. and Grimm, S. S. 1996. Comparison of two phenology models for predicting flowering and maturity date of soybean. *Crop Science*, 36, 1606–14.

Piper, L., Boote, J. and Jones, W. 1998. Evaluation and improvement of crop models using regional cultivar trial data. *Applied Engineering in Agriculture*, 14, 435.

Polson, D. E. 1972. Day neutrality in soybeans. *Crop Science*, 12, 773–6.

Raes, D., Steduto, P., Hsiao, T. C. and Fereres, E. 2009. AquaCrop—the FAO crop model to simulate yield response to water II. Main algorithms and soft ware description. *Agronomy Journal*, 101, 438–47.

Roberts, E. H. and Summerfield, R. J. 1987. Measurement and prediction of flowering in annual crops. In: Atherton, J. G. (Ed.), *Manipulation of Flowering*. Butterworths, London, UK, pp. 17–50.

Robertson, M. J., Carberry, P. S., Huth, N. I., Turpin, J. E., Probert, M. E., Poulton, P. L., Bell, M., Wright, G. C., Yeates, S. J. and Brinsmead, R. B. 2002. Simulation of growth and development of diverse legume species in APSIM. *Australian Journal of Agricultural Research*, 53, 429–46.

Rosenzweig, C., Jones, J. W., Hatfield, J. L., Ruane, A. C., Boote, K. J., Thorburn, P., Antle, J. M., Nelson, G. C., Porter, C., Janssen, S., Asseng, S., Basso, B., Ewert, F., Wallach, D., Baigorria, G. and Winter, J. M. 2013a. The Agricultural Model Intercomparison and Improvement Project (AgMIP): Protocols and pilot studies. *Agricultural and Forest Meteorology*, 170, 166–82.

Rosenzweig, C., Elliott, J., Deryng, D., Ruane, A. C., Muller, C., Arneth, A., Boote, K. J., Folberth, C., Glotter, M., Khabarov, N., Neumann, K., Piontek, F., Pugh, T. A. M., Schmid, E., Stehfest, E., Yang, H. and Jones, J. W. 2013b. Assessing agricultural risks of climate change in the 21st century in a global gridded crop model intercomparison. *Proceedings of the National Academy of Sciences*, 111(9), 3268–73.

Rotter, R. P., Carter, T. R., Olesen, J. and Porter, J. R. 2011. Crop-climate models need an overhaul. *Nature Climate Change*, 1, 175–7.

Ruiz-Nogueira, B., Boote, K. J. and Sau, F. 2001. Calibration and use of CROPGRO-soybean model for improving soybean management under rainfed conditions. *Agricultural Systems*, 68, 151–73.

Salmeron, M. and Purcell, L. C. 2016. Simplifying the prediction of phenology with the DSSAT-CROPGRO-soybean model based on relative maturity group and determinacy. *Agricultural Systems*, 148, 178–87.

Sau, F., Boote, K. J. and Ruiz-Nogueira, B. 1999. Evaluation and improvement of CROPGRO-soybean model for a cool environment in Galicia, Northwest Spain. *Field Crops Research*, 61, 273–91.

Seddigh, M., Jolliff, G. D. and Orf, J. H. 1989. Night temperature effects on soybean phenology. *Crop Science*, 29, 400–6.

Setiyono, T. D., Weiss, A., Specht, J., Bastidas, A. M., Cassman, K. G. and Dobermann, A. 2007. Understanding and modeling the effect of temperature and daylength on soybean phenology under high-yield conditions. *Field Crops Research*, 100, 257–71.

Setiyono, T., Cassman, K. G., Specht, J, Dobermann, A., Weiss, A., Yang, H. S., Conley, S. P., Robinson, A. P., Pedersen, P. and De Bruin, J. L. 2010a. Simulation of soybean growth and yield in near-optimal growth conditions. *Field Crops Research*, 119, 161–74.

Setiyono, T., Cassman, K. G., Specht, J., Weiss, A., Dobermann, A. and Yang, H. S. 2010b. SoySim: A simulation model for soybean growth and yield. University of Nebraska – Lincoln, Lincoln, NE.

Shanmugasundaran, S. and Tsou, S. C. S. 1978. Photoperiod and critical duration for flower induction in soybean. *Crop Science*, 18, 598–601.

Sinclair, T. R. 1986. Water and nitrogen limitations in soybean grain production I. Model development. *Field Crops Research*, 15, 125–41.

Sinclair, T. R., Muchow, R. C., Ludlow, M. M., Leach, G. J., Lawn, R. J. and Foale, M. A. 1987. Field and model analysis of the effect of water deficits on carbon and nitrogen accumulation by soybean, cowpea and black gram. *Field Crops Research*, 17, 121–40.

Sinclair, T. R., Kitani, S., Hinson, K., Bruniard, J. and Horie, T. 1991. Soybean flowering date: Linear and logistic models based on temperature and photoperiod. *Crop Science*, 31, 786–90.

Sinclair, T. R., Farias, J. R., Neumaier, N. and Nepomuceno, A. L. 2003. Modeling nitrogen accumulation and use by soybean. *Field Crops Research*, 81(2/3), 149–58

Soltani, A. and Sinclair, T. R. 2012. *Modeling Physiology of Crop Development, Growth and Yield*. CABI International, Wallingford, UK.

Steduto, P., Hsiao, T. C., Raes, D. and Fereres, E. 2009. AquaCrop—the FAO crop model to simulate yield response to water I. Concepts and underlying principles. *Agronomy Journal*, 101, 426–37.

Steinberg, R. A. and Garner, W. W.1936. Response of certain plants to length of day and temperature under controlled conditions. *Journal of Agricultural Research*, 52, 943–60.

Stewart, D. W., Cober, E. R. and Bernard, R. L. 2003. Modeling genetic effects on the photothermal response of soybean phenological development. *Agronomy Journal*, 95, 65–70.
Stockle, C. O., Martin, S. A. and Campbell, G. S. 1994. CropSyst, a cropping systems simulation model: Water/nitrogen budgets and crop yield. *Agricultural Systems*, 46, 335–59.
Stockle, C. O., Marcello, D. and Roger N. 2003. CropSyst, a cropping systems simulation model. *European Journal of Agronomy*, 18, 289–307.
Summerfield, R. J. and Minchin, F. R. 1976. An integrated strategy for daylength and temperature sensitivity screening of potentially tropic-adapted soybeans. In: Goodman, R. M. (Ed.), *Expanding the Use of Soybeans*. INTSOY Series No. 10. University of Illinois, Urbana, IL, pp. 186–91.
Summerfield, R. J. and Roberts, E. H. 1984. Soyabeans. In: Halevy, A. H. (Ed.), *A Handbook of Flowering*. CRC Press, Boca Raton, FL.
Summerfield, R. J. and Wilcox, J. R. 1978. Effects of photoperiod and air temperature on growth and yield of economic legumes. In: Summerfield, R. J., Bunting, A. H. (Eds), *Advances in Legume Sciences*. Royal Botanic Gardens, Kew, England, pp. 17–36.
Summerfield, R. J., Lawn, R. J., Qi, A., Ellis, R. H., Roberts, E. H., Chay, P. M., Brouwer, J. B., Rose, J. L., Shanmugasundaram, S., Yeates, S. J. and Sandover, S. 1993. Towards the reliable prediction of time to flowering in 6 annual crops. 2. Soybean (Glycine-max). *Experimental Agriculture*, 29, 253–89.
Summerfield, R. J., Asumadu, H., Ellis, R. H. and Qi, A. 1998. Characterization of the photoperiodic response of post-flowering development in maturity isolines of soyabean *Glycine max* (L.) Merrill 'Clark'. *Annals of Botany*, 82, 765–71.
Thomas, J. F. and Raper Jr., C. D. 1976. Photoperiodic control of seed filling for soybeans. *Crop Science*, 16, 667–72.
Thompson, J. A., Bernard, R. L. and Nelson, R. L. 1997. A third allele at the soybean dt1 locus. *Crop Science*, 37, 757–62.
Torrion, J., Setiyono, T., Cassman, K. G. and Specht, J. 2011. Soybean phenology simulation in the North-Central United States. *Agronomy Journal*, 103, 1661–7.
Upadhyay, A. P., Ellis, R. H., Summerfield, R. J., Roberts, E. H. and Qi, A. 1994. Characterization of photothermal flowering responses in maturity isolines of soyabean [*Glycine max* (L.) Merrill] cv. Clark. *Annals of Botany*, 74, 87–96.
van den Berg, M., Driessen, P. M. and Rabbinge, R. 2002. Water uptake in crop growth models for land use systems analysis: II. Comparison of three simple approaches. *Ecological Modelling*, 148, 233–50.
Wang, E and Engel, T. 1998. Simulation of phenological development of wheat crops. *Agricultural Systems*, 58, 1–24.
Wang, E., Robertson, M. J., Hammer, G. L., Carberry, P. S., Holzworth, D., Meinke, H., Chapman, S. C., Hargreaves, J. N. G., Huth, N. I. and McLean, G. 2002. Development of a generic crop model template in the cropping system model APSIM. *European Journal of Agronomy*, 18, 121–40.
Wang, F., Fraisse, C. W., Kitchen, N. R. and Sudduth, K. A. 2003. Site-specific evaluation of the CROPGRO-soybean model on Missouri claypan soils. *Agricultural Systems*, 76, 985–1005.
Wilkerson, G. G., Jones, J. W., Boote, K. J., Ingram, K. T. and Mishoe, J. W. 1983. Modeling soybean growth for crop management. *Transactions of the ASAE*, 26, 63–73.
Wilkerson, G. G., Jones, J. W., Boote, K. J. and Mishoe, J. W. 1985. SOYGRO V5.0: Soybean crop growth and yield model. Technical documentation. Agricultural Engineering Department, University of Florida, Gainesville, FL.
Wilkerson, G. G., Jones, J. W., Boote, K. J., Buol, G. S. 1989. Photoperiodically sensitive interval in time to flower of soybean. *Crop Science*, 29, 721–6.
Williams, J. R., Jones, C. A. and Dyke, P. T. 1984. A modeling approach to determining the relationship between erosion and soil productivity. *Transactions of the ASAE*, 27, 129–44.
Yang, H. S., Dobermann, A., Lindquist, J. L., Walters, D. T., Arkebauer, T. J. and Cassman, K. G. 2004. Hybrid-maize – a maize simulation model that combines two crop modeling approaches. *Field Crops Research*, 87, 131–54.

Yang, H. S., Grassini, P., Cassman, K. G., Aiken, R. M. and Coyne, P. I. 2017. Improvements to the Hybrid-Maize model for simulating maize yields in harsh rainfed environments. *Field Crops Research*, 204, 180–90.

Yin, X. and Struik, P. C. 2008. Applying modelling experiences from the past to shape crop systems biology: the need to converge crop physiology and functional genomics. *New Phytologist* 179(3), 629–42. doi:10.1111/j.1469-8137.2008.02424.x.

Yin, X., Kropff, M. J., McLaren, G. and Visperas, R. M. 1995. A nonlinear model for crop development as a function of temperature. *Agricultural and Forest Meteorology*, 77, 1–16.

Chapter 11

Defining and implementing best management practices in soybean production

D. Reynolds, Mississippi State University, USA

1 Introduction

This chapter addresses the sustainable production of soybean through the development of best management practices (BMPs). Before we can describe these practices, we must first understand what a BMP is. A BMP can be defined as the optimal way to achieve a desired goal within the resources available (*Black's Law Dictionary*). In our context, agricultural BMPs can be defined more specifically as 'management, agronomic and technological practices that can permit economic and viable production while achieving the least possible adverse impact on the environment' (Anon. 2002). BMPs are thus management, agronomic and technological practices that optimize the profitability of soybean production while minimizing adverse effects on the environment.

A review of the literature will quickly reveal that BMPs are available for just about any discipline you might imagine. In fact, many of the authors of the other discipline-specific chapters in this book have provided some examples of a BMP for their discipline area. However, it is imperative to note that soybean production BMPs are specific to geographic regions. For instance, a soybean production BMP designed for a state in the southern United States, such as Mississippi, may not be adequate for a state in the Midwest such as Iowa. Furthermore, one should be aware that production practices within a region may dictate differences in the adoption of BMPs. For example, soil type and environmental conditions may determine what type of tillage practice can be used.

http://dx.doi.org/10.19103/AS.2017.0034.11

A term related to BMP is 'good agricultural practices' (GAPs). These have been defined as 'practices that address environmental, economic and social sustainability for on-farm processes, and result in safe and quality food and non-food agricultural products' (FAO 2003). Whilst GAPs have a wider scope than BMPs do, both are concerned with the sustainable optimization of agricultural production. A sample of GAP guides specifically for soybean production help to identify the scope of GAP/BMPs for this crop. Four sources have been used to identify common elements of GAP in soybean production:

- The United States: Pioneer (2017)
- The United States: University of Missouri Extension Service (2017a,b)
- Kenya: Kenya Agricultural Research Institute (KARI 2013)
- Nigeria: N2Africa/Africa Soil Health Consortium (ASHC) (N2Africa 2014)

The common elements are summarized in Table 1.

The dynamic nature of BMPs for soybean production makes it extremely difficult to cite specific practices that will be the same across large geographic regions. However, as Table 1 suggests, there are some general practices that transcend geographic areas and are common to all soybean producers. Ultimately, development of BMPs should include decisions on the following components: tillage system, soil fertility, variety selection, agronomic practices and pest management. In this chapter, we will attempt to address BMPs common to all geographic areas but whose details may be location specific. In these cases, we will describe the practice and recommend that individuals consult agronomic professionals within their geographic region for specific details. Lastly, one should realize that the selection of a specific management practice in one aspect of production may influence or even dictate the BMP in other areas of production.

2 Tillage

Historically, tillage has been relied on as a means of mechanical weed control and seedbed preparation (Buhler and Gunsolus 1996; Lal et al. 2007; Popp et al. 2000; Tu et al. 2001).

Table 1 Common GAP elements

GAP	Pioneer	Missouri	KARI	N2N
Site selection and preparation (including tillage)	Y	Y	Y	Y
Fertilizer/soil health	Y	Y		Y
Variety/seed selection	Y	Y	Y	Y
Seed treatment (incl. inoculation)	Y	Y		Y
Planting	Y	Y	Y	Y
Irrigation		Y		
Weed control	Y	Y	Y	Y
Pest and disease management		Y	Y	Y
Harvesting	Y	Y	Y	Y
Post-harvest operations (e.g. storage)			Y	Y

Production systems utilizing frequent tillage are commonly referred to as conventional tillage systems. In addition to weed control, tillage may also affect water infiltration into the soil and surface runoff; however, these effects are usually short-lived following the tillage event (Benjamin 1993; Bronick and Lal 2005; Green et al. 2003; Lampurlanes and Cantero-Martinez 2006).

The advent of modern herbicides and more efficient planting equipment has allowed for the adoption of no-tillage or conservation tillage practices (Derpsch et al. 2010; Givens et al. 2009; Karayel 2009; Lanie et al. 1994). As the name implies, a true no-tillage system eliminates mechanical tillage and relies upon other technologies, mainly herbicides, for weed control prior to and following planting. In terms of seedbed preparation in no-tillage systems, many specialty planters and drills have been developed for planting into undisturbed soils that often have high levels of crop residue present from the preceding crop (Karayel 2009).

Definitions of conventional and no-tillage systems are much more distinct than definitions of minimum, reduced or conservation tillage systems. Simply put, minimum, reduced or conservation tillage systems are somewhere in between conventional and no-tillage systems. While they do not implement tillage on the entire field as with conventional tillage systems, they do, however, incorporate some sort of tillage and therefore cannot be coined no-tillage systems. Conservation tillage systems often rely on chemical weed control prior to and shortly after planting. In some areas, this system would be described as strip tillage as only a small strip of soil where the crop is planted is disturbed (Morrison 2002). Another example of conservation tillage is the stale-seedbed system often used in the southern United States (Heatherly 1999b; Heatherly and Elmore 1983). In this system, tillage for seedbed preparation is performed several months prior to planting and is left fallow. Prior to planting, existing weeds and/or cover crops are terminated with herbicides, thus leaving soil undisturbed (Oliver et al. 1993; Shaw 1996).

Which of these systems is best for soybean production? Much like we described previously, any one of these systems may be the BMP for producers in a given region. However, topography and consequently soil erosion often dictate which type of tillage system can be implemented. Highly erodible land (HEL) is defined by the United States Department of Agriculture-National Resources Conservation Service (USDA-NRCS) as land with soils that have an erodibility index of eight or more (NRCS 2017). Thus, HEL requires the use of a conservation or no-tillage system to prevent erosion. However, this is not the only case where no-tillage or conservation tillage cropping systems are used. Regardless of topography or erodibility, research has demonstrated many beneficial attributes to conservation and no-tillage systems (Derpsch et al. 2010).

Many factors would contribute to the determination of what the BMP practice, relative to tillage, might be. We have just described regulatory consideration, but other factors might include equipment availability and type, variety selection and pest pressure. While conservation and no-tillage systems offer yield advantages, these advantages often do not manifest for several years. Frequently, producers switch from conventional tillage to conservation tillage systems with unrealistic expectations and are disappointed with the results (DeLaune et al. 2012; Lal 1991; Logan et al. 1991; Rhoton 2000). Currently, there is high interest in the use of cover crops for not only the soil health benefits, but also for weed suppression, especially herbicide-resistant weeds (Ateh and Doll 1996; Liebel et al. 1992; Moore et al. 1994; Shilling et al. 1995; Teasdale et al. 1991).

3 Soil fertility

Regardless of the tillage system, soil fertility is extremely important in soybean production. Without proper soil testing and amendments, soybean yield potential can never be met. The first step in this component of a BMP is soil testing. Without knowing soil pH or nutrient levels, it is impossible to accurately amend soil and likewise optimize production (Mengel et al. 1987). Soil pH between 6.0 and 6.5 is ideal for soybean nutrient availability as shown in Fig. 1 (Johnson 1987; Truog 1947). The results of a soil test may indicate that nutrients are available in sufficient quantities for optimum production; however, the pH of the soil may preclude their availability to the plant (Truog 1947). Thus, in order to maximize utilization of soil nutrients, soil pH must be adjusted to the appropriate level.

An additional aspect of any fertility programme is accounting for nutrient removal as a function of crop harvest. Each metric ton of soybean seed contains approximately 55, 12, 20, 3.5 and 3 kg of nitrogen (N), phosphorus (P_2O_5), potassium (K_2O), magnesium (Mg) and sulphur (S), respectively (Phillips and Majumdar 2015). Typical nutrient removal levels for varying soybean yields common throughout the world are depicted in Table 2. Nutrient demand is greatest from the late vegetative growth stages through the beginning of seed development (Bender et al. 2015). Additionally, modern varieties with higher yield potential have a higher nutrient demand compared to older varieties (Bender et al. 2015). Many agricultural soils have nutrient reserves that do not require amendments of phosphorus or potassium for optimum yields. While these reserves may last several years, high-yielding varieties used today can often mine these nutrients down to insufficient

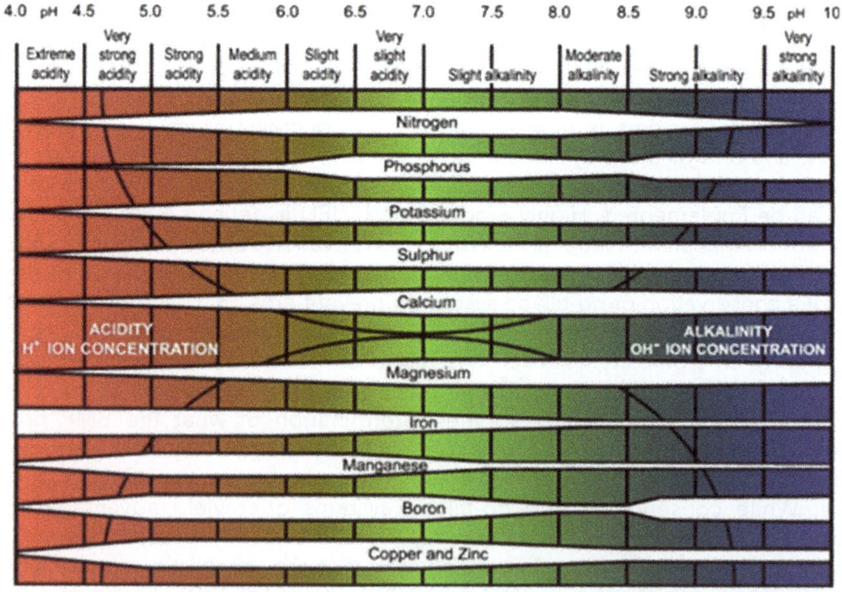

Figure 1 Availability of plant nutrients in relation to soil acidity. The availability of different pH bands is indicated by the width of the white bar: the wider the bar, the more available is the nutrient (redrawn for PDA from Truog (1947)). Soil reaction influence on availability of plant nutrients. *Soil Science Society of America Journal.*

Table 2 Kilograms of essential nutrients removed by a typical yielding soybean crop

	N	P_2O_5	K_2O	Mg	S
Metric tons/ha	kg/ha				
2	110	24	40	7	6
3	165	36	60	10.5	9
4	220	48	80	14	12

Source: Adapted from Phillips and Majumdar (2015).

levels (Bender et al. 2015). Therefore, it is important to have soil tested each year to determine nutrient levels and amend according to the test results.

In the presence of proper soil bacterium and environmental conditions, symbiotic bacteria can fix soybean's N requirements from the atmosphere (Hungria and Vargas 2000; Mengel et al. 1987; Zahran 1999). In some soybean-producing areas of the world that may undergo intense drought and heat, inoculation may be required each year as bacterium communities decline (Hungria and Vargas 2000; Zahran 1999). In fields where legumes have been produced in recent years, it is generally accepted that additional inoculation is not required; however, if legumes have not been produced in a field for three years, it is recommended to inoculate at planting (DeBruin et al. 2008; Ruiz et al. 2009; Salvagiotti et al. 2008).

4 Variety selection

Proper variety selection may arguably have the single biggest impact on a producer's yield potential. Even under optimum fertility and environmental conditions, not all soybean varieties will perform similarly. Consequently, most soybean-producing regions have geography-specific recommendations for variety selection. Many genotype-by-environment experiments are conducted throughout soybean production areas of the world (Gurmu et al. 2009; Sudarić et al. 2006; Yan et al. 2010). These experiments are designed to identify the variety with the greatest yield potential when grown under the environmental conditions of a given geographic area. Many factors contribute to these differences. Depending upon the latitude, maturity groups may differ significantly. Generally, it is accepted that in the northern hemisphere, the more northern latitudes will grow shorter-season maturity groups (I–II), while southern latitudes will grow longer-season maturity groups such as V–VII (Zhang et al. 2007). However, the adaptation of soybean with an indeterminate growth habit and the early season production system has resulted in the adoption of Maturity Group IV in areas that used to predominantly grow Maturity Groups V–VII (Heatherly 1999a; Salmeron et al. 2014). Earlier planting dates of these maturity groups in the south has been highly successful, resulting in higher soybean yields while facilitating earlier harvest in the fall. This approach maximizes growth and yield by capturing the more frequent early-season rainfall events during the vegetative and reproductive growth stages as well as allowing harvest prior to late-season rains that may be detrimental to yields and grain quality.

Depending on geographic region, soybean can be subject to a variety of diseases. Several seed treatments and foliar-applied fungicides are available to manage a variety of these pests. However, some diseases have no seed or foliar-applied fungicide options for management (Athow 1987). Often the best approach to managing these diseases would be the selection of a variety that has a natural genetic tolerance to this specific pest. A good example would be sudden death syndrome or stem canker where the selection of a variety with resistance to this organism would be the most effective management approach (Athow 1987).

Another important aspect of variety selection is that of a transgenic trait. In many soybean production areas, various herbicide and insecticide traits are available (Castle et al. 2006). These are discussed in more detail in the chapters related to entomology and weed science. Many factors contribute to the selected trait. One must consider the economics of adopting the use of transgenic crops to determine if they are cost-effective (Carpenter 2010; Givens et al. 2009). In areas of high adoption of a specific transgenic trait, like glyphosate resistance, producers will often adopt technologies in an attempt to mitigate any off-target deposition of the respective herbicide used by neighbouring producers. Also, the evolution of herbicide resistance in weed species has driven adoption of some transgenic traits. Ironically, it was the over-reliance on a single mode of action provided by some of these transgenic crops that led to the rapid development of resistance (Owen et al. 2005; Givens et al. 2009). In areas where glyphosate-resistant weed species have developed, producers have been forced to utilize more soil residual herbicides while in many cases also adopting glufosinate, dicamba or 2,4-D resistant soybean varieties (Byker et al. 2013; Johnson et al. 2010). The bottom line is that the use of multiple modes of action and multiple methods of weed control is imperative to steward these technologies from the development of species with resistance to multiple modes of action (Legleiter and Bradley 2008).

5 Agronomic practices

Planting dates vary by geographic region and may vary greatly within a given geographic area. In the mid-southern United States, planting may range from mid-April to late June (Salmeron et al. 2014). Generally, after the last spring frost, soybean can be planted. Planting dates have recently shifted dramatically in the mid-southern United States. Traditionally in this region, soybean Maturity Groups VI to VII were planted from May to July. By 2014, most of the soybean crop in Mississippi was composed of Maturity Groups IV and V most of which are planted from mid-April to mid-May. This shift is due to the adoption of the early planting soybean production system as described by Heatherly (1999a).

Row spacing is dependent upon many factors. Producers who have equipment set to wider rows (100 cm) for other crops such as cotton may produce soybean on the same width rows due to equipment availability. However, in some cases, soybean is planted in twin rows on these wider row spacings in an attempt to obtain quicker canopy closure for enhanced weed control and to maximize yields (Bruns 2011a,b; Bruns and Young 2012; Janovicek et al. 2006). In areas where raised beds are needed for furrow irrigation, the minimum row spacing is generally around 75 cm as it is difficult to create beds that are less than 75 cm wide. Consequently, when producers elect to plant on 38- or 19-cm row spacings, they are generally planted without raised beds (Bruns and Young 2012).

A great deal of research has been conducted to evaluate the optimum seeding rate and row-width combination for soybean. The ultimate goal in modern soybean production is to maximize yield while planting as low of a plant population as possible in order to conserve seed costs which can be quite high for transgenic varieties. Research has shown that plant populations can vary significantly without affecting yields (Heatherly et al. 1999; Lueschen and Hicks 1977). A seeding rate of approximately 320 000 seeds/ha will generally allow for optimum yield potential (Carpenter and Board 1997; Lueschen and Hicks 1977). Research has shown that seeding rates lower than this in some instances will yield as well as the higher seeding rates (Carpenter and Board 1997; Lueschen and Hicks 1977). The problem with attempting to lower the seeding rate to increase profitability is that any unexpected factor that might affect soybean establishment may result in unacceptable reductions in plant population and may result in the need to replant.

6 Pest management

The best management strategy for pests in particular is to have a proactive rather than a reactive strategy. A good point of reference would be that of historical records for a particular field or area. Additionally, this would include scouting for disease, insect and weed pests. Producers should include allowances in their production budgets for these services. Some pests are best controlled with prophylactic treatments, while others should be treated when the pest reaches an economic injury level. Economic injury levels are the point at which an economic loss will occur without treatment for the given pest (Bauer and Mortensen 1992; Ragsdale et al. 2007). Additionally, multiple modes of action should be used for all pesticides when possible. The over-reliance on a single mode of action leads to the development of resistant pests that further complicate their management (Norsworthy et al. 2012; Sparks and Nauen 2015). Each of the major disease, insect and weed pests is discussed in discipline-specific chapters.

As previously mentioned, the adoption of other production practices may greatly affect other areas of production. A good example of this is in the area of insect management where cover crops are being used. Termination of some cover crops in a short interval prior to planting could lead to increased insect pressure during emergence and early vegetative growth (Dunbar et al. 2016). Generally, if cover crops are terminated, a minimum of three weeks prior to planting, insect pressure should not be increased.

In an era of heightened environmental concerns and the possibility of severe crop injury due to off-target applications of some pesticides, pesticides should be applied with proper equipment under proper environmental conditions to help mitigate crop injury, yield loss and environmental exposure. Current areas of high concern include off-target deposition of synthetic-auxin herbicides such as dicamba to dicamba-sensitive soybean and the effect of several insecticides on pollinator species (Egan et al. 2014; Zeng et al. 2013). We must be good stewards of these technologies in order to sustain their availability to soybean producers around the world.

7 Development of best management practices

It should now be apparent that there is no 'one size fits all' scenario for the development of BMPs for soybean production. Many of the most important aspects of such a management plan have been outlined. It is important to be reminded that our definition of a BMP is management, agronomic and technological practices that optimize the profitability of soybean production while minimizing adverse effects to the environment. I would like to emphasize that producing the highest yield may neither be the most profitable nor minimize adverse effects to the environment.

In order to develop your BMPs, the first step is to determine the type of tillage system you will use (conventional, no-tillage, etc.) based upon the factors described above. This decision will likely impact several of your other production practices, particularly as it relates to weed control. This step may also impact the selection of row spacing and/or the use of a flat versus raised bed production systems. The second step is to properly sample and test your soil and amend as needed based upon your soil test results and your potential yield goal. The third step will be the selection of the best variety for your production system. This may be heavily influenced by pest pressures (both weed and disease), transgenic trait use or quite simply market availability. The germplasm of the variety selected may have a significant effect on your yield potential. The fourth step is the selection of your agronomic practices. As we discussed, these may be dictated by some of your initial choices like tillage, equipment availability or transgenic traits. These practices often overlap with the fifth step which is the selection and implementation of your pest management programme. This programme may include variety selection as well as seed treatments and foliar-applied treatments. Too often as researchers we lose sight of the complex interaction of all of these components of a BMP. They are truly dynamic, and adjustments in one area of the system often result in changes in other areas. The utilization of data specific to a geographic area will facilitate the development of the most effective production system for producers in that region allowing them to optimize profitability while minimizing adverse effects to the environment.

8 References

Anon. 2002. Manual of Best Management Practices (BMPs) for agriculture in New Hampshire. Agricultural Best Management Practices Task Force/USDA, Concord, NH. https://nerc.org/documents/manure_management/best_management_nh.pdf

Ateh, C. M. and J. D. Doll. 1996. Spring-planted winter rye (*Secale cereale*) as a living mulch to control weeds in soybean (*Glycine max*). *Weed Technology* 10: 347–53.

Athow, K. L. 1987. Fungal diseases. In: J. R. Wilcox (Ed.), *Soybeans: Improvement, Production, and Uses*, 2nd ed. (pp. 687–727). Madison, WI: American Society of Agronomy, Crop Science Society of America, Soil Science Society of America.

Bauer, T. A. and D. A. Mortensen. 1992. A comparison of economic and economic optimum thresholds for two annual weeds in soybeans. *Weed Technology* 6: 228–35.

Bender, R. R., J. W. Haegele and F. E. Below. 2015. Nutrient uptake, partitioning, and remobilization in modern soybean varieties. *Agronomy Journal* 107: 563–73.

Benjamin, J. G. 1993. Tillage effects on near-surface soil hydraulic properties. *Soil and Tillage Research* 26: 277–88.

Black's Law Dictionary. 2012. What is BEST MANAGEMENT PRACTICE (BMP)? Definition of BEST MANAGEMENT PRACTICE (BMP) (Black's Law Dictionary). *The Law Dictionary*. N.p. 12 October 2012. Web. 03 July 2017. http://thelawdictionary.org/best-management-practice-bmp/

Bronick, C. J. and R. Lal. 2005. Soil structure and management: A review. *Geoderma* 124: 3–22.

Bruns, H. A. 2011a. Comparisons of single-row and twin-row soybean production in the Mid-South. *Agronomy Journal* 103: 702–8.

Bruns, H. A. 2011b. Planting date, rate, and twin-row vs. single-row soybean in the Mid-South. *Agronomy Journal* 103: 1308–13.

Bruns, A. H. and L. D. Young. 2012. Raised seedbeds for soybean in twin rows increase yields over flat seedbeds. *Crop Management* 11(1). Online. doi: 10.1094/CM-2012-0712-01-RS.

Buhler, D. D. and J. L. Gunsolus. 1996. Effect of date of preplant tillage and planting on weed populations and mechanical weed control in soybean (*Glycine max*). *Weed Science* 44: 373–9.

Byker, H. P., N. Soltani, D. E. Robinson, F. J. Tardif, M. B. Lawton and P. H. Sikkema. 2013. Control of glyphosate-resistant horseweed (*Conyza canadensis*) with dicamba applied preplant and postemergence in dicamba-resistant soybean. *Weed Technology* 27: 492–96.

Carpenter, J. E. 2010. Peer-reviewed surveys indicate positive impact of commercialized GM crops. *Nature Biotechnology* 28(4): 319–21.

Carpenter, A. C. and J. E. Board. 1997. Branch yield components controlling soybean yield stability across plant populations. *Crop Science* 37: 885–91.

Castle, L. A., G. Wu and D. McElroy. 2006. Agricultural input traits: Past, present and future. *Current Opinion in Biotechnology* 17: 105–12.

De Bruin, J. L., P. Pedersen, S. P. Conley, J. M. Gaska, S. L. Naeve, J. E. Kurle, R. W. Elmore, L. J. Giesler and L. J. Abendroth. 2008. Probability of yield response to inoculants in fields with a history of soybean. *Crop Science* 50: 265–72.

DeLaune, P. B., J. W. Sij, S. C. Park and L. J. Krutz. 2012. Cotton production as affected by irrigation level and transitioning tillage systems. *Agronomy Journal* 104: 991–5.

Derpsch, R., T. Friedrich, A. Kassam and H. Li. 2010. Current status of adoption of no-till farming in the world and some of its main benefits. *International Journal of Agricultural and Biological Engineering* 3(1): 1–25.

Dunbar, M. W., M. E. O'Neal and A. J. Gassmann. 2016. Increased risk of insect injury to corn following rye cover crop. *Journal of Economic Entomology* 109: 1691–7.

Egan, J. F., K. M. Barlow and D. A. Mortensen. 2014. A meta-analysis on the effects of 2,4-D and dicamba drift on soybean and cotton. *Weed Science* 62: 193–206.

FAO. 2003. Development of a Framework for Good Agricultural Practices: Committee on Agriculture Seventeenth Session (COAG/2003/6), Rome, Italy.

Givens, W. A., D. R. Shaw, G. R. Kruger, W. G. Johnson, S. C. Weller, B. G. Young, R. G. Wilson, M. D. K. Owen and D. Jordan. 2009. Survey of tillage trends following the adoption of glyphosate-resistant crops. *Weed Technology*. 23: 150–5.

Green, T. R., L. R. Ahuja and J. G. Benjamin. 2003. Advances and challenges in predicting agricultural management effects on soil hydraulic properties. *Geoderma* 116: 3–27.

Gurmu, F., H. Mohammed and G. Alemaw. 2009. Genotype × environment interactions and stability of soybean for grain yield and nutrition quality. *African Crop Science Journal* 17: 87–99.

Heatherly, L. G. 1999a. Early soybean production system. In: L. G. Heatherly and H. F. Hodges (Eds), *Soybean Production in the Midsouth* (pp. 103–18). Boca Raton, FL: CRC Press.

Heatherly, L. G. 1999b. The stale seedbed planting system. In: L. G. Heatherly and H. F. Hodges (Eds), *Soybean Production in the Midsouth* (pp. 93–102). Boca Raton, FL: CRC Press.

Heatherly, L. G. and C. D. Elmore. 1983. Response of soybeans (*Glycine max*) to planting in untilled, weedy seedbed on clay soil. *Weed Science* 31: 93–9.

Heatherly, L. G., A. Blaine, H. F. Hodges, R. A. Wesley and N. Buehring. 1999. Variety selection, planting date, row spacing, and seeding rate. In: L. G. Heatherly and H. F. Hodges (Eds), *Soybean Production in the Midsouth* (pp. 41–51). Boca Raton, FL: CRC Press.

Hungria, M. and M. A. T. Vargas. 2000. Environmental factors affecting N_2 fixation in grain legumes in the tropics, with an emphasis on Brazil. *Field Crops Research* 65: 151–64.

Janovicek, K. J., W. Deen and T. J. Vyn. 2006. Soybean response to zone tillage, twin-row planting, and row spacing. *Agronomy Journal* 98: 800–7.

Johnson, R. R. 1987. Crop management. In: J. R. Wilcox (Ed.), *Soybeans: Improvement, Production, and Uses, Second Edition* (pp. 355–90). Madison, WI: American Society of Agronomy, Crop Science Society of America, Soil Science Society of America.

Johnson, B., B. Young, J. Matthews, P. Marquardt, C. Slack, K. Bradley, A. York, A. York, S. Culpepper, A. Hager, K. Al-Khatib, L. Steckel, M. Moechnig, M. Loux, M. Bernards and R. Smeda. 2010. Weed control in dicamba-resistant soybeans. *Crop Management* 9(1). Online. https://doi.org/10.1094/CM-2010-0920-01-RS

Karayel, K. 2009. Performance of a modified precision vacuum seeder for no-till sowing of maize and soybean. *Soil & Tillage Research* 104: 121–5.

KARI. 2013. Better soybean yields through good agricultural practices. http://africasoilhealth.cabi.org/wpcms/wp-content/uploads/2014/05/259-KARI-soybean-leaflet.pdf

Lal, R. 1991. Tillage and agricultural sustainability. *Soil and Tillage Research* 20: 133–46.

Lal, R., D. C. Reicosky and J. D. Hanson. 2007. Evolution of the plow over 10,000 years and the rationale for no-till farming. *Soil and Tillage Research* 93: 1–12.

Lampurlanes, J. and C. Cantero-Martinez. 2006. Hydraulic conductivity, residue cover and soil surface roughness under different tillage systems in semiarid conditions. *Soil and Tillage Research* 85: 13–26.

Lanie, A., J. Griffin, P. Vidrine and D. Reynolds 1994. Weed control with non-selective herbicides in soybean (*Glycine max*) stale seedbed culture. *Weed Technology* 8: 159–64.

Legleiter, T. R. and K. W. Bradley. 2008. Glyphosate and multiple herbicide resistance in common waterhemp (*Amaranthus rudis*) populations from Missouri. *Weed Science* 56: 582–7.

Liebl, R., F. W. Simmons, L. M. Wax and E. W. Stoller. 1992. Effect of rye (*Secale cereale*) mulch on weed control and soil moisture in soybean (*Glycine max*) *Weed Technology* 6: 838–46.

Logan, T. J., R. Lal and W. A. Dick. 1991. Tillage systems and soil properties in North America. *Soil and Tillage Research* 20: 241–70.

Lueschen, W. E. and D. R. Hicks. 1977. Influence of plant population on field performance of three soybean cultivars. *Agronomy Journal* 69: 390–3.

Mengel, D. B., W. Segars and G. W. Rehm. 1987. Soil fertility and liming. In: J. R. Wilcox (Ed.), *Soybeans: Improvement, Production, and Uses*, 2nd ed. (pp. 461–96) Madison, WI: American Society of Agronomy, Crop Science Society of America, Soil Science Society of America.

Moore, M. J., T. J. Gillespie and C. J. Swanton. 1994. Effect of cover crop mulches on weed emergence, weed biomass, and soybean (*Glycine max*) development. *Weed Technology* 8: 512–18.

Morrison Jr., J. E. 2002. Strip tillage for 'N–till' row crop production. *Applied Engineering in Agriculture* 18(3): 277–84.

Norsworthy, J. K., S. M. Ward, D. R. Shaw, R. S. Llewellyn, R. L. Nichols, T. M. Webster, K. W. Bradley, G. Frisvold, S. B. Powles, N. R. Burgos, W. W. Witt and M. Barrett. 2012. Reducing the risks of herbicide resistance: Best management practices and recommendations. *Weed Science* 60: 31–62.

NRCS. 2017. Conservation compliance on highly erodible land and wetlands. NRCS, New York, NY. https://www.nrcs.usda.gov/wps/portal/nrcs/detail/ny/programs/?cid=nrcs144p2_027057

N2Africa. 2014. Better soybean through good agricultural practices. http://www.n2africa.org/sites/n2africa.org/files/366%20N2Africa%20Nigeria%20soybean%20booklet_0.pdf

Oliver, L. R., T. E. Klingaman, M. McClelland and R. C. Bozsa. 1993. Herbicide systems in stale seedbed soybean (*Glycine max*) production. *Weed Technology* 7: 816–23.

Owen, M. D. K. and I. A. Zelaya. 2005. Herbicide-resistant crops and weed resistance to herbicides. *Pest Management Science* 61: 301–11.

Phillips, S. and K. Majumdar. 2015. Scientific principles supporting right rate. In: T. W. Bruulsema, P. E. Fixen and G. D. Sulewski (Eds), *4R Plant Nutrition – Metric Version* (pp. 206–49). Peachtree Corners, Georgia: International Plant Nutrition Institute.

Pioneer 2017. Maximizing soybean yields by improving agricultural practices. https://www.pioneer.com/home/site/us/agronomy/library/maximize-soybean-yield/

Popp, M. P., L. R. Oliver, C. R. Dillon, T. C. Keisling and P. M. Manning. 2000. Evaluation of seedbed preparation, planting method, and herbicide alternatives for dryland soybean production. *Agronomy Journal* 92: 1149–55.

Ragsdale, D. W., B. P. McCornack, R. C. Venette, B. D. Potter, I. V. MacRae, E. W. Hodgson, M. E. O'Neal, K. D. Johnson, R. J. O'Neil, C. D. DiFonzo, T. E. Hunt, P. A. Glogoza and E. M. Cullen. 2007. Economic threshold for soybean aphid (Hemiptera: Aphididae). *Journal of Economic Entomology* 100: 1258–67.

Rhoton, F. E. 2000. Influence of time on soil response to no-till practices. *Soil Science Society of America Journal* 64: 700–9.

Ruiz Diaz, D. A., P. Pedersen and J. E. Sawyer. 2009. Soybean response to inoculation and nitrogen following long-term grass pasture. *Crop Science* 49: 1058–62.

Salmeron, M., E. E. Gbur, F. M. Bourland, N. W. Buehring, L. Earnest, F. B. Fritschi, B. R. Golden, D. Hathcoat, J. Lofton, T. D. Miller, C. Neely, G. Shannon, T. K. Udeigwe, D. A. Verbree, E. D. Vories, W. J. Wiebold and L. C. Purcell. 2014. Soybean maturity group choices for early and late plantings in the Midsouth. *Agronomy Journal* 106: 1893–901.

Salvagiotti, F., K. G. Cassman, J. E. Specht, D. T. Walters, A. Weiss and A. Dobermann. 2008. Nitrogen uptake, fixation and response to fertilizer N in soybeans: A review. *Field Crops Research* 108: 1–13.

Shaw, D. R. 1996. Development of stale seedbed weed control programs for southern row crops. *Weed Science* 44: 413–16.

Shilling, D. G., B. J. Brecke, C. Hiebsch and G. MacDonald. 1995. Effect of soybean (*Glycine max*) cultivar, tillage, and rye (*Secale cereale*) mulch on sicklepod (*Senna obtusifolia*). *Weed Technology* 9: 339–42.

Sparks, T. C. and R. Nauen. 2015. IRAC: Mode of action classification and insecticide resistance management. *Pesticide Biochemistry and Physiology* 121: 122–8.

Sudarić, A., D. Šimić and M. Vratarić. 2006. Characterization of genotype by environment interactions in soybean breeding programmes of southeast Europe. *Plant Breeding* 125: 191–4.

Teasdale, J. R., C. E. Beste and W. E. Potts. 1991. Response of weeds to tillage and cover crop residue. *Weed Science* 39: 195–9.

Truog, E. 1947. Soil reaction influence on availability of plant nutrients. *Soil Science Society of America Journal* 11: 305–8.

Tu, M., C. Hurd and J. M. Randall and The Nature Conservancy. 2001. Weed control methods handbook: tools & techniques for use in natural areas. All U.S. Government Documents (Utah Regional Depository), Paper 533. http://digitalcommons.usu.edu/govdocs/533

University of Missouri Extension Service. 2017a. Soybean production in Missouri. http://extension.missouri.edu/p/G4410

University of Missouri Extension Service. 2017b. Missouri soybean handbook. http://extension.missouri.edu/p/M123

Yan, Z., J. G. Lauer, R. Borges and N. de Leon. 2010. Effects of genotype × environment interaction on agronomic traits in soybean. *Crop Science* 50: 696–702.

Zahran, H. H. 1999. Rhizobium-legume symbiosis and nitrogen fixation under severe conditions and in an arid climate. *Microbiology and Molecular Biology Reviews* 63: 968–89.

Zeng, G. M., M. Chen and Z. T. Zeng. 2013. Risks of neonicotinoid pesticides. *Science* 340(6139): 1403.

Zhang, L. X., Kyei-Boahen, S., Zhang, J., Zhang, M. H., Freeland, T. B., Watson Jr., C. E. and Liu, X. M. 2007. Modifications of optimum adaptation zones for soybean maturity groups in the USA. *Crop Management*. Online. doi:10.1094/CM-2007-0927-01-RS.

Chapter 12

Sustainable soybean production research and practice: contrasting case studies from three soybean production areas

Roger W. Elmore, Rodrigo Werle and Charles Wortmann, University of Nebraska, USA; Phinehas Tukamuhabwa, Makerere University, Uganda; and Nathan Mueller, University of Nebraska, USA

1 Introduction

A sustainable production system is one that (Wortmann et al., 2016): '... not only produces adequate food supplies but also ...

- generates profit for the farmer
- protects the integrity of the resource base
- improves quality of life of individuals and communities
- ensures the nutritional value and safety of food.'

Unfortunately, in the short term, farmers are rarely, if ever, paid for working to achieve the last three sustainability indicators. However, at times, US farmers are compensated through cost-sharing agreements to add infrastructure that promotes sustainability – for example, terraces or waterways – or to initially offset the costs of sustainable cropping systems, for example, cover crops. Unfortunately, this is not the norm in other soybean production areas of the world. Nevertheless, despite incentives for US farmers, they too have some way to go to attain long-term sustainability. In fact, we suggest they may have further to go in certain areas than farmers in other countries.

http://dx.doi.org/10.19103/AS.2017.0034.13

Good agricultural practices (GAPs) are an essential ingredient in achieving a sustainable agricultural system. 'GAPs for a specific crop need to balance the numerous dimensions of sustainable crop production. Through additive and interactive effects, as part of a set of practices…' (Wortmann et al., 2016). The GAPs contribute to all the four aspects of a sustainable system as well as producing healthy food.

Of course, GAPs vary within and among production systems. Improved production technology coupled with favourable weather resulted in increasing productivity in many areas such as in Nebraska and Brazil over the last three decades (Fig. 1). For example, from 1984 to 2013, yields increased at 48.6 and 46.5 kg ha^{-1} yr^{-1} in Nebraska and Brazil, respectively (USDA-NASS, 2016; FAO, 2016). Overall, US soybean yields increased at 29.1 kg ha^{-1} yr^{-1}. However, these yield increases are not consistent with yield trends in other parts of the world. For example, soybean yields in Eastern Africa have been at best stable, or maybe even declining (Fig. 1). Production has steadily increased in the US and Brazil, but has not changed much for Uganda during the past decade (Fig. 2).

Nevertheless, yield gaps between current cropping systems and potential yield exist in all three situations – these are documented for cereals in several countries, but not yet for soybean (GYGA, 2017). In addition, we recognize that GAPs must do more than increase soybean yield and production; they must also contribute to sustainable production systems.

Our objective here is to address the question of whether we produce soybean sustainably. To that end, we detail three case studies – Nebraska/U.S., Brazil and Uganda – to compare and contrast current production practices with research-based information on GAPs with an eye on production system sustainability and closing the yield gap. In other words, are

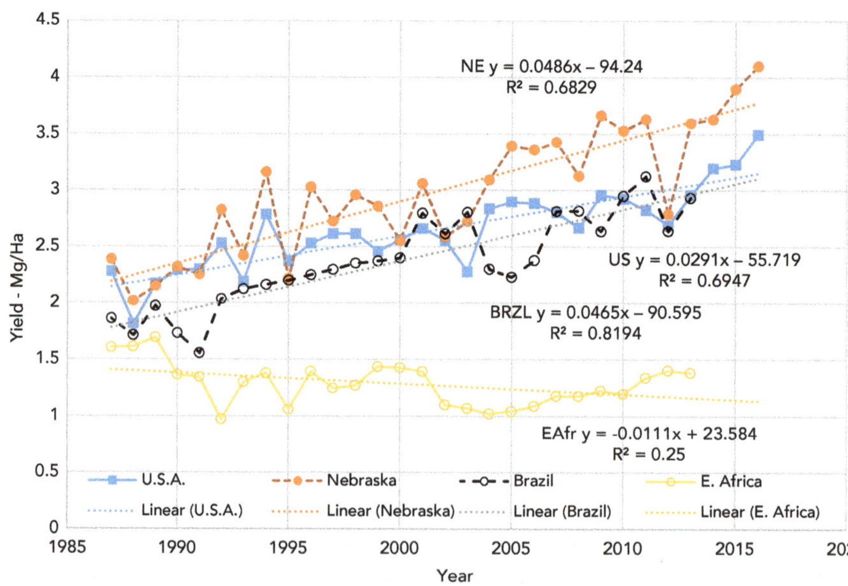

Figure 1 Soybean yields – 30-year trends: U.S. Nebraska, Brazil and East Africa. Data from USDA-NASS, 2016 and FAO, 2016.

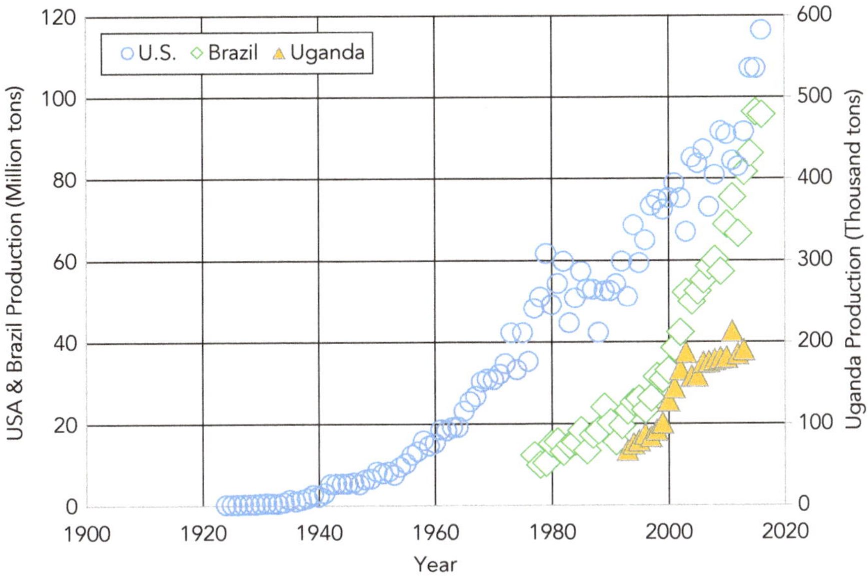

Figure 2 Annual soybean production for U.S. and Brazil (left axis) and Uganda (right axis). Data from USDA-NASS (U.S.), CONAB, www.conab.gov.br (Brazil), and FAO, 2016 (Uganda).

current practices not only 'good', – in the sense of improving productivity – but also, do they contribute to a sustainable production system based on the parameters listed above?

We have chosen three contrasting production areas as case studies to address our objective. In each case study, we document current research understandings and contemporary practices as well as we can. We also attempt to gage the general sustainability of the practices. For each case, we start by documenting important decisions at or before planting, then discuss soil management issues, and finally address pest management and site-specific management. We use Fig. 3 to visually show where GAPs work well in promoting sustainable production in each of the three systems, and to illustrate – in our opinion – where practices appear to lag behind current research.

2 Soybean production in Nebraska, USA

2.1 Decisions at or before planting

Nebraska lies at the western end of the corn/soybean belt in the Midwest US. Precipitation ranges from 760–890 mm in the southeast to 380–510 mm in the west (HPRCC, 2017). The state has about 18.3 million ha in farms, and produces 11.2% of the US corn and 7.3% of its soybean (USDA-NASS, 2017a). Relative to other states, it ranks third in corn production and fifth in soybean production. In 1929, Nebraska reported 810 ha planted; by 1938 – when yields were reported for the first time, the state averaged 0.8 MG ha^{-1}, and planted

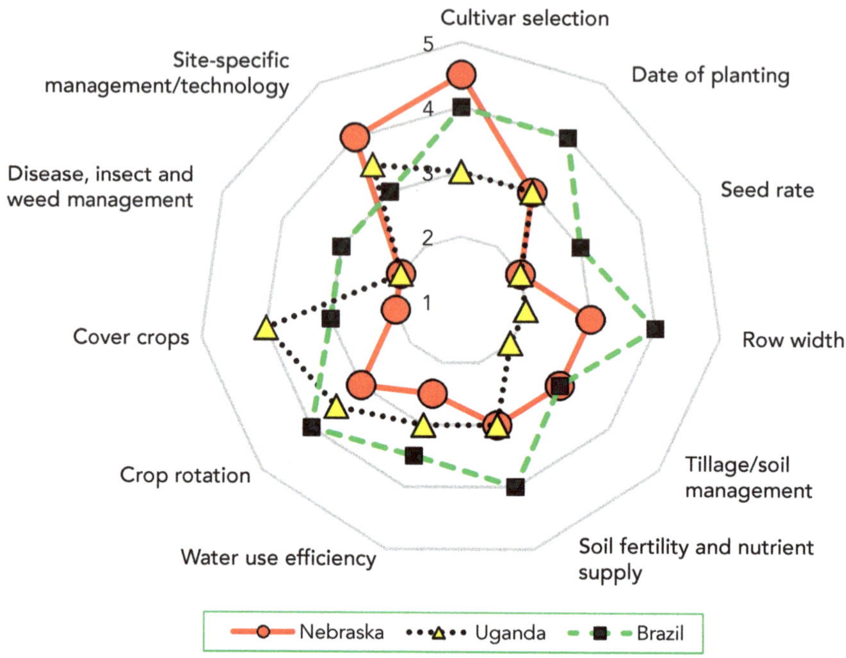

Figure 3 Evidence of sustainable GAPs among three soybean production areas. 1) No information; 2) limited or no evidence of sustainable GAPs; 3) some evidence of sustainable GAPs; 4) encouraging evidence of sustainable GAPs; 5) Perfect.

area had more than tripled to 2835 ha (USDA-NASS, 2017b). Production area dramatically increased beginning in the 1960s.

Nebraska is the most irrigated state in the United States with 3.4 million ha irrigated. Nearly half, 1.0 million ha, of the 2.1 million ha soybean were irrigated in 2016 (USDA-NASS, 2017b). To provide an example of the flexibility of Nebraska farmers in response to grain markets, 2.3 million ha were planted in 2017 (USDA-NASS, 2017b). Soybean consistently yields more than the national average in most years – Nebraska experienced record high yields in 2016, 4.2 MG ha^{-1} (Fig. 1) (USDA–NASS, 2017b). However, gaps between current and potential yields range from 10 to 30% (Grassini et al., 2015). Even in this highly productive system, there is room for improvement (Fig. 3).

Cultivar selection

How do US farmers select soybean cultivars for planting? A typical tongue-in-cheek response to this question is that they choose based on only three traits: yield, yield and yield! Obviously in our commodity-based production systems, where farmers are not necessarily paid for above-average quality, yield pays the bills. Nevertheless, other factors like maturity group (MG), standability – which often correlates with height and lodging, disease resistance and herbicide resistance traits – are also important (Heatherly and Elmore, 2004; Mueller et al., 2014).

Before the latter part of the twentieth century, most US farmers planted seed retained from their previous production (bin-run seed). Yet, keeping one's own grain for seed was not always the most productive practice. Dunphy and Ferguson (1991) reported a 0.13 Mg ha^{-1} yield increase with certified seed compared with bin-run seed based on over 200 comparisons. After the advent and rapid adoption of transgenic cultivars – first approved in 1994 (Fig. 4), planting bin-run seed of transgenic cultivars was illegal, thus 'bin-run' seed planting declined. However, with the original glyphosate gene patent expiration, farmers now are reconsidering the planting of seed from their harvests (Mason et al., 2016b, 2017). However, farmers must still ensure that cultivars they intend to plant are not covered by other patents or plant cultivar protection, separate from the Roundup Ready 1 patent.

Although yields continue to increase in both United States and Nebraska (Fig. 1), there is no evidence that building herbicide resistance in transgenic cultivars has on its own improved yields. Indeed, although herbicide resistance improves a plant's ability to withstand stress from weeds, it inherently does not improve yield potential. A plant with a transgenic gene for herbicide resistance will yield exactly the same as a plant with the same genetic makeup, but without the transgene in the absence of weed pressure – this assumes no yield drag associated with the transgene insertion process.

In a survey, Grassini et al. (2016) found that most Nebraska farmers know the optimum MG for soybean in their area; the range of maturities planted in each region was only about one-half of a MG unit – for example MG 3.0–3.5. Farmers also tended to plant slightly later MG for non-irrigated fields. The authors speculated that this helps non-irrigated soybean mitigate the impact of high temperatures – most likely stress during flowering.

There is much to gain or lose with cultivar selection. Mueller et al. (2014) suggest that proper cultivar selection can improve yields by up to 1.01 Mg ha^{-1} (Table 1). Today, seed companies offer many cultivar choices and bountiful performance data. Farmers would be wise to compare performance data from seed companies along with third-party testing programmes (e.g. Iowa State Univ., 2016; Regassa, 2016). Well-conducted cultivar comparisons aid in optimal cultivar selection.

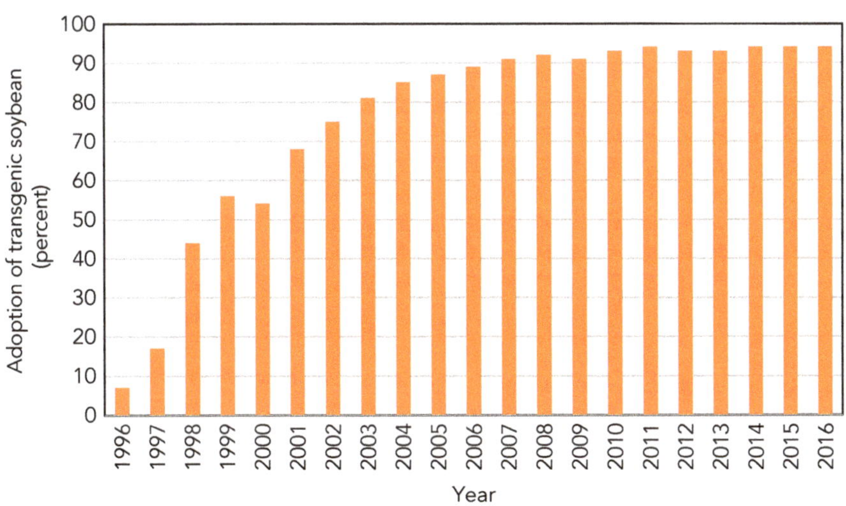

Figure 4 Transgenic soybean adoption: U.S. Data from USDA-NASS.

Table 1 Effect of various management factors on soybean yield. From Mueller et al., 2014

Rank	Management factor	Yield (Mg ha^{-1}) difference
1	Cultivar selection*	1.01
2	Planting date	0.54
3	Weed control**	0.54
4	Phosphorus fertility	0.34
6	Crop rotation	0.34
5	Row spacing***	0.20
6	Seeding rate	0.20

* Three-year average performance difference among cultivars from 2011 to 2013 in South Dakota State University trials.

** One week delay in weed control on 30-inch rows.

*** Multi-state study from 2009 to 2011 funded by the United Soybean Board, led by Seth Naeve, University of Minnesota.

Nebraska soybean farmers – and farmers across the United States – have ample opportunities to select excellent genetic materials. And, in general, they do so. Yields continue to increase with minimal increases in other inputs, thus promoting sustainable production (Fig. 3).

Date of planting

Nebraska farmers plant an average of 0.47 d yr^{-1} earlier in spring than 30 years ago – this is probably true for much of the Midwest US (Fig. 5) (Specht et al., 2013). Earlier planting promotes yield (Table 1) (Bastidas et al., 2008). Rees (2015) found that 50% of farmers reported planting soybean after they have finished planting maize and the remaining plant based on the calendar date: 40%, 7% and 3% plant in late April to the first week of May, mid-May and mid-late May, respectively. In a similar survey by Jasa and Rees (2016), 55% of responding farmers planted soybean after maize planting, and 38%, 6% and 2% in late April/first week of May, mid-May and late May, respectively.

Planting around 1 May increases final yields compared with earlier or later planting (Specht et al., 2013); more than 90% of farmers target that time frame. Grassini et al. (2016) found that Nebraska farmers lose between 34 and 50 kg d^{-1} if planting after 1 May, but that water supply is more important than planting date for rainfed production where water supply limits yield potential. Farmers balance weather conditions and the yield trade-offs associated with late-planting maize versus late-planting soybeans when choosing planting dates. Earlier planting usually improves yields without increasing expenses and because of earlier crop canopy closure reduces soil water evaporation and weed competition and may reduce herbicide inputs.

This GAP contributes to sustainability; however, Nebraska farmers could improve yields even more and thus, increase sustainability by placing a higher priority on planting earlier (Fig. 3) – albeit with a few trade-offs. No doubt soybean fits well into the Nebraska cropping system alongside maize. But, to plant soybean earlier may necessitate farmers delaying maize planting, which often reduces maize yields in Nebraska and elsewhere in the United States (Abendroth et al., 2017).

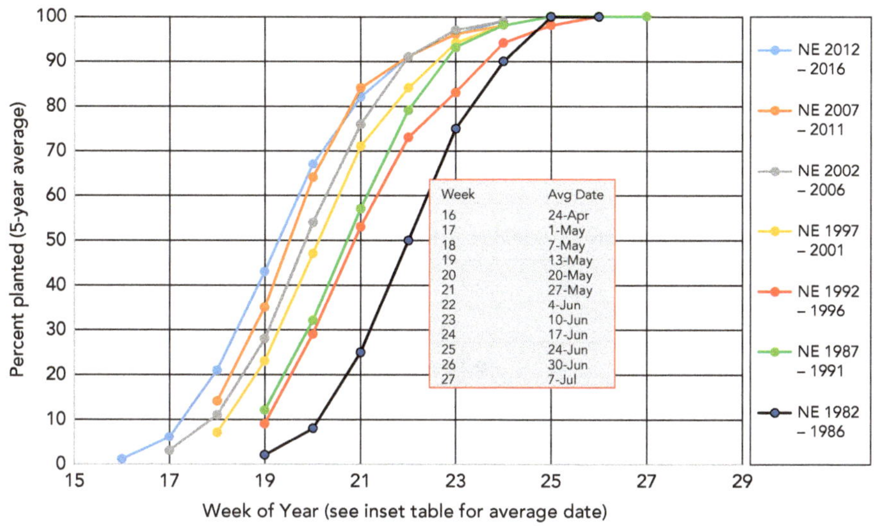

Figure 5 Soybean planting progress for Nebraska, 5-year averages. Data compiled from USDA-NASS, 2016.

Seeding rate

Another way to improve profits and, thus, sustainability is to reduce seeding rates (Table 1). Average US seeding rates appear to be declining by about 5000 seeds ha^{-1} yr^{-1} (Fig. 6), but there are opportunities to further reduce rates even more. Harvest plant populations of about 462 000 plants ha^{-1} maximized yields in an Iowa study, but greater than 95% maximum yield occurred with final populations as low as 258 600 plants ha^{-1}; the high seeding rates resulted in lost profit (De Bruin and Pedersen, 2008). Conley and Santini (2007) found that 95% of farmers with large farms (>400 ha) tended to plant less than 500 000 seed ha^{-1} while those with smaller farms (<40 ha), a fewer proportion of farmers – 77% – planted less than 500 000 seed ha^{-1}.

In response to 'What is your current soybean seeding rate?', 4%, 22%, 55% and 19% of the responding farmers reported planting over 494 000, 420 000–494 000, 346 000–420 000 and 272 000–346 000 seed ha^{-1}, respectively (Rees, 2015). Results were similar in 2016 (Jasa and Rees, 2016). Grassini et al. (2016) found that Nebraska farmers planted an average of 395 000 seed ha^{-1} with a range from 301 000 to 506 000.

The University of Nebraska's On-farm Research Network project included soybean seeding rates as a factor from 2006 to 2013 (UNL, 2013). The most profitable seeding rates ranged from 198 000 to 395 000 seed ha^{-1} with no site experiencing an increase in profit above 395 000 seed ha^{-1}. Grassini et al. (2016) suggest that a farmer's choice to over-plant may be related to the lower per unit cost of soybean relative to maize, a lack of knowledge about soybean responses to seeding rate and/or risk aversion. Farmers often over-plant soybean, reducing profit and, thus, sustainability (Fig. 3).

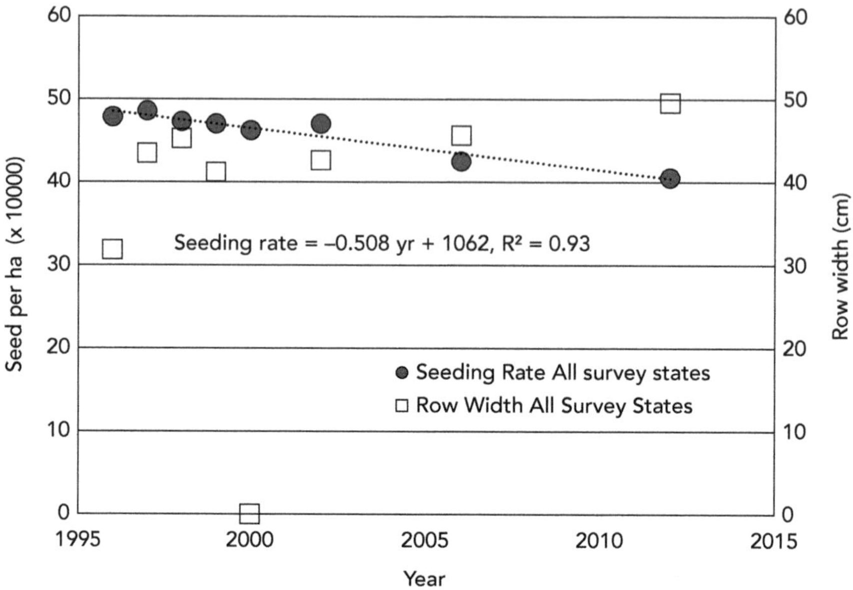

Figure 6 Soybean seeding rates and row spacing in the U.S. From ERS-ARMS, 2016.

Row width

Soybean response to row width varies with location and other factors. Interestingly, farmers appear to have increased row width for soybean (Fig. 6), although researchers often report advantages to narrow rows due in part to faster canopy closure (De Bruin and Pedersen, 2008; Hanna et al., 2008). In a 2013 and 2014 study at eight Nebraska locations, soybean yielded an average of 0.235 Mg ha^{-1} more in 38-cm compared to 76-cm row spacing (Giesler et al., 2016). In Wisconsin, Conley and Santini (2007) found that 92% of farmers with farms of >400 ha compared with 48% of farmers with <40 ha planted soybean with less than 51-cm row spacing – that is, larger farmers were more likely than smaller farmers to use narrow rows.

Economics and increased incidence of Sclerotinia stem rot (*Sclerotinia sclerotiorum*) and other diseases with narrow rows handicap adoption of narrow row production in states east of Nebraska (De Bruin and Pedersen, 2008). Narrow row soybean spacing promotes early canopy closure and higher yield, especially where the crop experiences severe soil water deficits early in the growing season – such as before mid-July and in areas with short growing seasons (Grassini et al., 2016). Moreover, narrow row spacing may reduce the likelihood of weed resurgence in soybeans due to faster canopy closure rate (Bradley, 2006). The combination of greater yield potential and faster canopy closure allows this GAP to contribute to sustainability, but increased adoption would promote this even more (Fig. 3). We recognize that a farmer's decision on row spacing includes not only yield potential, but also tillage system, other crops in the rotation, equipment and potential disease and weed management issues among others.

2.2 Soil, pest and site-specific management

Tillage and soil management

Soil management including tillage practices can dramatically affect sustainability. Knuteson et al. (2016) found that 91% of farmers in Illinois and Wisconsin use reduced and/or no-till practices to maintain a minimum of 30% soil cover. Reduced tillage decreases soil loss to erosion. In the same survey, 90% of farmers used management practices that would limit soil compaction to improve water infiltration, reduce erosion and improve crop productivity. Over 90% of the respondents used contour strips to reduce soil erosion and limit pesticide runoff. Soil sampling to ascertain soil nutrient levels was practised by 86% of the farmers.

In Iowa, soybean yields were not changed with conventional tillage compared with reduced tillage systems based on results from a 14-year crop rotation and tillage study in Southwest Iowa (Al-Kaisi and Kwaw-Mensah, 2015). Therefore, the additional cost of any tillage pass did not generate additional yield and income with reduced profit.

The adoption of no-till soybean production in Nebraska ranges widely among counties (Fig. 7). Reduced tillage prior to soybean planting would further improve farmer profitability and enhance sustainable soybean production (Fig. 3).

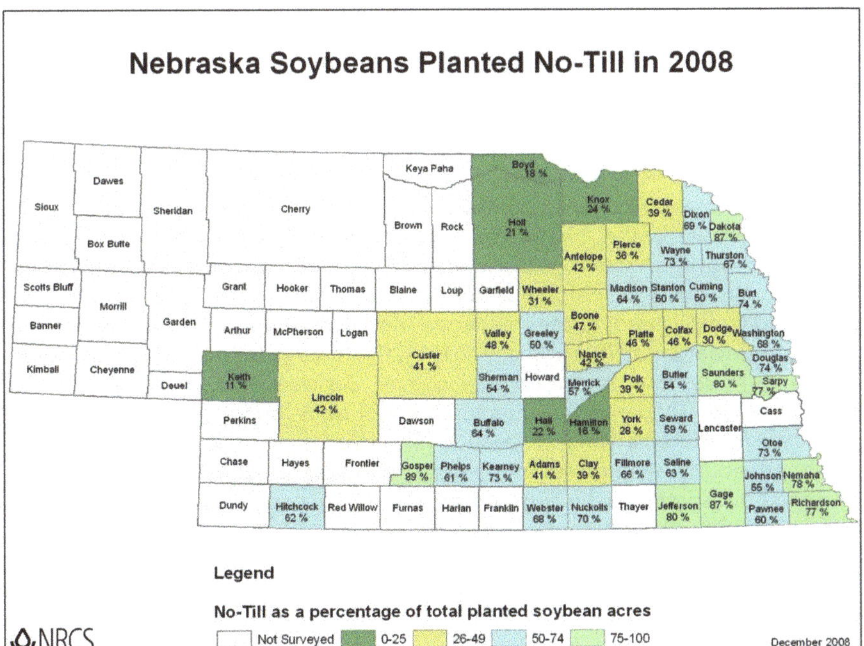

Figure 7 No-till soybeans in Nebraska, 2008. From NRCS, 2008.

Soil fertility and nutrient supply

Soybean has a low probability of yield response to P application if Bray-1 P is above 10 ppm, which is also true for maize following soybean in rotation; continuous maize requires a higher soil P level (Wortmann et al., 2008; Shaver et al., 2014). However, soybean is more sensitive to low soil pH than maize and pH should be kept well above 5.8 for the maize–soybean rotation (Shaver et al., 2014).

Soybean on average obtains about 65% of plant N through biological N fixation (Salvagiotti et al., 2008) and is efficient in recovering N from the soil, commonly leaving soil more depleted of nitrate-N compared with maize. Nitrogen application at soybean planting is recommended for High Plains production such as in western Nebraska, but not for other parts of Nebraska. Routine in-season application of N or N plus S, such as at the R4 growth stage, has little profitable effect (Wortmann et al., 2012). However, when equipped for fertigation, a low rate of N application during early grain fill may be profitable.

Soybean response to applied K and S is not common in Nebraska. Zinc application is recommended when diethylenetriamine pentaacetate (DTPA)–Zn is below 0.4 ppm.

Lime-induced iron deficiency chlorosis (IDC) can be detrimental to soybean yield on calcareous soils. Planting cultivars tolerant to IDC at relatively high within-row plant densities can be at least a partial solution to the iron deficiency problem (Shaver et al., 2014). Placement of Fe-chelated EDDHA in the seed furrow or as a seed coating is also effective in reducing IDC. Foliar application of Fe as a crop rescue treatment may not produce positive results.

Soybean inoculation with *Bradyrhizobium japonicum* improves yields by providing N fixation in areas where soybean has not been grown with inoculated seed or where *Bradyrhizobium* was severely reduced such as by prolonged flooding or sedimentation. Conley and Santini (2007) found that 17.8% of responding Indiana farmers reported using a seed inoculant. In a Wisconsin survey, Furseth et al. (2012) found that inoculants were used by 85% of farmers. According to Marburger et al. (2016), this inter-state difference '... is primarily because of the greater crop rotation diversity in Wisconsin compared with the mainly maize–soybean rotations in Indiana' (p. 402). There was virtually no inoculation effect on soybean grain yield for 28 Nebraska comparisons of inoculated with non-inoculated seed in replicated, farmer-managed trials (2.4 kg ha^{-1} increase with inoculation with a standard deviation of 90.4 kg ha^{-1}) (2001–2013 data compiled by Laura Thompson, Nebraska On-farm Research Network, pers. comm., September 2016).

De Bruin et al. (2010) in a multi-state summary of data found that '... inoculant offers limited success for either a yield increase or improved economic return on soils where soybean has previously been grown in the upper Midwest' (p. 265). Inoculant use is often, '... widespread and prophylactic because of low product cost, ease of application, and increasing soybean commodity prices' (Marburger et al., 2016).

Profits, and thus sustainability, could be improved, if farmers would carefully consider their situation before fertilizing and inoculating soybean. Although not all Nebraska farmers follow GAPs for soil fertility, doing so would result in greater sustainability (Fig. 3).

Water-use efficiency

Water is the most limiting resource for rainfed soybean production in Nebraska and the Great Plains. Grassini et al. (2016) in their survey of soybean farmers in three Nebraska

Natural Resource Districts found that water supply sets the upper limit on soybean yields in these dry environments. To produce soybean without stress, the crop needs to receive about 635 mL of water during the growing season. More water does not increase yield. Grassini et al. (2016) estimated that yields increase about 97 kg per 10 mL of water supply up to the 635 mL threshold. Nevertheless, many farmers apply more than 760 mm of water per year. Extension workers encourage farmers to carefully monitor soil water, precipitation and irrigation. Increased vigilance on irrigation applications would improve sustainability (Fig. 3).

Crop rotation

Knuteson et al. (2016) in a survey of Illinois and Wisconsin farmers found that 93% rotate crops (at least two years) and that 79% grow more than one agricultural crop on their farm per year. Rotating soybean with other crops is conducive to higher yields of both soybeans and the rotated crop (Mason et al., 2016a; Table 1). The maize–soybean rotation is the most common in Nebraska and throughout the Corn Belt. In a study conducted from 1986 through 2001, Wilhelm and Wortmann (2004) found the average rainfed soybean yield rotated to maize was 2.58 Mg ha^{-1} (with individual years ranging from 1.56 to 4.01 Mg ha^{-1}), whereas, continuous soybeans yielded on average 2.35 Mg ha^{-1} (with individual years ranging from 1.27 to 3.25 Mg ha^{-1}). Other crop-rotation studies in the Midwest suggest that soybean yields can be increased further (0.27 to 0.34 Mg ha^{-1}) with a longer rotation than maize–soybean, for example, maize–maize–soybean or wheat–maize–soybean (Nafziger et al., 2013, 2015). Crop rotation not only assists farmers with increasing yield potential, but also allows production with less fertilizer and pesticide use. Crop rotations break the life cycles of several troublesome diseases and insects and allow opportunities for diversified and more effective weed-management strategies. A rough estimate is that slightly more than half of Nebraska's maize was rotated to soybeans in 2016 (4.0×10^6 ha maize planted and 2.1×10^6 ha soybean; USDA-NASS, 2017b). Thus, increasing diversity through crop rotation would promote greater sustainability in Nebraska cropping systems (Fig. 3).

Cover crops

Cover crops can reduce soil erosion and build soil organic matter, when current soil organic matter is sufficiently low, and thus promotes sustainability. Cover crops are planted on 4.2 million ha in the United States annually (CTIC, 2014). Farmers using cover crops plant them on 24% of their cropland area (CTIC, 2014). Since 2009, cover crop area increased 30% yr^{-1} (CTIC, 2014) with 21 and 18% of the cover crop users planting before and after soybean, respectively, in a maize–soybean rotation. Integration of cover crops into both irrigated and rainfed maize–soybean cropping systems – ideally without yield loss for the primary crop yield – is much studied in Nebraska and other states (Kohler-Cole et al., 2016).

Drewnoski et al. (2015) found that about one-third of Nebraska farmers planted cover crops and concluded that farmers with larger farms were more likely to plant cover crops. The biggest limitations to adoption were 1) time and labour, 2) water use by cover crops and 3) costs versus returns. Common-sense situations for cover crops in soybean cropping systems include following maize silage and seed maize harvest in Nebraska; these two cropping systems accounted for 23% and 10% of cover crop area, respectively. However, seed maize and silage systems account for only small proportions of the maize–soybean production area.

By their nature, if cover crops do not reduce primary crop yields, systems with cover crops will be more sustainable. Nebraska farmers could increase sustainability by adopting more of this GAP – planting more cover crops (Fig. 1).

Disease, insect and weed management

Integrated pest (disease, weed and insect) management (IPM) philosophy is key to sustainable agriculture. Knuteson et al. (2016) found that 73%, 75% and 39% of farmers in Illinois and Wisconsin scouted for insects, weeds and diseases, respectively. More than 85% used scouting thresholds for pest management decision-making. Nearly all of the Illinois and Wisconsin farmers surveyed indicated they used cultural practices to manage disease, insect and weed problems. However, Grassini et al. (2016) in their poll of Nebraska farmers found that 1 in 5 irrigated soybean fields received an in-season foliar fungicide between R3 to beginning pod set even though there was no evidence of disease presence – not a sustainable practice. Seed treatments enhanced soybean stands with early planting dates. Between 70% and 80% of Nebraska farmers treated seed with either a fungicide, insecticide or both (Grassini et al., 2016). Fungicide and insecticide seed treatments are important in many Nebraska soybean production environments (Giesler and Ziems, 2008; Specht et al., 2015). The adoption of glyphosate-resistant (GR) soybean cultivars and over-reliance on glyphosate as a post-emergence herbicide option for weed control in Nebraska and much of the United States have led to the occurrence of glyphosate resistance in multiple weed species. In order to sustainably manage weeds, we recommend farmers use multiple effective modes of action at each application time, include active ingredients with soil residual activity in the tank-mix, spray herbicides in a timely fashion and rotate crops and herbicide modes of action as often as possible. Such integration of practices is also important to avoid or delay the development of resistance for other biotic constraints. The ongoing challenges of Western corn rootworm management is an example of a biotic challenge that can develop with reliance on simple crop rotations.

Greater use of GAPs for disease, insect and weed management would enhance sustainability in Nebraska (Fig. 3).

Site-specific management/technology

Technology drives sustainability in various ways. US farmers often adopt new management practices rapidly. Early in the twenty-first century, GR soybean occupied 90% of US soybean production area within 13 years of their introduction (Duke and Powles, 2009); in 2016, US farmers planted herbicide-resistant soybean on 94% of their fields (Fig. 4) (USDA-NASS, 2016). Soon after the introduction of GR soybean, although yields in weed-free environments were 5–10% less with GR soybean (Elmore et al., 2001), farmers rapidly adopted GR soybean because it initially allowed superior weed management plus GR technology was more environmentally benign than previous weed-management practices (Fig. 3) (Duke and Powles, 2009). Unfortunately, development of GR weed resistance now reduces the benefits of GR technology.

Knuteson et al. (2016) report that 48% of Wisconsin and Illinois farmers surveyed use Global Positioning System (GPS) technology on their farms. Schimmelpfennig and Ebel (2011) found that prior to 2006, precision agricultural practices and technology adoption rates were less than expected early on in their use; 40% used yield monitors and 'very few' farmers had adopted GPS maps and variable rate application technology. But, they found that both maize and soybean producers who used yield monitors had higher yields and lower fuel expenses than those who did not.

The use of GPS, auto-steer technology and automatic section control (sections or individual rows) helps reduce both skips and overlaps when planting, fertilizing and spraying. It also allows farmers to work in poor visibility conditions and experience less fatigue (Runge et al., 2014). The use of precision monitors along with sensors located on equipment enables farmers to more quickly notice equipment malfunctions and errors. Data collected during these operations can be used to improve input tracking, provide insight into observations seen in the field and help make data-driven management decisions.

Technology reduces inputs, increases yield and improves environmental stewardship and thus, sustainability, such as through reduced overapplication of pesticides and fertilizers. Thus these GAPs enhance sustainability in Nebraska (Fig. 3).

3 Soybean production in Uganda

3.1 Decisions at or before planting

Soybean was introduced into Uganda in 1908. In the 1940s and 1950s, the colonial government encouraged production to combat protein malnutrition among soldiers of the King's African Rifles and kwashiorkor among children. Production growth was slow but, with well-organized market demand, production tripled during the 20 years from 1993 to 2013 (Fig. 2 and 6) (FAO, 2016). Although the rate of increase in production may have slowed in recent years (Fig. 8) and soybean production area accounts for less than 2% of crop production land, Uganda is now third in soybean production for Africa, following South Africa and Nigeria. The increased production is primarily from planting more land area; yields are stable or declining and average now about 1.2 t ha^{-1} (Fig. 1). Tukamuhabwa et al. (2012) report that sub-Saharan farm yields range between 600 and 800 kg ha^{-1}, while research station yields have reached 3.5 MG ha^{-1}. A soybean yield gap assessment has not been developed for Uganda. However, for comparison, the gap between current rainfed maize yields and potential yield is around 50% (GYGA, 2017). However, considering that current farm soybean yields range around 1.2 MG ha^{-1}, an estimated yield gap could be closer to two-thirds.

Soybean production occurs throughout the 1100–1800 m above sea level elevation range of Uganda where soil water is adequate, but is more concentrated in the Northern Moist Farmlands at 1050–1200-m elevation (Wortmann and Eledu, 1999). The production area is within 1.5° of the equator and much of this area has two cropping seasons per year. Depending on rainfall amount and distribution, soybean may be produced in either season. In contrast to both Nebraska and Brazil which are both mechanized for essentially all management practices, soybeans in Uganda are harvested by breaking lower stems by hand and air-drying for later threshing.

Despite past and current efforts to improve yield, there are few indications of significant progress (Fig. 1). However, we will document several GAPs already in place and possible ways to move forward.

Cultivar selection

The Centre for Soybean Improvement and Development at Makerere University develops soybean cultivars that are 1) high yielding with average maturity of less than 120 d, 2) resistant to soybean rust disease and groundnut leaf miners, 3) resistant to lodging and pod shattering, 4) effectively nodulate with native rhizobia, 5) contain high-grain protein and oil, 6) set pods adequately high and 7) have end-user acceptance in terms of seed appearance and other traits. About 93% of the production is with cultivars released since 2004 (Table 2). Maksoy 2N, released in 2008, together with Maksoy 3N, released in 2010, account for 72% of soybean seed sales.

Seed supply is through both formal and informal means. The Centre for Soybean Improvement and Development at Makerere University provided about 190 tons of foundation seed over recent years to companies producing and marketing certified seed. In addition, farmers in five districts were trained in quality seed production, which they may market to other farmers or to organizations that facilitate seed supply to farmers. However, most of the seed sown is from the farmers' own harvest (known as bin-run seed in the United States). Planting higher quality seed of improved cultivars would increase sustainability (Fig. 3).

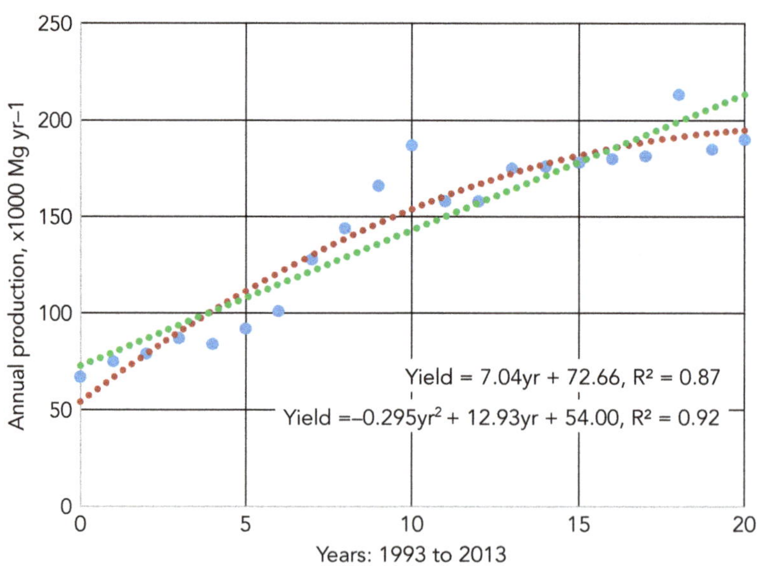

Figure 8 The increase in soybean production in Uganda over a 20-year period (with linear and quadratic trends) (from FAO, 2016).

Table 2 Nutrient uptake by soybeans

Plant part	N	P_2O_5	K_2O	Ca	Mg	S
	kg Mg^{-1}					
Grain	51	10	20	3	2	5.4
Crop residue	32	5.4	18	9.2	4.7	10
Total	83	15.4	38	12.2	6.7	15.4
% exported via grain	61	65	53	25	30	35

Source: EMPRAPA (2013).

Date of planting

March followed by September are the main months for planting soybean. Seed is placed in holes made with hand hoes for 96% of the soybean farmers. However, planting time is tied not to the calendar but rather to soil water availability considering other crops in the often diverse cropping systems, health of the household, success of earlier plantings and so on; much flexibility with planting date is needed to achieve sustainable GAP. Planting schedules depend on the situation. Onset of the rains is erratic. Early planting, while desired, often fails due to climate or biotic factors, and replanting or gap filling is important. Cropping systems are diverse and planting of different crops needs to be prioritized. Health problems often interfere with these labour-intensive production systems. Although this does not seem reasonable to some of us, flexibility in planting and other operations is essential to sustainability (Fig. 3) – a prescribed 3-week planting window will not work. Improved weather forecasts will undoubtedly aid in decision-making and promote greater sustainability.

Seeding rate and row width

Planting in 45-cm row spacing is recommended but intra- and inter-row spacing varies within and between fields. Available seed, whether from a seed company or that saved by farmers, is of variable quality for viability and seedling vigour. These traits are more difficult to achieve in the tropics, where seed is untreated or inadequately treated for the control of many potential biotic constraints that occur in the tropics. Relatively low and uneven populations are often achieved given the seeding rate and field conditions. Farmers depend heavily on soybean's ability to compensate for the increased area available per plant. More research and greater precision planting – with quality seed – will increase sustainability of these practices (Fig. 3).

3.2 Soil, pest and site-specific management

Tillage and soil management

Land is tilled before planting with very little reduced-tillage production, although research under similar growing conditions in Uganda with sorghum indicates an opportunity to increase yield by more than 10% with less cost by replacing pre-plant tillage with glyphosate application (Kaizzi et al., 2007; Nansamba et al., 2016). Adoption of reduced tillage systems could improve sustainability (Fig. 3).

Soil fertility and nutrient supply

Ugandan farmers use little fertilizer in producing soybeans. However, research results do indicate high profit potential from application of 10 to 15 kg ha^{-1} fertilizer P for soybean production with less response and yield potential in the relatively drought-prone Lake Kyoga basin compared with other production areas in Uganda (Fig. 9; Kaizzi et al, 2012). Soybean responses to applied N and K compared with P were small. Soybean fertilizer-use decision tools are available for all important production areas of Uganda. Promiscuous soybean cultivars developed in Uganda nodulate effectively with native rhizobia; inoculum use is negligible. Planting of non-promiscuous cultivars that depend on *B. japonicum* for effective nodulation would be relatively less sustainable. Ugandan farmers appear to produce soybean in a sustainable fashion in terms of soil fertility and nutrient supply, as well as in terms of water-use efficiency and crop rotations – see the following two sections (Fig. 3).

Water-use efficiency

Most Ugandan soybean production is rainfed. However, rainfall is limited and unpredictable and yields correspondingly are low; the frequency of occurrence and severity of soil water deficits are an important yield determinant.

Crop rotation

Soybean is mostly produced as a sole crop with less than 20% of the production area intercropped with maize or another crop. Crop rotation with several other crops is the norm and essential for sustainable production systems where biotic constraints may be present nearly all months of the year.

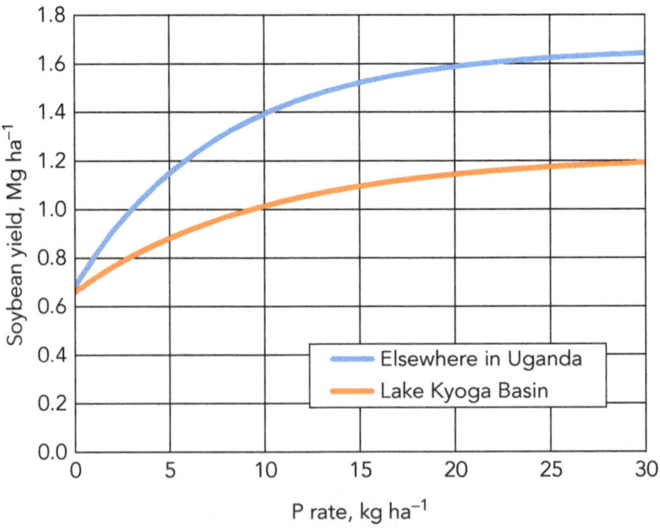

Figure 9 Representative response functions for soybean to applied P in Uganda (Kaizzi et al, 2012).

Cover crops

In most Ugandan production systems, soil cover develops with natural growth of weedy and non-weedy species following harvest of the previous crop with the result that most and maybe nearly all soybean is planted following a cover crop/grazed forage crop. This plant growth contributes much to erosion control, ecosystem diversity, nutrient cycling and return of organic material to the soil. This is positive in terms of sustainability (Fig. 3).

Disease, insect and weed management

There is little use of pesticides in soybean production. Weed control is mostly by weeding with hand hoes. Soybean rust (*Phakopsora pachyrhizi*) is the disease of greatest concern and managed primarily through planting resistant cultivars and crop rotation. Rust-resistant breeding lines in one set of studies yielded up to 2.4 MG ha^{-1} while the control cultivars yielded 0.8–1.2 MG ha^{-1} (Tukamuhabwa et al. 2012). Soybean rust evolution in overcoming breed resistance is a threat to sustainability, while development of herbicide-resistant weeds is not a threat. More focus on disease, weed and insect management could certainly improve yields and profit and thus sustainability, while careless use of pesticides contributes to reduced sustainability of human health (Fig. 3).

Site-specific management/technology

There is little if any thought about GPS or advanced technology in this setting. However, site-specific management does not necessarily depend on GPS or other related technology. Farmers are generally well aware of variation in soils across their fields. Given that land use decisions are commonly made on <0.5 ha basis in Uganda, farmers by tradition practice site-specific management. Another component of site-specific management is to intentionally farm disparate, non-contiguous pieces of land. These practices spread risk and this adds to sustainability, although production in small land parcels enhances opportunities for invasion of biotic constraints (Fig. 3).

4 Soybean production in Brazil

4.1 Decisions at or before planting

Brazil is second to the United States in soybean production globally (Fig. 2). Despite the differences in climatic conditions, soybean production systems in Brazil and the United States are very similar in terms of technology, mechanization and productivity, but contrasts dramatically with those of Uganda. Over the last 30 years, soybean production in Brazil has increased from from 8 million to 33 million ha, from 1.7 to 3 Mg ha^{-1} and from 10 000 000 to 95 000 000 Mg yr^{-1} (Fig. 2, 10 and 11; CONAB, 2016). The significant increase in area and total production is due to the expansion of soybean production in the Cerrado (Central-West, Brazil; Fig. 12) and adoption of sustainable production practices, such as no-till, crop rotation, cover crops and biotechnology, which allowed Brazil to become a major player in world soybean production. A soybean yield gap assessment has not been done for Brazil. However, for comparison, the gap between current rainfed maize yields

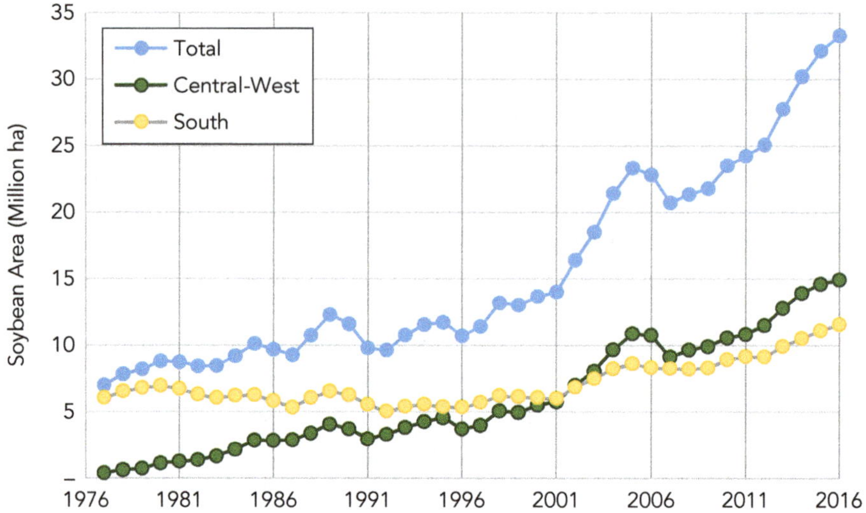

Figure 10 Soybean production area in Brazil. Source: CONAB, www.conab.gov.br.

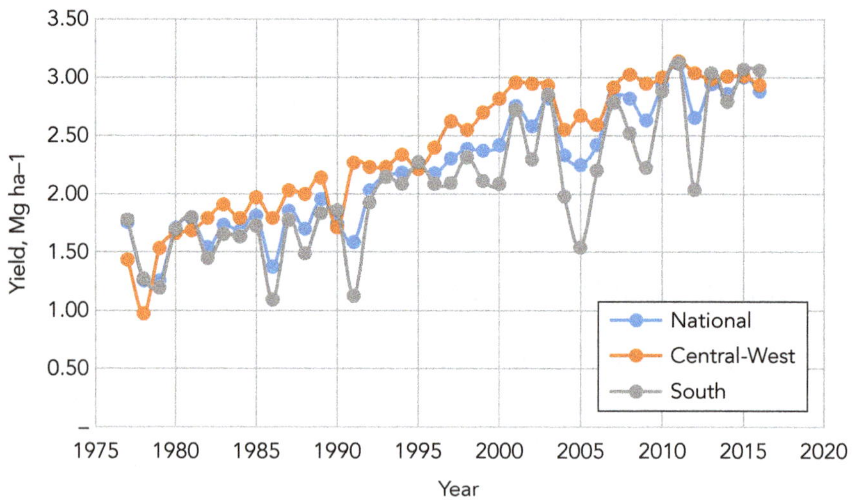

Figure 11 Mean soybean yields in Brazil. Source: CONAB, www.conab.gov.br.

and potential yield is around 30% (GYGA, 2017). That is similar to the 30–40% yield gap for rainfed soybean in neighbouring Argentina.

Soybeans were introduced in Brazil by Japanese immigrants in the early 1900s, but their adoption as a major crop occurred in 1970s due to the increase in international demand. Until the 1980s, the soybean production area was concentrated in subtropical southern

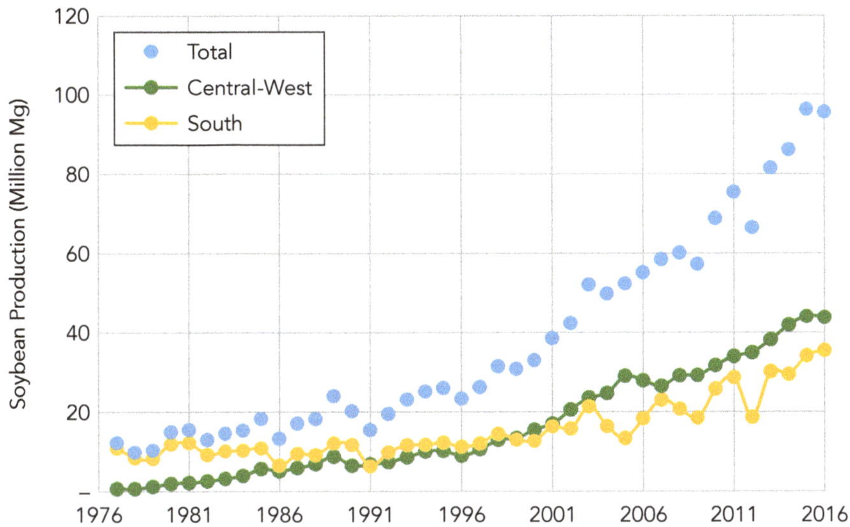

Figure 12 Soybean production in Brazil. Source: CONAB, www.conab.gov.br.

Brazil (Fig. 12). The 'boom' in soybean production in Brazil started when soybean became a viable crop option in the Cerrado. Until the 1960s, the Cerrado area of Brazil was considered inappropriate for row crop production because of infertile acid soils. EMBRAPA (Brazilian Agricultural Research Corporation) played a key role in expanding soybean production in the Cerrado by conducting research and educating farmers on the need for lime applications to correct low pH, fertilizer application to increase the soil P and K availability and planting of soybean cultivars adapted to tropical conditions (earlier soybean cultivars came from the United States with temperate climate adaptation). Increased soybean production in the Cerrado contributed to economic growth in central and northern parts of Brazil. Currently, the major challenge for soybean farmers in the central-west and northern areas of Brazil is the lack of infrastructure for grain storage and transportation, the development of which has been indefinitely delayed by current national political issues.

With the continual increase in world population and demand for higher quality food, there will be an increase in the demand for soybean that needs to be met by increased productivity and/or production area. Researchers and farmers cooperate to increase yield and Brazil has the opportunity to expand soybean production area including much of the 40 million hectares of degraded pasture that could potentially be converted into row crop production without any additional deforestation (Assunção and Chiavari, 2015). Therefore, Brazil has tremendous potential to assist the demands of a growing population.

Cultivar selection

Due to soybean sensitivity to photoperiod, different MGs are recommended for each part of the country (Fig. 13). Longer night lengths stimulate the crop to enter reproductive stages. If the crop has not developed enough prior flowering, it will not achieve its yield potential. Recommended MGs adapted to Brazil range from MG V to X. The closer to

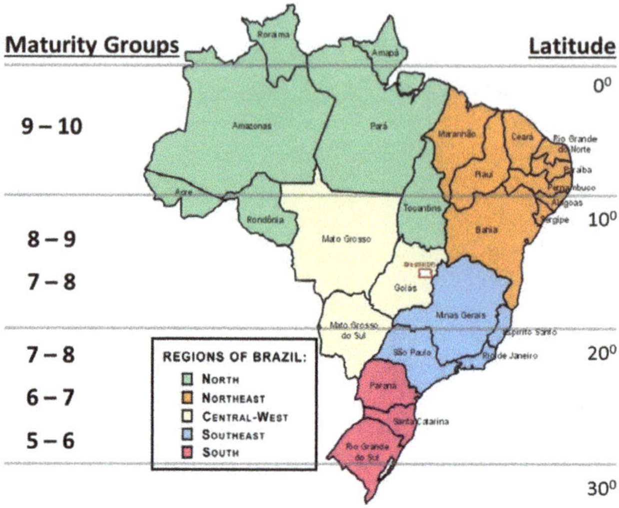

Figure 13 Distribution of maturity groups in Brazil adapted to each region. Adapted from Alliprandini et al. (2009).

the equator, the sooner the soybeans are likely to flower; therefore, cultivars with longer maturity are recommended. In the summer, days are longer in the southern part than closer to the equator, therefore earlier MGs are recommended for southern Brazil. Moreover, recommendations to farmers are to select for soybean cultivars tolerant to common diseases and nematodes, along with higher protein content and yield potential. Farmers have ready access to high-quality seed of high-yielding cultivars. Yields continue to increase with minimal increases in inputs – a sustainable situation (Fig. 3).

Date of planting

The optimal window for soybean planting in Brazil ranges from mid-October through November (EMBRAPA, 2013). Later MGs perform better when planted in October and shorter maturity cultivars present higher yield potential when planted in November. Planting later than November often results in higher incidence of insects (e.g. stinkbugs) and diseases (e.g. soybean rust) besides shortening the crop life cycle and yield potential. Improved sustainability correlates with choosing the best cultivar for the planting date – Brazilian farmers are following this GAP (Fig. 3).

Seeding rate

No yield difference has been observed from populations ranging from 200 000 to 500 000 plant ha^{-1} (EMBRAPA, 2013). Until the 1980s, 400 000 plant ha^{-1} was a common target population. With the advancement of planting techniques, cultivars, availability of more efficient post-emergence herbicides and seed treatments, current recommendations are to reduce population to 300 000 plants ha^{-1}, and 200 000–250 000 plants ha^{-1} for high fertility areas where lodging is a common issue. Moreover, with the increased seed costs due to high adoption of transgenic varieties (Fig. 14), farmers are adopting lower seeding

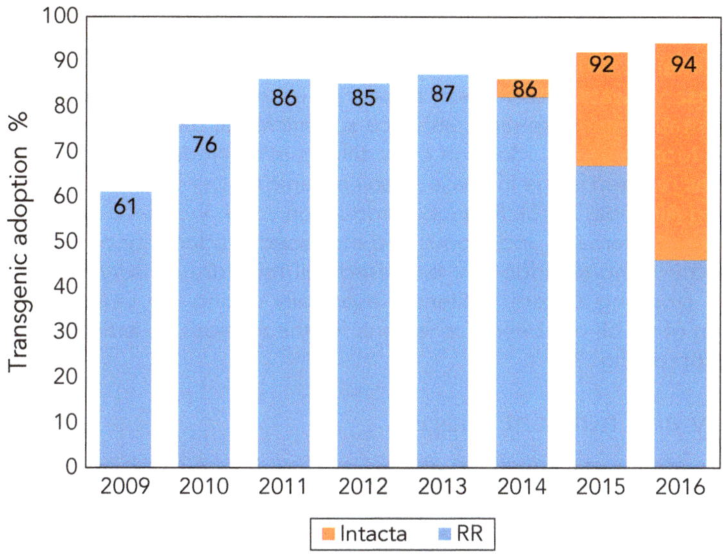

Figure 14 Per cent of soybean production area in Brazil planted to transgenic soybean. RR (glyphosate resistant); Intacta = RR trait plus insect pest resistance.

rates – a sustainable GAP – but more farmers could adopt lower seeding rates to improve sustainability even more (Fig. 3).

Row width

Research on row spacing shows that narrower spacing results in faster canopy closure, increases light interception and reduces weed competition (EMBRAPA, 2013). However, row spacing narrower than 40 cm does not allow late-season operations such as fungicide application for disease management (i.e. white mould, Asian soybean rust, etc.) without significant plant damage. Most equipment in Brazil is developed for 40–50-cm row spacing and farmers are advised to adopt row spacing of 40 cm or wider – they have adopted this GAP. Narrower rows reduce evaporation and early season weed pressure while improving yield potential – all promoting sustainable GAPs (Fig. 3).

4.2 Soil, pest and site-specific management

Tillage and soil management

Farmers' awareness of the importance of soil conservation strategies has increased tremendously over the past 50 years in Brazil. No-till, or 'Sistema de Plantio Direto' ('direct seeding system') as it is referred to, started in the 1970s after farmers realized that some of the European soil management practices, which included intensive tillage, were not appropriate for tropical environments. Brazilian growers learned from US pioneer farmers and researchers about no-till strategies and quickly adapted it to tropical conditions. Public–private partnerships were fundamental during the early stages of no-till development as nutrient recommendations and chemical control option for weeds needed development. Nowadays, approximately 80% of annual row crop production in Brazil is under no-till (Motter and Almeida, 2015). Besides improving productivity, no-till has assisted Brazilian farmers to preserve water, reduce soil losses via water and wind erosion, increase soil carbon sequestration and retention, enhance soil microbial diversity and reduce reliance on pesticides. Moreover, the adoption of no-till has led to reduced reliance on fossil fuel by 70% and has resulted in the increase of soil organic matter from over 1% to more than 3% (Motter and Almeida, 2015). Reduced production costs, increased machinery lifespan, reduced machinery demand and labour and increased productivity of marginal land are also associated with no-till adoption. Increased soil moisture conservation due to no-till has increased cropping system resilience, especially during dry years and in marginal land. Adoption of no-till has been a milestone for the increased sustainability of soybean production in Brazil (Fig. 3).

Soil fertility and nutrient supply

Brazilian soils are mostly highly weathered and of low organic matter and nutrient availability, but can be very productive with application of lime, P and other essential nutrients. Lime and P application ameliorate aluminium toxicity, which is otherwise a major limiting constraint of crop yield. The ideal pH for soybean is between 5 and 6; lime application decisions are based on soil test results. Nutrient application decisions need to consider crop demand (Table 2), expected yield and current soil test results. Fertilizer placement at planting is recommended beside or below the seed. Seed/seedling contact with fertilizer often reduces water absorption and could cause seedling death, especially when high rates of KCl are applied in furrow at planting, which reduces stands and yield potential.

Biological N fixation, in areas where soybeans have been cultivated and *B. japonicum* and/or *B. elkanii* are present, supplies 70–95% of the N demand by soybeans, whereas, the remaining 5–30% are supplied by soil organic N mineralization (Câmara, 2014). According to research, there is no need for application of N if crop rotation and retention of crop residue in the field are practised. Farmers, in general, follow GAPs for soil fertility and nutrients, which promotes sustainability (Fig. 3).

Water-use efficiency

The widespread adoption of no-till combined with appropriate cultivars has increased water-use efficiency of soybeans in Brazil. Soybeans are grown during the wet season in Brazil, and irrigation is not a common practice. According to the National Water Agency, less than 5% of soybean production area of Brazil is irrigated. Despite the benefits of irrigation systems, the occurrence of natural rainfall during the soybean production season in Brazil allows farmers to sustainably produce a good soybean crop without the need of supplemental water.

Crop rotation

Rotation is a common practice in row crop production in Brazil. Given the weather conditions and depending upon the region within the country, 3–5 crops can be grown in a two-year period. Soybeans are typically followed by 'safrinha' corn (short-season corn), which is then followed by wheat, oats, pearl millet or edible beans. Soybeans can be planted the next rainy season or sometimes cotton is grown depending on commodity prices and weather conditions. Crop rotation is a sustainable and common strategy in Brazil.

Cover crops

Despite the lack of data to support such a statement, there seems to be an increase in the adoption of cover and forage crops as part of Brazilian cropping systems. Oats, pearl millet and/or radishes are commonly grown as cover crops for soil conservation purposes and to generate additional biomass for no-till production in regions where there is not enough precipitation to grow a cash crop during the dry winter months. Crop–livestock integration, where a forage crop is planted for grazing during or after the main crop is harvested, is becoming a common strategy in the country, helping producers diversify their source of income and keep their systems productive throughout the year. In some instances, cover crops are also grown to assist with pest management, particularly to reduce plant pathogenic nematode populations.

Disease, insect, nematode and weed management

Due to the tropical/subtropical climate, disease and insect management is important in Brazil. The lack of severe winters permits year-around crop production, which allows insects and pathogens to continue their cycle and build up numbers/inoculum for subsequent crops. The adoption of transgenic Intacta soybean with insect pest resistance is an important IPM component. Diseases such as soybean rust (*Phakopsora pachyrhizi*) and white mould (*S. sclerotiorum*) demand constant scouting and fungicide application. *Meloidogyne* spp., *Heterodera glycines*, *Pratylenchus brachyurus* and *Rotylenculus reniformis* are troublesome nematode species in soybean production in Brazil. Crop rotation and resistant cultivars are the major strategies for nematode management. Weed management became somewhat easy and convenient after wide adoption of GR soybeans following official approval in 1998 (Fig. 12), but glyphosate resistance developed for some weed species including *Lolium multiflorum*, *Conyza bonariensis*, *Conyza canadensis* and *Digitaria insularis*. In order to sustainably manage insects, diseases and weeds, farmers are encouraged to rotate crops, integrate cultural and mechanical practices and alternate herbicide modes of action. Weed resistance management and IPM practices are sustainable GAPs but more could be done (Fig. 3).

Site-specific management/technology

Similar to the United States, Brazilian farmers are technology-driven. The advancements in genetics (i.e. drought, herbicide and insect tolerance traits) and machinery (e.g. precision planting, spraying and harvest) have been well received and adopted in the country. The high costs associated with new technologies and lack of government support have made these technologies less readily available to small-scale producers; however, these technologies are highly adopted by large-scale operations. The use of these technologies has helped producers increase sustainability of their cropping systems by adjusting seeding rates, cultivar selection and pesticide usage for overall enhanced crop production.

5 Summary and conclusions

Comparing and contrasting soybean management research recommendations and actual production practices in three drastically different production areas reveal the opportunities to sustainably increase both the quantity and quality of future soybean production. It is essential that global soybean production increases while embracing improved farmer profitability, environmental stewardship, quality of life for individuals and communities, all with safe and nutritious food sources.

This chapter addresses agronomic aspects of sustainable soybean production. Obviously, other aspects of the value chain are critical including transportation, governmental policies, food processing and storage, as well as human consumption and waste. Increased implementation of research-based agronomic recommendations have political, social, behavioural and economic constraints. For example, in Nebraska and Brazil, increased adoption of no-till planting and reduced seeding rates is more of a behavioural issue – rather than an issue such as lack of resources or inputs. In contrast, Ugandan farmers have low financial ability and access to mechanization, and both inputs and markets constrain sustainable soybean production. However, in all three cases, many opportunities exist to overcome these production constraints while improving sustainability.

6 Where to look for further information

Soybean farmers in all three countries cited in this chapter have indeed made advances in sustainable production practices in the last five or more decades. By incorporating what at the time were GAPs, yields at least in the United States and Brazil continued to increase – we say 'at the time were GAPs', because GAPs can and will change with improved knowledge of cropping systems and technology, climate variation, environmental impact awareness, as well as economic and political forces. However, we presented a case here – for all three countries – that inroads to sustainability have occurred, but there is a considerable way to go to keep up with the GAPs required for future sustainability. To maintain a handle on the pursuit of GAPs, consider these resources:

For information relative to Nebraska/US:

University of Nebraska Extension: http://cropwatch.unl.edu/

Iowa State University Extension Integrated Crop Management Newsletter: https://crops.extension.iastate.edu/cropnews

University of Minnesota Extension: http://www.extension.umn.edu/agriculture/soybean/

University of Wisconsin Extension: http://www.coolbean.info/

University of Illinois Extension: http://bulletin.ipm.illinois.edu/

Kansas State University Extension: http://www.agronomy.k-state.edu/extension/crop-production/soybeans/index.html

University of Missouri Extension: https://plantsciences.missouri.edu/grains/soybean/

Purdue University Extension: https://ag.purdue.edu/arp/Pages/purduesoybeancenter.aspx

Ohio State University Extension: https://agcrops.osu.edu/newsletter/corn-newsletter

University of Arkansas: https://www.uaex.edu/farm-ranch/crops-commercial-horticulture/soybean/

University of Kentucky: http://graincrops.ca.uky.edu/
American Soybean Association: https://soygrowers.com/

For information relative to Brazil:
EMBRAPA: https://www.embrapa.br/en/soja/apresentacao

For information relative to Uganda GAPs:
Makerere University: https://news.mak.ac.ug/

7 References

Abendroth, L. J., K. P. Woli, A. J. W. Myers and R. W. Elmore. 2017. Yield-based corn-planting date recommendation windows for Iowa. *Crop Forage Turfgrass Manag.* 3:1. Open access. doi:10.2134/cftm2017.02.001.

Al-Kaisi, M. and D. Kwaw-Mensah. 2015. Long-term tillage and crop rotation effects on soil carbon and soil productivity in southwest Iowa. http://farms.ag.iastate.edu/sites/default/files/LongTermTillageArmstrong_1.pdf

Alliprandini, L. F., C. Abatti, P. F. Bertagnolli, J. E. Cavassim, H. L. Gabe, A. Kurek, M. N. Matsumoto, M. A. R. de Oliveira, C. Pitol, L. C. Prado and C. Steckling. 2009. Understanding soybean maturity groups in Brazil: Environment, cultivar classification, and stability. *Crop Sci.* 49:801–8. doi:10.2135/cropsci2008.07.0390.

Assunção, J. and J. Chiavari. 2015. Towards efficient land use in Brazil. The new climate economy. The global commission of the economy and climate. http://climatepolicyinitiative.org/wp-content/uploads/2015/09/Towards-Efficient-Land-Use-in-Brazil.pdf. Accessed 10 March 2016.

Bastidas, A. M., T. D. Setiyono, A. Dobermann, K. G. Cassman, R. W. Elmore, G. L. Graef and J. E. Specht. 2008. Soybean sowing date: The vegetative, reproductive, and agronomic impacts. *Crop Sci.* 48: (2) 727–40.

Bradley, K. W. 2006. A review of the effects of row spacing on weed management in corn and soybeans. *Crop Manag.* doi:10.1094/CM-2006-0227-02-RV.

Câmara, G. M. S. 2014. Fixação biológica de nitrogênio em soja. IPNI. International Plant Nutrition Institute. Informações agronômicas n° 147.

CONAB. 2016. Companhia Nacional de Abastecimento. www.conab.gov.br

Conley, S. P. and J. Santini. 2007. Crop management practices in Indiana soybean production systems. *Crop Manag.*. doi:10.1094/CM-2007-0104-01-RS.

CTIC (Conservation Technology Information Center). 2014. Report of the 2013–14 Cover Crop Survey. Joint publication of the Conservation Technology Information Center and the North Central Region Sustainable Agriculture Research and Education Program. http://www.ctic.org/media/CoverCrops/CTIC_04_Cover_Crops_report.pdf.

De Bruin, J. L. and P. Pedersen. 2008. Effect of row spacing and seeding rate on soybean yield. *Agron. J.* 100:704–10.

De Bruin, J. L., P. Pedersen, S. P. Conley, J. M. Gaska, S. L. Naeve, J. E. Kurle, R. W. Elmore, L. J. Giesler and L. J. Abendroth. 2010. Probability of yield response to inoculants in fields with a history of soybean. *Crop Sci.* 50:265–72.

Drewnoski, M., N. Muller, R. Saner, P. Jasa, G. Zoubek, J. Rees, K. Pekarek, R. Elmore, D. Redfearn and G. Lesoing. 2015. Cover crop survey of Nebraska farmers. Univ. of Nebraska Extension. *CropWatch*. http://cropwatch.unl.edu/image/CW_News/2015/UNL-2015-Cover-Crop-Survey-Summary-2.pdf.

Duke, S. O. and S. B. Powles. 2009. Glyphosate-resistant crops and weeds: Now and in the future. *AgBioForum* 12 (3 and 4): 346–57. http://naldc.nal.usda.gov/download/40555/PDF.

Dunphy, J. and J. Ferguson. 1991. Field performance of farmer-saved and professionally-grown soybean seed lots. In: *Proceedings of the American Seed Trade Association Annual Meeting.*

Elmore, R. W., F. W. Roeth, L. A. Nelson, C. A. Shapiro, R. N. Klein, S. Z. Knezevic and A. Martin. 2001. Glyphosate resistant soybean cultivar yields compared to sister lines. *Agron. J.* 93:408–12.

EMBRAPA. 2013. Tecnologia de produção de soja – Região central do Brasil 2014. http://ainfo.cnptia.embrapa.br/digital/bitstream/item/95489/1/SP-16-online.pdf. Accessed 10 March 2016.

ERS. 2016. ARMS farm financial and crop production practices. http://www.ers.usda.gov/data-products/arms-farm-financial-and-crop-production-practices/tailored-reports-crop-production-practices.aspx.

FAO. 2016. FAOSTAT. http://faostat3.fao.org/home/E.

Furseth, B. J., S. P. Conley and J. M. Ané. 2012. Soybean response to rhizobia and seed-applied rhizobia inoculants in Wisconsin. *Crop Sci.* 52:339–44.

Gielser, L., T. Hunt, J. Miller, R. Seymour and B. Wright. 2016. Soybean production management strategies comparison. In: *2016 Soybean Management Field Days*. Univ. of Nebraska. Extension. http://ardc.unl.edu/2016SMFDBooklet-FINAL.pdf.

Gielser, L. J. and A. D. Ziems. 2008. Seed treatment fungicides for soybeans. NebGuide G1852. Univ. of Nebraska, Extension. Lincoln, NE.

Grassini, P., J. A. Torrion, H. S. Yang, J. Rees, Daryl Andersen, K. G. Cassman and J. E. Specht. 2015. Soybean yield gaps and water productivity in the western U.S. corn belt. *Field Crops Res.* 179:150–63.

Grassini, P., J. M. Rees and J. E. Specht. 2016. What does it take to produce 80+ bu/ac soybean routinely? Univ. of Nebraska Extension, draft NebGuide.

GYGA. 2017. Global yield gap analysis. http://www.yieldgap.org/

Hanna, S., S. P. Conley, G. E. Shaner and J. B. Santini. 2008. Fungicide application timing and row spacing effect on soybean canopy penetration and grain yield. *Agron. J.* 100(5):1488–92.

Heatherly, L. G. and R. W. Elmore. 2004. Managing inputs for peak production (Chapter 10). In: H. R. Boerma and J. E. Specht (Eds) *Soybeans: Improvement, Production, and Uses*, 3rd Ed. No. 16. ASA, CSSA, and SSSA, Madison, WI.

HPRCC. 2017. High Plains Regional Climate Center, 30-year climate normals. https://hprcc.unl.edu/onlinedataservices.php

Iowa State University. 2016. ICPT soybean reports. Iowa State Univ. and Iowa Crop Improvement Assoc. http://www.croptesting.iastate.edu/Soybean/Reports.aspx (16 August 2016).

Jasa, L. and J. Rees. 2016. Results of 2016 CropWatch soybean planting survey. Personal communication.

Kaizzi, C. K., J. Byalebeka, C. S. Wortmann and M. Mamo. 2007. Low input approaches for soil fertility management in semi-arid eastern Uganda. *Agron. J.* 99:847–53.

Kaizzi, C. K., J. Byalebeka, O. Semalulu, I. Alou, W. Zimwanguyizza, A. Nansamba, E. Odama, P. Musinguzi, P. Ebanyat, T. Hyuha, A. Kashura and C. Wortmann. 2012. Optimizing smallholder returns to fertilizer use: Bean, soybean and groundnut. *Field Crops Res.* 127:109–19.

Knuteson, D., S. Conley, P. Mitchell, F. Dong, A. J. Bussan, J. Wyman and M. Broeske. 2016. The national sustainable soybean initiative: A grower driven sustainability program to enhance US soybean production and markets. Univ. of Wisconsin Extension. http://www.coolbean.info/pdf/soybean_research/national_sustainability_initiative/SoybeanSurveyResults_2013_FINAL.pdf..

Kohler-Cole, K., R. Elmore, H. Blanco, C. Francis, D. Heeren, S. Irmak, C. Proctor, C. Shapiro, T. Shaver and M. Stockton. 2016. Biomass production of winter annual cover crops in corn and soybean. Univ. of NE Extension. *CropWatch*. http://cropwatch.unl.edu/2016/spring-biomass-production-winter-annual-cover-crops-corn-and-soybean

Marburger, D. A., S. Mourtzinis, J. M. Gaska and S. P. Conley. 2016. Do crop rotation and tillage influence seed-applied inoculant decisions? *Agron. J.* 108:402–6.

Mason, S., T. Galusha and Z. Kmail. 2016a. Soybean yield and nodulation response to crop history and inoculation. *Agron. J.* 108:309–12.

Mason, S. C., C. G. Walters, T. D. Galusha, R. K. Wilson and Z. Kmail. 2017. Planting Saved Roundup Ready 1 soybean seed implications on yield and profit. *Agron. J.* 109:729–34.

Mason, S. C., C. Walters and T. Galusha. 2016b. Saving Roundup Ready soybean seed. Cornhusker Economics. Univ. of Nebraska. http://agecon.unl.edu/cornhusker-economics/2016/saving-roundup-ready-soybean-seed.

Motter, P. and H. G. de Almeida. 2015. Plantio direto: A tecnologia que revolucionou a agricultura brasileira, 144p http://febrapdp.org.br/download/publicacoes/LIVRO_PLANTIO_DIRETO_WEB.pdf.

Mueller, N., R. Elmore and T. Regassa. 2014. Why soybean cultivar is critical to improving farm yield. Univ. of Nebraska Extension, *CropWatch*. http://cropwatch.unl.edu/why-soybean-cultivar-critical-improving-farm-yield (15 August 2016).

Nafziger, E., M. Johnson, B. Mansfield and A. Pelter. 2015. The effects of long-term rotation and tillage on corn, soybean, and wheat yields. University of Illinois: http://web.extension.illinois.edu/nwiardc/downloads/58547.pdf.

Nafziger, E., E. Adee, M. Johnson, B. Mansfield and A. Pelter. 2013. Crop rotation and corn and soybean yield. University of Illinois: http://web.extension.illinois.edu/nwiardc/downloads/48081.pdf.

Nansamba, A., K. C. Kaizzi, A. B. Twaha, P. Ebanyat and C. S. Wortmann. 2016. Reduced tillage, rotation, and soil fertility management for sorghum production in Uganda. *Agron. J.* 108:1–10.

NRCS. 2008. Nebraska soybeans planted no-till in 2008. Accessed November 2016. http://www.nrcs.usda.gov/Internet/FSE_DOCUMENTS/nrcs142p2_029359.pdf

Rees, Jennifer. 2015. Results of the CW soybean planting survey. Univ. of Nebraska Extension, *CropWatch*. http://cropwatch.unl.edu/results-cw-soybean-planting-survey.

Regassa, T. 2016. Soybean cultivar test results. Univ. of Nebr. Extension, *CropWatch*. http://cropwatch.unl.edu/cultivartest/soybean (15 August 2016).

Runge, M., J. Fulton, T. Griffin, S. Virk and A. Brook. 2014. Automatic section control technology for row crop planters. Alabama A&M & Aubern University Extension: http://www.aces.edu/pubs/docs/A/ANR-2217/ANR-2217.pdf.

Salvagiotti, F., A. Weiss, A. Dobermann, D. T. Walters, K. G. Cassman and J. E. Specht. 2008. Nitrogen uptake, fixation and response to fertilizer N in soybeans: A review. *Field Crops Res.* 108:1–13.

Schimmelpfennig, D. and R. Ebel. 2011. On the doorstep of the information age: Recent adoption of precision agriculture. EIB-80, U.S. Dept. of Agriculture, Economic Research Service. http://www.ers.usda.gov/media/81195/eib80_1_.pdf.

Shaver, T., R. Ferguson, E. Penas and W. Bart Stevens. 2014. Soybean (Chapter 16), pp. 116–20. In: T. Shaver (Ed.) *Nutrient Management for Agronomic Crops in Nebraska*. University of Nebraska-Lincoln Extension EC 155.

Specht, J., P. Grassini and J. Rees. 2013. Data show Nebraskans planting soybean earlier each year. Univ. of Nebraska, *CropWatch*. http://cropwatch.unl.edu/data-show-nebraskans-planting-soybean-earlier-each-year.

Specht, J., J. Rees, P. Grassini and N. Mueller. 2015. Soybean planting tips for 2015. Univ. of Nebraska Extension, *CropWatch*. http://cropwatch.unl.edu/soybean-planting-tips-optimal-yield-2015.

Tukamuhabwa, P., H. K. Oloka, T. Sengooba and P. Kabayi. 2012. *Euphytica* 183:1. doi:10.1007/s10681-011-0404-3.

UNL. 2013. Nebraska on-farm research: Soybean population studies (2006–2013). Univ. of Nebraska Extension. https://docs.google.com/presentation/d/1Vyu-Kt9-DaMleKjMeXvGksaduWWjLe4zeZlqM-yflDk/pub?start=falseandloop=falseanddelayms=3000andslide=id.g3db2340d5_5_75.

USDA-NASS. 2016. Acreage. ISSN: 1949-1522 http://usda.mannlib.cornell.edu/usda/current/Acre/Acre-06-30-2016.pdf.

USDA-NASS. 2017a. News Release, Nebraska Rank in U.S. Agriculture, May 2017. https://www.nass.usda.gov/Statistics_by_State/Nebraska/Publications/Rank_and_Agribusiness/rank2017.pdf.

USDA-NASS. 2017b. Quick Stats, May 2017. https://www.quickstats.nass.usda.gov.

Wilhelm, W. W. and C. Wortmann. 2004. Tillage and rotation interactions for corn and soybean grain yield as affected by precipitation and air temperature. *Agron. J.* 96: 425–32.

Wortmann, C. S. and C. A. Eledu. 1999. *Uganda's Agroecological Zones: A Guide for Planners and Policy Makers*. CIAT, Kampala, Uganda.

Wortmann, C. S., A. R. Dobermann, R. B. Ferguson, G. W. Hergert, C. A. Shapiro, D. D. Tarkalson, D. T. Walters. 2008. High yield corn response to applied phosphorus, potassium, and sulfur in Nebraska. *Agron. J.* 101:546–55.

Wortmann, C. S., C. A. Shapiro, R. Ferguson, and M. Mainz. 2012. Irrigated soybean has a small response to nitrogen applied during early reproductive growth. *J. Crop Manage.* http://www.plantmanagementnetwork.org/sub/cm/research/2012/nitrogen/.

Wortmann, C. S., P. Grassini and R. W. Elmore. 2016. Optimizing maize-based cropping systems: Sustainability, good agricultural practices (GAP) and yield goals. In: Watson, D. (Ed.) *Achieving Sustainable Cultivation of Maize Volume 2: Cultivation Techniques, Pest and Disease Control*. Burleigh Dodds Science Publishing, Cambridge, UK (ISBN: 978 1 78676 012 8; www.bdspublishing.com).

Chapter 13

Supporting smallholders in soybean cultivation: the example of Zimbabwe

Byron Zamasiya and Kefasi Nyikahadzoi, University of Zimbabwe, Zimbabwe

1 Introduction

Low crop productivity due to soil fertility decline is a major challenge constraining maize-based smallholder farming systems in sub-Saharan Africa. The decline in soil fertility is largely due to continuous cropping without soil replenishment and reduced N_2 fixation (Dakora and Keya, 1997). Most of the soils in this region contain low nutrient levels, for instance P levels range between 5 and 15 mg kg^{-1} and total N ranges between 0.02% and 0.07% (Sanchez, 2002). The situation is exacerbated by the presence of dysfunctional institutions, climate change and exorbitant costs of inorganic fertilisers following the removal of subsidies. Although most of the smallholder farmers are in need of mineral fertilisers to boost cereal crop productivity, over 60% are unable to afford the high prices (Morris et al., 2007). This has increased food, income and nutrition insecurity among smallholder farmers.

Grain legumes are seen as having the potential to replenish soil nitrogen and provide the much needed protein in maize-based diets (Chianu et al., 2009). In Zimbabwe, soybean (*Glycine max* (L.) Merrill) has emerged as a key crop that has the potential to help improve soil fertility, food and income security among smallholder farmers (Zamasiya et al., 2014; Chianu et al., 2009). Soybean is renowned for its high propensity to fix atmospheric nitrogen in symbiosis with introduced or indigenous *Rhizobia* (Smaling et al., 2008; Pompi et al., 1998). On-farm experiments show that maize grain yield after soybean crop would reach 1.5 tonnes ha^{-1} compared to 0.5 tonnes ha^{-1} for maize grown after maize

http://dx.doi.org/10.19103/AS.2017.0034.14

(Kasasa et al., 1999). Soybean has a high protein level of approximately 40% and 20% oil compared to other legumes such as cowpea (*Vigna unguiculata* L.), groundnut (*Arachis hypogaea* L.) and common bean (*Phaseolus vulgaris* L.) (Mpepereki et al., 2000). Soybean is perceived to be more resistant to pests and diseases than other legume crops (Giller et al., 2011; Mpepereki et al., 2000). These attributes make it an ideal crop for supplementing protein-deficient maize-based diets among smallholder farmers. Despite the nutritional, income and biological nitrogen fixation (BNF) benefits of soybean, its adoption among smallholder farming systems in sub-Saharan Africa is still low. Studies note that the majority of smallholder farmers grow soybean on small plots (Mapfumo et al., 2001) with grain yield barely exceeding 500 kg ha^{-1} (Mafongoya et al., 2006). The low adoption is attributed to biophysical, social and economic conditions prevailing in developing countries.

To promote the realisation of these benefits, government, research institutions, development institutions and non-governmental organisations embarked on collaborative programmes that supported smallholder farmers in the cultivation, processing and marketing of soybeans. Despite the immense contributions of these programmes on smallholder cultivation of soybeans, there is very scanty literature to display their impact, lessons learnt and challenges. Studies have concentrated on documenting biophysical researches on soybean biomass and grain yield under different fertiliser blends and farmer management practices (Kanonge et al., 2015; Mpepereki et al., 2000; Mabika and Mariga, 1996; Mpepereki et al., 1996a). Other studies have also dwelled on demonstrating the role of *Rhizobium* inoculation on soybean grain yield, biomass and N_2 production (Zengeni et al., 2006; Mpepereki et al., 2000). The concentration of research on the biophysical side of soybean production has left an information gap for policy making on how to enhance the cultivation of soybean among smallholder farmers. As such, this chapter seeks to add to the existing literature on soybean production and marketing by focusing on the opportunities and constraints for soybean cultivation by smallholder farmers using the example of Zimbabwe.

2 Background to soybean production in Zimbabwe

Large-scale soybean production in Zimbabwe started in the 1970s with the majority of the output coming from commercial farmers (Technoserve, 2011; Rusike et al., 2000). The major soybean producing areas are Mashonaland Central, Mashonaland West and Mashonaland East provinces of Zimbabwe. These areas are spread around Harare, the major processing centre. Production among the commercial farmers was highly mechanised and they averagely produced 95% of the national output while smallholder farmers accounted for 5% (Estehuizen, 2011). The commercial farmers had easy access to inputs, financial capital, irrigation services and well-developed marketing services (Madanzi et al., 2012). Smallholder farmers cultivated soybean as a side crop on small patches of land. The low yields among smallholder farmers are attributed to lack of access to inoculant, low usage of inorganic fertilisers, poor quality germplasm and lack of access to certified seed in appropriate packs (Rusike et al., 2000; Mabika and Mariga, 1996), poor seed quality and lack of nutrient resources (Mapfumo et al., 2001). Unlike the commercial farmers, the majority of smallholder farmers sell their produce to itinerant traders at farm gate prices.

Figure 1 shows the contribution of smallholder and commercial farmers to national soybean production in Zimbabwe. Following the land reform exercise, contribution of commercial farmers fell to 65% while the contribution of smallholder farmers rose to 35% (Technoserve, 2011). The decline is attributed to the disruption of agricultural production

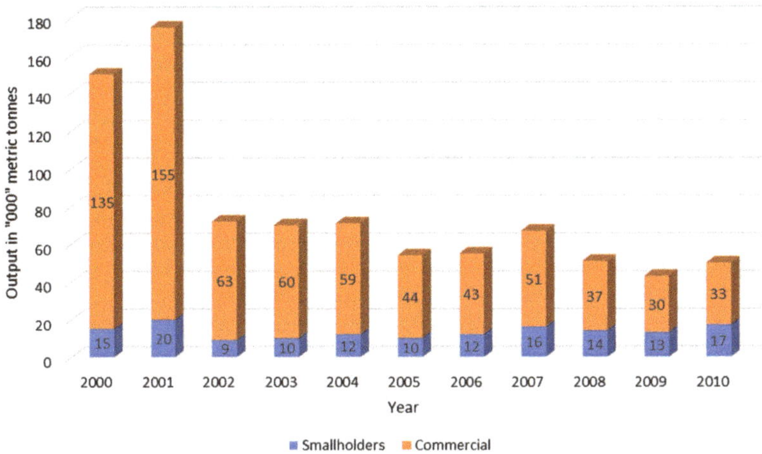

Figure 1 Soybean production by smallholder and commercial farmers in Zimbabwe. Source: Technoserve, 2011.

on commercial farms and hyperinflation which made it expensive to access credit. This decline has left Zimbabwe importing 70% of its soybean requirements from India, South Africa, Malawi and Zambia.

Zimbabwe has four soybean processing plants with a combined capacity of 460 000 metric tonnes. These are Surface Investments (180 000 metric tonnes), Olivine (120 000 metric tonnes), National Foods (80 000 metric tonnes) and United Refineries (80 000 metric tonnes). In 2011, Surface Investments was operating at full capacity while Olivine was operating at 10% due to low soybean supply (Technoserve, 2011). Surface Investments acquired processing technology that can process at US$40–50/metric tonne against an industry average of US$120–140/metric tonne. United Refineries and National Foods plants were not in operation. Although the country has a processing capacity of 460 000 metric tonnes, smallholder and commercial farmers have only managed to produce 30% of national demand with the deficit being filled by imports from Malawi, Zambia and South Africa (Technoserve, 2011). This supply deficit has reduced capacity utilisation to about 16% at the country's major soybean processing plant (Technoserve, 2011). Despite the huge supply deficit, smallholder farmers have not been able to step in and fill the supply gap (Zamasiya et al., 2014; Technoserve, 2011). The booming poultry, piggery and aquaculture production sectors where soybean is used to produce feeds and soybean cake is an opportunity to increase the production of soybean by smallholder farmers (Zamasiya et al., 2014; Technoserve, 2011). Despite these opportunities for increasing soybean production, the contribution of smallholder producers to national soybean output has remained low.

3 Review of selected soybean promotion projects in Zimbabwe

Since 1986, the government of Zimbabwe's research and development partners embarked on collaborative projects that sought to promote soybean cultivation and utilisation among smallholder farmers. Some of the projects include the Hurungwe Soybean Project, the Zimbabwe Soybean Promotion Taskforce, Africare's Soybean and Market Linkages Project and the Putting Nitrogen Fixation to Work for Smallholder Farmers in Africa (N2Africa project). The projects reported in this chapter were selected based on the number of farmers reached and the geographical spread of the projects. A common approach by all these research-led soybean promotion projects is that they showcased production technologies, rhizobial inoculants and varietal testing using different fertiliser blends through field-based farmer- and researcher-managed demonstrations (Mapfumo, 2009). Research institutions provided technical backstopping to communities and extension staff through training.

3.1 Hurungwe district soybean project

The Hurungwe District Soybean project was initiated by the Ministry of Community Development and Women Affairs in August 1986 (Whingwiri, 1996). The purpose of the project was to promote the cultivation and utilisation of soybeans in smallholder farmers' diets. At that time, Hurungwe district accounted for the majority of the national share of soybeans that was produced by smallholder farmers (Rusike et al., 2000). The farmers in the district were already cultivating soybeans using a local variety known as Hurungwe Special. Prior to the project, extension workers had little knowledge of the crop and farmers used retained or carry over seed (Whingwiri, 1996). None of the farmers used inoculants before planting or inorganic fertilisers. Soybean yields in the district ranged between 500 and 600 kg/ha (Whingwiri, 1996). The project introduced soybean demonstrations hosted by farmers. The demonstrations showcased the use of Rhizobium japonicum inoculant and inorganic fertilisers namely single super phosphate (18% P_2O_5) and ammonium nitrate (33.5% nitrogen). Furthermore, the demonstrations showcased the yield responses of Hurungwe Special and a new variety Roan under different fertiliser and management regimes. Agricultural extension officers were trained on how to use the inoculant, and grow and harvest soybean. Community development workers were also trained on how to process soybeans for household consumption in different foods. In the 1988/9 farming season, the project stopped demonstrations and initiated a competition among farmers to grow at least 0.5 ha of soybeans. For the first time in their farming history, the majority of the smallholder farmers in Hurungwe district purchased certified soybean seed and Japonicum inoculant. Whingwiri (1996) noted that the area under soybean cultivation increased by approximately 400% from 306 ha in the 1986/7 farming season to 1300 ha in the 1988/9 farming season.

3.2 Zimbabwe Soybean Promotion Taskforce project

The Zimbabwe Soybean Promotion Taskforce was established in February 1996 with financial support from the Kellogg Foundation. The Taskforce was formed as a result of a workshop convened by the University of Zimbabwe and attended by research institutions, agro-industrial processors and extension to discuss why the adoption of soybean by smallholder farmers was very low (Mpepereki et al., 1996a). Its membership was drawn from the University of Zimbabwe, Zimbabwe Farmers Union, Department of Research and Specialist Services, Department of Agricultural Technical and Extension Services (AGRITEX) and Olivine Industries (an agroenterprise processor for soybean) and Commercial Farmers Union (Giller et al., 2011). The Taskforce's main purpose was to promote the adoption of soybeans by smallholder farmers in Zimbabwe. The project focused on training smallholder farmers on soybean production technologies, processing, marketing and agronomic evaluation of specific and promiscuous nodulating soybean varieties (Rusike et al., 2000). In the 1996/7 farming season, the project targeted five farmers per village from 11 communal areas (Giller et al., 2011) that were within a radius of 250 km from Harare. These communal areas were in Natural Farming Region II and III where the climatic conditions are favourable for soybean production. The radius of 250 km ensured proximity to the main soybean processors in Harare. The target farmers were selected on the basis of having previous experience in growing soybean or interest in the crop. The taskforce trained extension officers and smallholder farmers from the 11 clusters on the importance of soybean production, input acquisition, processing, utilisation and marketing (Giller et al., 2011). Field-based farmer-managed demonstrations were established to showcase soybean cultivation technologies, use of rhizobia inoculant and varietal performance under different fertiliser regimes. Each demonstration site was 0.4 ha in size and had four different soybean cultivars. Soybean seeds planted on each demonstration were inoculated and a seed rate of 80 kg/ha was used. Lime and compound L (N: P: K: S 5:8:8:8) were applied at rates of 500 kg/ha and 125 kg/ha (Rusike et al., 2000), respectively. At the end of the first season, smallholder farmers sold 28 tonnes of soybeans grains to Atrax Commodities (Rusike et al., 2000). Other unrecorded sales were also made to Cairns Food and National Foods.

The 1997/8 farming season was marked by a sharp rise in demand for soybean by smallholder farmers. Smallholder farmers preferred an input pack comprising 10 kg of seed, 1 sachet *Rhizobium* inoculant, 25 kg of Compound L fertiliser and 50 kg of lime. Due to the non-availability of the inputs at a local level, the Taskforce had to facilitate the repackaging of 50 kg soybean seed packets into 10 kg packs and also provided transport. The cost of the input packet was Z$200 (US$20) but was subsidised to Z$150 (US$15)[1] by input suppliers (Rusike et al., 2000). Since the demand for the input package outstripped supply, some smallholder farmers had to use retained seeds and planted without inoculant, lime and Compound L fertiliser. Between 1996 and 2000, the demand for *Rhizobium* inoculant rose from 60 000 to over a 100 000 sachets (Giller et.al., 2011). The Taskforce assisted smallholder farmers with market information and facilitated a marketing arrangement with Olivine Industries in which the company paid Z$4500/tonne for 270 tonnes of soybean grain (Rusike et al., 2000). This price was lucrative when compared to the Grain Marketing Board's Z$3000/tonne. However, smallholder farmers received an effective price of Z$4000/tonne after paying for handling and transport services. In the last season of the project (1998/9), the project was working with 10 000 smallholder farmers. Area under soybean cultivation increased from 22 ha in 1996/7 to 4000 ha in 1998/9 season. Sales rose from 18.2 tonnes to 700 tonnes. Olivine Industries tested the quality of the soybeans and upgraded them from grade D to B. The upgrading also carried a price premium for the farmers (Giller et al., 2011; Rusike et al., 2000). This shows that barring the challenges, the soybean taskforce was very effective in promoting soybean cultivation by smallholder farmers. Although the project was a huge success in promoting soybean cultivation by smallholder farmers, the subsequent land reform and economic crisis led to sharp declines in production.

1. Based on an exchange rate of US$1:Z$10.

3.3 Africare Zimbabwe's Soybean Market Linkages Project

Africare Zimbabwe implemented a Soybean Market Linkages Project in Rushinga, Mt Darwin, Shamva and Bindura districts of Zimbabwe. The project was implemented for five cropping seasons, that is, from January 2002 to December 2007. Its objective was to improve rural farmers' access to markets, value chain development and nutrition. Community-based facilitators and extension officers trained over 5320 smallholder farmers in soybean production, marketing and processing using the training of trainers' (ToTs) approach (Africare, 2010). The ToTs approach was meant for lead farmers who would then cascade the training to follower farmers by demonstrating best practices, enhancing farmer-to-farmer exchange, monitoring the performance of trainees and participatory evaluation of the project. Further, the project boosted soybean production through promoting innovative seed multiplication techniques. The project assisted in the establishment of farmer information centres for the dissemination of research and marketing information. Through a revolving loan fund from Africare, project beneficiaries established three oil pressing plants in Shamva and Bindura districts. The plants enabled local value addition and improved the incomes and nutrition of over 2 000 households. Project beneficiaries also received training in market research and marketing. The purpose of the marketing committees was to ensure that smallholder soybean farmers obtained high farm gate prices by selling directly to processors. Moreover, the project promoted soybean grain marketing locally and regionally through farmer-led commodity associations.

3.4 Putting Nitrogen Fixation to Work for Smallholder Farmers in Africa (N2Africa) project

The N2Africa project is a large-scale project focused on increasing the production of grain legumes and the inputs from N_2 fixation in smallholder farming systems in sub-Saharan Africa (Giller et al., 2013). The N2Africa project was implemented in Zimbabwe, DR Congo, Ghana, Kenya, Malawi, Mozambique, Nigeria and Rwanda from 2011 to 2015 and then expanded into Ethiopia, Uganda, Tanzania, Liberia and Sierra Leone. The objective of this research and development partnership project was to develop, disseminate and promote appropriate N_2 fixation technologies for smallholder farmers focusing on soybeans (*Glycine max* (L.) Merrill), groundnuts (*Arachis hypogaea* L.), cowpeas (*Vigna unguiculata* (L.) Walp) and common beans (*Phaseolus vulgaris* L.). The project used partnerships with agricultural and research institutions in delivering the public good. In Zimbabwe, the project was implemented in seven districts using the lead farmer approach. In each district, lead farmers were identified with the help of agricultural extension officers and implementing partners. These lead farmers attended ToTs sessions on legume agronomy, inoculation, scouting and post-harvest handling. They then recruited 15 follower farmers each onto whom they imparted the skills. The project provided 5 kg of soybean seeds, one sachet of *Bradyrhizobium* and the requisite basal compound fertilisers to each lead farmer. Each follower farmer received 1 kg of seeds, and no compound fertiliser. The follower farmers were supposed to learn from the lead farmers and try out the legume production technologies on their pieces of land. Agro-dealers in the target districts received training from the project staff on the importance of selling certified seeds and on how to use the soybean inoculant. This assisted the agro-dealers in selling legume seed, storing the inoculant and advising farmers on how to use the inoculant. In the 4-year period, the project targeted 25 000 direct beneficiaries. The volumes of seed availed to smallholder farmers did not allow for market participation during the project lifespan.

3.5 Lessons emerging from previous projects

The key emerging lessons from these selected projects are that, with improved access to quality germplasm in the right packages, smallholder farmers can increase adoption of soybean. Availability of quality germplasm has to be followed up with training of extension officers, smallholder farmers and agro-dealers on soybean agronomy and processing. This model has high potential for adoption of soybean as agro-dealers can act as distribution points for seed, chemicals, fertilisers and inoculants. The training of smallholder farmers in agronomy, pest and diseases control is crucial for making smallholder farmers shift from subsistence to commercial production. During the harvesting period, agro-dealers can act as agencies for the soybean processors as they are strategically located in business centres in smallholder farming communities. Further, linkages to reliable output markets is a necessary condition for increased cultivation and marketing of soybean by smallholder farmers. Sustainable linkages could involve training of smallholder farmers in market research and marketing. It could also involve facilitating contractual arrangements involving farmers, fertiliser companies, seed companies and soybean processors. Research institutions can then provide support services in terms of technology dissemination through participatory action research of new varieties, mineral fertiliser blends and rhizobial strains. Such an agreement would guarantee access to good quality seed, inoculant at an uninflated price, chemicals, knowledge and a reliable market for smallholder farmers.

4 Constraints to soybean cultivation in Zimbabwe

Despite its potential contribution to soil fertility, food, nutritional and income security, the cultivation of soybean by smallholder farmers is threatened by a number of constraints which include biophysical and institutional issues.

One of the major challenges facing smallholder soybean producers relates to soil health issues (Rusike et al., 2000). The majority of the soils in most smallholder farming communities are acidic and inherently deficient in phosphorous and organic matter (Mtambanengwe and Mapfumo, 2009; Mabika and Mariga, 1996). Studies note that although soybean demands high levels of phosphorous, most smallholder farmers apply suboptimal levels of organic fertilisers to their maize in a maize–soybean rotation (Mpepereki et al., 2000; Shumba et al., 1992). Soybean varieties produced in Zimbabwe have a potential of yielding 3–4 tonnes per hectare; however, they would require extra top dressing with fertilisers rich in nitrogen (Giller et al., 2011). As such, continued growth of soybean without the requisite fertilisers gradually depletes soil fertility. However, the lack of financial resources to purchase inorganic fertilisers limits the potential for smallholder farmers to grow soybeans.

Lack of quality nutrient resources is also a major challenge affecting soybean grain yield in smallholder farming systems (Mtambanengwe and Mapfumo, 2009). Smallholder farmers allocate limited nutrient sources such as compost, woodland litter and cattle manure to their soybeans. They also apply limited inorganic fertilisers to their cereal crops, thus leaving the soybeans to benefit from residual fertility (Mapfumo et al., 2001). Studies note that the poor application of nutrient resources is driven by resource scarcity (Mtambanengwe and Mapfumo, 2009). When they are applied, the organic nutrient resources are of poor quality. The lack of quality nutrient resources is likely to suppress soybean grain yield for smallholder farmers and hinder its adoption. With low yields, smallholder farmers will remain subsistence oriented.

Although soybean is less susceptible to diseases unlike other legumes, studies note that the crop is susceptible to diseases, such as soybean rust, frogeye leaf spot, red and bacterial blight (Murithi et al., 2016; Seedco, 2010). Globally, soybean rust has been singled out as a major threat to grain yield (Dean et al., 2012) with yield losses of 10–90% recorded across the globe (Oloka et al., 2008). For instance, all soybean varieties (SC Siesta, SC Safari, SC Serenade and SC Santa)[2] produced by Seedco in Zimbabwe are highly susceptible to soybean rust. This implies that smallholder farmers have to control the disease by spraying fungicides two or three times. However, the purchase of fungicides is a constraint to intensification of soybean cultivation by smallholders who are already resource constrained.

Although supplementing legumes with inorganic basal fertilisers has been shown to double the yield, and increase plant growth and N_2 fixation compared with unfertilised controls, the high cost of inorganic basal compound fertilisers is a major impediment to intensification of soybean cultivation by smallholder farmers (Ndakidemi et al., 2006). For instance, in Zimbabwe, a bag of compound D fertiliser (NPK) costs US$29 in Harare and US$42 when it reaches the smallholder farmers. Given that a 20kg bucket of maize grain averagely costs US$3 in the smallholder farming communities, the cost of a bag of fertiliser is thus equivalent to 14 buckets of maize grain. This high conversion rate implies a low incentive to use mineral fertilisers by smallholder farmers. Considering the low maize grain yields in communal farmers, it is therefore difficult for farmers to sacrifice 14 buckets of maize for just one bag of fertiliser.

Studies note that soybean has the potential to fix 40–120 kgN/ha (Mabika and Mariga, 1996). The nitrogen is very useful for the subsequent maize crop in a maize–soybean rotation system. However, smallholder farmers do not apply inoculant to their soybean due to its unavailability at the time of planting, lack of knowledge on how to use the it, lack of refrigeration and overpricing (Zamasiya et al., 2014; Chianu et al., 2009; Whingwiri, 1996; Mabika and Mariga, 1996). As reported by Africare (2010), use of rhizobial inoculant has the potential to reduce production cost, especially in sandy soils. Failure to use inoculants limits the potential of soybean to fix atmospheric nitrogen and subsequently reduces soybean grain yield (Technoserve, 2011). Although *Rhizobium* inoculant is produced at $3.20 and retailed at $5 per sachet (Wommer et al., 2013), most smallholder farmers access the same sachet at double the recommended price (Zamasiya et al., 2014). The high cost prohibits smallholder farmers from using this yield-enhancing technology in soybean production.

Smallholder farmers also lack access to improved seeds from local markets (Rusike et al., 2013; Mudimu, 1996). In Zimbabwe, the majority of smallholder soybean farmers access seeds from donations with only 43.4% accessing from seed distributors and stockists (Mujaju and Jonga, 2015). This could be due to non-availability of the seed in appropriate packages and quantities (Rusike et al., 2000; Whingwiri, 1996), lack of knowledge of

2. http://seeds.seedco.co/profiles/investor/fullpage.asp?BzID=2050&to=cp&Nav=0&LangID=1&s=0&ID=12988.

the existence of improved varieties (Mabika and Mariga, 1996) and lack of appropriate varieties. Smallholder farmers who travelled to Harare from the Soybean Promotion Taskforce project sites could not access the inputs from traditional suppliers (Rusike et al., 2000). The Soybean Promotion Taskforce had to repack the improved seeds from 50 kg into 10 kg seed packs. Although Seedco packages soybean in 10 kg and 25 kg packets recommended as appropriate by farmers, access to seed by smallholder farmers is still low (Mujaji and Jonga, 2015). Due to failure to access improved soybean seed, smallholder farmers end up using retained seeds beyond the supplier-recommended recycling periods of 4 years. Continued use of recycled seeds may render them susceptible to diseases. Sadly, some of the seeds were retained as early as 1960s (Mabika and Mariga, 1996).

Soybeans are a capital-intensive crop that requires farmers to purchase rhizobial inoculant, improved seed and basal fertilisers (Technoserve, 2011; Mudimu, 1996). The absence of agricultural credit from formal lending institutions at concessionary lending rates implies that the crop has to compete for working capital with food security crops such as maize. Since maize is the staple food crop in Zimbabwe, most farmers end up using carry over seed and plant the soybeans without inoculant and/or basal fertilisers. The soybean crop is left to benefit from residual fertility (Mapfumo et al., 2001). Although fertiliser companies such as Zimbabwe Fertilizer Company offered fertilisers on credit to smallholder soybean producers in Goromonzi district in the 2012/13 farming season, the facility was made possible with a credit guarantee from HELP Germany. This implies that without collateral security or a third party guaranteeing the credit, smallholder farmers cannot access agricultural credit at concessionary rates. Mpepereki et al. (2000) noted that smallholder farmers shifted to promiscuous soybean varieties when the prices of improved seed and basal fertilisers increased. As such, the lack of access to agricultural credit constrains soybean production.

A key challenge facing soybean cultivation by smallholder farmers in Zimbabwe is the lack of continued research and extension support beyond the technology transfer stages of soybean promotion programmes. As noted by Zamasiya et al. (2014), extension officers play a pivotal role in disseminating agricultural technologies and information to farmers. They also act as the interface between researchers and farmers. The boom in soybean cultivation during project implementation is a direct result of access to research and extension services (Kanonge et al., 2015; Africare, 2010; Mpepereki et al., 2000). In the majority of research-led development projects, as soon as the projects ended, the umbilical cord between communities and research services is severed. Contact with extension is reduced due to lack of financial resources (Rusike et al., 2000; Mudimu, 1996). Without continued access to technology transfer services from extension and research services, soybean production will fall as farmers due to lack of access to support services. Mabika and Mariga (1996) observed that without access to extension services, smallholder farmers' yield for soybean is low. For instance, smallholder farmers in Hurungwe district used a low plant population of 100 000–150 000 plants per ha as opposed to the recommended practice of 300 000 plants per ha.

The role played by stakeholders in the different projects in linking soybean farmers to markets highlights the importance of access to reliable markets. For instance, the Africare project had to link the smallholder farmers through commodity associations to local and international buyers for soybean grain. Moreover, the Soybean Promotion Taskforce facilitated a marketing arrangement for the project beneficiaries with Olivine Industries (Rusike et al., 2000). In most cases, the project implementer's facilitator role in linking soybean producers to reliable markets discontinued upon the completion of the project's lifespan. The lack of a reliable market and access to price information would cause most smallholders to stop growing soybeans. Without access to market information smallholder farmers are less likely to participate in soybean marketing (Zamasiya et al., 2014). This implies that for smallholder farmers to grow soybeans, they need access to price information and reliable markets.

Studies note that soybean is a fairly new crop to most smallholder farmers and is not regarded as part of the traditional diet (Technoserve, 2011; Rusike et al., 2000). The majority of smallholder farmers are not aware of how this crop can be processed and thus incorporated into local foods. Low local demand for soybean foods was attributed to unfamiliarity with the taste (Mudimu, 1996). This implies that despite the nutritional benefits of soybean in a household's diet, the lack of knowledge on processing constrained its uptake. Without proper knowledge, most farmers end up seeing soybean as a cash crop only and not as a cash and food crop.

5 Opportunities for soybean cultivation in Zimbabwe

This research notes that despite the aforementioned constraints, there exist significant opportunities for supporting soybean cultivation by smallholder farmers in Zimbabwe. These opportunities include taping into supply deficit, boom in livestock industry and combining inorganic and organic fertilisers.

As noted by Opperman and Varia (2011) and Giller et al. (2011), Zimbabwe has an estimated processing capacity of 460 000 metric tonnes and requires an annual supply of 125 000 metric tonnes of soybean. However, local production from both commercial and smallholder farmers cannot satisfy the demand. The collapse of the commercial sector which accounted for over 90% of the national soybean output and the supply deficit implies that there is potential demand for soybeans from National Foods, United Refineries and Olivine Industries. This has also been aided by the placing of a ban by the Ministry of Agriculture on the importation of genetically modified soybeans from South Africa. Studies note that South Africa started growing genetically engineered soybean in 2001. South African government promulgated a legislation which allowed farmers to grow genetically engineered soybeans. Studies noted that in 2008, about 88% of land under soybean had genetically engineered varieties (African Centre for Biosafety, 2010). With a guaranteed market, there is potential for smallholder farmers to increase their production and tap into the underutilised process capacity.

Both smallholder and commercial farmers in Zimbabwe produce soybeans at a unit cost of US$300 per metric tonne compared to an import parity of US$730 per metric tonne (Technoserve, 2011). From a cost perspective, it is costly for local soybean processors to import soybean from South Africa, Malawi and India. Further, Zimbabwe has sophisticated processing plants which do so at a cost of US$40–$50 per metric tonne compared to regional costs of US$120–US$140 per metric tonne (Technoserve, 2011). This suggests that Zimbabwe has a competitive advantage against other regional producers in soybean processing. As such, it makes economic sense for processors in Zimbabwe to enter into contractual arrangements with local smallholder farmers and support them with inputs (seed and basal fertilisers). As such, there is room to organise the dispersed smallholder farmers into production hubs and support them to address local soybean demand. Development programs should give targeted support to medium and high resource endowed farmers as resources-constrained farmers are risk averse and have fewer resources to invest (Franke et al., 2014).

The boom in the poultry, pig and aquaculture industry is expected to increase demand for locally produced soybean as the crop is used in the manufacturing of stock feeds and soybean cake. It is estimated that by 2020, the poultry industry will require 220 000 metric tonnes per year, and the piggery and aquaculture industries would require 28 000 metric tonnes each per annum (Technoserve, 2011). The rise in soybean demand by these three sectors is underpinned by low per unit cost, a rise in incomes and a preference for white protein over beef. Given that current production is below 110 000 metric tonnes, there is potential for smallholder farmers to intensify cultivation of soybeans since the market is guaranteed.

Further, soybean is a crop that has high potential for income generation compared to other crops. The Zimbabwe Farmers Union noted that in 2012, most smallholder farmers in Mashonaland Provinces shifted to soybean due to the high spot prices that were paid by buyers. This view was also corroborated by Woomer et al. (2013) who noted that marginal returns were higher for soybean compared to other legumes. Farmers who participated in field-based demonstrations and diagnostic surveys noted that soybean is a very high valued crop in terms of income generation (Mudimu, 1996). This implies that if the farmers are imparted with proper agronomic knowledge on soybean varieties, inoculant use and fertilisers, there is huge potential for increased production. As such, the participation of smallholder farmers in field-based demonstrations will increase opportunities for intensification of cultivation of soybean.

The associated benefits of soybean cultivation such as BNF and soil fertility improvement from biomass make it a very good crop for rotation purposes in maize–legume rotation. Studies note that soybean has the potential to fix 40–120 kgN/ha (Mabika and Mariga, 1996) and produce 50–90% of its N_2 requirements (Mapfumo et al., 2001). Cultivation of soybean using the appropriate technologies and varieties will thus result in smallholder farmers making substantial cost savings on purchases of inorganic fertilisers for maize crop. Given the high cost of inorganic fertilisers and the declining soil fertility in most smallholder communities (Mapfumo et al., 2000), there is high potential for adoption of soybeans by smallholder farmers.

Although soil fertility remains a major challenge for soybean-producing smallholder farmers due to expensive mineral fertilisers and scarcity of nutrient resources, there is an opportunity to intensify soybean cultivation through combining organic and inorganic fertilisers. Mtambanengwe and Mapfumo (2009) observed that without use of inorganic fertilisers, smallholder farmers are less likely to significantly increase their grain yields. Organic nutrient resources have a great potential to build soil organic matter and increase pH and nutrient availability while mineral fertilisers readily release nutrients (Vanlauwe et al., 2010). Results from on-farm experiments show that soybean yields reached 1900 kg ha^{-1} with basal NPK compound and woodland litter (Kanonge et al., 2015). This implies that there is room to increase intensification of soybean by smallholder farmers through promoting a combination of organic and inorganic nutrient resources such as litter and basal compound fertilisers.

Southern Africa is prone to cyclical droughts which leave women and children vulnerable to food and nutrition security. UNICEF and World Food Programme purchase corn–soya blend which they use as part of feeding programmes. These humanitarian organisations use the corn–soya blend to feed vulnerable children and nursing mothers (Mutithi et al., 2016). The high demand of soybean for producing corn–soybean blend represents an opportunity for improving incomes (Lubungu et al., 2013) of smallholder farmers through intensification of soybean.

6 Conclusion and recommendations

This chapter has provided a review of collaborative projects that promoted soybean cultivation by smallholder farmers. The research has highlighted the constraints and opportunities for supporting smallholder farmers in soybean cultivation using the example of Zimbabwe. The research notes that there is high potential for soybean production in Zimbabwe due to the huge supply deficit in national production driven by the boom in livestock industry and consumers' preference for white meat. Despite the success of the soybean promotion projects, lack of access to production inputs (inoculants, improved germplasm and inorganic fertilisers), post-technology transfer research and extension services, agricultural credit facilities, price information and reliable markets constrain soybean production by smallholder farmers. Despite these challenges, Zimbabwe has opportunities for increased cultivation of soybeans by smallholder farmers. If adopted, there is potential to improve the food, nutrition and income security of smallholder farmers in Zimbabwe.

This research recommends that there is need for government, development partners and agricultural research institutions to train extension officers and smallholder farmers on soybean production technologies, inoculant use, soybean agronomy and processing. The training of smallholder farmers is pivotal for shifting smallholder farmers' production orientation from subsistence to commercial. There is also need for financial market institutions, fertiliser, soybean seed producers and soybean processors to establish a production to marketing system which allows smallholder farmers to access inputs on credit and then pay upon selling. A typical example is the one adopted by cotton merchants in Zimbabwe, where farmers are given the requisite inputs on credit and pay back when selling. This system will guarantee ready markets for input producers, reliable output markets for smallholder farmers and consistency of supply to soybean grain processors. However, the Ministry of Agriculture Mechanisation and Irrigation Development would need to organise farmers into commodity associations for ease of coordination and credit guaranteeing. The development of promiscuous and high nodulating soybean varieties will promote the cultivation of soybeans by smallholder farmers. There is also need for government and its development partners to promote consumption of soybean foods through training of farmers in home-based soybean food processing. Adoption of soybean-based foodstuffs would lead to increased cultivation of soybeans by smallholder farmers.

7 Where to look for further information

http://knowledgecenter.cimmyt.org/cgi-bin/koha/opac-detail.pl?biblionumber=4284&shelfbrowse_itemnumber=5167
http://www.n2africa.org/content/background-n2africa
http://www.partnership-africa.org/content/innovation-zimbabwe-africare
http://www.surfacewilmar.co.zw

8 References

African Centre for Biosafety (May 2010). Tracing Segregation and Labelling of Genetically Modified Products in South Africa. A position paper on the implementation of the Consumer Protection Act and mandatory labelling of GM food. http://www.biosafetyafrica.org.za.

Africare (2010). Innovation in Zimbabwe. President's corner.

Ajeigbe, H. A. and DashielL, K. (2010). Participatory research extension approach: N2 Africa extension method. Based on Elis-Jones, J., Schulz, S., Chikonye, D., de Haan, N., Kormawa, P., and Adezwa, D. (2005). Participatory Research and Exyension approaches: A guide for researchers and extension workers for involving farmers in research and development. IITA: Ibadan, Nigeria and Silsoe Research Institute: UK, p. 52.

Chianu, J. N., Ohiokpehai, O., Vanlauwe, B., Adesina, A., De Groote, H. and Sanginga, N. (2009). Promoting a versatile but yet minor crop: Soybean in the farming systems of Kenya. *Journal of Sustainable Development in Africa*, 10(4): 1–21.

Dakora, F. D. and Keya, S. O. (1997). Contribution of legume nitrogen fixation to sustainable agriculture in sub-Saharan Africa. *Soil Biology and Biochemistry*, 29: 809–17.

Dean, R., Van Kan, J. A., Pretorius, Z. A. and et al. (2012). The top 10 fungal pathogens in molecular plant pathology. *Molecular Plant Pathology*, 13: 413–30.

Franke, A. C., van den Brand, G. J. and Giller, K. E. 2014. Which farmers benefit most from sustainable intensification? An ex-ante impact assessment of expanding grain legume production in Malawi. *European Journal of Agronomy*, 58: 28–38.

Giller, K. E., Murwira, M. S., Dhliwayo, D. K. C., Mafongoya, P. L. and Mpepereki, S. (2011). Soyabeans and sustainable agriculture in southern Africa. *International Journal of Agricultural Sustainability*, 9(1): 50–8

Giller, K.E., Abaidoo, R., Baijuka, F., Bala, A., Boahen, S., Dashiell, K., Kantengwa, P., Sanginga, J.M., Sanginga, N., Simmons, A., Turner, A., de Wolf, J., Woomer, P., Vanlauwe, B. (2013). Putting Nitrogen to work for smallholder farmers in Africa. In: Vanlauwe, B., van Asten, P., Blomme, G. (eds). *Agro-Ecological intensification of agricultural systems in the African Highlands*. Routledge Taylor and Francis Group: New York, 156–74.

Estehuizen, D. (2011). Zimbabwe GAIN Annual Report. Global Agricultural Information Network. USDA Foreign Exchange Service, 1–12.

Kanonge, G., Mtambanengwe, F., Manzeke, G., Nezomba, H. and Mapfumo, P. (2015). Assessing the potential benefits of organic and mineral fertiliser combinations on legume productivity under smallholder management in Zimbabwe. *South African Journal of Plant and Soil*, 32(4): 241–48. DOI:10.1080/02571862.2015.1053156.

Kasasa, P., Mpepereki, S., Musiyiwa, K., Makonese, F. and Giller, K. E. (1999). Residual nitrogen benefits of promiscuous soyabeans to maize under field conditions. *African Crop Science Journal*, 7: 375–82.

Lubungu, M., Burke, J. and Sitko, J. (2013). Analysis of the soya bean value chain in Zambia's eastern province. IAPRI Working Paper 74.

Mabika, V. and Mariga I. K. (1996). An overview of Soyabean Research in smallholder farming sector of Zimbabwe. In *Soyabean in the smallholder cropping systems in Zimbabwe*. Mpepereki, S., Giller, K. E. and Makonese, F. (Eds). Government Printers: Zimbabwe, pp. 12–17.

Madanzi, T, Chiduza, C., Kageler, S. J. R. and Muziri, T. (2012). Effects of different plant populations on yield of different soybean (Glycine max (L) Merrill) varieties in a smallholder sector in Zimbabwe. *Journal of Agronomy*, 11(1): 9–16.

Mafongoya, P. L., Bationo, A., Kihara, J. and Waswa, B. S. (2006). Appropriate technologies to replenish soil fertility in southern Africa. *Nutrient Cycling in Agroecosystems*, 76: 137–51.

Mapfumo, P., Campbell, B. M., Mpepereki, S. and Mafongoya, P. (2000). Legumes in soil fertility management: The case of pigeonpea in smallholder farming systems of Zimbabwe. *African Crop Science Journal*, 9(4), 629–44.

Mapfumo, P. and Giller, K. E. (2001). *Soil fertility Management Strategies and Practices by Smallholder Farmers in Semi-arid Areas of Zimbabwe*. ICRISAT, FAO: Bulawayo, Zimbabwe.

Mapfumo, P. (2009). Integrating Sustainable soil fertility management innovations in staple cereal systems and other value chains to enhance livelihoods and environmental systems in Southern Africa Soil Fertility Consortium for Southern Africa. A SOFECSA Task Force Annual Technical Report for 2008 prepared for the Forum for Agricultural Research in Africa (FARA) Sub Saharan Africa Challenge Program (SSACP). CIMMYT: Harare, Zimbabwe.

Morris, M., Kelly, V., Kopicki, R. and Byerlee, D. (2007). *Fertilizer Use in African Agriculture: Lessons Learned and Good Practice Guidelines*. The World Bank: Washington, DC.

Mpepereki, S., Makonese, F. and Giller, K. E. (1996a). Exploiting promiscuous nodulation as a vehicle for integrating soybean (Glycine max [L.] Merrill) into smallholder cropping systemsof Zimbabwe. In *Soyabeans in Smallholder Cropping Systems of Zimbabwe*. Mpepereki, S., Makonese, F. and Giller, K. E. (Eds). SoilFertNet/CIMMYT: Harare, Zimbabwe, pp. 77–84.

Mpepereki, S., Makonese, F. and Giller, K. E. (Eds). 1996b. Soyabeans in Smallholder Cropping Systems of Zimbabwe. SoilFertNet/CIMMYT: Harare, Zimbabwe, p. 87.

Mpepereki, S., Javaheri, F., Davis, P. and Giller, K. E. (2000). Soyabeans and sustainable agriculture Promiscuous soyabeans in southern Africa. *Field Crops Research* 65, 137–49.

Mudimu, G. D. (1996). Socioeconomic factors in soybean production, consumption and marketing in Hurungwe communal lands: A case study of institutions, technology and market linkages. In *Soyabean in the Smallholder Cropping Systems in Zimbabwe*. Mpepereki, S., Giller, K. E. and Makonese, F. (Eds). Soil Fertility Network for Maize Based Smallholder Farming Systems, University of Zimbabwe: Harare Zimbabwe, 26–32.

Mujaju, C. and Jonga, M. (2015). The African Seed Access Index Country Report on Zimbabwe. Zimbabwe Brief. http://tasai.org/wp-content/themes/tasai2016/img/briefs_Zimbabwe_TASAI_Policy_Brief_2015.pdf

Murithi, H. M., Beed, F., Tukamuhabwa, P., Thomma B. P. H. J. and Joosten, M. H. A. J. (2016). Soybean production in eastern and southern Africa and threat of yield loss due to soybean rust caused by Phakopsora pachyrhizi. *Plant Pathology*, 65, 176–88.

Ndakidemi, P. A., Dakora, F. D., Nkonya, E. M., Ringo, D. and Mansoor, H. (2006). Yield and economic benefits of common bean (Phaseolus vulgaris) and soybean (Glycine max) inoculation in northern Tanzania. *Australian Journal of Experimental Agriculture*, 46, 571–7.

Okogun, J. A. and Sanginga, N. (2003). Can introduced and indigenous rhizobia strains compete for nodule formation by promiscuous soybean in the moist savanna agroecological zone of Nigeria? *Journal of Biology and Fertility of Soils*, 38, 26–31.

Oloka, H. K., Tukamuhabwa, P., Sengooba, T. and Shanmugasundram, S. (2008). Reaction of exotic soybean germplasm to Phakopsora pachyrhizi in Uganda. *Plant Disease*, 92, 1493–6.

Opperman, C. and Varia, N. (2011). Technical Report: Soybean Value Chain. United States Agency for International Development Southern Africa Trade Hub. Botswana.

Pompi, I., Mpepereki, S. and Gwata, E. (1998). Zimbabwe Soybean Promotion Taskforce: Objectives, Achievements and Agenda for the Future. In Waddington S. R., Murwirwa H. K., Kumwenda J. D. T., Hikwa D. and Tagwira F. *Proc Soil Fertility Network Results and Planning Workshop*, 7–11 July 1997: Mutare, Zimbabwe. Soil Fertility Network. University of Zimbabwe: Harare Zimbabwe, pp. 271–73.

Rusike, J., Sukume, C., Doward, A., Mpepereki, S. and Giller, K. E. (2000). *The Economic Potential for Smallholder Soyabean Production in Zimbabwe*. Soil Fertility Network Special Publication. University of Zimbabwe: Harare Zimbabwe.

Seed Co (2011). *Agronomy Manual*. Seedco Pvt Limited: Harare, Zimbabwe.

Shumba, E. M., Chisenga, M. M. and Ndebele, C. (1992). Contribution of fertilizer and management practices to the grain yield of maize in semi-arid areas of Zimbabwe. *Zimbabwe Journal of Agricultural Research*, 30(2), 147–3.

Smaling, E. M. A., Roscoe, R., Lesschen, J. P., Bouwman, A. F. and Comunello, F. (2008). From forest to waste: Assessment of the Brazilian soybean chain, using nitrogen as a marker. *Agric. Ecosyst. Environ.*, 128(3), 185–97. http://dx.doi.org/10.1016/j.agee.2008.06.005.

Technoserve (2011). *Southern Africa Regional Soybean Roadmap, Zimbabwe Value Chain Analysis*. Technoserve. http://www.technoserve.org/files/downloads/technoserve-bmgf-regional-report.pdf.

Vanlauwe, B., Bationo, A., Chianu, J., Giller, K. E., Merckx, R., Mokwunye, U., Ohiokpehai, O., Pypers, P., Tabo, R., Shepherd, K. D., Smaling, E. M. A., Woomer, P. L., Sanginga, N. (2010). Integrated soil fertility management: Operational definition and consequences for implementation and dissemination. *Outlook on Agriculture* 39: 17–24.

Woomer, P. L., Balume, I., Afullah, W., Zamasiya, B., Mahamadi, D. and Chataika, B. (2013). Report on Market Analysis of inoculant production and use. Putting Nitrogen Fixation to work for smallholder farmers in Africa. www.N2Africa.org, 22. http://www.n2africa.org/sites/n2africa.org/files/N2Africa_Market%20Analysis%20of%20Inoculant%20Production%20and%20Use_0.pdf.

Zamasiya, B., Mango, N., Nyikahadzoi, K., and Siziba, S. (2014). Determinants of soybean market participation by smallholder farmers in Zimbabwe. *Journal of Development and Agricultural Economics*, 6(2): 49–58.

Zengeni, R., S. Mpepereki and K. E. Giller, (2006). Manure and soil properties affect survival and persistence of soybean nodulating rhizobia in smallholder soils of Zimbabwe. *Applied Soil Ecolology*, 32: 232–42.

Index

Lightning Source UK Ltd.
Milton Keynes UK
UKHW02n1411310518
323519UK00001B/3/P

9 781786 761125